Fundamentals of Semiconductor Materials and Devices

Fundamentals of Semiconductor Materials and Devices

Adrian Kitai

Departments of Engineering Physics and
Materials Science and Engineering,
McMaster University, Hamilton, ON, Canada

WILEY

Registered Office(s)
John Wiley & Sons, Inc., 111 River Street, Hoboken, NJ 07030, USA
John Wiley & Sons Ltd, The Atrium, Southern Gate, Chichester, West Sussex, PO19 8SQ, UK

For details of our global editorial offices, customer services, and more information about Wiley products visit us at www.wiley.com.

Wiley also publishes its books in a variety of electronic formats and by print-on-demand. Some content that appears in standard print versions of this book may not be available in other formats.

A catalogue record for this book is available from the Library of Congress

Hardback ISBN: 9781119891406; ePub ISBN: 9781119891420; ePDF ISBN: 9781119891413

Cover Image: © Redlio Designs/Getty Images
Cover Design: Wiley

Set in 9.5/12.5pt STIXTwoText by Integra Software Services Pvt. Ltd, Pondicherry, India
Printed and bound by CPI Group (UK) Ltd, Croydon, CR0 4YY

C9781119891406_061023

Contents

Acknowledgments

I would like to thank McMaster University colleagues Oleg Rubel and Randy Dumont and Wiley staff Jenny Cossham, Skyler Van Valkenburgh, Martin Preuss, Monica Chandrasekar, Dilip Varma and Christy Michael who played important roles in providing advice, expertise, general guidance, editing, and support.

Additional expertise from Mark Tuckerman (computational chemistry) and Nicolas Heumann (physics) is gratefully acknowledged.

I need to thank the students in the senior level course I teach at McMaster University for their invaluable assistance in finding errors in the manuscript.

Finally, this book would not have been possible without the patient support of my family.

Preface

The traditional, discrete silos of materials science, electrical engineering, engineering physics, physics, and chemistry are being increasingly blended due to the expanding needs and expectations of semiconductor materials and devices in both industry and academia.

This book is designed to introduce the topic of semiconductor materials and devices at a foundational level and to a broad audience. Only knowledge of first- and second-year-undergraduate-level math and science is assumed. This is a third- or fourth-year-undergraduate or graduate-level textbook.

The book begins with introductory chapters on quantum and solid-state physics followed by an introduction to practical semiconductor materials and basic p-n junction theory.

Next, to prepare for optoelectronic devices, radiation theory covering the very important concepts of the exciton, the exchange interaction, singlets/triplets, and the joint density of states is introduced. Associated coverage includes accelerating charges, a quantum treatment of the dipole radiator, the dipole matrix element, and Forster/ Dexter radiative energy transfer processes.

The electrical and optical properties of photodiodes, solar cells, and light emitting diodes follow together with an introduction to bipolar junction transistors and junction field effect transistors.

The silicon MOSFET (metal oxide field effect transistor) is presented next in detail. Coverage starts with MOS band theory and the metal oxide capacitor. Coverage of the MOSFET is extended to the nanoscale and a series of new nanoscale concepts are introduced in context. These concepts include details of bulk and interface charge trapping, coulombic shielding and the Debye length, the two-dimensional regime for channel charge, effective work functions, surface mobility, electron tunneling, Fowler Nordheim current, and saturated carrier velocity.

Key nanoscale fabrication strategies for MOSFET devices are covered including ion implantation, self-aligned structures, deep UV lithography, rapid thermal annealing, high oxidation potential, high K dielectric material (HfO_2), and high thermal budget electrode materials.

Content then moves into truly nanoscale concepts introduced via the quantum dot and progresses to the small organic molecule size range. Optical properties and the physics of quantum confinement in quantum dots are covered with an effective mass model to explain an increase in energy gap. Limitations of the effective mass model are explained as well as the need for a Rydberg term. The exciton radius relative to the quantum dot radius is a key new concept which can lead to an additional energy correction for the calculation of energy gap.

Vibrational aspects including the Stokes shift in quantum dot emission and absorption spectra are then understood by presenting the Frank-Condon principle. A configurational concept is introduced starting from a basic bi-atomic bonding model and is then generalized to the widely used configurational coordinate axis applicable to multi-atomic systems including quantum dots. The quantum harmonic oscillator is introduced in this context. Auger processes and biexcitons are treated in some detail and the biocompatibility of quantum dots is introduced.

Molecular electronics is treated next since the molecular length scale constitutes a natural progression toward smaller structures. This requires the introduction of organic semiconductor molecules and both oligomeric and polymeric materials. A range of well-known organic oligomers is presented. Sigma and pi bonds are introduced and the concepts of conjugated bonding leading to intramolecular electron transport through pi bands is covered together with intermolecular transport and the latter's dependence on molecular packing in both crystalline and amorphous molecular materials. The highest occupied molecular orbital and the lowest unoccupied molecular orbital are introduced and shown to define the effective energy gap in molecular semiconductors.

Functional molecule side groups are discussed in terms of modifying emission and absorption spectra and as a means to allow for solubility and low-cost solution deposition of oligomers in contrast to vapor deposition via vacuum evaporation.

The carrier mobility and electron affinity values of important molecular materials are discussed and tabulated. The low carrier mobilities and their strong influence on organic device design are described. Band models of a series of both polymeric and oligomeric organic light emitting devices are presented showing the importance and relevance of electron affinity. Electron and hole transport within these devices leading to the formation of excitons is described.

The molecular exciton and the associated energy transfer processes are now applied to molecular optoelectronic devices. A good understanding of the exciton is needed for both organic light emitting diodes and organic solar cells. Additional concepts are introduced including phosphorescent organic molecules, organometallic molecule spin–orbit coupling and thermally stimulated delayed fluorescence. Both homojunction and heterojunction organic solar cells are introduced as well as the bulk heterojunction leading to substantial improvements in organic solar cell performance.

Finally, materials and devices are considered in which at least one dimension exists at length scales down to the angstrom range. At the smallest length scales, two-dimensional atomically thin semiconductors are introduced via the Transition Metal Dichalcogenides (TMDCs). These materials naturally form semiconductors having well-defined energy gaps in a single dichalcogenide layer. There are no dangling bonds since bonding between layers in analogous bulk materials is via van der Walls forces only. Details of known atomic structures are presented.

Due to the limitations of bulk band theory for the smallest structures, the linear combination of atomic orbitals and density functional theory are introduced and examples are provided. Data is presented including the progression in band gaps and band diagram dispersion curves from bulk to single layers of TMDCs exhibiting both indirect gap and direct gap behavior. Examples of prototype TMDC transistors and their characteristics are presented. Coulomb scattering is shown to limit carrier mobility in practice.

One-dimensional semiconductor devices are introduced firstly by a description of the finFET which is an extension of the MOSFET. The finFET is then further extended to the concept of a true nanowire FET that is a type of gate-all-around FET. Fabrication steps for gate-all-around FETs are presented.

This book is uniquely comprehensive compared to existing books. It covers concepts at an introductory level but also highlights their relevance to a wide range of semiconductor materials and devices from the bulk to the nanoscale. Sitting at the interface between engineering and physics, the book employs math and physics throughout, but always ensures that the presentation is relevant to real applications. The math is sufficiently simple to allow a wide spectrum of scientists and engineers to follow the material. This book contains pedagogically useful questions and exercises throughout.

About the Companion Website

This book is accompanied by a companion website which includes a number of resources created by author for students and instructors that you will find helpful.

www.wiley.com/go/kitai_fundamentals

The website includes the following resource for each chapter:

- PowerPoint slides

Introduction to Quantum Mechanics

Objectives

1. Review the classical electron and motivate the need for a quantum mechanical model.
2. Present experimental evidence for the photon as a fundamental constituent of electromagnetic radiation.

Fundamentals of Semiconductor Materials and Devices, First Edition. Adrian Kitai.
© 2023 John Wiley & Sons Ltd. Published 2023 by John Wiley & Sons Ltd.
Companion Website: www.wiley.com/go/kitai_fundamentals

3. Introduce quantum mechanical relationships based on experimental results and illustrate these using examples.
4. Introduce expectation values for important measureable quantities based on the uncertainty principle.
5. Motivate and define the wavefunction as a means of describing particles.
6. Present Schrödinger's equation and its solutions for practical problems relevant to semiconductor materials and devices.
7. Introduce spin and the associated magnetic properties of electrons.
8. Present the postulates of quantum mechanics.
9. Introduce the Pauli exclusion principle and give an example of its application.

1.1   INTRODUCTION

The study of semiconductor materials and devices relies on the electronic properties of solid state materials and hence a fundamental understanding of the behavior of electrons in solids.

Electrons are responsible for electrical properties and optical properties in metals, insulators, inorganic semiconductors, and organic semiconductors. These materials form the basis of an astonishing variety of electronic components and devices.

The electronics age in which we are immersed would not be possible without the ability to grow these materials, control their electronic properties, and finally fabricate structured devices using them which yield specific electronic and optical functionality.

Electron behavior in solids requires an understanding of the electron that includes the quantum mechanical description; however, we will start with the classical electron.

1.2   THE CLASSICAL ELECTRON

We describe the electron as a particle having mass

$$m = 9.11 \times 10^{-31} \text{ kg}$$

and negative charge of magnitude

$$q = 1.602 \times 10^{-19} \text{ coul}$$

If an external electric field $\varepsilon(x, y, z)$ is present in three-dimensional space and an electron experiences this external electric field, the magnitude of the force on the electron is

$$F = q\varepsilon$$

The direction of the force is opposite to the direction of the external electric field due to the negative charge on the electron. If ε is expressed in volts per meter (V/m) then F will have units of newtons.

If an electron accelerates through a distance d from point A to point B in vacuum due to a uniform external electric field ε it will gain kinetic energy ΔE in which

$$\Delta E = Fd = q\varepsilon d \qquad (1.1a)$$

This kinetic energy ΔE gained by the electron may be expressed in joules within the Meters-Kilograms-Seconds (MKS) unit system. We can also say that the electron at point A has a potential energy U that is higher than its potential energy at point B. Since total energy is conserved,

$$|\Delta U| = |\Delta E| \qquad (1.1b)$$

There exists an electric potential $V(x,y,z)$ defined in units of the volt at any position in three-dimensional space associated with an external electric field. We obtain the spatially dependent potential energy $U(x,y,z)$ for an electron in terms of this electric potential as

$$U(x,y,z) = -qV(x,y,z)$$

We also define the *electron-volt*, another commonly used energy unit. By definition, one electron-volt in kinetic energy is gained by an electron if the electron accelerates in an electric field between two points in space whose difference in electric potential ΔV is one volt.

EXAMPLE 1.1

Find the relationship between two commonly used units of energy, namely the electron-volt and the joule.

Consider a uniform external electric field in which $\varepsilon(x) = 1\,V/m$. If an electron accelerates in vacuum in this uniform external electric field between two points separated by one meter and therefore having a potential difference of one volt, then from Equation 1.1a it gains kinetic energy expressed in joules of

$$\Delta E = Fd = q\varepsilon d = 1.602 \times 10^{-19} \text{ coulombs} \times \frac{1V}{m} \times 1m = 1.602 \times 10^{-19}\,J$$

But, by definition, one electron-volt in kinetic energy is gained by an electron if the electron accelerates in an electric field between two points in space whose difference in electric potential ΔV is one volt, and we have therefore shown that the conversion between the joule and the electron-volt is

$$1eV = 1.602 \times 10^{-19} \text{ J}$$

If an external magnetic field $\boldsymbol{B}$ is present, the force on an electron depends on the charge q on the electron as well as the component of electron velocity $\boldsymbol{v}$ perpendicular to the magnetic field which we shall denote $\boldsymbol{v}_\perp$. This force, called the *Lorentz Force*, may be expressed as $\boldsymbol{F} = -q(\boldsymbol{v}_\perp \times \boldsymbol{B})$. The force is perpendicular to both the velocity component of the electron and to the magnetic field vector. The Lorentz force enables the electric motor and the electric generator.

This classical description of the electron generally served the needs of the vacuum tube electronics era and the electric motor/generator industry in the first half of the twentieth century.

In the second half of the twentieth century, the electronics industry migrated from vacuum tube devices to solid state devices once the transistor was invented at Bell Laboratories in 1954. The understanding of the electrical properties of semiconductor materials from which transistors are made could not be achieved using a classical description of the electron. Fortunately the field of *quantum mechanics*, which was developing over the span of about 50 years before the invention of the transistor, allowed physicists to model and understand electron behavior in solids.

In this chapter we will motivate quantum mechanics by way of a few examples. The classical description of the electron is shown to be unable to explain some simple observed phenomena, and we will then introduce and apply the quantum-mechanical description which has proven to work very successfully.

1.3   TWO-SLIT ELECTRON EXPERIMENT

One of the most remarkable illustrations of how strangely electrons can behave is illustrated in Figure 1.1. Consider a beam of electrons arriving at a pair of narrow, closely spaced slits formed in a solid. Assume that the electrons arrive at the slits randomly in a beam having a width much wider than the slit dimensions. Most of the electrons hit the solid, but a few electrons pass through the slits and then hit a screen placed behind the slits as shown.

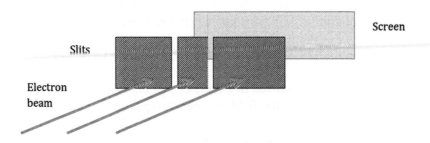

FIGURE 1.1   Electron beam emitted by an electron source is incident on narrow slits with a screen situated behind the slits.

If the screen could detect where the electrons arrived by counting them, we would expect a result as shown in Figure 1.2.

In practice, a screen pattern as shown in Figure 1.3 is obtained. This result is impossible to derive using the classical description of an electron.

It does become readily explainable, however, if we assume the electrons have a wavelike nature. If light waves, rather than particles, are incident on the slits then there are particular positions on the screen at which the waves from the two slits cancel out. This is because they are out of phase. At other positions on the screen the waves add together because they are in-phase. This pattern is the well-known interference pattern generated by light traveling through a pair of slits. Interestingly we do not know which slit a particular electron passes through. If we attempt to experimentally determine which slit an electron is passing through we immediately disrupt the experiment and the interference pattern disappears. We could say that the electron somehow goes through both slits. Remarkably, the same interference pattern builds up slowly and is observed even if electrons are emitted from the electron source and arrive at the screen one at a time. We are forced to interpret these results as a very fundamental property of small particles such as electrons.

We will now look at how the two-slit experiment for electrons may actually be performed. It was done in the 1920s by Davisson and Germer. It turns out that very narrow slits are required to be able to observe the electron behaving as a wave due to

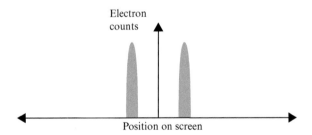

FIGURE 1.2   Classically expected result of two-slit experiment.

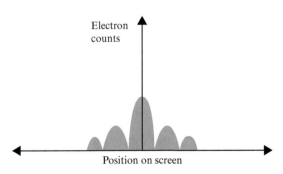

FIGURE 1.3   Result of two-slit experiment. Notice that a wavelike electron is required to cause this pattern. If light waves rather than electrons were used then a similar plot would result except the vertical axis would be a measure of the light intensity instead.

the small wavelength of electrons. Fabricated slits having the required very small dimensions are not practical, but Davisson and Germer realized that the atomic planes of a crystal can replace slits. By a process of electron reflection, rows of atoms belonging to adjacent atomic planes on the surface of a crystal act like tiny reflectors that effectively form two beams of reflected electrons that then reach a screen and form an interference pattern similar to that shown in Figure 1.3.

Their method is shown in Figure 1.4. The angle between the incident electron beam and each reflected electron beam is θ. The spacing between surface atoms belonging to adjacent atomic planes is d. The path length difference between the two beam paths shown is $d\sin\theta$. A maximum on the screen is observed when

$$d\sin\theta = n\lambda \tag{1.2a}$$

or an integer number of wavelengths. Here, n is an integer and λ is the wavelength of the waves. A minimum occurs when

$$d\sin\theta = \left(\frac{2n+1}{2}\right)\lambda \tag{1.2b}$$

which is an odd number of half wavelengths causing wave cancellation.

In order to determine the wavelength of the apparent electron wave we can solve Equations 1.2a and 1.2b for λ. We have the appropriate values of θ; however, we need to know d. Using X-ray diffraction and Bragg's law we can obtain d. Note that Bragg's law is also based on wave interference except that the waves are X-rays.

The results that Davisson and Germer obtained were quite startling. The calculated values of λ were on the order of angstroms, where 1 angstrom $\left(\text{Å}\right)$ is one tenth of a nanometer. This is much smaller than the wavelength of light which is on the order of thousands of angstroms, and it explains why regular slits used in optical

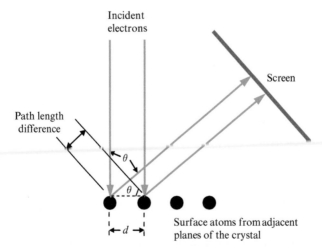

FIGURE 1.4   Davisson-Germer experiment showing electrons reflected off adjacent crystalline planes. Path length difference is $d\sin\theta$.

experiments are much too large to observe electron waves. But more importantly *the measured values of λ actually depended on the incident velocity v or momentum mv of the electrons* used in the experiment. Increasing the electron momentum by accelerating electrons through a higher potential difference before they reached the crystal caused λ to decrease, and decreasing the electron momentum caused λ to increase. By experimentally determining λ for a range of values of incident electron momentum, the following relationship was determined:

$$\lambda = \frac{h}{p} \qquad (1.3a)$$

This is known as the *de Broglie equation*, because de Boglie postulated this relationship before it was validated experimentally. Here p is the magnitude of electron momentum, and $h = 6.63 \times 10^{-34}$ Js is a constant known as *Planck's constant*. In an alternative form of the equation we define $\hbar$, pronounced *h-bar* to be $\hbar = \frac{h}{2\pi}$ and we define k, the *wavenumber* to be $k = \frac{2\pi}{\lambda}$. Now we can write the de Broglie equation as

$$p = \frac{h}{\lambda} = \hbar k \qquad (1.3b)$$

Note that p is the magnitude of the momentum vector **p** and k is the magnitude of *wavevector* **k**. The significance of wavevectors will become clearer in Chapter 2.

EXAMPLE 1.2

An electron is accelerated through a potential difference $\Delta V = 10,000$ volts.

a) Find the electron energy in both joules and electron-volts.
b) Find the electron wavelength.

Solution

a) Assume the initial kinetic energy of the electron was negligible before it was accelerated. The final energy is

$$E = q\Delta V = 1.6 \times 10^{-19} \text{ coul} \times 10,000\text{V} = 1.6 \times 10^{-15} \text{J}$$

To express this energy in electron-volts,

$$E = 1.6 \times 10^{-15} \text{J} \times \frac{1\text{eV}}{1.6 \times 10^{-19} \text{J}} = 10^4 \text{eV}$$

b) From Equation 1.3b

$$\lambda = \frac{h}{p} = \frac{h}{\sqrt{2mE}} = \frac{6.63 \times 10^{-34} \text{ Js}}{\sqrt{2 \times 9.11 \times 10^{-31} \text{kg} \times 1.6 \times 10^{-15} \text{J}}}$$
$$= 1.23 \times 10^{-11} \text{m} = 0.123\text{Å}$$

1.4   THE PHOTOELECTRIC EFFECT

About 30 years before Davisson and Germer discovered and measured electron wavelengths, another important experiment had been undertaken by Heinrich Hertz. In 1887 Hertz was investigating what happens when light is incident on a metal. He found that electrons in the metal can be liberated by the light. It takes a certain amount of energy to release an electron from a metal into vacuum. This energy is called the *workfunction* Φ, and the magnitude of Φ depends on the metal.

If the metal is placed in a vacuum chamber, the liberated electrons are free to travel away from the metal and they can be collected by a collector electrode also located in the vacuum chamber shown in Figure 1.5. This is known as the *photoelectric effect*.

By carefully measuring the resulting electric current I flowing through the vacuum, scientists in the last decades of the nineteenth century were able to make the following statements:

1. The electric current increases with increasing light intensity.
2. The color or wavelength of the light is important also. If monochromatic light is used, there is a particular critical wavelength above which no electrons are released from the metal even if the light intensity is increased.
3. This critical wavelength depends on the type of metal employed. Metals composed of atoms having low ionization energies such as cerium and calcium have larger critical wavelengths. Metals composed of atoms having higher ionization energies such as gold or aluminum require smaller critical wavelengths.
4. If light having a wavelength equal to the critical wavelength is used then the electrons leaving the metal surface have no initial kinetic energy and collecting these electrons requires that the voltage V must be positive to accelerate the electrons away from the metal and toward the second electrode.
5. If light having wavelengths smaller than the critical wavelength is used then the electrons do have some initial kinetic energy. Now, even if V is negative so as to retard the flow of electrons from the metal to the electrode, some

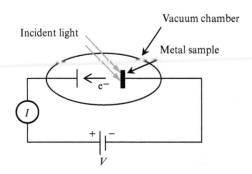

FIGURE 1.5   The photoelectric experiment. The current I flowing through the external circuit is the same as the vacuum current.

electrons may be collected. There is, however, a maximum negative voltage of magnitude V_{max} for which electrons may be collected. As the wavelength of the light is further decreased, V_{max} increases (the maximum voltage becomes more negative).

6. If very low intensity monochromatic light with a wavelength smaller than the critical wavelength is used then individual electrons are measured rather than a continuous electron current. Suppose at time $t = 0$ there is no illumination, and then at time $t > 0$ very low intensity light is turned on. The first electron to be emitted may occur virtually as soon as the light is turned on, or it may take a finite amount of time to be emitted after the light is turned on. There is no way to predict this amount of time in advance.

Einstein won the Nobel Prize for his conclusions based on these observations. He concluded the following:

1. Light is composed of particle-like entities or *wave-packets* commonly called *photons*.
2. The intensity of a source of light is determined by the photon flux density, or the number of photons being emitted per second per unit area.
3. Electrons are emitted only if the incident photons each have enough energy to overcome the metal's workfunction.
4. The photon energy of an individual photon of monochromatic light is determined by the color (wavelength) of the light.
5. Photons of a known energy are randomly emitted from monochromatic light sources, and we can never precisely know when the next photon will be emitted.
6. Normally we do not notice these photons because there are a very large number of photons in a light beam. If, however, the intensity of the light is low enough then the photons become noticeable and light becomes *granular*.
7. If light having photon energy larger than the energy needed to overcome the workfunction is used then the excess photon energy causes a finite initial kinetic energy of the escaped electrons

By carefully measuring the critical photon wavelength λ_c as a function of V_{max}, the relationship between photon energy and photon wavelength can be determined: The initial kinetic energy E_k of the electron leaving the metal can be sufficient to overcome a retarding (negative) potential difference V between the metal and the electrode. Since the kinetic energy lost by the electron as it moves against this retarding potential difference V is $E_k = |qV|$ we can deduce the minimum required photon energy E using the energy equation $E = \Phi + qV_{max}$. This experimentally observed relationship is

$$E = h\frac{c}{\lambda}$$

where c is the velocity of light. Since the frequency of the photon ν is given by

$$\lambda = \frac{c}{\nu}$$

we obtain the following relationship between photon energy and photon frequency:

$$E = h\nu = \hbar\omega \tag{1.4}$$

where $\omega = 2\pi\nu$. Note that Planck's constant h is found when looking at either the wavelike properties of electrons in Equation 1.3b or the particle-like properties of light in Equation 1.4. This *wave-particle duality* forms the basis for quantum mechanics.

Equation 1.4 which arose from the photoelectric effect defines the energy of a photon. In addition, Equation 1.3b which arose from the Davisson and Germer experiment applies to both electrons *and* photons. This means that a photon having a known wavelength carries a specific momentum $p = \frac{h}{\lambda}$ even though a photon has no mass. The existence of photon momentum is experimentally proven since light-induced pressure can be measured on an illuminated surface.

In summary, electromagnetic waves exist as photons which also have particle-like properties such as momentum and energy, and particles such as electrons also have wavelike properties such as wavelength.

EXAMPLE 1.3

Ultraviolet light with wavelength 190 nm is incident on a metal sample inside a vacuum envelope containing an additional collector electrode. The collector electrode potential relative to the sample potential is defined by potential difference V as shown in Figure 1.5. Photoelectric current is observed if $V \geq -1.4\text{V}$ and ceases if $V \leq -1.4\text{V}$. Identify the metal using the following table.

Metal	Aluminum	Nickel	Calcium	Cesium
Workfunction (eV)	4.1	5.1	2.9	2.1

Solution

190 eV photons have energy $E = h\nu = \dfrac{hc}{\lambda} = \dfrac{6.63 \times 10^{-34}\,\text{Js} \times 3 \times 10^{8}\,\text{ms}^{-1}}{190 \times 10^{-9}\,\text{m}} = 1.05 \times 10^{-18}\,\text{J}$

Expressed in electron-volts, this photon energy is $\dfrac{1.05 \times 10^{-18}\,\text{J}}{1.6 \times 10^{-19}\,\text{JeV}^{-1}} = 6.56\text{eV}$

The potential difference V will decelerate photoelectrons released from the metal surface if $V < 0$. This results in an energy loss of the photoelectrons and if $V = -1.4\text{volts}$, the energy loss will be 1.4eV. The photons therefore supply enough energy to overcome both the metal workfunction Φ and the energy loss due to deceleration. Hence the predicted metal workfunction is $\Phi = 6.56\text{eV} - 1.4\text{eV} = 5.16\text{eV}$. The metal having the closest workfunction match is nickel.

1.5   WAVE-PACKETS AND UNCERTAINTY

Uncertainty in the precise position of a particle is embedded in its quantum mechanical wave description. The concept of a wave-packet introduced in Section 1.4 for light is important since it is applicable to both photons and particles such as electrons.

A wave-packet is illustrated in Figure 1.6 showing that the wave-packet has a finite size. A wave-packet can be analyzed into, or synthesized from, a set of component sinusoidal waves, each having a distinct wavelength, with phases and amplitudes such that they interfere constructively only over a small region of space to yield the wave-packet, and destructively elsewhere. This set of component sinusoidal waves of distinct wavelengths added to yield an arbitrary function is an example of a Fourier series.

The uncertainty in the position of a particle described using a wave-packet may be approximated as Δx as indicated in Figure 1.6. The uncertainty depends on the number of component sinusoidal waves being added together in a Fourier series: If only one component sinusoidal wave is present, the wave-packet is infinitely long, and the uncertainty in position is infinite. In this case, the wavelength of the particle is precisely known, but its position is not defined. As the number of component sinusoidal wave components of the wave-packet approaches infinity, the uncertainty Δx of the position of the wave-packet may drop and we say that the wave-packet becomes increasingly localized.

An interesting question now arises: If the wave-packet is analyzed into, or composed from, a number of component sinusoidal waves, can we define the precise wavelength of the wave-packet? It is apparent that as more component sinusoidal waves, each having a distinct wavelength, are added together the uncertainty of the wavelength associated with the wave-packet will become larger. From Equation 1.3b, the uncertainty in wavelength results in an uncertainty in momentum p, and we write this momentum uncertainty as Δp.

By doing the appropriate Fourier series calculation (see Appendix 2), the relationship between Δx and Δp can be shown to satisfy the following condition:

$$\Delta x \Delta p \geq \frac{\hbar}{2} \tag{1.5}$$

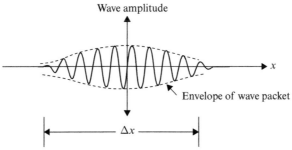

FIGURE 1.6   Wave-packet. The envelope of the wave-packet is also shown. More insight about the meaning of the wave amplitude for a particle such as an electron will become apparent in Section 1.6 in which the concept of a probability amplitude is introduced.

As Δx is reduced there will be an inevitable increase in Δp. This is known as the *Uncertainty Principle*. We cannot precisely and simultaneously determine the position and the momentum of a particle. If the particle is an electron we know less and less about the electron's momentum as we determine its position more and more precisely.

Since a photon is also described in terms of a wave-packet, the concept of uncertainty applies to photons as well. As the location of a photon becomes more precise, the wavelength or frequency of the photon becomes less well defined. Photons always travel with velocity of light c in vacuum. The exact arrival time t of a photon at a specific location is uncertain due to the uncertainty in position. For the photoelectric effect described in Section 1.4, the exact arrival time of a photon at a metal was observed not to be predictable for the case of monochromatic photons for which $E = h\nu = h\dfrac{c}{\lambda}$ is accurately known. If we allow some uncertainty in the photon frequency $\Delta\omega$ the energy uncertainty $\Delta E = \hbar\Delta\omega$ of the photon becomes finite, but then we can know more about the arrival time. The resulting relationship which may be calculated by the same approach as presented in Appendix 2 may be written as $\Delta E \Delta t \geq \dfrac{\hbar}{2}$. This type of uncertainty relationship is useful in time-dependent problems and, like the derivation of uncertainty for particles such as electrons, it results from a Fourier transform: The frequency spectrum $\Delta\omega$ of a pulse in the time domain becomes wider as the pulse width Δt becomes narrower.

A wave travels with velocity $v = f\lambda = \omega / k$. Note that we refer to this as a *phase velocity* because it refers to the velocity of a point on the wave that has a given phase, for example the crest of the wave. For a traveling wave-packet, however, *the velocity of the particle described using the wave-packet is not necessarily the same as the phase velocity of the individual waves making up the wave-packet.* The velocity of the particle is actually determined using the velocity of the wave-packet envelope shown in Figure 1.6. The velocity of propagation of this envelope is called the *group velocity* v_g because the envelope is formed by the Fourier sum of a group of waves.

When photons travel through media other than vacuum, *dispersion* can exist. Consider the case of a photon having energy uncertainty $\Delta E = \hbar\Delta\omega$ due to its wave-packet description. In the case of this photon traveling through vacuum, the group velocity and the phase velocity are identical to each other and equal to the speed of light c. This is known as a *dispersion-free* photon for which the wave-packet remains intact as it travels. But if a photon travels through a medium other than vacuum there is often finite dispersion in which some Fourier components of the photon wave-packet travel slightly faster or slightly slower that other components of the wave-packet, and the photon wave-packet broadens spatially as it travels. For example photons traveling through optical fibers typically suffer dispersion which limits the ultimate temporal resolution of the fiber system.

It is very useful to plot ω versus k for the given medium in which the photon travels. If a straight line is obtained then $v = \omega / k$ is a constant and the velocity of each Fourier component of the photon's wave-packet is identical. This is dispersion-free propagation. In general, however, a straight line will not be observed and dispersion exits.

In Appendix 3 we analyze the velocity of a wave-packet composed of a series of waves. It is shown that the wave-packet travels with velocity

$$v_g = \frac{d\omega}{dk}$$

This is valid for both photons and particles such as electrons. For wave-packets of particles, however, we can further state that

$$v_g = \frac{d\omega}{dk} = \frac{1}{\hbar}\frac{dE}{dk}$$

This relationship will be important in Chapter 2 to determine the velocity of electrons in crystalline solids.

1.6 THE WAVEFUNCTION

Based on what we have observed up to this point, the following four points more completely describe the properties of an electron in contrast to the description of the classical electron of Section 1.2:

1. The electron has mass m.
2. The electron has charge q.
3. The electron has wave properties with wavelength $\lambda = \dfrac{h}{p}$.
4. The exact position and momentum of an electron cannot be measured simultaneously.

Quantum mechanics provides an effective mathematical description of particles such as electrons that was motivated by the above observations. A *wavefunction* ψ is used to describe the particle and ψ may also be referred to as a *probability amplitude*. In general, ψ is a complex number which is a function of space and time. Using cartesian spatial coordinates, $\psi = \psi(x,y,z,t)$. We could also use other coordinates such as spherical polar coordinates in which case we would write $\psi = \psi(r,\theta,\phi,t)$.

The use of complex numbers is very important for wavefunctions because it allows them to represent waves as will be seen in Section 1.7.

Although ψ is a complex number and is therefore not a real, measureable or *observable* quantity, the quantity $\psi^*\psi = |\psi|^2$ where ψ^* is the complex conjugate of ψ, is an observable and must be a real number. $|\psi|^2$ is referred to as a *probability density*. At any time t, using cartesian coordinates, the probability of the particle being in volume element dxdydz at location (x,y,z) will be $|\psi(x,y,z)|^2$ dxdydz. If a particle exists, then it must be somewhere is space and we can write

$$\int_{all\ space} |\psi|^2\ dV = \int_{-\infty}^{\infty}\int_{-\infty}^{\infty}\int_{-\infty}^{\infty} |\psi|^2\ dxdydz = 1 \tag{1.6}$$

The wavefunction, therefore, fundamentally recognizes the attribute of uncertainty and simultaneously is able to represent a wave. We cannot precisely define the position of the particle; however, we can determine the probability of it being in a specific region. Equation 1.6 is referred to as the normalization condition for a wavefunction and a wavefunction that satisfies this equation is a *normalized* wavefunction.

In order to give the particle we are trying to describe the attributes of a wave, the form of ψ is a mathematical wave expression such as the sinusoidal function used in Example 1.4.

EXAMPLE 1.4

A time-independent wavefunction $\psi(x)$ is defined in one dimension along the *x*-axis.

Such that:

if $x < -\dfrac{2\pi}{a}$ or $x > \dfrac{2\pi}{a}$ then $\psi(x) = 0$

and if $-\dfrac{2\pi}{a} \leq x \leq \dfrac{2\pi}{a}$ then $\psi(x) = A\sin(ax)$

Normalize this wavefunction by determining the appropriate value of coefficient A.

Solution

$$\int_{-\infty}^{\infty} |\psi|^2 \, dx = A^2 \int_{-\frac{2\pi}{a}}^{\frac{2\pi}{a}} (\sin(ax))^2 \, dx = \frac{A^2}{2} \int_{-\frac{2\pi}{a}}^{\frac{2\pi}{a}} (1 - \cos(2ax)) \, dx$$

$$= \frac{A^2}{2}\left(\frac{4\pi}{a}\right) = 1$$

Hence

$$A = \sqrt{\frac{a}{2\pi}}$$

Now the normalized wavefunction is:

for $x < -\dfrac{2\pi}{u}$ and $x > \dfrac{2\pi}{a}$ $\psi(x) = 0$

and for $-\dfrac{2\pi}{a} \leq x \leq \dfrac{2\pi}{a}$ $\psi(x) = \sqrt{\dfrac{a}{2\pi}}\sin(ax)$

1.7   THE SCHRÖDINGER EQUATION

The most fundamental law in classical physics also applies to quantum physics. This is the conservation of energy. More specifically, this can be expressed as follows: The total energy of a closed system is fixed and is the sum of the internal potential and kinetic energies.

Building on the emerging understanding of particles we have outlined in this chapter and through the remarkable insights of Erwin Schrödinger, in 1925 the following wave equation, called the *Schrödinger Equation*, was postulated:

$$-\frac{\hbar^2}{2m}\nabla^2\psi(x,y,z)+U(x,y,z)\psi(x,y,z)=i\hbar\frac{\partial\psi}{\partial t} \tag{1.7}$$

$U(x,y,z)$ represents the potential energy in the electric field in which the particle of mass m exists and the equation allows the particle's wavefunction $\psi(x,y,z)$ to be found. The first term is associated with the kinetic energy of the particle. The second term is associated with the potential energy of the particle, and the right-hand side of the equation is associated with the total energy E of the particle. Once ψ is known, the particle's position, energy, and momentum can be determined either as specific values or as spatial distribution functions consistent with the uncertainty principle. For time-varying systems, a possible time evolution of the particle's properties may also be described.

This equation is applicable to a range of particles including electrons and protons; we are most often interested in the electrical properties of materials and we will therefore focus now on the electron. In Chapter 7 it will be applied to the proton and to atomic nuclei.

By solving Schrödinger's equation for an electron in a few simple scenarios we will be able to appreciate the utility of the equation. In addition, a better understanding will be gained about the quantum mechanical wavefunction-based description of particles.

Let us propose a solution to Equation 1.7 having the form

$$\psi(x,y,z,t)=\psi(x,y,z)T(t) \tag{1.8}$$

Note that we have separated the solution into two parts, one for spatial dependence and one for time dependence. Now substituting Equation 1.8 into Equation 1.7 and dividing by $\psi(x,y,z)T(t)$ we obtain

$$-\frac{\hbar^2}{2m}\frac{\nabla^2\psi(x,y,z)}{\psi(x,y,z)}+U(x,y,z)=i\hbar\frac{1}{T(t)}\frac{dT(t)}{dt}$$

Since the left side is a function of independent variables x,y,z only and the right side is is a function of independent variable t only, the only way for the equality to hold for both arbitrary spatial locations and for arbitrary moments in time is for both sides of the equation to be equal to a constant that we will call E.

The resulting equations are

$$i\hbar \frac{dT(t)}{dt} = ET(t) \tag{1.9}$$

and

$$-\frac{\hbar^2}{2m}\nabla^2 \psi(x, y, z) + U(x, y, z)\psi(x, y, z) = E\psi(x, y, z) \tag{1.10}$$

Equations. 1.9 and 1.10 are the result of the just-described method known as *the separation of variables* applied to Equation 1.7. Equation 1.9 is easy to solve and has solution

$$T(t) = Ae^{-i\frac{E}{\hbar}t} \tag{1.11a}$$

If we now identify E with the energy of the electron and use Equation 1.4 we obtain

$$E = h\nu = \hbar\omega$$

and therefore

$$T(t) = Ae^{-i\omega t} \tag{1.11b}$$

which represents the expected time-dependence of a wave having frequency ω.

Equation 1.10 is known as the *Time-Independent Schrödinger Equation* and it is useful for a wide variety of steady state (time-independent) situations as illustrated in Examples 1.5 to 1.8.

EXAMPLE 1.5

Consider a one-dimensional problem in which the only spatial coordinate is the x-axis. Assume that the potential energy of the electron for all values of x is zero.

a) Find the wavefunctions.
b) Find the probability density.
c) Can the wavefunctions be normalized?

Solution

a) This absence of potential energy for all values of x implies that there is no net force acting on the electron. This situation may be applicable if no electric field is present. Now Equation 1.10 becomes

$$-\frac{\hbar^2}{2m}\frac{d^2\psi(x)}{dx^2} = E\psi(x) \tag{1.12}$$

The general solution to this differential equation is

$$\psi(x) = Ae^{ikx} + Be^{-ikx} \tag{1.13}$$

where

$$k = \frac{\sqrt{2mE}}{\hbar}$$

k is the wavenumber. In the complex plane $e^{\pm i2\pi}$ represents one complete cycle of a wave of wavelength $\lambda = 2\pi$ and therefore for one complete cycle of the wave, $kx = k\lambda = 2\pi$. It follows that

$$k = \frac{2\pi}{\lambda}$$

consistent with the de Broglie equation.

From Equations 1.11b and 1.13 the overall solution to the Schrödinger equation is now

$$\psi(x,t) = \left(Ae^{ikx} + Be^{-ikx}\right)e^{-i\omega t} = Ae^{i(kx-\omega t)} + Be^{-i(kx+\omega t)}$$

This is the general equation of two traveling plane waves with components moving in both the positive and negative x-directions described by the first term and the second term, respectively. Note that each term represents a phasor in the complex plane that rotates either clockwise or counterclockwise as we move along an x-axis or a time axis.

a) We can now calculate the probability density

$$|\psi|^2 = \psi^*\psi = \left(Ae^{-i(kx-\omega t)} + Be^{i(kx+\omega t)}\right)\left(Ae^{i(kx-\omega t)} + Be^{-i(kx+\omega t)}\right)$$
$$= A^2 + B^2 + ABe^{2ikx} + ABe^{-2ikx} = A^2 + B^2 + 2AB\cos 2kx$$

Note that $|\psi|^2$ is a real number and it represents the probability distribution function of the electron.

If we consider the wavefunction to be a plane wave traveling in the positive x-direction then with reference to Equation 1.13 we conclude that $B = 0$ and now $|\psi|^2 = A^2$ which is a constant.

b) In order to normalize the wavefunction to satisfy Equation 1.6 we require that

$$\int_{-\infty}^{\infty} |\psi|^2 \, dx = 1$$

This requires that $|A| \to 0$ and there is a vanishingly small probability of finding the electron in any differential length dx along the x-axis. This is consistent with the infinite length of the x-axis and the fact that the wavefunction is describing a single electron over this infinite length.

EXAMPLE 1.6

Assume the solutions to Example 1.5 have the property $|A|=|B|$.

a) Find and sketch the probability density function.
b) Discuss normalization of the wavefunctions.

Solution

a) The equality $|A|=|B|$ is a necessary condition for a standing wave. If we let
$C = A^2 + B^2 = 2A^2 = 2AB$ then the probability density function becomes

$$|\psi|^2 = C(1+\cos 2kx)$$

This can be plotted:

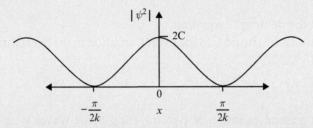

b) Once again, in order to normalize the wavefunction we require that $C \to 0$ and there is a vanishingly small probability of finding the electron in any differential length dx along the x-axis.

EXAMPLE 1.7

Discuss the uncertainty in position and the uncertainty in momentum for the electron in Examples 1.5 and 1.6.

Solution

For Example 1.5 there is a traveling wave solution with a velocity in the positive x-direction. This wave has momentum $p = \dfrac{h}{\lambda}$. Although there is no uncertainty in momentum, this does not violate the uncertainty principle of Equation 1.5 because Δx, or the length of the x-axis, is infinite.

In Example 1.6 there is uncertainty in momentum. This is because momentum p may be positive or negative, i.e. $p = \pm\dfrac{h}{\lambda}$, depending on the direction of each of the two waves forming the standing wave. We can approximate the uncertainty in momentum as $\Delta p \cong 2\dfrac{h}{\lambda}$. Note that any finite value of wavelenght λ is valid and therefore any finite value of Δp is possible. This again does not violate the uncertainty principle of Equation 1.5 because Δx is infinite.

EXAMPLE 1.8

a) Find the energy of the electron for Example 1.5.
b) Show that the result of solving the Schrödinger equation is consistent with Equation 1.3.

Solution

a) The Schrödinger equation also tells us the energy of the electron. If we substitute Equation 1.13 into Equation 1.12 we have

$$-\frac{\hbar^2}{2m}\frac{d^2}{dx^2}\left(Ae^{ikx} + Be^{-ikx}\right) = E\psi(x) \tag{1.14}$$

Differentiating and solving for E we obtain

$$E = \frac{\hbar^2 k^2}{2m} \tag{1.15}$$

The energy of the electron is determined by the electron wavenumber k. It is important to note that the electron energy depends on the electron wavenumber, and since there is no restriction on wavenumber there is also no restriction on possible values of electron energy.

b) Since we have assumed $U = 0$, all the electron's energy is kinetic energy $E = \dfrac{p^2}{2m}$ and hence comparing to Equation 1.15 we see that $p = \hbar k = \dfrac{h}{\lambda}$. This is Equation 1.3 and therefore the de Broglie equation is implicitly embedded in Schrödinger's equation.

1.8   THE ELECTRON IN A ONE-DIMENSIONAL WELL

In practical situations electrons are not free to move infinite distances along an axis and we will now consider the case of an electron that is free to move over a finite portion of the x-axis only. Beyond this range the electron will encounter potential barriers that limit its movement.

Consider a one-dimensional steady state or time-invariant problem in which an electron is free to move around in the *one-dimensional potential well* illustrated in Figure 1.7. The potential energy of an electron is zero inside the well and Φ outside the well. This model can be thought of in the context of a hypothetical solid having one dimension a with vacuum outside the solid. The electron potential energy is zero inside the solid and the height of the potential well is equal to the *workfunction* Φ of the solid.

To determine the wavefunctions of the electron we again make use of the time-independent Schrödinger Equation 1.10

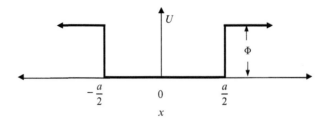

FIGURE 1.7   Potential well with zero potential for $-\frac{a}{2} \le x \le \frac{a}{2}$ and a potential of Φ for $x < -\frac{a}{2}$ or $x > \frac{a}{2}$. In a hypothetical solid we can consider the wall height of the well to be equal to the workfunction of the solid.

$$-\frac{\hbar^2}{2m}\frac{d^2\psi(x)}{dx^2}+U(x)\psi(x)= E\psi(x) \tag{1.16}$$

Inside the well, $U=0$ and hence

$$-\frac{\hbar^2}{2m}\frac{d^2\psi(x)}{dx^2}= E\psi(x)$$

The general solution of this from Equation 1.13 is $\psi(x)= \alpha e^{ikx} + \beta e^{-ikx}$ with $k = \dfrac{\sqrt{2mE}}{\hbar}$. Note that these terms represent components of traveling waves moving in opposite directions. Since there is no preferred traveling wave direction due to the symmetry of the potential well about the origin we can conclude that these waves form a standing wave, i.e. $|\alpha|=|\beta|$. Since there is no time dependence, these solution are referred to as *stationary states* or *eigenstates*. An *eigenenergy* is associated with an eigenstate.

There are two possibilities:

If $\alpha = \beta$ we obtain an even function

$$\psi(x)= \alpha\left(e^{ikx} +e^{-ikx}\right)= A\cos kx \tag{1.17a}$$

or if $\alpha = -\beta$ we obtain an odd function

$$\psi(x)= \alpha\left(e^{ikx} -e^{-ikx}\right)= A\sin kx \tag{1.17b}$$

Note that coefficient A is an imaginary number in Equation 1.17b. This is not a difficulty since the observable $|\psi|^2 = \psi^*\psi$ is always a real number.

In regions where the potential is equal to Φ we again apply Equation 1.16. We will assume that the potential step in Figure 1.7 satisfies the condition $\Phi > E$. Equation 1.16 may be written

$$\frac{\hbar^2}{2m}\frac{d^2\psi(x)}{dx^2} = (\Phi - E)\psi(x)$$

Since $(\Phi - E)$ is positive, the general solution is

$$\psi(x) = Ce^{\gamma x} + De^{-\gamma x}$$

where

$$\gamma = \frac{\sqrt{2m(\Phi - E)}}{\hbar}$$

To further simplify $\psi(x)$ we note that for $x \leq -\frac{a}{2}$ it follows that $D = 0$ to eliminate the physically impossible solution in which $\psi(x)$ rises exponentially as x goes more negative. Similarly for $x \geq \frac{a}{2}$ it follows that $C = 0$.

Hence for $x \leq -\frac{a}{2}$ we have

$$\psi(x) = Ce^{\gamma x} \tag{1.18a}$$

and for $x \geq \frac{a}{2}$,

$$\psi(x) = De^{-\gamma x} \tag{1.18b}$$

We can now apply boundary conditions at $x = -\frac{a}{2}$ and at $x = \frac{a}{2}$. These boundary conditions require that both the wavefunction ψ and its slope $\frac{d\psi}{dx}$ are continuous at the boundaries. In the absence of this condition the second derivative of the wavefunction $\frac{d^2\psi}{dx^2}$ would become infinite and the Schrödinger equation could not be satisfied.

In the even case, for ψ to be continuous at $x = \frac{a}{2}$, we obtain, using Equations 1.17a and 1.18b,

$$D\exp\left(-\frac{\gamma a}{2}\right) - A\cos\left(\frac{ka}{2}\right) = 0 \tag{1.19a}$$

and for $\frac{d\psi}{dx}$ to be continuous at $x = \frac{a}{2}$, by differentiating Equations 1.17a) and 1.18b), we further obtain

$$D\gamma\exp\left(-\frac{\gamma a}{2}\right) - Ak\sin\left(\frac{ka}{2}\right) = 0 \tag{1.19b}$$

To obtain simultaneous solutions for D and A in Equation 1.19, the determinant of the matrix formed from the coefficients must be zero. Thus,

$$\begin{vmatrix} \exp\left(-\dfrac{\gamma a}{2}\right) & -\cos\left(\dfrac{ka}{2}\right) \\ \gamma\exp\left(-\dfrac{\gamma a}{2}\right) & -k\sin\left(\dfrac{ka}{2}\right) \end{vmatrix} = 0$$

This may be simplified to

$$k\tan\left(\frac{ka}{2}\right) = \gamma \tag{1.20}$$

Only discrete values of k and γ are allowed due to the periodicity of the tangent function.

In the odd case for ψ to be continuous at $x = \dfrac{a}{2}$ we obtain, using Equations 1.17b and 1.18b,

$$D\exp\left(-\frac{\gamma a}{2}\right) - A\sin\left(\frac{ka}{2}\right) = 0 \tag{1.21a}$$

and for $\dfrac{d\psi}{dx}$ to be continuous at $x = \dfrac{a}{2}$, by differentiating Equations 1.17a and 1.18b, we further obtain

$$D\gamma\exp\left(-\frac{\gamma a}{2}\right) + Ak\cos\left(\frac{ka}{2}\right) = 0 \tag{1.21b}$$

To obtain simultaneous solutions to D and A in the two Equations 1.21 the determinant of the matrix formed from the coefficients must be zero. Thus,

$$\begin{vmatrix} \exp\left(-\dfrac{\gamma a}{2}\right) & -\sin\left(\dfrac{ka}{2}\right) \\ \gamma\exp\left(-\dfrac{\gamma a}{2}\right) & k\cos\left(\dfrac{ka}{2}\right) \end{vmatrix} = 0$$

This may be simplified to

$$-\gamma\tan\left(\frac{ka}{2}\right) = k \tag{1.22}$$

Again, only discrete values of k and γ are allowed.

EXAMPLE 1.9

Assume a potential well with $a = 12\times10^{-10}$ m and $\Phi = 5$ eV.

a) Find the allowed electron energy values.
b) Find and graph the allowed wavefunctions.

c) Estimate the uncertainties in position and momentum.

d) Show that uncertainty is consistent with the uncertainty principle.

Solution

a) For the even solutions, since $k = \dfrac{\sqrt{2mE}}{\hbar}$ and $\gamma = \dfrac{\sqrt{2m(\Phi - E)}}{\hbar}$, we use Equation 1.20 to obtain

$$\tan\left(\frac{a\sqrt{2mE}}{2\hbar}\right) = \sqrt{\frac{\Phi - E}{E}} \tag{1.23}$$

Similarly for the odd solutions, we use Equation 1.22 to obtain

$$\tan\left(\frac{a\sqrt{2mE}}{2\hbar}\right) = -\sqrt{\frac{E}{\Phi - E}} \tag{1.24}$$

After substituting the values for a, Φ, $\hbar$, and m, we now have Equation 1.23 or 1.24 with only one unknown, the electron energy E. The solutions may be obtained by plotting both sides of each equation as a function of E and looking for intersections of the curves as shown below.

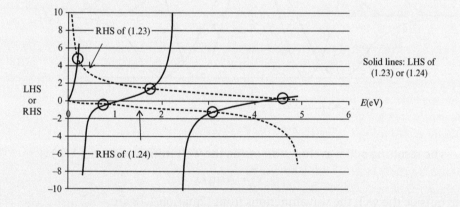

The only five possible solutions to the Schrödinger equation (circled) are found at the intersections of the curves representing the two sides of Equation 1.23 corresponding to the even case (three solutions) and Equation 1.24 corresponding to the odd case (two solutions).

The five possible values of energy E corresponding to the intersections of the curves are shown in the table below. Each energy is labeled using a *quantum number n* that refers to the eigenenergy corresponding to each of the five eigenstates. The lowest possible electron eigenenergy is called the *ground state* energy level $E_1 = 0.20\text{eV}$. If there is only one electron in the energy well then this electron will remain in the ground state unless it receives enough energy to jump into one of the four possible *excited states*.

(continued)

(*continued*)

Quantum number n	n = 1	n = 2	n = 3	n = 4	n = 5
Energy level E_n	E_1	E_2	E_3	E_4	E_5
Energy	0.20 eV	0.80 eV	1.75 eV	3.05 eV	4.55 eV

b) The resulting even wavefunction inside the well from Equation 1.17a is now

$$\psi(x) = A \cos kx$$

and outside the well from Equations 1.18 we have

$$\psi(x) = De^{\pm \gamma x}$$

The three possible even wavefunctions resulting from Equations 1.17a and 1.18 are shown below. Note that we have not normalized these wavefunctions.

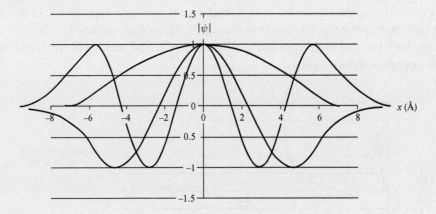

The resulting odd wavefunction inside the well from Equation 1.17b) is

$$\psi(x) = A \sin kx$$

and outside the well the wavefunctions from Equation 1.18 are

$$\psi(x) = De^{\pm \gamma x}$$

The two possible odd wavefunctions resulting from Equations 1.17b and 1.18 are plotted below. The wavefunctions are not normalized. Note that due to the anti-symmetry of these wavefunctions C and D in Equation 1.18 have opposite signs but equal magnitudes.

c) The uncertainty in position for any of the eigenstates is roughly equal to the width of the potential well or $12\,\text{Å}$.

Possible momentum values for any particular solution may be obtained from the electron energy corresponding to the solution. The solution having the lowest

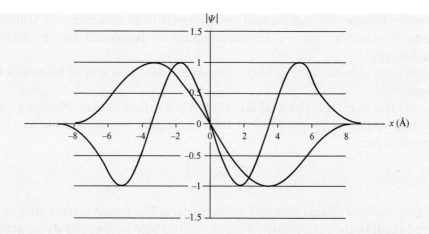

energy is $E_1 = 0.20\text{eV}$. Momentum in this state can be positive or negative and its values are given by

$$p = \pm\sqrt{2mE} = \pm\sqrt{2\times9.11\times10^{-31}\,\text{Kg}\times0.20\text{eV}\times1.6\times10^{-19}\,\text{JeV}^{-1}}$$
$$= \pm2.41\times10^{-25}\,\text{Kgms}^{-1}$$

Uncertainty in momentum can be estimated as $\Delta p = 2\times2.41\times10^{-25}\,\text{Kgms}^{-1}$ $= 4.82\times10^{-25}\,\text{Kgms}^{-1}$

Now, $\Delta x\Delta p = 12\times10^{-10}\,\text{m}\times4.82\times10^{-25}\,\text{Kgms}^{-1} = 5.78\times10^{-34}\,\text{Kgm}^2\text{s}^{-1}$
$$= 5.78\times10^{-34}\,\text{Js}$$

The uncertainty principle from Equation 1.5 requires that

$$\Delta x\Delta p \geq \frac{\hbar}{2} = \frac{h}{4\pi} = \frac{6.63\times10^{-34}\,\text{Js}}{4\times3.14} = 5.27\times10^{-35}\,\text{Js}$$

Our estimate of uncertainty satisfies the uncertainty principle.

A limiting case of Example 1.9 is an *infinite potential well* for which $\Phi = \infty$. See Problem 1.6. This is used in Chapter 2 to develop the behavior of electrons in semiconductors.

1.9   THE HYDROGEN ATOM

An important and widely applicable system involves a spherically symmetric potential energy experienced by an electron at radius r from a positive charge Q. The positive charge can be the nucleus of an atom such as a proton in a hydrogen atom in which case $Q = q$. Other sources of positive charge are possible and will be discussed in subsequent chapters.

In order to solve for the allowed energy levels of the electron, we will apply Schrödinger's equation and we therefore require an expression for the electron's potential energy U.

Gauss's law may be used to obtain this expression. It is one of Maxwell's four equations.

Gauss's law states that the total flux of the electric field passing through an arbitrary closed surface is proportional to the charge enclosed by the surface. In its mathematical form,

$$\oint \varepsilon \cdot d\mathbf{s} = \frac{Q}{\epsilon_0}$$

where Q is the total charge enclosed by the surface. The closed surface integral has integrand equal to the dot product of ε, an electric field vector, and $d\mathbf{s}$, a vector of magnitude equal to an infinitesimal portion of the area of the closed surface and directed normal to, and outward from, the closed surface portion. The closed surface is known as a *Gaussian surface* and ϵ_0 is the electric permittivity of vacuum.

In the case of a point charge Q located at the center of a spherical Gaussian surface, the integral is simple to calculate due to the symmetry of the problem. In this case,

$$\oint \varepsilon \cdot d\mathbf{s} = \varepsilon \oint d\mathbf{s} = 4\pi r^2 \varepsilon$$

The dot product of the vectors in the integrand becomes the scalar product of ε and $d\mathbf{s}$ and by symmetry, ε is a constant over the spherical Gaussian surface. Applying Gauss's law and solving for ε we obtain

$$\varepsilon = \frac{Q}{4\pi\epsilon_0 r^2} \tag{1.25}$$

which is Coulomb's law because $F = q\varepsilon$.

An electron far away from a proton will accelerate toward the proton. The gain in the electron's kinetic energy ΔE as it accelerates over distance d corresponds to an equivalent loss ΔU in its potential energy. The potential energy of the electron at infinite separation from the proton is defined as zero.

The potential energy of the electron at a finite separation from the proton can be obtained using Equations 1.1a.and 1.1b Since ε is a function of separation, the change in potential energy as the electron moves a distance $d = \infty - r$ from infinite separation to a distance r from the proton requires integrating over the distance traveled where ε comes from Equation 1.25. Therefore

$$U(r_0) = \Delta U = -q\varepsilon d = -\frac{qQ}{4\pi\epsilon_0} \int_\infty^r \frac{1}{(r')^2} dr' = -\frac{qQ}{4\pi\epsilon_0 r} = -\frac{q^2}{4\pi\epsilon_0 r} \tag{1.26}$$

To manage the use of Schrödinger's equation, we will assume that the central charge Q does not move under the influence of the motion of the electron. In the case of a hydrogen atom, the error created by this assumption is small because the proton mass is about three orders of magnitude heavier than the electron mass. We will be able to resolve this error later.

Thus far, the examples in this chapter of the application of Schrödinger's equation have been one dimensional. A three-dimensional form is shown in Equation 1.10 but is not directly useable here because it is written in Cartesian coordinates whereas the hydrogen atom is inherently more suited to a spherical geometry.

As a result, both the potential energy $U(r,\theta,\phi)$ and the resulting wavefunctions $\psi(r,\theta,\phi)$ need to be expressed using spherical polar coordinates, and Equation 1.10 needs to be re-cast in spherical polar coordinates.

We will now obtain a solution to this problem for the case where the wavefunction is spherically symmetric and therefore independent of coordinates θ and ϕ. In this case the wavefunction is a function of r only.

The first term in Schrödinger's equation in three dimensions is

$$-\frac{\hbar^2}{2m}\nabla^2\psi = -\frac{\hbar^2}{2m}\left(\frac{\partial^2\psi}{\partial x^2}+\frac{\partial^2\psi}{\partial y^2}+\frac{\partial^2\psi}{\partial z^2}\right)$$

The radius coordinate r in terms of cartesian coordinates x, y, z is

$$r=\sqrt{x^2+y^2+z^2}$$

Now,

$$\frac{\partial\psi}{\partial x}=\frac{\partial\psi}{\partial r}\frac{\partial r}{\partial x}=\frac{x}{\sqrt{x^2+y^2+z^2}}\frac{\partial\psi}{\partial r}=\frac{x}{r}\frac{\partial\psi}{\partial r}$$

and

$$\frac{\partial^2\psi}{\partial x^2}=\frac{\partial}{\partial x}\left(\frac{x}{r}\frac{\partial\psi}{\partial r}\right)=\frac{1}{r}\frac{\partial\psi}{\partial r}+x\frac{\partial}{\partial x}\left(\frac{1}{r}\frac{\partial\psi}{\partial r}\right)=\frac{1}{r}\frac{\partial\psi}{\partial r}+x\frac{\partial r}{\partial x}\frac{\partial}{\partial r}\left(\frac{1}{r}\frac{\partial\psi}{\partial r}\right)$$

$$=\frac{1}{r}\frac{\partial\psi}{\partial r}+\frac{x^2}{r}\frac{\partial}{\partial r}\left(\frac{1}{r}\frac{\partial\psi}{\partial r}\right)$$

Repeating the above for the y and z coordinates and combining,

$$\nabla^2\psi=\frac{3}{r}\frac{\partial\psi}{\partial r}+\frac{x^2+y^2+z^2}{r}\frac{\partial}{\partial r}\left(\frac{1}{r}\frac{\partial\psi}{\partial r}\right)=\frac{3}{r}\frac{\partial\psi}{\partial r}+r\frac{\partial}{\partial r}\left(\frac{1}{r}\frac{\partial\psi}{\partial r}\right)$$

$$=\frac{2}{r}\frac{\partial\psi}{\partial r}+\frac{\partial^2\psi}{\partial r^2}=\frac{1}{r}\frac{\partial^2}{\partial r^2}(r\psi)$$

Using Equation 1.10, we can now rewrite Schrödinger's equation as

$$-\frac{\hbar^2}{2mr}\frac{\partial^2}{\partial r^2}[r\psi(r)]+U(r)\psi(r)=E\psi(r) \tag{1.27}$$

which is the form of the time-independent Schrödinger equation required for spherically symmetric potentials and spherically symmetric wavefunctions.
With $U(r)$ from Equation 1.26 we have

$$-\frac{\hbar^2}{2mr}\frac{\partial^2}{\partial r^2}\left[r\psi(r)\right]-\frac{q^2}{4\pi\epsilon_0 r}\psi(r)-E\psi(r)=0 \tag{1.28}$$

This differential equation in $\psi(r)$ is not particularly simple to solve. The simplest type of solution that one might consider, given the second order differential equation, is an exponential function of radius r.

We will try $\psi(r)=Ae^{-\alpha r}$ and determine if it satisfies the equation by substitution. We would expect a negative exponent to avoid a physically unreasonable result. The left-hand side is

$$-\frac{\hbar^2}{2mr}\frac{\partial^2}{\partial r^2}\left[rAe^{-\alpha r}\right]-\frac{q^2}{4\pi\epsilon_0 r}Ae^{-\alpha r}-EAe^{-\alpha r} \tag{1.29}$$

After taking the second derivative, collecting terms, and simplifying, (see Problem 1.12) the left- hand side becomes

$$\left(\frac{\hbar^2\alpha}{m}-\frac{q^2}{4\pi\epsilon_0}\right)\frac{1}{r}-\left(\frac{\hbar^2\alpha^2}{2m}+E\right) \tag{1.30}$$

This expression will only be zero for all values of r if the $\frac{1}{r}$ term is eliminated which requires that

$$\frac{\hbar^2\alpha}{m}-\frac{q^2}{4\pi\epsilon_0}=0$$

and therefore

$$\alpha=\frac{mq^2}{4\pi\epsilon_0\hbar^2}$$

In addition, the remainder of Equation 1.30 must be zero to satisfy Equation 1.28 and we require that

$$\frac{\hbar^2\alpha^2}{2m}+E=0$$

The validity of proposed solution is now confirmed for one specific value of α. The eigenenergy is

$$E=-\frac{\hbar^2\alpha^2}{2m}=-\frac{mq^4}{8\epsilon_0^2 h^2}=-13.6\text{eV}$$

The magnitude of this well-known energy is known as the *Rydberg constant* which represents the amount of energy required to ionize a hydrogen atom or equivalently to separate the electron and proton infinitely far apart.

The associated wavefunction is found by substituting the appropriate value of α into the exponentially decaying function $\psi(r) = Ae^{-\alpha r}$ resulting in

$$\psi(r) = Ae^{-\frac{mq^2}{4\pi\epsilon_0 \hbar^2}r}$$

The electron wavefunction amplitude decreases exponentially as radius r increases. The radius a_0 at which the wavefunction amplitude drops to $\frac{1}{e}$ of its maximum amplitude is obtained by setting the magnitude of the exponent of $\psi(r)$ to one yielding

$$a_0 = \frac{4\pi\epsilon_0 \hbar^2}{mq^2} = 0.53\text{Å}$$

This value of radius is defined as the *Bohr radius*.

The solution we found for the hydrogen atom is known as the *ground state* of the atom. It is called an s orbital because it is spherically symmetric. There are other valid solutions to Schrödinger's equation for the hydrogen atom having eigenenergies that are smaller in magnitude than the Rydberg constant. These represent *excited states* of the hydrogen atom in which the electron and the proton are, on average, further apart and less tightly bound. Some of these other solutions result in wavefunctions that are also spherically symmetric and they are also denoted as s orbitals. In addition, there are solutions that are not spherically symmetric and that are functions of both r and θ. Finally there are solutions that depend on all three coordinates r, θ, and ϕ. The orbitals that are not spherically symmetric are the well-known p, d, and f orbitals. The complete set of solutions is derived in well-known textbooks on quantum mechanics. See Problem 1.11.

Ground state wavefunction $\psi(r) = Ae^{-\frac{mq^2}{4\pi\epsilon_0 \hbar^2}r} = Ae^{-\frac{r}{a_0}}$ can be normalized to determine the value of A. The integral that must be used for normalization must correctly treat differential volume elements in spherical polar coordinates. Hence

$$\int_0^{2\pi} \int_0^{\pi} \sin\theta \int_0^{\infty} r^2 |\psi(r)|^2 \, dr d\theta d\phi = \int_0^{\infty} 4\pi r^2 A^2 e^{-2\frac{r}{a_0}} \, dr = 1 \qquad (1.31)$$

After solving for A, the normalized wavefunction is (see Problem 1.13)

$$\psi(r) = \frac{1}{\sqrt{\pi}} a_0^{-\frac{3}{2}} e^{-\frac{r}{a_0}} \qquad (1.32)$$

A small error in our analysis of the hydrogen atom is caused by the motion of the nucleus of the atom under the influence of the Coulomb force exerted by the electron. This error can be removed by considering the center of mass of the atom and then replacing the electron mass m with μ, where μ is referred to as a *reduced mass*. The derivation of the reduced mass covered in Appendix 4 is straightforward Newtonian physics. The result is

$$\frac{1}{\mu} = \frac{1}{m} + \frac{1}{M}$$

where M is the mass of the proton.

The atomic radius a now deviates from the Bohr radius a_0 and becomes

$$a_{\text{reduced mass}} = \frac{4\pi\epsilon_0\hbar^2}{\mu q^2} \tag{1.33}$$

Also, the binding energy of the ground state electron of the hydrogen atom deviates in magnitude from the Rydberg constant and becomes

$$E_{\text{reduced mass}} = -\frac{\mu q^4}{8\epsilon_0^2 h^2} \tag{1.34}$$

In the hydrogen atom, proton mass is more than three orders of magnitude larger than electron mass and therefore $\mu \cong m$. However, in some atom-like entities, the mass of the positive charge can be significantly lighter than the proton mass and the reduced mass correction becomes more significant.

1.10   ELECTRON TRANSMISSION AND REFLECTION AT POTENTIAL ENERGY STEP

Consider now the influence of a potential step on the propagation of a beam of electrons. See Figure 1.8 showing a potential step of height U_0 which may be positive or negative. Let us assume that a beam of electrons with kinetic energy $E > U_0$, moving from left to right, is incident upon the potential step at position $x = 0$.

The wavefunctions of the electrons are given by solutions of the time-independent Schrödinger equation. For $x \le 0$,

$$\psi_1(x) = Ae^{ik_1x} + Be^{-ik_1x}$$

where

$$k_1 = \sqrt{\frac{2mE}{\hbar^2}}$$

Wavefunctions ψ_2 will describe electrons that are transmitted across the step. Since the wavefunctions have energies $E > U_0$ we obtain

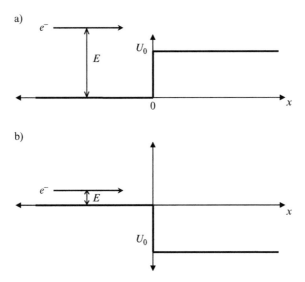

FIGURE 1.8 a) A beam of electrons with kinetic energy E moving from left to right is incident upon a potential step of height $U_0 > 0$ at $= 0$. A transmitted electron flux exists if $E > U_0$. b) If $U_0 < 0$ the condition $E > U_0$ is valid for all values of E.

$$\psi_2 = Ce^{ik_2x}$$

where

$$k_2 = \sqrt{\frac{2m(E-U_0)}{\hbar^2}}$$

Since both ψ and $\dfrac{d\psi}{dx}$ must be continuous across the boundary at $x = 0$ we can write

$$Ae^{ik_1x}\Big|_{x=0} + Be^{-ik_1x}\Big|_{x=0} = Ce^{ik_2x}\Big|_{x=0}$$

or

$$A + B = C \tag{1.35}$$

and

$$ik_1 Ae^{ik_1x}\Big|_{x=0} - ik_1 Be^{-ik_1x}\Big|_{x=0} = ik_2 Ce^{ik_2x}\Big|_{x=0}$$

or

$$Ak_1 - Bk_1 = Ck_2 \tag{1.36}$$

Using Equations 1.35 and 1.36 the result can now be written as

$$\frac{B}{A} = \frac{k_1 - k_2}{k_1 + k_2}$$

and

$$\frac{C}{A} = \frac{2k_1}{k_1 + k_2}$$

Consider the electron beam having wavefunction Ae^{ikx}. The flux of these electrons is proportional to the product of the probability density function and the electron momentum $k = p/\hbar$. We now define *the reflection coefficient* R as a ratio of fluxes between incident and reflected waves or

$$R = \frac{B^2 k_1}{A^2 k_1} = \left(\frac{k_1 - k_2}{k_1 + k_2}\right)^2$$

and the *transmission coefficient* T as a ratio of fluxes between incident and transmitted waves or $T = \dfrac{C^2 k_2}{A^2 k_1} = \dfrac{4k_1 k_2}{(k_1 + k_2)^2}$

Note that $R + T = 1$ as expected.

The reflection coefficient R will approach 1 if $k_1 \ll k_2$ or if $k_2 \ll k_1$. We will refer to these results when we describe electrons passing through potential barriers in semiconductor diode devices in Chapter 3.

1.11   SPIN

Electrons and other particles possess another important characteristic called *spin*. In an experiment performed by Stern and Gerlach in the 1920s, a beam of silver atoms was evaporated from solid silver in a vacuum furnace and then passed through a magnetic field as shown in Figure 1.9. The magnetic field was a converging field in which the lines of magnetic field are more dense near one pole of the magnet.

Consider a magnetic dipole caused by an orbiting electron. This is similar to the magnetic dipole formed by a current-carrying wire in the shape of a coil wound around a core. In Figure 1.10 the silver atom is regarded as such a magnetic dipole. The converging field lines are also shown between the two distinctly shaped magnetic poles. A Lorentz force $F = -q(v \times B)$ is directed outward on the electron as it orbits. There is a component of this force $F_\perp$ that causes a net translational force on the magnetic dipole toward the upper magnet pole. If the magnetic field had no convergence then there would be no translational force.

Using silver atoms, a net magnetic dipole within every silver atom is detected in the Stern-Gerlach experiment: The results are shown on the screen in Figure 1.9. Remarkably, atoms arrive at the screen in only two spatially discrete zones. This is not expected classically since there was no pre-alignment of the magnetic dipole direction of the silver atoms, and yet all the atoms arriving at the screen appear, with equal probability, to have entered the magnets with a magnetic dipole direction either pointing so as to cause a force $F_\perp$ toward the north pole or a force $-F_\perp$ toward the south pole. It is as if only two orientations of magnetic dipole, commonly referred to

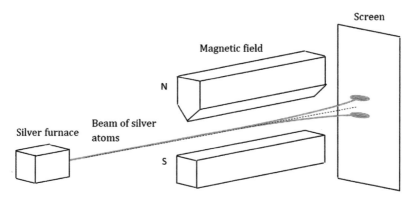

FIGURE 1.9 The Stern-Gerlach experiment in which silver atoms are vaporized in a furnace and sent through a converging magnetic field. The north and south poles of the magnet have distinct shapes which give the field lines a higher density near the north pole.

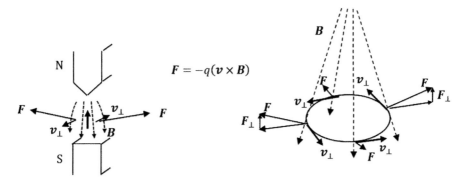

FIGURE 1.10 A magnetic dipole formed by an orbiting electron passing through a converging magnetic field. The magnetic field causes a Lorentz force on the electron as shown in the detailed drawing on the right. For each electron position there is a force $F_\perp$ that acts normal to the plane of electron orbit and toward the north pole of the magnet.

as *spin up* and *spin down* are allowed in spite of the random orientation of silver atoms leaving the oven. Another way to state this is to say that the classically expected continuum of magnetic dipole directions is simply not observable.

In Section 1.3 electrons passing through two slits interfere with each other just as if they are waves. If we try to look at the path of an individual electron to see which slit it actually passes through we interfere with the system and destroy the interference pattern. We are forced to assume that each electron somehow passes through both slits. The measurement to look at the electron path, no matter how carefully performed, disturbs the electron. In the Stern-Gerlach experiment the act of measuring the magnetic dipole direction actually determines the only possible directions of the magnetic dipole and all other possible directions simply cease to exist. If we attempt to determine just how the random silver atom orientations emerging from the furnace are reduced to just two orientations we will similarly disturb the experiment no matter how carefully we make the determination and we will again observe only two orientations of the magnetic dipole that depend on the orientation of the apparatus used to make the measurement. We conclude that quantum mechanics is very definite about what discrete states may and may not be observed.

Although silver atoms contain 47 electrons it turns out that only the one outermost electron in the silver atom is responsible for this magnetic dipole. This can be confirmed by repeating the Stern-Gerlach experiment with many smaller atoms such as hydrogen atoms which also exhibit magnetic dipole behaviour.

Also surprising is the observation that orbital motion of the electron in either a silver atom or a hydrogen atom is not responsible for the deflection observed in the Stern-Gerlach experiment. Instead the electron itself is seen to intrinsically constitute a magnetic dipole. We say that the electron has a *magnetic dipole moment* μ_s. In fact, isolated electrons also exhibit only two possible orientations of magnetic dipole moment.

If we calculate the strength of the magnetic dipole moment of an electron required to explain the observed deflection in the Stern-Gerlach experiment it turns out that it is equal to a fundamental quantity called the *Bohr magneton* μ_b given by

$$\mu_b = \frac{e\hbar}{2m}$$

where m is the mass of the electron. We say that the electron has "spin" even though this is not physically accurate terminology. A quantum number s associated with spin is given the value $s = 1/2$ and as a result we can write for the electron

$$\mu_s = g_s \mu_b m_s$$

Here g_s is called the *spin g factor* and is equal to 2. m_s is the *secondary spin quantum number* and is equal in magnitude to the spin quantum number s. Hence $m_s = \pm 1/2$ to denote the two possible directions of the quantized spin magnetic moment.

In most multi-electron atoms with odd numbers of electrons such as silver, all the electrons have $m_s = \pm 1/2$ but the electrons occur in pairs having $m_s = +1/2$ and $m_s = -1/2$ and their spin magnetic moments therefore cancel out except for the final electron which determines the net spin magnetic moment. Exceptions to this include transition metals and rare-earth elements with partly filled inner shells. Since these atoms have incomplete pairings of inner shell electrons there can be a higher value of net spin.

In addition *orbital angular momentum* which occurs due to the quantum states of electrons surrounding the nucleus of the atom may contribute to the overall atomic magnetic moment. The net orbital angular momentum of multi-electron atoms is often zero because of cancellation of electron magnetic moments but transition metals and rare earth elements often have a net orbital angular momentum and hence a net magnetic moment due to orbital angular momentum. Some atomic states do not have orbital angular momentum. The ground state of hydrogen atom is an example of this; however, excited states of the hydrogen atom may have orbital angular momentum due to the dependence of their wavefunctions on coordinates θ and ϕ.

Practical permanent magnetic materials consist of rare earth and transition metal alloys and compounds.

The Schrödinger equation does explain orbital angular momentum but it does not predict the existence of a spin magnetic moment for particles such as electrons. The existence of spin was proved by Dirac in 1928 by applying Einstein's theory of relativity to the Schrödinger equation.

1.12   THE PAULI EXCLUSION PRINCIPLE

If a single electron exists in an energy well such as the well of Example 1.9 then it will normally occupy the ground state of the energy well and will have quantum number $n = 1$. If a second electron is added to the well it can also occupy the ground state of the energy well but the spins of the two electrons will point in opposite directions. A third electron will be forced to exist in the $n = 2$ state.

By studying the data concerning the energy levels of electrons in atoms, Pauli in 1928 found the following principle which applies to atoms, molecules, and any other multi-electron system such as a potential well containing more than one electron: *Each electron must have a unique set of quantum numbers and be in a unique quantum state.*

Table 1.1 shows the allowed quantum states for the quantum well of Example 1.9. As more electrons are added to the well they will occupy the higher excited states, each having a unique set of quantum numbers up to the tenth and final electron that can be accommodated in the well.

This table is relevant to a one-dimensional potential well. In a three-dimensional potential well known as a *quantum box* the result is similar. Solving the Schrödinger equation for this three-dimensional well will result in three quantum numbers n_x, n_y, and n_z. Since each electron also has a spin quantum number m_s there are a total of four quantum numbers for each electron. The three-dimensional case will be covered in Chapter 2 because it is essential for understanding electronic properties of solids.

In atoms, four quantum numbers result. The first three, normally labeled n, l, and m_l arise from solving the Schrödinger equation, and the spin quantum number m_s is the fourth.

TABLE 1.1   The allowed quantum states for the one-dimensional energy well of Example 1.9. Interactions between electrons have been neglected for simplicity.

Electron count	1 and 2	3 and 4	5 and 6	7 and 8	9 and 10
Quantum number	Ground state	First excited state	Second excited state	Third excited state	Fourth excited state
n	1	2	3	4	5
m_s	+1/2 −1/2	+1/2 −1/2	+1/2 −1/2	+1/2 −1/2	+1/2 −1/2

Since electrons interact with each other in multi-electron systems such as atoms and molecules it is necessary to include electron–electron interactions in our understanding of these systems. In Section 3.6, a more complete analysis of these effects is discussed. This is important for organic electronic devices.

One family of particles called *fermions* includes electrons, protons, neutrons, positrons, and muons. All fermions obey the Pauli exclusion principle and have spin $= 1/2$.

Another family of particles called *bosons* includes photons and alpha particles. These particles have integer spin values and do not obey the Pauli exclusion principle. In addition *phonons* or lattice vibrations are *quasiparticles* that are also bosons.

The concepts of allowable electron states and the Pauli exclusion principle will be further developed in Chapter 4 since they explain important fundamentals of organic electronic materials and devices.

1.13   OPERATORS AND THE POSTULATES OF QUANTUM MECHANICS

The time-dependent Schrödinger equation introduced in Section 1.7 can be rewritten by making use quantum mechanical *operators*. The equation now takes the form

$$\hat{T}\psi + \hat{U}\psi = \hat{E}\psi$$

Here, $\hat{T} = -\dfrac{\hbar^2}{2m}\nabla^2$ is the *kinetic energy operator* that operates on the wavefunction to extract the kinetic energy. In one dimension, $\hat{T} = -\dfrac{\hbar^2}{2m}\dfrac{\partial^2}{\partial x^2}$. The *potential energy operator* is $\hat{U}$ where $\hat{U} = U$. In this case the operator simply multiplies the wavefunction by potential energy which is a scalar quantity. $\hat{E}$ is the *energy operator* $\hat{E} = i\hbar\dfrac{\partial\psi}{\partial t}$ which operates on the wavefunction to extract the total energy.

A very widely used operator known as the *Hamiltonian* is $\hat{H}$ is defined as $\hat{H} = \hat{T} + \hat{U}$ which results in the very compact form of the time-dependent Schrödinger equation

$$\hat{H}\psi = \hat{E}\psi$$

The time-independent Schrödinger equation may also be written using operators. Equation 1.10 becomes, in cartesian coordinates,

$$\hat{H}\psi(x,y,z) = E\psi(x,y,z)$$

The wavefunction ψ, after being acted on by the Hamiltonian operator $\hat{H}$, yields the eigenenergy E multiplied by the wavefunction. The type of equation in which an operator acts on a wavefunction yielding a real number multiplying the wavefunction is called an *eigenequation*. Wavefunctions that are valid solutions to the Schrödinger equation

$$\hat{H}\psi(x,y,z) = E\psi(x,y,z)$$

are called *eigenfunctions* and the resulting values of E are the corresponding *eigenvalues*. Since these eigenvalues are observables they must be real numbers.

At this point we can present the *postulates of quantum mechanics* which are listed below. The wide range of operators and observables falls under the umbrella of these postulates.

Postulate 1. The state of a quantum mechanical system is completely specified by a function $\psi(\mathbf{r},t)$ that depends on the coordinates of the particle(s) and on time. This function, called the wavefunction or probability amplitude, has the important property that $\psi^*(\mathbf{r},t)\psi(\mathbf{r},t)d\mathbf{r}$ is the probability that the particle lies in the volume element $d\mathbf{r}$ located at $\mathbf{r}$ at time t.[1]

For the case of a single particle, the probability of finding it *somewhere* is 1, so that we have the normalization condition

$$\int_{-\infty}^{\infty} \psi^*(\mathbf{r},t)\psi(\mathbf{r},t)d\mathbf{r} = 1$$

It is customary to also normalize many-particle wavefunctions to 1. Each wavefunction must be single-valued, continuous, and finite.

Postulate 2. To every observable in classical mechanics there corresponds a linear, Hermitian operator $\hat{A}$ in quantum mechanics. (See Section 1.14.)

Some common operators occurring in quantum mechanics are collected in Table 1.2.

TABLE 1.2 Physical observables and their corresponding quantum operators (single particle).

Observable		Operator	
Name	Symbol	Symbol	Operation
Position	r	$\hat{r}$	Multiply by r
Momentum	p	$\hat{p}$	$-i\hbar\left(\hat{i}\dfrac{\partial}{\partial x}+\hat{j}\dfrac{\partial}{\partial y}+\hat{k}\dfrac{\partial}{\partial z}\right)$
Kinetic energy	T	$\hat{T}$	$-\dfrac{\hbar^2}{2m}\left(\dfrac{\partial^2}{\partial x^2}+\dfrac{\partial^2}{\partial y^2}+\dfrac{\partial^2}{\partial z^2}\right)$
Potential energy	$U(r)$	$\hat{U}(r)$	Multiply by $U(r)$
Total energy	E	$\hat{H}$	$-\dfrac{\hbar^2}{2m}\left(\dfrac{\partial^2}{\partial x^2}+\dfrac{\partial^2}{\partial y^2}+\dfrac{\partial^2}{\partial z^2}\right)+U(r)$
Angular momentum	L_x	$\hat{L}_x$	$-i\hbar\left(y\dfrac{\partial}{\partial z}-z\dfrac{\partial}{\partial y}\right)\hat{i}$
	L_y	$\hat{L}_y$	$-i\hbar\left(z\dfrac{\partial}{\partial x}-x\dfrac{\partial}{\partial z}\right)\hat{j}$
	L_z	$\hat{L}_z$	$-i\hbar\left(x\dfrac{\partial}{\partial y}-y\dfrac{\partial}{\partial x}\right)\hat{k}$

Postulate 3. In any measurement of the observable associated with operator $\hat{A}$, the only values that will ever be observed are the eigenvalues a which satisfy the eigenvalue equation

$$\hat{A}\psi = a\psi$$

It is possible to have a continuum of eigenvalues in the case of unbound states. If the system is in an eigenstate of $\hat{A}$ with eigenvalue a, then any measurement of the observable A will yield a.

Although measurements must always yield an eigenvalue, the state does not have to be in an eigenstate of $\hat{A}$ *initially*. An arbitrary state can be expanded in the complete set of eigenvalues of $\hat{A}$ where $\hat{A}\psi_i = a_i\psi_i$ and $\psi = \sum_i^n c_i\psi_i$ where n may go to infinity. In this case we only know that the measurement of A will yield *one* of the values a_i but we don't know which one. However, we do know the *probability* that eigenvalue a_i will occur – it is the absolute value squared of the coefficient or $|c_i|^2$ provided ψ is normalized.

After measurement of ψ yields some eigenvalue a_i, the wavefunction immediately collapses into the corresponding eigenstate ψ_i. Measurement affects the state of the system.

Postulate 4. If a system is in a state described by a normalized wavefunction ψ, then the average or expectation value of the observable corresponding to $\hat{A}$ is given by

$$\langle A \rangle = \int_{-\infty}^{\infty} \psi^* \hat{A}\psi dr = \langle \psi | \hat{A} | \psi \rangle$$

See Section 1.14.

Postulate 5. The wavefunction or state function of a system evolves in time according to the time-dependent Schrödinger equation

$$\widehat{H}\psi(r,t) = i\hbar \frac{\partial \psi}{\partial t}$$

Postulate 6. The total wavefunction must be antisymmetric with respect to the interchange of all coordinates of one fermion with those of another. Electronic spin must be included in this set of coordinates. This postulate will be explained in Chapter 4 in the context of the *exchange interaction*.

1.14   EXPECTATION VALUES AND HERMITIAN OPERATORS

Although there is uncertainty in the position of an electron in a one-dimensional potential well, it is still possible to obtain an *expectation value* of position for an electron. Since $|\psi|^2$ is the probability density function we can determine the average value or expectation value for the position of an electron by first normalizing the wavefunction to ensure that $\int_{-\infty}^{+\infty} |\psi|^2 dx = 1$. Then we can determine the expectation value of the position x of the electron by conventional statistical methods regarding $|\psi|^2$ as a probability distribution function and finding the average value of variable x by integrating the product $|\psi|^2 x$ over the x-axis. Hence we obtain $\langle x \rangle = \int_{-\infty}^{+\infty} |\psi|^2 \, x \mathrm{d}x$ where $\langle x \rangle$ is used to denote the expectation value of variable x. Since $|\psi|^2 = \psi^* \psi$ we can also write this using the complex conjugate of the wavefunction as

$$\langle x \rangle = \int_{-\infty}^{+\infty} \psi^* x \psi \mathrm{d}x$$

Using a common shorthand notation known as *Dirac notation* or *bra-ket notation* we would write

$$\langle x \rangle = \int_{-\infty}^{+\infty} \psi^* x \psi \mathrm{d}x = \langle \psi | x | \psi \rangle$$

It is also possible to determine expectation values of other variables such as the expectation value of momentum or the expectation value of energy of an electron by making use of operators that extract the relevant quantity from the wavefunction. For example, the expectation value of the momentum of the electron in the solutions to the Schrödinger equation in Example 1.9 is zero because the solutions originate from waves traveling in both directions with equal probabilities. The expectation value of momentum is found from the general expression

$$\langle p \rangle = \int_{-\infty}^{+\infty} \psi^* \hat{p} \psi \mathrm{d}x = \langle \psi | \hat{p} | \psi \rangle$$

where $\hat{p}$ is the *momentum operator*. The expectation value of energy of a wavefunction is

$$\langle E \rangle = \int_{-\infty}^{+\infty} \psi^* \hat{H} \psi \mathrm{d}x = \langle \psi | \hat{H} | \psi \rangle$$

The concept of expectation value is particularly useful in cases where the wavefunction of the electron is in a *superposition* state which is not an eigenstate of an eigenequation.

See Problems 1.7 and 1.9. Use will be made of expectation values of superposition states in Chapters 4 and 9.

An eigenequation operator that yields real eigenvalues is called a *Hermitian* operator. The Hamiltonian $\hat{H}$ is an example of a Hermitian operator. There are other valid Hermitian operators in quantum mechanics that allow the observable quantity that is being sought to be determined from the relevant eigenequation. These eigenequations allow physical quantities such as position, momentum, angular momentum, spin, and energy to be determined from the valid eigenequation.

Consider an operator $\hat{A}$ where $\psi_a(r)$ and $\psi_{a'}(r)$ are eigenfunctions of $\hat{A}$. In order for $\hat{A}$ to be a Hermitian operator it must satisfy the following relation:

$$\langle \psi_a | \hat{A} | \psi_{a'} \rangle = \langle \psi_{a'} | \hat{A} | \psi_a \rangle^* \tag{1.37}$$

See Problem 1.14. Any two wavefunctions of a Hermitian operator are *orthogonal*. Consider $\psi_a(x)$ and $\psi_{a'}(x)$, the two eigenstates of Hermitian operator $\hat{A}$, and their corresponding real eigenvalues a and a' respectively. Thus the two relevant eigenequations are $\hat{A}\psi_a = a\psi_a$ and $\hat{A}\psi_{a'} = a'\psi_{a'}$. Multiplying the complex conjugate of the first eigenequation by $\psi_{a'}$ and integrating over all space, and then rewriting the result in Dirac notation we obtain

$$\int_{-\infty}^{\infty} \psi_{a'} \left(\hat{A}\psi_a \right)^* dr = a \int_{-\infty}^{\infty} \psi_a^* \psi_{a'} dr = \left\langle \psi_{a'} \left| \hat{A} \right| \psi_a \right\rangle^*$$

Multiplying the second eigenequation by ψ_a^* and integrating over all space, and then rewriting the result in Dirac notation we obtain

$$\int_{-\infty}^{\infty} \psi_a^* \hat{A}\psi_{a'} dr = a' \int_{-\infty}^{\infty} \psi_a^* \psi_{a'} dr = \left\langle \psi_a \left| \hat{A} \right| \psi_{a'} \right\rangle$$

From Equation 1.37, the two equations are equal and therefore

$$\left(a - a'\right) \int_{-\infty}^{\infty} \psi_a^* \psi_{a'} dr = 0$$

Provided the two eigenvalues are distinct, this proves that the eigenstates are orthogonal functions. Functions ψ_a and $\psi_{a'}$ therefore satisfy the well-known requirement for two functions to be orthogonal, namely

$$\int_{-\infty}^{\infty} \psi_a^* \psi_{a'} dr = 0$$

1.15   SUMMARY

1.1 Our classical understanding of the electron includes its charge $q = 1.6 \times 10^{-19}$ coul, its force in an electric field $F = q\varepsilon$, and its energy gain upon acceleration across a potential difference V of $E = qV$. The electron-volt (eV) is a useful unit equivalent to 1.6×10^{-19} J. In a magnetic field the force on an electron traveling with velocity component $\boldsymbol{v}_\perp$ perpendicular to the magnetic field $\boldsymbol{B}$ is given by the Lorentz force $\boldsymbol{F} = -q\left(\boldsymbol{v}_\perp \times \boldsymbol{B}\right)$ which is perpendicular to both the velocity and the magnetic field direction.

1.2 If a beam of electrons is passed through a set of two slits and detected using a screen, the spatial pattern of the electrons can only be explained using wave theory in which the electrons are treated as waves rather than particles. A quantitative analysis of the wave nature of the electrons may be determined by carefully examining the interference pattern of the electron waves using classical wave theory to derive wavelength λ. It is found that λ depends on the momentum $p = mv$ of the electrons according to the de Broglie equation $\lambda = \dfrac{h}{p}$ where Planck's constant h is experimentally found to be $h = 6.63 \times 10^{-34}$ Js.

This equation may also be written in the form $p = \hbar k$ where k is a wave-number.

1.3 If a beam of monochromatic radiation is incident upon a metal sample in vacuum the photoelectric effect is observed. Here, electrons gain enough energy from the light to overcome the workfunction of the metal. Once free, these electrons can travel through the vacuum and be picked up by a second electrode. The measured photocurrent can be used to determine the energy supplied by the monochromatic light. This energy is experimentally found to be $E = h\nu = h\dfrac{c}{\lambda}$ where ν and λ are the frequency and wavelength, respectively, of the light. The concept of the photon arises from this experiment.

1.4 The Heizenberg uncertainty principle states that $\Delta x \Delta p \geq \dfrac{\hbar}{2}$. Here, Δx and Δp refer to the uncertainty in position and momentum of a particle, respectively. It is not possible to simultaneously measure both the precise position and momentum of a particle such as an electron.

1.5 The wavefunction ψ is introduced to describe particles such as electrons. This description inherently embodies probability to address the required attributes of both spatial uncertainty and momentum uncertainty. In additon ψ possesses wavelike properties and is a complex number. The wavefunction ψ may also be referred to as a probability amplitude. The spatial probability density of a particle may be obtained from ψ by determining $|\psi|^2$. We require that $\displaystyle\int_{\text{all space}} |\psi|^2 \, dV = 1$ to ensure the existence of the particle being described.

1.6 The Schrödinger equation $-\dfrac{\hbar^2}{2m}\nabla^2\psi + U\psi = i\hbar\dfrac{\partial\psi}{\partial t}$ is a differential equation that allows us to determine the specific form of the wavefunction ψ describing a particle with potential energy given by $U(x, y, z)$. It also allows the determination of the energy and momentum of the particle. This equation is a highly successful postulate of quantum mechanics. The simplest types of solutions to the Schrödinger equation describe particles such as electrons that are free to move along an axis in one dimension. Any energy and momentum is a possible solution to the equation.

1.7 For an electron in a one-dimensional potential well, the solutions to the Schrödinger equation become very specific. Only discrete solutions are possible due to the requirement that we satisfy boundary conditions. A set of quantum numbers is used to label a corresponding set of wavefunctions and energies of the electron in the well. The amplitude of ψ and the probability density $|\psi|^2$ may be plotted on an axis. The uncertainty in position and momentum resulting from these solutions are shown to satisfy the requirements of the uncertainty principle.

1.8 The hydrogen atom ground state ionization energy can be calculated by solving the Schrödinger equation for an electron in the spherically symmetric potential of a positively charged nucleus. The result is 13.6 eV which is named the Rydberg constant. The Bohr radius of this electron is 0.53 Å. The masses of both the nucleus and the electron may be used to calculate a reduced mass. The reduced mass can be used to correct for the nonstationary nucleus.

1.9 Electrons having energy $E > U_0$ encountering a potential step of height U_0 may be transmitted or reflected at the potential step. Step height U_0 may be positive or negative.

1.10 Expectation values or average values for position may be calculated using the weighted integral $x = \int_{-\infty}^{+\infty} |\psi|^2 x\, dx$ in the one-dimensional case. It is also possible to obtain expectation values for momentum and energy. Hermitian operators yield real eigenvalues and the eigenfunctions of Hermitian operators are orthogonal. The Hamiltonian is a Hermitian operator.

1.11 Although not predicted by the Schrödinger equation unless we include relativistic effects, particles such as electrons possess a property called spin. Spin results in a magnetic moment inherent to the particle. The fundamental unit of this magnetic moment is the Bohr magneton $\mu_b = \dfrac{e\hbar}{2m}$. Only two values of spin are observed which correspond to spin quantum numbers denoted $m_s = \pm 1/2$, and the electron's magnetic moment is $\mu_s = g_s \mu_0 m_s$ The classically expected continuum of spin directions is not observed.

1.12 The Pauli exclusion principle states that only one electron can exist in one quantum state. In systems containing multiple electrons there will be a set of quantum states. Each quantum state will have a unique set of quantum numbers. In a three-dimensional system such as an atom or a quantum box there will be four quantum numbers, one for each spatial dimension and one for spin direction. Fermions obey the Pauli exclusion principle but bosons do not.

PROBLEMS

1.1 In the Davisson-Germer experiment electrons are reflected off two atomic planes that terminate at the surface of a crystalline solid. The distance between the two planes measured at the crystal surface is 2Å. The electrons used are accelerated through a potential difference of 250 V.
   (a) Find the first three angles θ between the incident and reflected electrons that give a maximum electron intensity on the detection screen.
   (b) Find the first three angles θ between the incident and reflected electrons that give a minimum electron intensity on the detection screen.
   (c) Explain why larger slits used for light wave interference experiments would not be useful for electron interference experiments. What slit dimensions are used for light wave interference experiments?

1.2 It is difficult to accept the observation that regardless of how small the electron current becomes in the Davisson-Germer experiment, even to the point of having electrons passing through the apparatus one at a time, there is still an interference pattern on the screen. Do an internet search and read discussions of how physicists attempt to rationalize this. Write a two page summary of your findings.

1.3 Find the wavelengths of the following particles:
  (a) An electron having kinetic energy of 1×10^{-19} J.
  (b) A proton having kinetic energy of 1×10^{-19} J.
  (c) An electron traveling at 1% of the speed of light
  (d) A proton traveling at 1% of the speed of light.

1.4 The photoelectric effect is observed by shining blue monochromatic light having wavelength 450nm onto a sample of cesium. The collector electrode is biased to a voltage V relative to the cesium potential.
  (a) Use the internet to look up the workfunction of cesium.
  (b) Find the most negative value of V, the retarding potential, at which a photocurrent may be observed.
  (c) What is the longest wavelength of radiation that would allow photocurrent from the cesium to be observed if $V = 0$? In what part of the electromagnetic spectrum is this radiation?
  (d) What element has the smallest workfunction? Find several metals that have among the largest workfunctions. Use references or the internet.
  (e) Frequently there is more than one workfunction value for a given metal. Look at reference data and explain reasons for this.

1.5 The wavefunction of an electron in one dimension is given as follows:
  For $x\leq0$ $\psi=0$
  For $0<x<10$ $\psi=A\sin 5\pi x$
  For $x\geq10$ $\psi=0$
  (a) Normalize the wavefunction by determining the value of A.
  (b) What is the probability of finding the electron in the range $3\leq x\leq4$?
  (c) What is the probability of finding the electron in the range $3\leq x\leq3.1$?
  (d) Make a plot of $|\psi|^2$ as a function of x.

1.6 The solution of Example 1.9 may be simplified in the limiting case in which $\Phi=\infty$.
  (a) Starting from Equation 1.16, show that the allowed energy values in this limiting case are $E_n = \dfrac{hn^2}{8ma^2}$
  where n is a quantum number. Hint: The wavefunctions must be zero outside the well. Also the boundary condition that $\dfrac{d\psi}{dx}$ is continuous is no longer relevant since the Schrödinger equation now includes an infinite value of Φ.
  (b) Determine the even and odd wavefunctions.
  (c) Is there a restriction on the values of quantum number n?
  (d) Plot $|\psi|^2$ as a function of x for each of the 5 wavefunctions of d).

1.7 Calculate the expectation value of position and the expectation value of momentum for the following wavefunction. Hint: For momentum, use the integral form of expectation values as defined in Section 1.14 and write the wavefunction as a superposition state

$$x < -\frac{4\pi}{a} \quad \psi(x) = 0$$

$$-\frac{4\pi}{a} \le x \le \frac{2\pi}{a} \quad \psi(x) = A\sin(ax)$$

$$x > \frac{2\pi}{a} \quad \psi(x) = 0$$

$$\psi(x) = A\exp(ikx)$$

1.8   Normalize the wavefunctions that are solutions to the problem in Example 1.9.

1.9   Due to a supply of some extra energy we find that an electron in the potential well of Problem 1.6 can be found in the two lowest energy levels E_1 and E_2 that are each a solution to the Schrödinger equation rather than being just in the ground state. Since this electron is in any one of two quantum states ψ_n that are each a solution to the Schrödinger equation, we call this a *superposition state* ψ_s which is written

$$\psi_s = a\psi_1 + b\psi_2$$

Note that the superposition state will also be a valid solution to the Schrödinger equation since the equation is a linear differential equation.

(a) Normalize each of ψ_1 and ψ_2.

(b) Normalize ψ_s.

(c) Find coefficients a and b such that the probability of the electron having energy E_1 is twice as large as the probability that it has energy E_2. Ensure that ψ_1, ψ_2 and ψ_s are all normalized.

(d) Find the expectation value of the energy of the electron.

(e) Search the internet for an interactive simulation of superposition states of an electron in an infinite square well and confirm that your result is consistent with this.

1.10   The concept of observable electron spin magnetic moment having only two possible orientations, namely spin up and spin down, is difficult to accept because we rightly expect that the silver atoms in the type of experiment performed by Stern and Gerlach are randomly oriented when they emerge from the furnace. Do an internet search and read discussions of how physicists attempt to rationalize our inability to explain how all but two of these orientations "disappear." Write a two page summary of your findings.

1.11   The hydrogen atom contains one electron that exists in the potential energy $U(r)$ caused by a positively charged nucleus. The Schrödinger equation can be solved using spherical polar coordinates for a complete set of electron wavefunctions and eigenenergies as a function of r, θ and ϕ rather than just the ground solution as presented in Section 1.9. Find a suitable textboook on atomic quantum mechanics (see, for example, Suggestions for Further Reading) that contains the solution to this problem and summarize the method and the results in 3–4 pages. Highlight the labeling of these orbitals as s, p, d, and f orbitals.

1.12 Show that Equation 1.30 follows from Equation 1.29.

1.13 Show that Equation 1.32 is obtained by performing the integral of Equation 1.31.

1.14 Show that if an operator $\hat{A}$ satisfies Equation 1.37 it must have real eigenvalues.

NOTE

1. In this context, the symbol dr is used to denote a volume element in space

SUGGESTIONS FOR FURTHER READING

1. Thornton, S. and Rex, A. (2013). *Modern Physics for Scientists and Engineers*, 4e. Cengage Learning.

2. Solymar, I. and Walsh, D. (2004). *Electrical Properties of Materials*, 7e. Oxford University Press.

3. Griffiths, D.J. (2017). *Quantum Mechanics*. Cambridge University Press.

Semiconductor Physics

CONTENTS

Objectives

1. Understand semiconductor band theory and its relevance to semiconductor devices.
2. Obtain a qualitative understanding of how bands depend on semiconductor materials.
3. Introduce the concept of the Fermi energy.
4. Introduce the concept of the mobile hole in semiconductors.
5. Derive the number of mobile electrons and holes in semiconductor bands.
6. Obtain expressions for the conductivity of semiconductor material based on the electron and hole concentrations and mobilities.
7. Introduce the concepts of doped semiconductors and the resulting electrical characteristics.
8. Understand the concept of excess, nonequilibrium carriers generated by either illumination or by current flow due to an external electric power source.
9. Introduce the physics of traps and carrier recombination and generation.
10. Introduce alloy semiconductors and the distinction between direct gap and indirect gap semiconductors.

2.1 INTRODUCTION

A fundamental understanding of electron behavior in crystalline solids is available using the *band theory of solids*. This theory explains a number of fundamental attributes of electrons in solids including:

(i) concentrations of charge carriers in semiconductors;
(ii) electrical conductivity in metals and semiconductors;
(iii) optical properties such as absorption and photoluminescence;
(iv) properties associated with junctions and surfaces of semiconductors and metals.

The aim of this chapter is to present the theory of the band model, and then to exploit it to describe the important electronic properties of semiconductors.

2.2   THE BAND THEORY OF SOLIDS

There are several ways of explaining the existence of energy bands in crystalline solids. The simplest picture is to consider a single atom with its set of discrete energy levels for its electrons. The electrons occupy quantum states with quantum numbers n, l, m, and s denoting the energy level, orbital, and spin state of the electrons. Now if a number N of identical atoms is brought together in very close proximity as in a crystal, there is some degree of spatial overlap of the outer electron orbitals. This means that there is a chance that any pair of these outer electrons from adjacent atoms could trade places. The Pauli exclusion principle, however, requires that each electron occupy a unique energy state. Satisfying the Pauli exclusion principle becomes an issue because electrons that trade places effectively occupy new, *spatially extended* energy states.

In fact, since outer electrons from adjacent atoms may trade places, outer electrons from *all* the atoms may effectively trade places with each other and therefore a set of outermost electrons from the N atoms all appear to share a energy state that extends through the entire crystal. The Pauli exclusion principle can only be satisfied if these electrons occupy a set of distinct, spatially extended energy states. This leads to a set of slightly different energy levels for the electrons that all originated from the same atomic orbital. We say that the atomic orbital splits into an *energy band* containing a large but finite set of electron states with closely spaced energy levels. Additional energy bands will exist if there is some degree of spatial overlap of the atomic electrons in lower-lying atomic orbitals. This results in a set of energy bands in the crystal. Electrons in the lowest-lying atomic orbitals will remain virtually unaltered since there is virtually no spatial overlap of these electrons in the crystal.

Figure 2.1 shows a one-dimensional crystal lattice having lattice constant a on the x-axis. The potential energy $U(x)$ experienced by an electron in the crystal is principally controlled by atomic nuclear charges, bound atomic electrons, and electrons

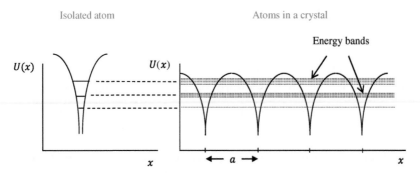

FIGURE 2.1   The energy levels of a single atom are shown on the left. Once these atoms form a crystal, their proximity causes energy level splitting. The resulting sets of closely-spaced energy levels are known as energy bands shown on the right. The electrons in the crystal exist in a periodic potential energy $U(x)$ that has a period equal to the lattice constant a as shown.

involved in bonds. For example, in a covalently bonded crystal, positively charged atomic sites provide potential valleys to a mobile electron and negatively charged regions where covalent bonding electrons are concentrated provide potential barriers. Because the lattice potential energy repeats according to the lattice constant, $U(x)$ is known as a *periodic potential energy*.

The picture we have presented is conceptually a very useful one and it suggests that electrical conductivity may arise in a crystal due to the formation of spatially extended electron states. It does not, however, directly allow us to quantify and understand important aspects of these electrons. We need to understand the number and the behavior of the electrons that move about in the solid to determine the electrical properties of semiconductors and other materials in semiconductor devices.

The quantitative description of these spatially extended electrons requires the use of wavefunctions that describe their spatial distribution as well as their energy and momentum. These wavefunctions may be obtained by applying Schrödinger's equation to the electrons in a periodic potential energy. The following sections present the resulting *band theory* of crystalline solids and the results.

2.3 BLOCH FUNCTIONS

There is an important constraint on the wavefunctions that can exist for an electron in a periodic potential. This constraint comes about from the consideration of the periodicity of the crystal. We will now assume an infinitely large crystal. Later in Section 2.7 we will look at the effects of finite crystal dimensions.

It is very difficult to calculate full wavefunction solutions to Schrödinger's equation in a periodic potential energy. We will therefore start by assuming that $\psi(x)$ is a valid solution to Schrödinger's equation for a periodic potential energy $U(x)$ in one dimension with lattice constant a such as that shown in Figure 2.1. Since we do not have an expression for $\psi(x)$, we will write it in terms of a Fourier series that expresses an arbitrary wavefunction. A summation of component plane waves is made, each wave having a unique wavelength. Therefore

$$\psi(x) = \sum_n a_n e^{ik_n x} \qquad (2.1)$$

Note that this Fourier series was used in Appendix 2 for waves, but we can use it for wavefunctions also. The component waves are $e^{ik_n x}$ and the Fourier coefficients are a_n. From Example 1.5 each component wave component is the spatial part of a traveling plane wave with time dependence given by $e^{-i\omega t}$. We will focus on the spatial part of the wavefunction.

We can also express an arbitrary periodic potential energy $U(x)$ such as that shown in Figure 2.1 by a Fourier series and we can write

$$U(x) = \sum_m U_m e^{iK_m x} \qquad (2.2)$$

in which the Fourier coefficients are U_m. Since $U(x)$ is periodic with period equal to lattice constant a, we know that the wave components of this Fourier series must have specific wavelengths λ_m such that the lattice constant a is an integer multiple of λ_m. Hence

$$\lambda_m = \frac{a}{m}, \quad m = 1, 2, 3 \ldots$$

which means that wavenumber K_m is restricted to values $K_m = \frac{2\pi}{\lambda_m} = \frac{2\pi m}{a}$. The fundamental Fourier component of $U(x)$ exists when $m = 1$.

Note that we have omitted the term for $m = 0$, which means that we are neglecting to include non-zero average potential energy. This is justifiable since in Section 2.4 it will become clear that we are interested in a relative, rather than absolute, energy scale.

The Fourier series expressions in Equations 2.1 and 2.2 may now be substituted into the time-independent Schrödinger equation (Equation 1.10)

$$-\frac{\hbar^2}{2m}\frac{d^2\psi(x)}{dx^2} + U(x)\psi(x) = E\psi(x)$$

and we obtain

$$-\frac{\hbar^2}{2m}\frac{d^2}{dx^2}\left[\sum_{n_1} a_{n_1} e^{ik_{n_1}x}\right] + \left[\sum_m U_m e^{iK_m x}\right]\left[\sum_{n_2} a_{n_2} e^{ik_{n_2}x}\right] = E\sum_{n_3} a_{n_3} e^{ik_{n_3}x}$$

or

$$\frac{\hbar^2}{2m}\sum_{n_1} a_{n_1} k_{n_1}^2 e^{ik_{n_1}x} + \left[\sum_m U_m e^{iK_m x}\right]\left[\sum_{n_2} a_{n_2} e^{ik_{n_2}x}\right] = E\sum_{n_3} a_{n_3} e^{ik_{n_3}x}$$

and finally

$$\frac{\hbar^2}{2m}\sum_{n_1} a_{n_1} k_{n_1}^2 e^{ik_{n_1}x} + \sum_{m,n_2} a_{n_2} U_m e^{i(K_m + k_{n_2})x} = E\sum_{n_3} a_{n_3} e^{ik_{n_3}x} \tag{2.3}$$

There are three summations, each of which represents a Fourier series of component waves. We are free to assign to each Fourier series its own independent index and, to make this clear, we have labeled the index variables n with subscripts 1, 2, and 3 on the three summations.

Let us examine a specific term in the first summation that could give us a solution to Equation 2.3: Select a specific value for k_{n_1}. We now require that

$$k_{n_1} = K_m + k_{n_2} = k_{n_3} \tag{2.4}$$

because we cannot express a wave of infinite spatial extent having a given wavenumber in terms of a Fourier series of waves having other wavenumbers that do not include the given wavenumber. This is because the terms in a Fourier series are *orthogonal* to each other.

The second summation in Equation 2.3 is interesting due to the double index. It comprises a Fourier series of waves $e^{i(K_m + k_{n_2})x}$. In order for Equation 2.4 to be satisfied, we need to choose a value of $n_2 \neq n_1 = n_3$.

The smallest value of K_m is $\dfrac{2\pi}{a}$ for $m = 1$. In this case, we could satisfy Equation 2.4 by making $k_{n_2} = k_{n_1} - \dfrac{2\pi}{a}$. The next available term in the Fourier series for $U(x)$ has $K_m = \dfrac{4\pi}{a}$ and Equation 2.4 could be satisfied if $k_{n_2} = k_{n_1} - \dfrac{4\pi}{a}$. As we extend this analysis to higher order terms of $U(x)$ we discover that adjacent terms of $\psi(x)$ must have wavenumbers that are separated by $\dfrac{2\pi}{a}$ and we realize that allowable forms of $\psi(x)$ are restricted to those that may be written

$$\psi_n(x) = \sum_m a_{n-m} e^{i(k - K_m)x} = e^{ikx} \sum_m a_{n-m} e^{-iK_m x}$$

where the first term in the summation contains $K_m = K_1 = \dfrac{2\pi}{a}$ and $n = n_1 = n_3$. $K_1 = \dfrac{2\pi}{a}$ is the fundamental wavenumber of the Fourier series representation of $U(x)$, and adjacent terms have wavenumbers that are separated by $\dfrac{2\pi}{a}$. We are free to choose any value of k and we have therefore removed the subscript from k. This leads directly to the result that the allowed wavefunctions for the periodic potential must be of the form

$$\psi_n(x) = u_n(x)e^{ikx} \tag{2.5}$$

where

$$u_n(x) = \sum_m a_{n-m} e^{-iK_m x} \tag{2.6}$$

Summation $u_n(x)$ must have the same periodicity as the periodic potential $U(x)$ because it contains wave components that have the same wavenumbers as the periodic potential and importantly, the first term in the summation of terms of $u_n(x)$ has the fundamental wavenumber of the Fourier series representation of $U(x)$. We can now formally state *Bloch's theorem*: Electrons in a periodic potential energy $U(x)$ must have wavefunctions called *Bloch functions* expressed as $\psi(x) = u(x)e^{ikx}$ where $u(x)$ has the same periodicity as the periodic potential energy.

A Bloch function contains the spatial part of a traveling wave e^{ikx} modulated by $u(x)$. The wave may travel in either direction along the x-axis depending on the sign of wavenumber k. The choice of k for the wavenumber is consistent with the notion of a *crystal wavenumber* for an electron traveling through a periodic potential in a crystal.

The Bloch function is very general in that the form of the periodic potential is not specified and the crystal is assumed to be infinite in length. Our understanding of

Bloch functions will improve as we further investigate the properties of electrons in periodic potentials in Sections 2.4 and 2.5.

We have derived Bloch's theorem in one dimension. The derivation in three dimensions is very similar. See Suggestions for Further Reading.

2.4   THE KRONIG–PENNEY MODEL

The *Kronig–Penney model* builds on Bloch functions and is able to further explain the essential features of band theory. First, consider an electron that exists in a specific one-dimensional periodic potential energy $U(x)$. The periodic potential energy can be approximated by a series of regions having zero potential energy separated by potential energy barriers of height U_0 and width b as shown in Figure 2.2, resulting in a periodic potential energy with period $a+b$. We will assume an infinite number of potential energy barriers extending over the range $-\infty < x < \infty$. We associate $a+b$ with the lattice constant of the crystal. Note that the electric potential energy in a real crystal such as that represented in Figure 2.1 is different from the idealized shape of this periodic potential energy; however, the result turns out to be relevant in any case, and Schrödinger's equation is much easier to solve starting from the $U(x)$ of Figure 2.2.

We now take $U(x)$ from Figure 2.2 as well as the Bloch functions from Equation 2.5 and apply them to the time-independent Schrödinger equation (Equation 1.10) in one dimension. Hence,

$$-\frac{\hbar^2}{2m}\frac{d^2}{dx^2}\psi(x)+U(x)\psi(x)=E\psi(x)$$

or

$$-\frac{\hbar^2}{2m}\frac{d^2}{dx^2}\left[u(x)e^{ikx}\right]+U(x)u(x)e^{ikx}-Eu(x)e^{ikx}=0$$

Now, consider regions along the x-axis for which $U(x)=0$. The solutions in these regions will be designated $\psi_1 = u_1(x)e^{ikx}$ and we have

$$-\frac{\hbar^2}{2m}\frac{d^2}{dx^2}\left[u_1(x)e^{ikx}\right]-Eu_1(x)e^{ikx}=0$$

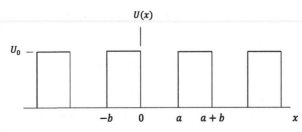

FIGURE 2.2   Periodic one-dimensional potential energy $U(x)$ used in the Kronig–Penney model.

or

$$-\frac{\hbar^2}{2m}\frac{d}{dx}\left[e^{ikx}\frac{du_1(x)}{dx}+u_1(x)ike^{ikx}\right]-Eu_1(x)e^{ikx}=0$$

and hence

$$-\frac{\hbar^2}{2m}\left[e^{ikx}\frac{d^2u_1(x)}{dx^2}+ik\frac{du_1(x)}{dx}e^{ikx}+ik\frac{du_1(x)}{dx}e^{ikx}-k^2u_1(x)e^{ikx}\right]-Eu_1(x)e^{ikx}=0$$

If we define $K^2=\dfrac{2mE}{\hbar^2}$ then this can be written

$$\frac{d^2u_1(x)}{dx^2}+2ik\frac{du_1(x)}{dx}-\left(k^2-K^2\right)u_1(x)=0 \tag{2.7}$$

Note that our use of variable K here is distinct from its use in Section 2.3. Now, consider regions along the x-axis in which $U(x)=U_0$. The solutions in these regions will be designated $\psi_2=u_2(x)e^{ikx}$ and, using the same approach, we obtain

$$\frac{d^2u_2(x)}{dx^2}+2ik\frac{du_2(x)}{dx}-\left(k^2-\beta^2\right)u_2(x)=0 \tag{2.8}$$

where we define $\beta^2=K^2-\dfrac{2mU_0}{\hbar^2}=\dfrac{2m(E-U_0)}{\hbar^2}$. Note that if $E>U_0$ then β is a real number and if $E<U_0$ then β is an imaginary number.

Soutions to Equations 2.7 and 2.8 are

$$u_1(x)=Ae^{i(K-k)x}+Be^{-i(K+k)x}\ \text{for}\ 0\le x\le a \tag{2.9}$$

and

$$u_2(x)=Ce^{i(\beta-k)x}+De^{-i(\beta+k)x}\ \text{for}\ -b\le x\le 0 \tag{2.10}$$

respectively. Note that we have been focusing on selected regions of the x-axis. To satisfy Schrödinger's equation for all values of x, both ψ_1 and ψ_2, and hence u_1 and u_2, must be continuous, and their first derivatives must also be continuous. At $x=0$ we require that $u_1=u_2$ and that $\dfrac{du_1}{dx}=\dfrac{du_2}{dx}$. From Equations 2.9 and 2.10, this gives us

$$A+B-C-D=0$$

and

$$(K-k)A-(K+k)B-(\beta-k)C+(\beta+k)D=0$$

In addition, the Bloch theorem requires that $u(x)$ is periodic with period equal to that of $U(x)$. Hence we can state that $u_1(a) = u_2(-b)$ and that $\left.\dfrac{du_1}{dx}\right|_{x=a} = \left.\dfrac{du_2}{dx}\right|_{x=-b}$. From Equations 2.9 and 2.10 this gives us

$$Ae^{i(K-k)a} + Be^{-i(K+k)a} - Ce^{-i(\beta-k)b} - De^{i(\beta+k)b} = 0 \tag{2.11}$$

and

$$(K-k)Ae^{i(K-k)a} - (K+k)Be^{-i(K+k)a} - (\beta-k)Ce^{-i(\beta-k)b} + (\beta+k)De^{i(\beta+k)b} = 0 \tag{2.12}$$

Equations 2.9, 2.10, 2.11 and 2.12 constitute four equations with four unknowns A, B, C, and D. Solutions exist only if the determinant of the coefficients of A, B, C and D is zero (Cramer's rule). See Problem 2.1. The result is that

$$\frac{-(K^2 + \beta^2)}{2K\beta}(\sin Ka)(\sin \beta b) + (\cos Ka)(\cos \beta b) = \cos k(a+b) \tag{2.13}$$

We will now choose the cases in which total electron energy $E < U_0$ for which β is an imaginary number. This is seen to best represent Figure 2.1 in which $U(x)$ is larger than E in the relevant portions of the x-axis. Let us define $\beta = i\gamma$ where γ is a real number. Equation 2.13 may now be re-written

$$\frac{\gamma^2 - K^2}{2K\gamma}(\sin Ka)(\sinh \gamma b) + (\cos Ka)(\cosh \gamma b) = \cos k(a+b) \tag{2.14}$$

This may be simplified if the limit $b \to 0$ and $U_0 \to \infty$ is taken such that bU_0 is constant (see Problem 2.2). We now define

$$P = \frac{mU_0 ab}{\hbar^2}$$

Since $\gamma \gg K$ and $\gamma b \ll 1$, using Equation 2.14 we obtain

$$\cos ka = P\frac{\sin Ka}{Ka} + \cos Ka \tag{2.15}$$

Here, k is the wavenumber of the electron in the periodic potential and

$$K = \frac{1}{\hbar}\sqrt{2mE} \tag{2.16}$$

which means that K is a term associated with the electron's energy.

There are two limiting cases. In the absence of a periodic potential, $P = 0$ and $\cos ka = \cos Ka$. This implies that all values of momentum are possible and there is no restriction on the wavenumber consistent with Example 1.8.

If $P \to \infty$ then by examining Equation 2.15, solutions only exist if $\dfrac{\sin Ka}{Ka} = 0$

which means that $Ka = ka = n\pi$ where n is a non-zero integer. Only specific values of

k are possible. This turns out to be consistent with the allowed values of k for an electron in a potential well of width a having infinite walls. This will be presented in Example 2.1.

The interesting results lie between these limiting cases. Equation 2.15 only has solutions if its right-hand side is between -1 and $+1$, which restricts the possible values of Ka. The right-hand side is plotted as a function of Ka in Figure 2.3.

Since K and E are related by Equation 2.15, these allowed ranges of Ka actually describe *energy bands* (allowed ranges of E) separated by *energy gaps* or *bandgaps* (forbidden ranges of E). Ka may be re-plotted on an energy axis, which is related to the Ka axis by the square root relationship of Equation 2.15. It is convenient to view E on a vertical axis and k on a horizontal axis as shown in Figure 2.4. Note that $k = \dfrac{n\pi}{a}$, with n an integer, at the highest and lowest points of each energy band where the left side of Equation 2.13 is equal to ± 1. These critical values of k occur at *Brillouin zone boundaries*, and the remaining values of k lie within *Brillouin zones*. Figure 2.4 clearly shows the energy bands and energy gaps. Also, it is seen that k may be positive or negative to denote the direction of motion of an electron along the x-axis. For this reason we should view k as a wavevector $\mathbf{k}$, although we will continue to refer to it as a wavenumber with both positive and negative values which is adequate for a one-dimensional analysis. Note that only three of many energy bands are illustrated.

Let us again consider an electron not inside a periodic potential, but instead in a one-dimensional space with zero potential energy. Solving Equation 1.10 for a free electron with $U = 0$ as shown in Example 1.5 yields the solution

$$\psi(x) = Ae^{ikx} + Be^{-ikx}$$

where

$$E = \frac{\hbar^2 k^2}{2m} \qquad (2.17)$$

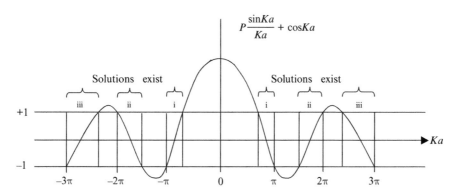

FIGURE 2.3 Graph of right-hand side of Equation 2.15 as a function of P for $P = 2$.

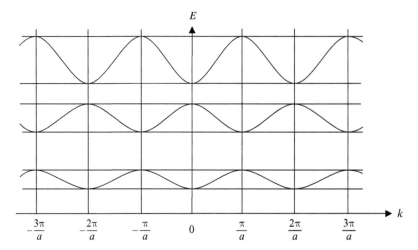

FIGURE 2.4 Plot of E versus k showing how k varies within each energy band and the existence of energy bands and energy gaps. The vertical lines at $k = n\dfrac{\pi}{a}$ are Brillouin zone boundaries. The first Brillouin zone extends from $-\dfrac{\pi}{a} > k > \dfrac{\pi}{a}$ and the second Brillouin zone includes both $-\dfrac{2\pi}{a} > k > -\dfrac{\pi}{a}$ (negative wavenumbers) and $\dfrac{\pi}{a} < k < \dfrac{2\pi}{a}$ (positive wavenumber).

This parabolic E versus k relationship may be plotted superimposed on the curves from Figure 2.4. The result is shown in Figure 2.5.

Taking the limit $P \to 0$, and combining Equations 2.15 and 2.16, we obtain:

$$E = \frac{\hbar^2 k^2}{2m}$$

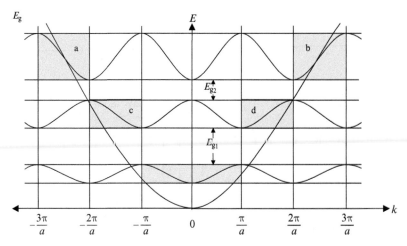

FIGURE 2.5 Plot of E versus k comparing the result of the Kronig–Penney model to the free electron parabolic result.

which is identical to Equation 2.17. This means that the dependence of E on k in Figure 2.5 will become parabolic if the amplitude of the periodic potential energy is reduced to zero. This is consistent with a free electron.

The important relationship between the parabola and the Kronig–Penney model is evident if we look at the solutions to Equation 2.15 within the shaded regions in Figure 2.5. Portions of the free electron parabola have been broken up by energy gaps and distorted in shape. If the periodic potential energy is weakened the solutions to Equation 2.15 more closely resemble the free electron parabola.

We refer to the plots of E versus k in the shaded regions of Figure 2.5 as dispersion relations for electrons in a periodic potential energy. The concept of a dispersion relation was introduced in Section 1.5.

At this point, we can draw some very useful conclusions based on the results of the Kronig-Penney model. The size of the energy gaps increases as the periodic potential energy increases in amplitude in a crystalline solid. Hence:

(i) The periodic potential energies and energy gaps are larger in amplitude for crystalline semiconductors that have small atoms since there are then fewer atomically bound electrons to screen the point charges of the nuclei of the atoms. In contrast, periodic potential energies and energy gaps are smaller in amplitude for crystalline semiconductors that have large atoms with more electron screening.

(ii) The periodic potential energies and hence energy gaps increase in amplitude for ionic semiconductors compared to covalent semiconductors since the ionic character of the crystal bonding increases the localization of positive and negative charges along the x-axis. This will be illustrated in Section 2.11 for some real semiconductors.

To extend our understanding of energy bands we now need to turn to another picture of electron behavior in a crystal.

2.5   THE BRAGG MODEL

Since electrons behave like waves, they will exhibit the behavior of waves that undergo reflections. Notice that in a crystal with lattice constant a, the Brillouin zone boundaries occur at wavenumbers

$$k = \frac{n\pi}{a} = \frac{2\pi}{\lambda}$$

and therefore

$$2a = n\lambda$$

The well-known Bragg condition relevant to electromagnetic waves (X-rays) that undergo strong reflections when incident on a crystal with lattice constant a is

$$2a \sin\theta = n\lambda$$

Now, if the electron is treated as such a wave incident at $\theta = 90°$

$$2a = n\lambda$$

which is precisely the case at Brillouin zone boundaries. We therefore make the following observation: Brillouin zone boundaries occur at the electron wavelengths that satisfy the requirement for strong reflections from crystal lattice planes according to the Bragg condition. This is not really a surprise since both electrons and photons exhibit wave properties as discussed in detail in Chapter 1.

The free electron parabola in Figure 2.5 is most similar to the Kronig–Penney model well away from Brillouin zone boundaries; however, as we approach Brillouin zone boundaries, strong deviations take place and energy gaps are observed.

There is therefore a fundamental connection between the Bragg condition and the formation of energy gaps. Electrons that satisfy the Bragg condition for a given value of n actually exist as standing waves at a corresponding Brillouin zone boundary. For these electrons, reflections will occur equally for electrons traveling in both directions of the x-axis. Provided electrons have wavelengths not close to the Bragg condition, they interact relatively weakly with the crystal lattice and behave more like free electrons.

The E versus k dependence immediately above and immediately below a particular energy gap is contained in four of the shaded regions in Figure 2.5 that we discussed in Section 2.4. The relevant shaded regions for E_{g2} in Figure 2.5 are labeled a, b, c, and d. These four regions are redrawn in Figure 2.6. Each region is shifted along the k axis by an amount $k = \pm\dfrac{2\pi}{a}$. Energy gap E_{g2} occurs at $k = \pm\dfrac{2\pi}{a}$. Since this is a standing wave condition with electron velocity and electron momentum $p = \hbar k$ equal to zero, E_{g2} is redrawn at $k = 0$ in Figure 2.6. Figure 2.6 is known as a *reduced zone scheme*, and only the first Brillouin zone is shown.

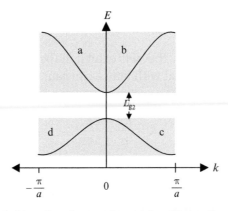

FIGURE 2.6   Plot of E versus k in reduced zone scheme taken from regions a, b, c, and d in Figure 2.5.

Translating the shaded regions along the k-axis by $k = \pm\dfrac{2\pi}{a}$ is permitted for the following reason: Revisiting the Bloch function $\psi(x) = u(x)e^{ikx}$ of Equation 2.5, we note that since $u(x)$ is periodic we can write $\psi(x+a) = u(x)e^{ikx}e^{ika}$ where a is the unit cell length. If $k = \dfrac{2\pi}{a}$ then $e^{ika} = e^{i2\pi} = 1$ and hence $\psi(x+a) = \psi(x)$. No physical change has occurred to the description of the electron. Hence, a shift of $\dfrac{2\pi}{a}$ along the k-axis is really a consequence of a shift in reference location within an infinite crystal by one unit cell of length a on an x-axis. This explains the repeating patterns in Figures 2.4 and 2.5. In fact, any k-value outside of the first Brillouin zone may be shifted back into the first Brillouin zone by moving it an integer number of shifts of $\pm\dfrac{2\pi}{a}$. We conclude that the reduced zone scheme limiting electron wavenumbers to the range $-\dfrac{\pi}{a} \le k \le \dfrac{\pi}{a}$ is valid.

We can now understand that the lower energy band of Figure 2.6 within shaded regions **c** and **d** shares the same k values as the upper band in regions **a** and **b**. The wavefunctions and therefore the energy ranges for the two bands differ, however. This is because each discrete energy value corresponding to each associated energy band for each given value of k corresponds to a discrete summation shown in Equation 2.6. The form of the periodic function $u(x)$ is distinct for each energy band. Figure 2.6 shows only two of many possible energy bands.

In summary, we can now view electrons moving in a periodic potential as being analogous to free electrons, but with dispersion curves having shapes shown in Figure 2.6 that differ from the free electron parabola. In Section 2.6, the treatment of these electrons will be simplified to extend our analogy to free electrons even further.

2.6   EFFECTIVE MASS IN THREE DIMENSIONS

The concept of *effective mass m** will now be introduced. It allows us to quantify electron behavior. Effective mass changes in a peculiar fashion near Brillouin zone boundaries, and generally it is not the same as the free electron mass m. It is easy to understand that the effective acceleration of an electron in a crystal due to an applied electric field will depend strongly on the nature of the reflections of electron waves off crystal planes. Rather than trying to calculate the specific reflections for each electron, we instead modify the mass of the electron to account for its observed willingness to accelerate in the presence of an applied force.

In Chapter 1, Example 1.5 stated the free electron relationship

$$E(k) = \frac{\hbar^2 k^2}{2m}$$

Upon taking the second derivative with respect to k

$$\frac{d^2 E(k)}{dk^2} = \frac{\hbar^2}{m}$$

is obtained, which can be solved for m resulting in

$$m = \frac{\hbar^2}{\dfrac{d^2 E(k)}{dk^2}}$$

Note that the electron mass is inversely proportional to the curvature of the free electron $E(k)$ parabola shown in Figure 2.5.

If the electron is in a periodic potential then, according to the Kronig Penney model, it will have an E versus k dependence as shown in Figure 2.6. Provided we restrict our attention to electrons that are close to the top or the bottom of an energy gap, it is very convenient to approximate the dependence of E versus k as parabolic with the general equation

$$E(k) = \frac{\hbar^2 k^2}{2m^*} + E' \tag{2.18}$$

where E' and m^* are constants. Note that m^* may be positive or negative. Again taking the second derivative of E with respect to k and solving for m^* we obtain

$$m^* = \frac{\hbar^2}{\dfrac{d^2 E(k)}{dk^2}}$$

where m^*, defined as the *effective mass*, is determined by the curvature of the specific parabola in Figure 2.6. This curvature, in turn, depends on the details of the periodic potential energy of the specific crystal in which the electron travels. If an electron exists in one of these parabolic bands, its group velocity v_g as discussed in Section 1.5 is

$$v_g = \frac{1}{\hbar} \frac{dE}{dk}$$

Note that the group velocity falls to zero at the Brillouin zone boundaries where the slope of the E versus k graph is zero. This is consistent with the case of a standing wave.

In general, m^* differs from free electron mass m due to the influence of the periodic potential. It is interesting to note that m^* may be negative for certain values of k. This may be understood physically: if an electron that is close to the Bragg condition is accelerated slightly by an applied force it may then move even closer to the Bragg condition, reflect more strongly off the lattice planes, and effectively accelerate in the direction opposite to the applied force.

An extension of the effective mass concept enables the treatment of three-dimensional semiconductors very near the bottom of a conduction band or the top of a valence band.

For the purposes of this book, the *parabolic, isotropic dispersion relation* will be assumed. This approximation states that the effective mass can be considered a constant that is direction independent. In Sections 2.2 to 2.5 the treatment of band theory has intentionally been one-dimensional to illustrate the important band concepts without the complexity of attempting this in three dimensions. Three-dimensional band modeling becomes specific to a given semiconductor material and this requires details of its lattice structure. In some materials, electron behavior is not isotropic. In this case the speed of an electron may depend on its direction, and it will accelerate to a different degree depending on the direction of the force. Sub-bands that involve spin considerations can also give rise to more than one effective mass. See Section 2.11. However, even in these materials (which include silicon), a suitably averaged effective mass is usually sufficient and the parabolic, isotropic dispersion relation is used. A key justification for this is carrier scattering to be discussed in Section 2.15 that causes randomization of the direction of carrier flow.

In summary, according to the parabolic, isotropic dispersion relation, the only required mathematical distinction between a free electron state in a potential $U = 0$, and an electron state at or near a band edge, is the use of the appropriate effective mass in place of the free electron mass.

2.7 NUMBER OF STATES IN A BAND

The curves in Figure 2.6 are misleading in that electron states in real crystals are discrete and only a finite number of states exist within each energy band. This means that the curves should be regarded as many closely-spaced points that represent quantum states.

Up until now, we have assumed crystals having a periodic potential energy of infinite length along the x-axis which is physically impossible. In order to determine the number of states in a band we need to consider a semiconductor crystal of finite length L. We start by approximating the crystal as a potential well of length L with potential energy $U = 0$ inside the well.

The problem of an electron in a potential well was analyzed in Section 1.8, but we will now assume infinite potential energy boundaries. See Example 2.1.

EXAMPLE 2.1

An electron is inside a potential well of length L with infinite walls and zero potential energy in the well. The well is shown below.

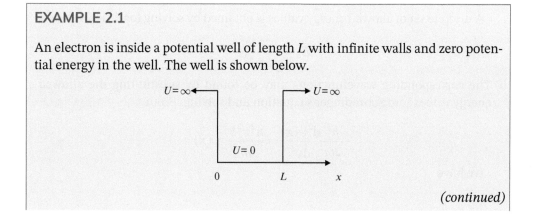

(continued)

(continued)

a) Find the allowed energy levels in the box.
b) Find the wavefunctions of these electrons.

Solution

a) Inside the well, from Schrödinger's equation, we can substitute $U(x) = 0$ and we
   obtain

$$-\frac{\hbar^2}{2m}\frac{d^2\psi(x)}{dx^2} = E\psi(x)$$

Solutions are of the form

$$\psi(x) = A\exp\left(\frac{i\sqrt{2mE}}{\hbar}x\right) + B\exp\left(\frac{-i\sqrt{2mE}}{\hbar}x\right)$$

In regions where $U = \infty$ the wavefunction is zero. In order to avoid discontinu-
ities in the wavefunction we satisfy boundary conditions at $x = 0$ and at $x = L$ and
require that $\psi(0) = 0$ and $\psi(L) = 0$. These boundary conditions can be written

$$0 = A + B \quad \text{or} \quad B = -A$$

and

$$0 = A\exp\left(\frac{i\sqrt{2mE}}{\hbar}L\right) + B\exp\left(\frac{-i\sqrt{2mE}}{\hbar}L\right)$$

$$= A\left(\exp\left(\frac{i\sqrt{2mE}}{\hbar}L\right) - \exp\left(\frac{-i\sqrt{2mE}}{\hbar}L\right)\right) = C\sin\left(\frac{\sqrt{2mE}}{\hbar}L\right)$$

where C is a constant.

We can regard this result as the superposition of two electron waves traveling
in opposite directions, reflecting off the infinite walls, and forming a standing
wave. Now $\sin\theta$ is zero provided $\theta = n\pi$ where n is an integer and hence

$$\frac{\sqrt{2mE}}{\hbar}L = n\pi$$

A discrete set of allowed energy values is obtained by solving for E. The result is

$$E_n = \frac{n^2\pi^2\hbar^2}{2mL^2}$$

b) The corresponding wavefunctions may be found by substituting the allowed
   energy values into Schrödinger's equation and solving. From

$$-\frac{\hbar^2}{2m}\frac{d^2\psi(x)}{dx^2} = \frac{n^2\pi^2\hbar^2}{2mL^2}\psi(x)$$

we have

$$\frac{d^2\psi(x)}{dx^2} = -\frac{n^2\pi^2}{L^2}\psi(x)$$

and hence

$$\psi_n(x) = A\ \sin\ kx = \sin\left(\frac{n\pi}{L}x\right)$$

From Example 2.1 we obtain wavefunction

$$\psi_n(x) = A\sin\left(\frac{n\pi}{L}x\right) \qquad (2.19)$$

where n is a quantum number and the associated wavenumbers are

$$k = \frac{n\pi}{L}, n = 1,2,3\ldots$$

As n increases we will eventually reach the k value corresponding to the Brillouin zone boundary as defined from the band model

$$k = \frac{\pi}{a}$$

This will occur when

$$\frac{n\pi}{L} = \frac{\pi}{a}$$

and therefore $n = \dfrac{L}{a}$ where n is the number of available states in a band. Note that n is the macroscopic length of the semiconductor crystal divided by the unit cell dimension. This is simply the number of unit cells in the crystal which we shall define as N. Since electrons have an additional quantum number s (spin quantum number) that may be either $\frac{1}{2}$ or $-\frac{1}{2}$, the maximum number of electrons that can occupy an energy band becomes $n_b = 2N$.

In Section 2.10 we will introduce a three-dimensional model of electron behavior in a semiconductor crystal and it will be possible to show that the same result $n_b = 2N$ is also valid in a three-dimensional crystal. See Problem 2.3.

2.8 BAND FILLING

The existence of $2N$ electron states in a band does not determine the actual number of electrons in the band. At low temperatures, the electrons will occupy the lowest allowed energy levels, and in a semiconductor like silicon, which has 14 electrons per atom, several low-lying energy bands will be filled. In addition, the highest occupied energy band will be full, and then the next energy band will be empty. This occurs because silicon has an even number of valence electrons per unit cell, and when there are N unit cells, there will be the correct number of electrons to fill the $2N$ states in the highest occupied energy band. A similar argument occurs for germanium as well as carbon (diamond) although diamond is an insulator due to its large energy gap.

Compound semiconductors such as GaAs and other III-V semiconductors as well as CdS and other II-VI semiconductors exhibit the same result: The total number of electrons per unit cell is even, and at very low temperatures in a semiconductor the highest occupied band is filled and the next higher band is empty.

In many other crystalline solids this is not the case. For example Group III elements Al, Ga, and In have an odd number of electrons per unit cell, resulting in the highest occupied band being half filled since the $2N$ states in this band will only have N electrons to fill them. These are metals. Figure 2.7 illustrates the cases we have described, showing the electron filling picture in semiconductors, insulators, and metals.

In Figure 2.7a the lowest empty band is separated from the highest filled band by an energy gap E_g that, in semiconductors, is typically in the range from less than 1eV to between 3eV and 4eV. A completely filled energy band will not result in electrical conductivity because for each electron with a positive wavenumber k there will be one having wavenumber $-k$. The result is a net electron wavenumber of zero within the band. There is no net electron momentum and hence no net electron flux even if an electric field is applied to the material.

Electrons may be promoted across the energy gap E_g by thermal or optical energy, in which case the filled band is no longer completely full and the empty band is no longer completely empty, and now electrical conduction occurs.

Insulators (Figure 2.7b), typically have E_g in the range from about 4 eV to over 6 eV. In these materials it is difficult to promote electrons across the energy gap.

In metals, Figure 2.7c shows a partly filled energy band as the highest occupied band. The energy gap has almost no influence on electrical properties whereas

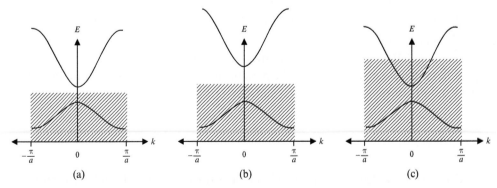

FIGURE 2.7   The filling of the energy bands in (a) semiconductors, (b) insulators, and (c) metals at temperatures approaching 0 K. Available electron states in the hatched regions are filled with electrons and the energy states at higher energies are empty. In band gaps there are no energy states and therefore, although the filling of available energy states is shown hatched, no electrons can exist.

occupied and vacant electron states within this partly filled band are significant: strong electron conduction takes place in metals because empty states exist in the highest occupied band, and electrons may be promoted very easily into higher energy states within this band such that a net electron momentum is produced by a non-zero net wavenumber. A very small applied electric field is enough to promote some electrons into the higher energy states that impart the net momentum to the electrons within the band and electron flow or electric current results.

2.9   FERMI ENERGY AND HOLES

We have seen that partly filled energy bands are of particular interest since they can give rise to electric currents. In semiconductors, a number of electrons may exist near the bottom of the lowest normally empty band due to electrons promoted across the energy gap from the highest normally filled band by thermal or optical excitation. The highest normally empty band is named the *conduction band* because a net electron flux or flow may be obtained in this band. The band directly below the conduction band is almost full; however, because there are empty states near the top of this band, it also exhibits conduction and is named the *valence band*. The electrons that occupy the valence band are valence electrons which exist in covalent bonds in a semiconductor such as silicon or partly covalent bonds in compound semiconductors that have mixed covalent/ionic bonding character.

Figure 2.8 shows the room temperature picture of a semiconductor in thermal equilibrium. Rather than an abrupt point along the energy axis defining the boundary between regions of occupied and unoccupied electron states, there is a more gradual transition between these regions along the energy axis. We define an imaginary horizontal line at energy E_F, called the *Fermi energy*. The Fermi energy represents an energy above which the probability of electron states being filled is under 50%, and below which the probability of electron states being filled is over 50%. We call the

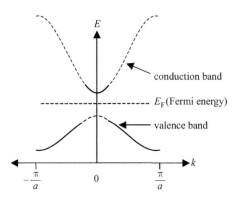

FIGURE 2.8   Room temperature semiconductor showing the partial filling of the conduction band and partial emptying of the valence band. Valence band holes are formed due to electrons being promoted across the energy gap. The Fermi energy lies between the bands. Solid lines represent energy states that have a significant chance of being filled.

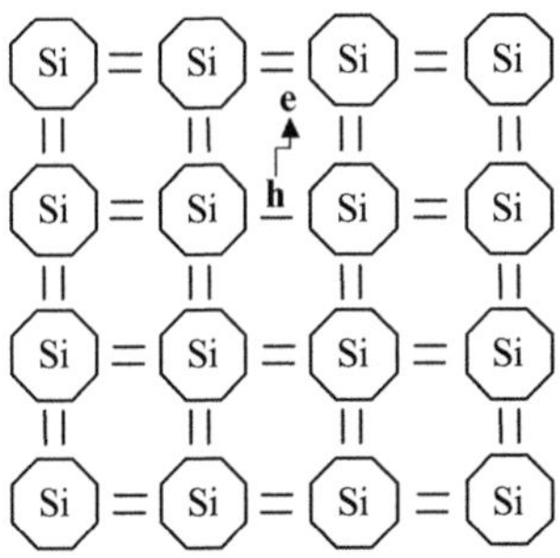

FIGURE 2.9   Silicon atoms have four covalent bonds as shown. Although silicon bonds are tetrahedral, they are illustrated in two dimensions for simplicity. Each bond requires two electrons, and an electron may be excited across the energy gap to result in both a hole in the valence band and an electron in the conduction band that are free to move independently of each other.

empty states in the valence band *holes*. Both valence band holes and conduction band electrons contribute to electrical conductivity.

In a semiconductor we can illustrate the valence band using Figure 2.9, which shows a simplified two-dimensional view of silicon atoms bonded covalently. Each covalent bond requires two electrons. The electrons in each bond are not unique to a given bond, and are shared between all the covalent bonds in the crystal, which means that the electron wavefunctions extend spatially throughout the crystal as described by Bloch functions. A valence electron can be thermally or optically excited and may leave a bond to form an *electron-hole pair* (EHP). The energy required for this is the bandgap energy of the semiconductor. Note that there are multiple bandgaps in crystalline materials as discussed in Section 2.4, but the technologically important bandgap is the one lying between the valence band and the conduction band, and we will henceforth use this definition for the bandgap of a given semiconductor. Once the electron leaves a covalent bond a hole is created. Since valence electrons form extended states, the hole is likewise shared among bonds and is able to move through the crystal. At the same time the electron that was excited enters the conduction band and is also able to move through the crystal resulting in two independent charge carriers.

In order to calculate the conductivity arising from a particular energy band, we need to know the number of electrons n per unit volume of semiconductor, and the number of holes p per unit volume of semiconductor resulting from the excitation of electrons across the energy gap E_g. In the special case of a pure or *intrinsic* semiconductor, we can write the carrier concentrations as n_i and p_i such that $n_i = p_i$

2.10   CARRIER CONCENTRATION

The determination of n and p requires us to examine the states in the conduction band that have a significant probability of being occupied by an electron, and the states in the valence band that have a significant probability of being occupied by a hole, and for each state we need to determine the probability of occupancy to give an appropriate weighting to the state.

A constant effective mass for the electrons or holes in a given energy band will be assumed. In real semiconductor materials the relevant band states are either near the top of the valence band or near the bottom of the conduction band as illustrated in Figure 2.8. In both cases the band shape may be approximated by a parabola, which yields a constant curvature and hence a constant effective mass as expressed in Equation 2.18.

The probability of occupancy by an electron in each band state depends strongly on energy and is given by the *Fermi–Dirac distribution function* which may be derived from Boltzmann statistics.

Consider a crystal lattice having lattice vibrations, or *phonons*, that transfer energy to electrons in the crystal. These electrons that occupy quantum states can also transfer energy back to the lattice, and thermal equilibrium will be established.

Consider an electron in a crystal that may occupy lower and higher energy states E_1^e and E_2^e respectively, and a lattice phonon that may occupy lower and higher energy states E_1^p and E_2^p respectively. Assume this electron makes a transition from energy E_1^e to E_2^e by accepting energy from the lattice phonon while the phonon makes a transition from E_2^p to E_1^p. For conservation of energy,

$$E_2^e - E_1^e = E_2^p - E_1^p \tag{2.20}$$

The probability of these transitions occurring can now be analyzed. Let $p(E^e)$ be the probability that the electron occupies a state having energy E^e. Let $p(E^p)$ be the probability that the phonon occupies an energy state having energy E^p. For a system in thermal equilibrium the probability of an electron transition from E_1^e to E_2^e is the same as the probability of a transition from E_2^e to E_1^e, and we can write

$$p(E_2^p)p(E_1^e)(1-p(E_2^e)) = p(E_1^p)p(E_2^e)(1-p(E_1^e)) \tag{2.21}$$

because the probability that an electron makes a transition from E_1^e to E_2^e is proportional to the terms on the left-hand side in which the phonon at E_2^p must be available and the electron at E_1^e must be available. In addition, the electron state at E_2^e must be vacant because electrons, unlike phonons, must obey the Pauli exclusion principle, which allows only one electron per quantum state. Similarly the probability that the electron makes a transition from E_2^e to E_1^e is proportional to the terms on the right-hand side.

From Boltzmann statistics (see Appendix 5) for phonons or lattice vibrations we use the Boltzmann distribution function:

$$p(E) \propto \exp\left(-\frac{E}{kT}\right) \tag{2.22}$$

Combining Equations 2.21 and 2.22 we obtain

$$\exp\left(-\frac{E_2^p}{kT}\right)p\left(E_1^e\right)\left(1-p\left(E_2^e\right)\right)=\exp\left(-\frac{E_1^p}{kT}\right)p\left(E_2^e\right)\left(1-p\left(E_1^e\right)\right)$$

which may be written

$$p\left(E_1^e\right)\left(1-p\left(E_2^e\right)\right)=\exp\left(\frac{E_2^p-E_1^p}{kT}\right)p\left(E_2^e\right)\left(1-p\left(E_1^e\right)\right)$$

Using Equation 2.20 this can be expressed entirely in terms of electron energy levels as

$$p\left(E_1^e\right)\left(1-p\left(E_2^e\right)\right)=\exp\left(\frac{E_2^e-E_1^e}{kT}\right)p\left(E_2^e\right)\left(1-p\left(E_1^e\right)\right)$$

Rearranging this we obtain

$$\frac{p\left(E_1^e\right)}{1-p\left(E_1^e\right)}\exp\left(\frac{E_1^e}{kT}\right)=\frac{p\left(E_2^e\right)}{1-p\left(E_2^e\right)}\exp\left(\frac{E_2^e}{kT}\right) \tag{2.23}$$

The left side of this equation is a function only of the initial electron energy level and the right side is a function only of the final electron energy level. Since the equation must always hold and the initial and final energies may be chosen arbitrarily we must conclude that both sides of the equation are equal to an energy-independent quantity, which could be a function of the remaining variable T. Let this function be $f(T)$. Hence using either the left side or the right side of the equation we can write

$$\frac{p(E)}{1-p(E)}\exp\left(\frac{E}{kT}\right)=f(T)$$

where E represents the electron energy level.

Solving for $p(E)$ we obtain

$$p(E)=\frac{1}{1+\dfrac{1}{f(T)}\exp\left(\dfrac{E}{kT}\right)} \tag{2.24}$$

The Fermi energy E_F can now be defined as the energy level at which $p(E)=\dfrac{1}{2}$ and hence

$$\frac{1}{f(T)}\exp\left(\frac{E_F}{kT}\right)=1$$

or

$$\frac{1}{f(T)}=\exp\left(\frac{-E_F}{kT}\right)$$

Under equilibrium conditions the final form of the probability of occupancy at temperature T for an electron state having energy E is now obtained by substituting this into Equation 2.24 to obtain

$$F(E) = \frac{1}{1 + \exp\left(\dfrac{E - E_F}{kT}\right)} \tag{2.25}$$

where $F(E)$ is used in place of $p(E)$ to indicate that this is the Fermi–Dirac distribution function. This function is graphed in Figure 2.10.

$F(E)$ is 0.5 at $E = E_F$ provided $T > 0$ K, and at high temperatures the transition becomes more gradual due to increased thermal activation of electrons from lower energy levels to higher energy levels. Figure 2.10 shows $F(E)$ plotted beside a semiconductor band diagram with the energy axis in the vertical direction. The bottom of the conduction band is at E_c and the top of the valence band is at E_v. At E_F there are no electron states since it is in the energy gap; however, above E_c and below E_v the values of $F(E)$ indicate the probability of electron occupancy in the bands. In the valence band the probability for a hole to exist at any energy level is $1 - F(E)$.

In Section 2.7 we found the total number of electron states in an energy band; however, the *distribution* of available energy levels in an energy band was not examined. We now require more detailed information about this distribution within the energy band. This is achieved by finding the *density of states* function $D(E)$, which gives the number of available energy states per unit volume over a differential energy range. It is needed in order to calculate the number of electrons or holes in an energy band. Knowing both the density of available energy states in a band as well as the probability of occupancy of these states in the band is essential. Once we have all this information we can obtain the total number of electrons or holes in a band.

The one-dimensional physics in Section 2.7 now needs to be extended into three dimensions. The three-dimensional density of states function may be derived by solving Schrödinger's equation for an infinite-walled *potential box* in which the

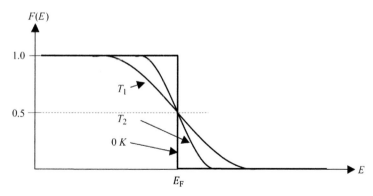

FIGURE 2.10 Plot of the Fermi–Dirac distribution function $F(E)$, which gives the probability of occupancy by an electron of an energy state having energy E. The plot is shown for two temperatures $T_1 > T_2$ as well as for 0 K. At absolute zero, the function becomes a step function.

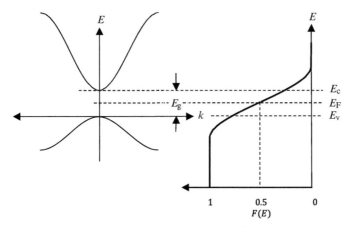

FIGURE 2.11   A semiconductor band diagram is plotted along with the Fermi–Dirac distribution function. This shows the probability of occupancy of electron states in the conduction band as well as the valence band. Hole energies increase in the negative direction along the energy axis. The hole having the lowest possible energy occurs at the top of the valence band. This occurs because by convention the energy axis represents electron energies and not hole energies. The origin of the energy axis is located at E_v for convenience.

wavefunctions must be expressed in three dimensions. In three dimensions, Schrödinger's equation is

$$\frac{-\hbar^2}{2m}\left(\frac{\partial^2}{\partial x^2}+\frac{\partial^2}{\partial y^2}+\frac{\partial^2}{\partial z^2}\right)\psi\left(x,y,z\right)+U\left(x,y,z\right)\psi\left(x,y,z\right)=E\psi\left(x,y,z\right)$$

Firstly, consider a box of dimensions a, b, and c in three-dimensional space in which $U=0$ inside the box when $0<x<a, 0<y<b, 0<z<c$. Outside the box, assume $U=\infty$.
Inside the box using Schrödinger's equation:

$$\frac{-\hbar^2}{2m}\left(\frac{\partial^2}{\partial x^2}+\frac{\partial^2}{\partial y^2}+\frac{\partial^2}{\partial z^2}\right)\psi\left(x,y,z\right)=E\psi\left(x,y,z\right) \tag{2.26}$$

If we let $\psi\left(x,y,z\right)=X(x)Y(y)Z(z)$ then upon substitution into Equation 2.26 and after dividing by $\psi(x,y,z)$ we obtain:

$$-\frac{\hbar^2}{2m}\left(\frac{1}{X(x)}\frac{d^2X(x)}{dx^2}+\frac{1}{Y(y)}\frac{d^2Y(y)}{dy^2}+\frac{1}{Z(z)}\frac{d^2Z(z)}{dz^2}\right)=E$$

Since each term contains an independent variable, we can apply *separation of variables* and conclude that each term is equal to a constant that is independent of x, y, and z.
Now, we have three equations

$$\frac{1}{X(x)}\frac{d^2X(x)}{dx^2}=-C_1 \tag{2.27a}$$

$$\frac{1}{Y(y)}\frac{d^2Y(y)}{dy^2}=-C_2 \tag{2.27b}$$

and

$$\frac{1}{Z(z)}\frac{d^2Z(z)}{dz^2}=-C_3 \tag{2.27c}$$

where

$$E=\frac{\hbar^2}{2m}(C_1+C_2+C_3) \tag{2.28}$$

The general solution to Equation 2.27a is

$$X(x)=A_1\exp(ikx)+A_2\exp(-ikx) \tag{2.29}$$

To satisfy boundary conditions such that $X(x)=0$ at $x=0$ and at $x=a$ we obtain

$$X(x)=A\sin k_x x$$

where

$$k_x=\frac{n_x\pi}{a}$$

with n_x a positive integer quantum number and

$$C_1=\left(\frac{n_x\pi}{a}\right)^2$$

Note that each sinusoidal solution of $X(x)$ comes from Equation 2.29 which represents two traveling plane waves. This is consistent with the formation of a standing wave from two oppositely-directed traveling waves.

Repeating a similar procedure for Equations 2.27b and 2.27c, and using Equation 2.28 we obtain:

$$\psi(x,y,z)=X(x)Y(y)Z(z)=ABC\sin\left(k_x x\right)\sin\left(k_y y\right)\sin\left(k_z z\right)$$

and

$$E=\frac{\hbar^2}{2m}\left(k_x^2+k_y^2+k_z^2\right)=\frac{\hbar^2\pi^2}{2m}\left[\left(\frac{n_x}{a}\right)^2+\left(\frac{n_y}{b}\right)^2+\left(\frac{n_z}{c}\right)^2\right] \tag{2.30}$$

If more than one electron is put into the box at zero kelvin the available energy states will be filled such that the lowest energy states are filled first.

We now need to determine how many electrons can occupy a specific energy range in the box. It is very helpful to define a three-dimensional space with

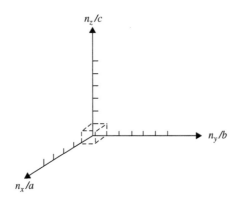

FIGURE 2.12 Reciprocal space lattice. A cell in this space is shown, which is the volume associated with one lattice point. The cell has dimensions $\frac{1}{a}, \frac{1}{b}, \frac{1}{c}$ and volume $\frac{1}{abc}$. The axes all have the same reciprocal length units, but the relative spacing between reciprocal space lattice points along each axis depends on the relative values of a, b, and c.

coordinates $\frac{n_x}{a}, \frac{n_y}{b}$ and $\frac{n_z}{c}$. In this three-dimensional space there are discrete points that are defined by these coordinates with integer values of n_x, n_y, and n_z in what is referred to as a *reciprocal space lattice*, which is shown in Figure 2.12.

From Equation 2.30 it is seen that a spherical shell in reciprocal space represents an equal energy surface because the general form of this equation is that of a sphere in reciprocal space. The number of reciprocal lattice points that are contained inside the positive octant of a sphere having a volume corresponding to a specific energy E will be the number of states having energy smaller than E. The number of electrons is actually twice the number of these points because electrons have an additional quantum number s for spin and $s = \pm\frac{1}{2}$. The positive octant of the sphere is illustrated in Figure 2.13.

Rearranging Equation 2.30 where r is the sphere radius in reciprocal space we obtain

$$\left(\frac{n_x}{a}\right)^2 + \left(\frac{n_y}{b}\right)^2 + \left(\frac{n_z}{c}\right)^2 = \frac{2mE}{\hbar^2\pi^2} = r^2 \tag{2.31}$$

The number of reciprocal lattice points inside the sphere is the volume of the sphere divided by the volume associated with each lattice point shown in Figure 2.12.

The volume of the sphere is

$$V = \frac{4}{3}\pi r^3$$

Using Equation 2.31 we obtain

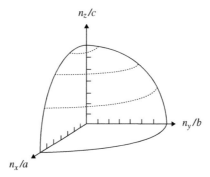

FIGURE 2.13 The positive octant of a sphere in reciprocal space corresponding to an equal energy surface. The number of electron states below this energy is twice the number of reciprocal lattice points inside the positive octant of the ellipsoid.

$$V = \frac{4}{3}\pi\left(\frac{2mE}{\hbar^2\pi^2}\right)^{\frac{3}{2}}$$

If the volume of the sphere is much larger than the volume associated with one lattice point then, including spin, the number of electrons having energy less than E approaches two times one-eighth of the volume of the positive octant of the sphere divided by the volume associated with one lattice point, or:

$$\text{number of electrons} = \frac{2\left(\frac{1}{8}\right)\frac{4}{3}\pi\left(\frac{2mE}{\hbar^2\pi^2}\right)^{\frac{3}{2}}}{\frac{1}{abc}}$$

We define $n(E)$ to be the number of electrons *per unit volume* of the box and therefore

$$n(E) = 2\left(\frac{1}{8}\right)\frac{4}{3}\pi\left(\frac{2mE}{\hbar^2\pi^2}\right)^{\frac{3}{2}}$$

We also define $D(E)$ to be the *density of states* function where

$$D(E) = \frac{dn(E)}{d(E)}$$

and finally we obtain

$$D(E) = \frac{\pi}{2}\left(\frac{2m}{\hbar^2\pi^2}\right)^{\frac{3}{2}} E^{\frac{1}{2}} \tag{2.32}$$

This form of the density of states function is valid for a box having $U = 0$ inside the box. In an energy band, however, $U(x)$ is a periodic function and the density of states function must be modified. Rather than the parabolic E versus k dispersion relation of Equation 2.17 for free electrons in which the electron mass is m, we must use the appropriate effective mass for an electron near the bottom or the top of an energy band as illustrated in Figure 2.8. This is as a result of the parabolic, isotropic dispersion relation. See Section 2.6.

The origin of the energy axis in Equation 2.32 must coincide with the top or the bottom of an energy band to be consistent with the relevant bandgap. It is important to remember that the density of states function is based on a density of available states in reciprocal space, and that for a specific differential range of k-values the corresponding range of energies along the energy axis is dependent upon the slope of the E versus k graph. The slope of E versus k in a parabolic band, and hence the density of states, is controlled by the effective mass.

Consequently, the density of states function in a conduction band is given by Equation 2.32, provided m_e^* is used in place of m, where m_e^* is the effective mass of electrons near the bottom of the conduction band. The point $E = 0$ should refer to the bottom of the band. We now have

$$D(E) = \frac{\pi}{2}\left(\frac{2m_e^*}{\hbar^2\pi^2}\right)^{\frac{3}{2}} E^{\frac{1}{2}} \qquad (2.33a)$$

Since E_v is defined as zero as in Figure 2.11 for convenience, the conduction band starts at $E_c = E_g$. $D(E - E_g)$ tells us the number of energy states available per differential range of energy within the conduction band, and we obtain

$$D(E - E_g) = \frac{\pi}{2}\left(\frac{2m_e^*}{\hbar^2\pi^2}\right)^{\frac{3}{2}}(E - E_g)^{\frac{1}{2}} \qquad (2.33b)$$

The total number of electrons per unit volume in the band is now given by

$$n = \int_{E_g}^{E_{max}} D(E - E_g)F(E)dE \qquad (2.34)$$

where E_{max} is the highest energy level in the energy band that needs to be considered as higher energy levels have a negligible chance of being occupied.

From Equation 2.25, since $E \geq E_g$ and $E_g - E_F \gg kT$, we can use the Boltzmann approximation:

$$F(E) \simeq \exp\left(-\frac{E - E_F}{kT}\right) \qquad (2.35)$$

Hence from Equations 2.33b, 2.34, and 2.35,

$$n = \frac{\pi}{2}\left(\frac{2m_e^*}{\hbar^2\pi^2}\right)^{\frac{3}{2}} \exp\left(\frac{E_F}{kT}\right)\int_{E_g}^{E_{max}}(E - E_g)^{\frac{1}{2}}\exp\left(-\frac{E}{kT}\right)dE$$

$$= \frac{\pi}{2}\left(\frac{2m_e^*}{\hbar^2\pi^2}\right)^{\frac{3}{2}} \exp\left(-\frac{E_g - E_F}{kT}\right)\int_{0}^{\infty}\sqrt{E}\exp\left(-\frac{E}{kT}\right)dE$$

The integral may be solved analytically provided the upper limit of the integral is allowed to be infinity. Using the infinite limit is allowable because the integrand becomes negligible for $E \gg kT$.

From standard integral tables and because $E_c = E_g$ we obtain

$$n_0 = N_c \exp\left[\frac{-(E_c - E_F)}{kT}\right] \tag{2.36a}$$

where

$$N_c = 2\left(\frac{2\pi m_e^* kT}{h^2}\right)^{\frac{3}{2}} \tag{2.36b}$$

The subscript on n indicates that equilibrium conditions apply. The validity of Equations 2.36a and 2.36b is maintained regardless of the choice of the origin on the energy axis since the quantity for determining the electron concentration is the *energy difference* between the conduction band edge and the Fermi energy.

The same procedure may be applied to the valence band. In this case we calculate the number of holes p in the valence band. The density of states function must be written as $D(-E)$ since from Figure 2.11 energy E is negative in the valence band and hole energy increases as we move in the negative direction along the energy axis. We can define a hole effective mass m_h^* and based on Equation 2.32 we obtain

$$D(-E) = \frac{\pi}{2}\left(\frac{2m_h^*}{\hbar^2\pi^2}\right)^{\frac{3}{2}}(-E)^{\frac{1}{2}}$$

The probability of the existence of a hole is $1 - F(E)$, and from Equation 2.25 if $E_F - E \gg kT$ we obtain

$$1 - F(E) \cong \exp\left(\frac{E - E_F}{kT}\right)$$

and now

$$p = \int_{0}^{-E_{max}} D(-E)(1 - F(E))dE$$

In an analogous manner to that described for the conduction band, we therefore obtain

$$p_0 = N_v \exp\left(\frac{-(E_F - E_v)}{kT}\right) \tag{2.37a}$$

where

$$N_v = 2\left(\frac{2\pi m_h^* kT}{h^2}\right)^{\frac{3}{2}} \tag{2.37b}$$

and m_h^*, the hole effective mass, is a positive quantity.

Equation 2.37a shows that the important quantity for the calculation of hole concentration is the *energy difference* between the Fermi energy and the valence band edge. Again the subscript on p indicates that equilibrium conditions apply.

We can now determine the position of the Fermi level and will again set $E_v = 0$ for convenience as illustrated in Figure 2.11. Since $n_i = p_i$ for an intrinsic semiconductor we equate Equations 2.36a and 2.37a and obtain

$$N_c \exp\left(\frac{-(E_g - E_F)}{kT}\right) = N_v \exp\left(\frac{-E_F}{kT}\right)$$

or

$$E_F = \frac{E_g}{2} + \frac{kT}{2}\ln\frac{N_v}{N_c} \tag{2.38}$$

The second term on the right side of Equation 2.38 is generally much smaller than $\frac{E_g}{2}$ and therefore the Fermi energy lies approximately in the middle of the energy gap.

From Equations 2.36a and 2.37a

$$np = N_c N_v \exp\left(\frac{-E_g}{kT}\right) \tag{2.39a}$$

and for an intrinsic semiconductor with $n_i = p_i$

$$n_i = p_i = \sqrt{N_c N_v} \exp\left(\frac{-E_g}{2kT}\right) \tag{2.39b}$$

which is a useful expression for carrier concentration as it is independent of E_F. See Example 2.2.

EXAMPLE 2.2

a) Calculate $n_i = p_i$ for silicon at room temperature and compare with the commonly accepted value.
b) Calculate $n_i = p_i$ for gallium arsenide at room temperature.

Solution

a) Using Appendix 6 to obtain silicon values $m_e^* = 1.08\,m$ and $E_g = 1.11$ eV,

$$N_c = 2\left(\frac{2\pi m_e^* kT}{h^2}\right)^{\frac{3}{2}} = 2\left(\frac{2\pi \times \left(1.08 \times 9.11 \times 10^{-31}\,\text{kg}\right) \times \left(0.026 \times 1.6 \times 10^{-19}\,\text{J}\right)}{\left(6.625 \times 10^{-34}\,\text{J s}\right)^2}\right)^{\frac{3}{2}}$$

$$= 2.84 \times 10^{25}\,\text{m}^{-3} = 2.84 \times 10^{19}\,\text{cm}^{-3}$$

and $m_h^* = 0.56\,m$

$$N_v = 2\left(\frac{2\pi m_h^* kT}{h^2}\right)^{\frac{3}{2}} = 2\left(\frac{2\pi \times \left(0.56 \times 9.11 \times 10^{-31}\,\text{kg}\right) \times \left(0.026 \times 1.6 \times 10^{-19}\,\text{J}\right)}{\left(6.625 \times 10^{-34}\,\text{J s}\right)^2}\right)^{\frac{3}{2}}$$

$$= 1.06 \times 10^{25}\,\text{m}^{-3} = 1.06 \times 10^{19}\,\text{cm}^{-3}$$

Now,

$$n_i = p_i = \sqrt{N_c N_v}\, \exp\left(\frac{-E_g}{2kT}\right) = \sqrt{2.84 \times 10^{19} \times 1.06 \times 10^{19}}$$

$$\times \exp\left(\frac{-1.11\text{eV}}{2 \times 0.026\,\text{eV}}\right) = 9.31 \times 10^9\,\text{cm}^{-3}$$

The commonly accepted value is $n_i = p_i = 1.5 \times 10^{10}$ cm^{-3}. The discrepancy relates mainly to three-dimensional aspects of the effective mass value, and the method and temperature at which effective mass is measured. We will continue to use the commonly accepted carrier concentration unless otherwise noted.

b) For GaAs from Appendix 6, $m_e^* = 0.067\,m$ and $E_g = 1.43$ eV. Hence

$$N_c = 2\left(\frac{2\pi m_e^* kT}{h^2}\right)^{\frac{3}{2}} = 2\left(\frac{2\pi \times \left(0.067 \times 9.11 \times 10^{-31}\,\text{kg}\right) \times \left(0.026 \times 1.6 \times 10^{-19}\,\text{J}\right)}{\left(6.625 \times 10^{-34}\,\text{J s}\right)^2}\right)^{\frac{3}{2}}$$

$$= 4.38 \times 10^{23}\,\text{m}^{-3} = 4.38 \times 10^{17}\,\text{cm}^{-3}$$

and

$$N_v = 2\left(\frac{2\pi m_h^* kT}{h^2}\right)^{\frac{3}{2}} = 2\left(\frac{2\pi \times \left(0.48 \times 9.11 \times 10^{-31}\,\text{kg}\right) \times \left(0.026 \times 1.6 \times 10^{-19}\,\text{J}\right)}{\left(6.625 \times 10^{-34}\,\text{J s}\right)^2}\right)^{\frac{3}{2}}$$

$$= 8.4 \times 10^{24}\,\text{m}^{-3} = 8.4 \times 10^{18}\,\text{cm}^{-3}$$

(continued)

(continued)

A much lower room temperature intrinsic carrier concentration in GaAs with bandgap of 1.43 eV compared to that in silicon with bandgap of 1.11 eV is expected. This is due to the negative exponential dependence of carrier concentration on bandgap.

$$n_i = p_i = \sqrt{N_c N_v} \exp\left(\frac{-E_g}{2kT}\right) = \sqrt{4.38 \times 10^{17} \times 8.4 \times 10^{18}} \exp\left(\frac{-1.43\,eV}{2 \times 0.026\,eV}\right)$$

$$= 2.65 \times 10^6\,cm^{-3}$$

2.11 SEMICONDUCTOR MATERIALS

The relationship between carrier concentration and E_g has now been established and we can look at examples of real semiconductors. A portion of the periodic table showing elements from which many important semiconductors are made is shown in Figure 2.14, together with a list of selected semiconductors and their energy gaps. Note that there are the Group IV semiconductors silicon and germanium, a number of III-V compound semiconductors having two elements, one from Group III and one from Group V respectively, and a number of II-VI compound semiconductors having elements from Group II and Group VI respectively.

A number of interesting observations may now be made. In Group IV crystals, the bonding is purely covalent. Carbon (diamond) is an insulator because it has an energy gap of 6 eV. The *energy gap decreases with atomic size* as we look down the Group IV column from C to Si to Ge and to Sn. Actually Sn behaves like a metal. Since its energy gap is very small, it turns out that the valence band and conduction band effectively overlap when a three-dimensional model of the crystal is considered rather than the one-dimensional model we have discussed. This guarantees some filled states in the conduction band and empty states in the valence band regardless of temperature. Sn is properly referred to as a *semi-metal* (its conductivity is considerably lower than

Group	II	III	IV	V	VI
		B	C	N	O
Element		Al	Si	P	S
	Zn	Ga	Ge	As	Se
	Cd	In	Sn	Sb	Te

Group	IV	IV	IV	IV	III-V	III-V	III-V	III-V	III-V	III-V	II-VI	II-VI
Element(s)	C	Si	Ge	Sn	GaN	AlP	GaP	AlAs	GaAs	InSb	ZnSe	CdTe
Energy gap (eV)	6	1.11	0.67	0	3.4	2.45	2.26	2.16	1.43	0.18	2.7	1.58

FIGURE 2.14 A portion of the periodic table containing some selected semiconductors composed of elements in Groups II to VI.

metals like copper or silver). We can understand this Group IV trend of decreasing energy gaps since the periodic potential of heavy elements will be weaker than that of lighter elements due to electron screening as described in Section 2.5.

As with Group IV materials, the energy gaps of III-V semiconductors decrease as we go down the periodic table from AlP to GaP to AlAs to GaAs and to InSb. The energy gaps of II-VI semiconductors behave in the same manner as illustrated by ZnSe and CdTe. Again, electron screening increases for heavier elements.

If we compare the energy gaps of a set of semiconductors composed of elements from the same row of the periodic table but with increasingly ionic bonding such as Ge, GaAs, and ZnSe, another trend becomes clear: *Energy gaps increase as the degree of ionic character becomes stronger.* The degree of ionic bond character increases the magnitude of the periodic potential and hence the energy gap.

The carrier concentration as a function of temperature according to Equation 2.39b is plotted for three semiconductors in Figure 2.15. Increasing energy gaps result in lower carrier concentrations at a given temperature.

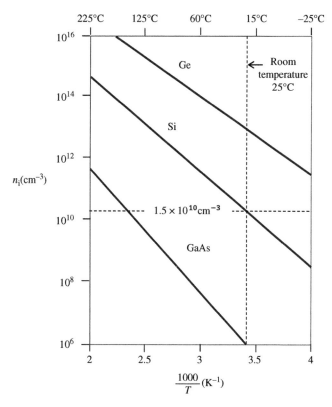

FIGURE 2.15 Plot of commonly accepted values of n as a function of $\frac{1}{T}$ for intrinsic germanium ($E_g = 0.7\text{eV}$), silicon ($E_g = 1.1\text{eV}$), and gallium arsenide ($E_g = 1.43\text{eV}$).

2.12   SEMICONDUCTOR BAND DIAGRAMS

The semiconductors in Figure 2.14 crystallize in either cubic or hexagonal structures. Figure 2.16a shows the *diamond* structure of silicon, germanium (and carbon), which is cubic.

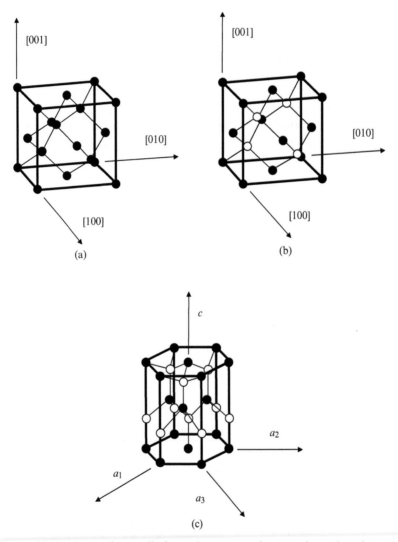

FIGURE 2.16   (a) The diamond unit cell of crystal structures of C, Si, and Ge. The cubic unit cell contains eight atoms. Each atom has four nearest neighbors in a tetrahedral arrangement. Within each unit cell, four atoms are arranged at the cube corners and at the face centers in a face-centered cubic (FCC) sublattice, the other four atoms are arranged in another FCC sublattice that is offset by a translation along one quarter of the body diagonal of the unit cell. (b) The zincblende unit cell contains four "A" atoms (black) and four "B" atoms (white). The "A" atoms form an FCC sublattice and the "B" atoms form another FCC sublattice that is offset by a translation along one quarter of the body diagonal of the unit cell. (c) The hexagonal wurtzite unit cell contains six "A" atoms and six "B" atoms. The "A" atoms form a hexagonal close-packed (HCP) sublattice and the "B" atoms form another HCP sublattice that is offset by a translation along the vertical axis of the hexagonal unit cell. Each atom is tetrahedrally bonded to four nearest neighbors. A vertical axis in the unit cell is called the *c*-axis.

Figure 2.16b shows the *zincblende* structure of a set of III-V and II-VI semiconductors, which is also cubic. Figure 2.16c shows the *hexagonal* structure of some additional compound semiconductors.

These three structures have features in common. Each atom has four nearest neighbors in a tetrahedral arrangement. Some crystals exhibit distortions from the ideal 109.47-degree tetrahedral bond angle; however, since all the compounds have directional covalent bonding to some extent, bond angles do not vary widely. Both the cubic (111) planes and the wurtzite (1000) planes normal to the *c*-axis have close-packed hexagonal atomic arrangements.

The energy gap and effective mass values for a given semiconductor are not sufficient information for optoelectronic applications. We need to re-examine the energy band diagrams for real materials in more detail.

The Kronig–Penney model involves several approximations. A one-dimensional periodic potential instead of a three-dimensional periodic potential is used. The periodic potential is simplified, and does not actually replicate the atomic potentials in real semiconductor crystals. For example, silicon has a diamond crystal structure with silicon atoms as shown in Figure 2.16a. Not only are three dimensions required, but also there is more than one atom per unit cell.

In addition, charges associated with individual atoms in compound semiconductors depend on the degree of ionic character in the bonding. This will affect the detailed shape of the periodic potential. Effects of electron shielding have not been accurately modeled. There are also other influences from electron spin and orbital angular momentum that influence energy bands in real crystals.

E versus *k* diagrams for various directions in a semiconductor crystal are often presented since the one-dimensional periodic potentials vary with direction. Although full three-dimensional band modeling is beyond the scope of this book the results for cubic crystals of silicon, germanium, gallium arsenide, gallium phosphide, gallium nitride, and cadmium telluride as well as for wurtzite GaN are shown in Figure 2.17a–g. For cubic crystals these figures show the band shape for an electron traveling in the [111] crystal direction on the left side and for the [100] direction on the right side. It is clear that the periodic potential experienced by an electron traveling in various directions changes: the value of a in $u_k(x) = u_k(x + a)$ appropriate for use in the Bloch function (Equation 2.5) for the [100] direction is the edge length of the cubic unit cell of the crystal. For the [111] direction a must be modified to be the distance between the relevant atomic planes normal to the body diagonal of the unit cell. For wurtzite crystals the two directions shown are the [0001] direction along the *c*-axis and the $\langle 1100 \rangle$ directions along the *a*-axes.

Note that there are multiple valence bands that overlap or almost overlap with each other rather than a single valence band. These are *sub-bands* for holes, which are due to spin–orbit interactions that modify the band state energies for electrons in the valence band. The sub-bands are approximately parabolic near their maxima. Because the curvatures of these sub-bands vary, they give rise to what are referred to as *heavy holes* and *light holes* with m^* as described by Equation 2.18. There are also *split-off* bands with energy maxima below the valence band edge.

2.13   DIRECT GAP AND INDIRECT GAP SEMICONDUCTORS

In Figure 2.17 the conduction bands often exhibit two energy minima rather than one minimum. Each local minimum can be approximated by a parabola whose curvature will determine the effective mass of the relevant electrons.

Referring to Figure 2.17c, we can see that the bandgap of GaAs is 1.43 eV where the valence band maximum and conduction band minimum coincide at $k = 0$. This occurs because the overall minimum of the conduction band is positioned at the same value of k as the valence band maximum and this results in a *direct gap* semiconductor. In Figure 2.17 GaAs, GaN, and CdTe are direct gap semiconductors. In contrast to GaAs, Si in Figure 2.17a has a valence band maximum at a different value of k

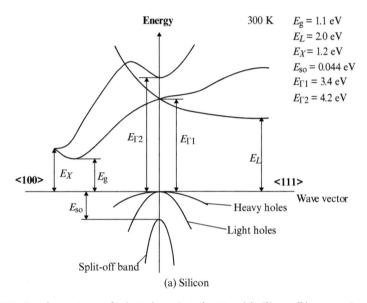

(a) Silicon

FIGURE 2.17   Band structures of selected semiconductors. (a) silicon, (b) germanium, (c) GaAs, (d) GaP, (e) cubic GaN, (f) CdTe, and (g) wurtzite GaN. Note that GaN is normally wurtzite. Cubic GaN is not an equilibrium phase at atmospheric pressure; however, it can be prepared at high pressure and it is stable once grown. Note that symbols are used to describe various band features. Γ denotes the point where $k = 0$. X and L denote the Brillouin zone boundaries in the $\langle 100 \rangle$ and $\langle 111 \rangle$ directions respectively in a cubic semiconductor. In (g) k_x and k_z denote the a and c directions, respectively, in a hexagonal semiconductor. See Figure 2.16c. Using the horizontal axes to depict two crystal directions saves drawing an additional figure; it is unnecessary to show the complete drawing for each k-direction since the positive and negative k-axes for a given k-direction are symmetrical. There are also energy gaps shown that are larger than the actual energy gap; the actual energy gap is the smallest gap. These band diagrams are the result of both measurements and modeling results. In some cases the energy gap values differ slightly from the values in Appendix 6. (a–d) After Levinstein, M., Rumyantsev, S., and Shur, M., Handbook Series on Semiconductor Parameters vol. 1. ISBN 9810229348. World Scientific, 1996 (e, g) After Morkoc, H., Handbook of Nitride Semiconductors and Devices, Vol. 1, ISBN 978-3-527-40837-5. Wiley 2008 (f) After Chadov, S., et al., Tunable multifunctional topological insulators in ternary Heusler compounds, Nature Materials 9, 541–545. 2010.

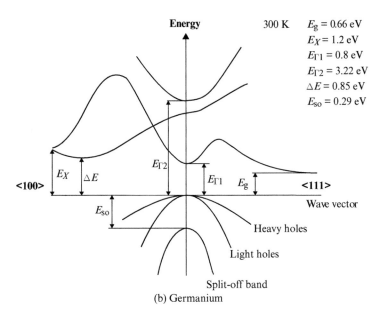

300 K $E_g = 0.66$ eV
$E_X = 1.2$ eV
$E_{\Gamma1} = 0.8$ eV
$E_{\Gamma2} = 3.22$ eV
$\Delta E = 0.85$ eV
$E_{so} = 0.29$ eV

(b) Germanium

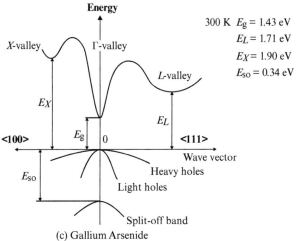

300 K $E_g = 1.43$ eV
$E_L = 1.71$ eV
$E_X = 1.90$ eV
$E_{so} = 0.34$ eV

(c) Gallium Arsenide

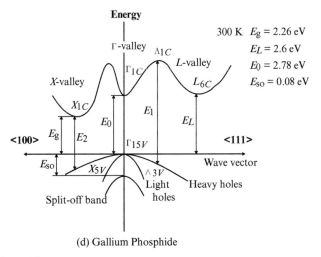

300 K $E_g = 2.26$ eV
$E_L = 2.6$ eV
$E_0 = 2.78$ eV
$E_{so} = 0.08$ eV

(d) Gallium Phosphide

FIGURE 2.17   (Cont'd)

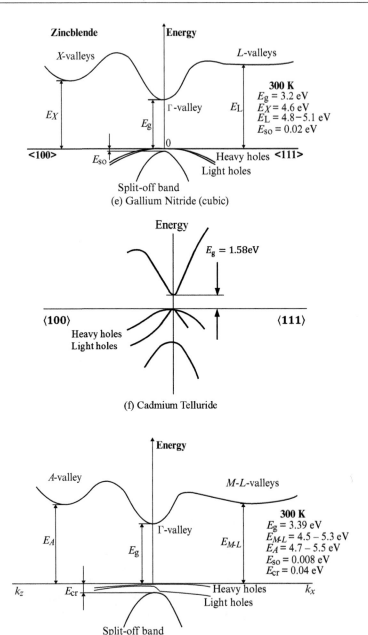

Zincblende

X-valleys

Energy

L-valleys

Γ-valley

E_X

E_g

E_L

300 K
$E_g = 3.2$ eV
$E_X = 4.6$ eV
$E_L = 4.8-5.1$ eV
$E_{so} = 0.02$ eV

<100>

E_{so}

Heavy holes <111>
Light holes

Split-off band
(e) Gallium Nitride (cubic)

Energy

$E_g = 1.58$eV

⟨100⟩

Heavy holes
Light holes

⟨111⟩

(f) Cadmium Telluride

Energy

A-valley

M-L-valleys

Γ-valley

E_A

E_g

$E_{M\text{-}L}$

300 K
$E_g = 3.39$ eV
$E_{M\text{-}L} = 4.5 - 5.3$ eV
$E_A = 4.7 - 5.5$ eV
$E_{so} = 0.008$ eV
$E_{cr} = 0.04$ eV

k_z

E_{cr}

Heavy holes k_x
Light holes

Split-off band
(g) Gallium Nitride (wurtzite)

FIGURE 2.17   (Cont'd)

than the conduction band minimum. That means that the energy gap of 1.1 eV is not determined by the separation between bands at $k = 0$, but rather by the distance between the overall conduction band minimum and valence band maximum. This results in an *indirect gap* semiconductor. Another indirect gap semiconductor in Figure 2.17 is the III-V material GaP.

The distinction between direct and indirect gap semiconductors is of particular significance for photosensitive and LED devices because processes involving photons occur in both cases, and photon absorption and generation properties differ considerably between these two semiconductor types.

An *electron-hole pair* (EHP) may be created if a photon is absorbed by a semiconductor and causes an electron in the valence band to be excited into the conduction band. For example, photon absorption in silicon can occur if the photon energy matches or exceeds the bandgap energy of 1.11 eV. Since silicon is an indirect gap semiconductor, however, there is a shift along the k-axis for the electron that leaves the top of the valence band and then occupies the bottom of the conduction band. In Section 2.4 we noted that $p = \hbar k$ and therefore a shift in momentum results. The shift is considerable as seen in Figure 2.17a, and it is almost the distance from the center of the Brillouin zone at $k = 0$ to the zone boundary at $k = \dfrac{\pi}{a}$ yielding a momentum shift of

$$\Delta p \simeq \hbar \frac{\pi}{a} \qquad (2.40)$$

During the creation of an EHP both energy and momentum must be conserved. Energy is conserved since the photon energy $\hbar \omega$ satisfies the condition $\hbar \omega = E_g$. Photon momentum $p = \dfrac{h}{\lambda}$ is very small, however, and is unable to provide momentum conservation. This is discussed further in Section 4.5. This means that a lattice vibration, or *phonon*, is required to take part in the EHP generation process. The magnitudes of phonon momenta cover a wide range in crystals and a phonon with the required momentum may not be available to the EHP process. This limits the rate of EHP generation, and photons that are not absorbed continue to propagate through the silicon.

If electromagnetic radiation propagates through a semiconductor we quantify absorption using an *absorption coefficient* α, which determines the resulting intensity of radiation by the exponential relationship

$$\frac{I(x)}{I_0} = e^{-\alpha x}$$

where I_0 is the initial radiation intensity and $I(x)$ is the intensity after propagating through the semiconductor over a distance x. Efficient crystalline silicon solar cells are generally at least $\simeq 100\,\mu\mathrm{m}$ thick for this reason due to their relatively low absorption coefficient. In contrast, GaAs (Figure 2.16c), is a direct gap semiconductor and has a much higher value of α (see Section 5.3). The thickness of GaAs required for sunlight absorption is only $\simeq 1\,\mu\mathrm{m}$. The value of α is an important parameter in PV semiconductors since sunlight that is not absorbed will not contribute to electric power generation. It is interesting to note that in spite of this difficulty silicon has historically been the most important solar cell material owing to its large cost advantage over GaAs.

In LEDs the process is reversed. EHPs *recombine* and give rise to photons, which are emitted as radiation. The wavelength range of this radiation may be in the infrared, the visible, or the ultraviolet parts of the electromagnetic spectrum, and is dependent on the semiconductor energy gap. Silicon is a poor material for LEDs because for an EHP recombination to create a photon, a phonon needs to be involved to achieve momentum conservation. The probability for this to occur is therefore much smaller and competing mechanisms for electron-hole pair recombination become important. These are known as *non-radiative recombination* events (see Sections 2.16 and 2.20). In contrast to silicon, GaAs can be used for high-efficiency LEDs and GaAs was used for the first practical LED devices due to its direct gap.

2.14   EXTRINSIC SEMICONDUCTORS

The incorporation of very small concentrations of impurities, referred to as *doping*, produces semiconductors that are called *extrinsic* to distinguish them from intrinsic semiconductors. Control of both the electron and hole concentrations over many orders of magnitude is enabled.

Consider the addition of a Group V atom such as phosphorus to a silicon crystal as shown together with a band diagram in Figure 2.18. This results in an *n-type semiconductor*. The phosphorus atom substitutes for a silicon atom and is called a *donor*; it introduces a new spatially localized energy level called the *donor level* E_d.

Because phosphorus has one more electron than silicon this donor electron is not required for valence bonding, is only loosely bound to the phosphorus, and can easily be excited into the conduction band. The energy required for this is $E_c - E_d$ and is

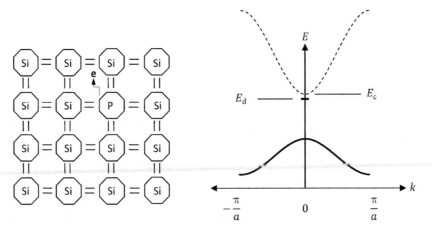

FIGURE 2.18   The substitution of a phosphorus atom in silicon (donor atom) results in a weakly bound extra electron occupying new energy level E_d that is not required to complete the covalent bonds in the crystal. It requires only a small energy $E_c - E_d$ to be excited into the conduction band, resulting in a positively charged donor ion and an extra electron in the conduction band.

referred to as the *donor binding energy*. If the donor electron has entered the conduction band, it is no longer spatially localized and the donor becomes a positively charged ion. The donor binding energy may be calculated by considering the well-known hydrogen energy quantum states in which the ionization energy for a hydrogen atom is given from Equation 1.34 by

$$E_{\text{Rydberg}} = \frac{\mu q^4}{8\varepsilon_0^2 h^2} = 13.6\text{eV} \tag{2.41}$$

and the Bohr radius is given by Equation 1.33 as

$$a_0 = \frac{4\pi\epsilon_0 \hbar^2}{\mu q^2} = 0.529\text{Å} \tag{2.42}$$

For the current purpose, two variables in Equations 2.41 and 2.42 must be changed. Whereas the hydrogen electron moves in a vacuum, the donor is surrounded by semiconductor atoms, which requires us to modify the dielectric constant from the free space value ϵ_0 to the appropriate value for silicon by multiplying by the relative dielectric constant ϵ_r. In addition, the free electron mass m must be changed to the effective mass m_e^*. The reduced mass μ is almost the same as m_e^*. This results in a binding energy from Equation 2.41 that is much smaller compared to the hydrogen atom, and an atomic radius from Equation 2.42 that is much larger compared to the Bohr radius. For n-type dopants in silicon the measured values of binding energy are approximately 0.05 eV compared to 13.6 eV for the Rydberg constant, and an atomic radius is obtained that is approximately an order of magnitude larger than the Bohr radius. Since the atomic radius is now one or more lattice constants, we can justify the use of the bulk silicon constants we have used in place of vacuum constants.

Consider now the substitution of a Group III atom such as aluminum for a silicon atom as illustrated in Figure 2.19. This creates a *p-type semiconductor*. The aluminum atom is called an *acceptor* and it introduces a new spatially localized energy level called the *acceptor level* E_a. Because aluminum has one fewer electron than silicon it can accept an electron from another valence bond elsewhere in the silicon, which results in a hole in the valence band. The energy required for this is $E_a - E_v$ and is referred to as the *acceptor binding energy*. If an electron has been accepted, the resulting hole is no longer spatially localized and the acceptor becomes a negatively charged ion. The binding energy may be estimated in a manner analogous to donor binding energies.

The introduction of either donors or acceptors influences the concentrations of charge carriers, and we need to be able to calculate these concentrations. The position of the Fermi level changes when dopant atoms are added, and it is no longer true that $n = p$; however, the Fermi–Dirac function $F(E)$ still applies. A very useful expression becomes the product of electron and hole concentrations in a given semiconductor.

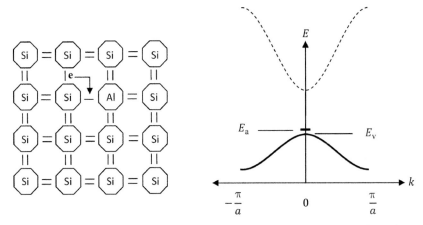

FIGURE 2.19   The substitution of an aluminum atom in silicon (acceptor atom) results in an incomplete valence bond for the aluminum atom. An extra electron may be transferred to fill this bond from another valence bond in the crystal. The spatially localized energy level now occupied by this extra electron at E_a is slightly higher in energy than the valence band. This transfer requires only a small energy $E_a - E_v$ and results in a negatively charged acceptor ion and an extra hole in the valence band.

For intrinsic material, we have calculated $n_i\, p_i$ and we obtained Equation 2.39a; however, Equations 2.36a and 2.37a are still valid and we can also conclude that

$$n_0 p_0 = n_i^2 = p_i^2 = N_c N_v \exp\left(\frac{-E_g}{kT}\right) \tag{2.43}$$

which is independent of E_F, and therefore is also applicable to extrinsic semiconductors. Here n_0 and p_0 refer to the equilibrium carrier concentrations in the doped semiconductor.

We now examine the *intermediate temperature condition* where the following apply:

(a) The ambient temperature is high enough to ionize virtually all the donors or acceptors.

(b) The concentration of the dopant is much higher than the intrinsic carrier concentration because the ambient temperature is not high enough to directly excite a large number of electron-hole pairs.

Under these circumstances, there are two cases. For donor doping in an n-type semiconductor we can conclude that

$$n_0 \simeq N_d \tag{2.44}$$

and combining Equations 2.43 and 2.44 we obtain

$$p_0 = \frac{n_i^2}{N_d} \tag{2.45}$$

where N_d is the donor concentration in donor atoms per unit volume of the semiconductor. For acceptor doping in a p-type semiconductor we have

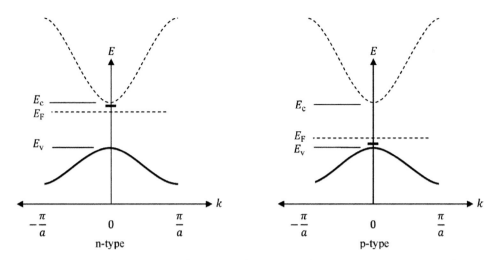

FIGURE 2.20 The band diagrams for n-type silicon with a donor doping concentration of 1×10^{17} cm^{-3} and p-type silicon with an acceptor doping concentration of 1×10^{17} cm^{-3}. Note that the Fermi energy rises to the upper part of the energy gap for n-type doping and drops to the lower part of the energy gap for p-type doping.

$$p_0 \simeq N_a \qquad (2.46)$$

and we obtain

$$n_0 = \frac{n_i^2}{N_a} \qquad (2.47)$$

The Fermi energy levels will change upon doping, and may be calculated from Equations 2.36a and 2.37a. In the case of n-type silicon the Fermi level will lie closer to the conduction band. In the case of p-type silicon the Fermi level will lie closer to the valence band (see Figure 2.20). In Example 2.3 we calculate some specific values of the Fermi energy position.

Consider n-type silicon at room temperature. The mobile electrons in the n-type silicon are called *majority carriers*, and the mobile holes are called *minority carriers*. We can also consider p-type silicon with mobile holes called majority carriers and mobile electrons called minority carriers.

EXAMPLE 2.3

Assume a silicon sample at room temperature.

a) Calculate the separation between E_c and E_F for n-type silicon having a phosphorus impurity concentration of 1×10^{17} cm^{-3}. Find both electron and hole concentrations.

b) Calculate the separation between E_c and E_F for p-type silicon having an aluminum impurity concentration of 1×10^{17} cm^{-3}. Find both electron and hole concentrations.

(continued)

(continued)

Solution

a) Using the intermediate temperature approximation $n_0 \cong 1 \times 10^{17}$ cm^{-3}. From Example 2.2 $n_i = 1.5 \times 10^{10}$ cm^{-3} and hence

$$p_0 = \frac{n_i^2}{n_0} = \frac{\left(1.5 \times 10^{10}\,\text{cm}^{-3}\right)^2}{1 \times 10^{17}\,\text{cm}^{-3}} = 2.25 \times 10^{3}\,\text{cm}^{-3}$$

and

$$n_0 = N_c \exp\left(\frac{-\left(E_c - E_F\right)}{kT}\right)$$

Solving for $E_c - E_F$,

$$E_c - E_F = kT \ln \frac{N_c}{n_0} = 0.026\,\text{eV} \ln\left(\frac{2.84 \times 10^{19}\,\text{cm}^{-3}}{1 \times 10^{17}\,\text{cm}^{-3}}\right) = 0.15\,\text{eV}$$

b) Using the intermediate temperature approximation, $p_0 \cong 1 \times 10^{17}$ cm^{-3} and hence

$$n_0 = \frac{n_i^2}{p_0} = \frac{\left(1.5 \times 10^{10}\,\text{cm}^{-3}\right)^2}{1 \times 10^{17}\,\text{cm}^{-3}} = 2.25 \times 10^{3}\,\text{cm}^{-3}$$

and

$$p_0 = N_v \exp\left(\frac{-\left(E_F - E_v\right)}{kT}\right)$$

Solving for $E_F - E_v$

$$E_F - E_v = kT \ln \frac{N_v}{p_0} = 0.026\,\text{eV} \ln\left(\frac{1.06 \times 10^{19}\,\text{cm}^{-3}}{1 \times 10^{17}\,\text{cm}^{-3}}\right) = 0.12\,\text{eV}$$

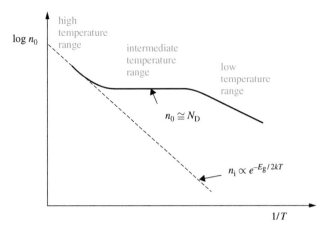

FIGURE 2.21   Carrier concentration as a function of temperature for an n-type extrinsic semiconductor. In the high-temperature range, carrier concentration is intrinsic-like and the Fermi energy is approximately mid-gap. In the intermediate-temperature range carrier concentration is controlled by the impurity concentration, virtually all the dopant atoms are ionized, and the Fermi energy is located above mid-gap. At low temperatures there is not enough thermal energy to completely ionize the dopant atoms and the Fermi energy migrates to a position between E_d and E_c as temperature drops. A similar diagram could be prepared for a p-type extrinsic semiconductor.

At low ambient temperatures, limited thermal energy causes only partial donor ionization and carrier concentration drops as temperature drops. At high ambient temperatures, the intrinsic electron-hole pair concentration may be significant and may exceed the doping concentration. In this case the semiconductor carrier concentrations can be similar to intrinsic material. These cases are illustrated in Figure 2.21. Of particular technological importance is the intermediate temperature range since the carrier concentrations are relatively independent of temperature and therefore semiconductor devices can operate over wide temperature ranges without significant variation in carrier concentrations.

2.15   CARRIER TRANSPORT IN SEMICONDUCTORS

The electrical conductivity of semiconductors is controlled by the concentrations of both holes and electrons as well as their ability to flow in a specific direction under the influence of an electric field. The flow of carriers is limited by scattering events in which carriers having a high instantaneous velocity frequently scatter off lattice vibrations (phonons), defects and impurities, and we can denote a *scattering time* or characteristic mean time between scattering events for this, referred to as τ. The resulting net flow velocity or *drift velocity* of a stream of carriers is much lower than their instantaneous velocity. The experimental evidence for this is summarized by Ohm's law, or

$$J = \sigma \varepsilon$$

which is a collision-limited flow equation that relates the current flow to the applied electric field. To understand this we consider Figure 2.22 showing the flow of electrons in a solid cylinder of cross-sectional area A in the positive x direction.

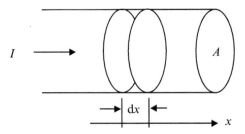

FIGURE 2.22   Current (I) flows along a solid semiconductor rod of cross-sectional area A.

If the carrier concentration is n and each carrier carrying charge $-q$ moves a distance dx in time dt then the amount of charge dQ passing across a given plane in the cylinder in time dt is $dQ = -nqAdx$. The carrier drift velocity is given by $\bar{v} = \dfrac{dx}{dt}$, and we can conclude that the current is

$$I = \frac{dQ}{dt} = \frac{-nqAdx}{dt} = -nqA\bar{v}$$

We also define the current density $J = \dfrac{I}{A}$ and hence we obtain Ohm's law or

$$J = -nq\bar{v} = \sigma\varepsilon$$

This is known as the *drift current density* and it requires the existence of an electric field. Note that since current is defined by convention as the flow of positive change, a negative sign is required when we consider electron drift velocity.

The application of an electric field can also be viewed using energy band diagrams. The well-known electrostatic relationship between electric field and electric potential and energy is given by

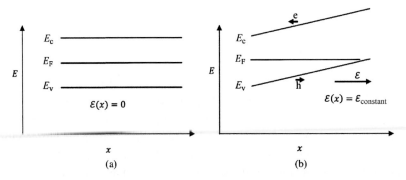

FIGURE 2.23   Spatial dependence of energy bands in an intrinsic semiconductor. If there is no electric field (a) the bands are horizontal and electron and hole energies are independent of location within the semiconductor. If an electric field ε is present inside the semiconductor the bands tilt. For an electric field pointing to the right, (b) electrons in the conduction band drift to the left, which decreases their potential energy. Holes in the valence band drift to the right, which decreases their potential energy. This reversed direction for hole energies is described in Figure 2.11.

$$\varepsilon(x) = -\frac{dV}{dx} = \frac{1}{q}\frac{dE}{dx} \qquad (2.48)$$

which states that an electric field causes a gradient in electric potential V and in addition an electric field causes a gradient in the potential energy E of a charged particle having charge q.

We can represent the conduction and valence bands in an applied electric field by showing the situation where the conduction and valence bands are separated by the energy gap. In Figure 2.8 this occurs at $k = 0$. Spatial dependence is indicated by using the x-axis to show the position in the x direction of the semiconductor as in Figure 2.23. Under equilibrium conditions, and provided there is no electric field present, the bands are horizontal lines. If a constant electric field is present the energy bands must tilt since from Equation 2.48 there will be a constant gradient in energy and the carriers in each band will experience a force F of magnitude $q\varepsilon$ in the directions shown and will travel so as to lower their potential energies. *The Fermi energy does not tilt*, since the electric field does not change the thermodynamic equilibrium.

We can now describe the flow of electrons. Since $\bar{v} \propto -\varepsilon$ we define

$$\bar{v} = -\mu\varepsilon$$

where μ is the carrier *mobility*, and we also conclude from Ohm's law that

$$\sigma = nq\mu$$

In order to confirm the validity of Ohm's law we can start with Newton's law of motion for an electron in an electric field

$$\frac{F}{m^*} = -\frac{q\varepsilon}{m^*} = \frac{d\bar{v}}{dt}$$

The treatment of carrier collisions requires adding the well-known *damping term* $\dfrac{\bar{v}}{\tau}$

where τ is the scattering time that results in a terminal velocity. This can be pictured by the example of a terminal velocity reached by a ping-pong ball falling in air. We now have

$$\frac{F}{m^*} = -\frac{q\varepsilon}{m^*} = \frac{d\bar{v}}{dt} + \frac{\bar{v}}{\tau} \qquad (2.49)$$

We can demonstrate the validity of the equation in steady state where $\dfrac{d\bar{v}}{dt} = 0$ and hence

$$\bar{v} = -\frac{q\tau\varepsilon}{m^*}$$

which is consistent with Ohm's law in which $|\bar{v}|$ is proportional to ε, and we see that

$$\mu = \frac{q\tau}{m^*}$$

This is understandable since a longer carrier scattering time τ will increase carrier mobility and a smaller carrier effective mass will increase average carrier velocity. In addition, we can examine the case where $\varepsilon = 0$. Now from Equation 2.49 we obtain

$$\frac{d\bar{v}}{dt} + \frac{\bar{v}}{\tau} = 0$$

which has solution $\bar{v}(t) = \bar{v}(0)\exp\left(-\frac{t}{\tau}\right)$. Carrier drift velocity will decay upon removal of the electric field with characteristic time constant equal to the scattering time τ.

In order to consider the contribution of both electrons and holes, we write the total drift current as

$$J_{\text{drift}} = J_\text{n} + J_\text{p} = q\left(n\mu_\text{n} + p\mu_\text{p}\right)\varepsilon \tag{2.50}$$

where

$$\mu_\text{n} = \frac{q\tau}{m_\text{n}^*}$$

and

$$\mu_\text{p} = \frac{q\tau}{m_\text{p}^*}$$

Separate mobility values μ_n and μ_p are needed for electrons and holes since they flow in different bands and may have different effective masses m_n^* and m_p^* respectively. The valence band has negative curvature, and Equation 2.18 shows that valence band electrons have negative effective mass; however, to ensure that μ_h is a positive quantity we define the effective mass of holes m_h^* to be a *positive* quantity equal in magnitude to this negative effective mass. Values of effective mass used to calculate μ_n and μ_p may differ from those in Appendix 6 due to three-dimensional considerations. See Suggestions for Further Reading at the end of this chapter.

Scattering time τ is determined from the combination of phonon scattering events, impurity scattering events and defect scattering events. Since the combined scattering rate $f = 1/\tau$ is measured in scattering events per second we can sum the scattering rates as follows

$$f = f_{\text{phonon}} + f_{\text{impurity}} + f_{\text{defect}}$$

where $f_{\text{phonon}} = 1/\tau_{\text{phonon}}$, $f_{\text{impurity}} = 1/\tau_{\text{impurity}}$ and $f_{\text{defect}} = 1/\tau_{\text{defect}}$, representing the scattering rates due to phonon scattering events, impurity scattering events and defect scattering events, respectively. We can also express this in terms of the average scattering times for each scattering type:

$$\frac{1}{\tau} = \frac{1}{\tau_{\text{phonon}}} + \frac{1}{\tau_{\text{impuriny}}} + \frac{1}{\tau_{\text{defect}}}$$

Contributions of each scattering type can be experimentally determined. For example, the phonon scattering contribution can be analyzed by observing changes in mobility as temperature is changed. Increasing temperatures decrease carrier mobility.

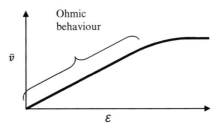

FIGURE 2.24 Dependence of drift velocity magnitude on electric field for a semiconductor.

The validity of Ohm's law has a limit. If the electric field is large, carrier velocity magnitudes saturate as shown in Figure 2.24 and will no longer be linearly proportional to the electric field. This occurs because energetic electrons transfer more energy to lattice vibrations. The magnitude of the electric field that results in saturation effects depends on the semiconductor.

Mobility values for a range of semiconductors are shown in Appendix 6. These are tabulated for intrinsic materials at room temperature; however, defects and impurities as well as higher temperatures have a substantial effect on mobility values since they decrease scattering times. Both undesirable impurities as well as intentionally introduced dopant atoms will cause scattering times and mobility values to decrease.

2.16 EQUILIBRIUM AND NONEQUILIBRIUM DYNAMICS

The carrier concentrations we have been discussing until now are *equilibrium* concentrations, and are in thermodynamic equilibrium with the semiconductor material. In equilibrium, both EHP generation and EHP recombination occur simultaneously; however, the net EHP concentration remains constant. We can express this using rate constants defined as follows:

$$G_{th} = \text{thermal EHP generation rate}\left(\text{EHP cm}^{-3}\,\text{s}^{-1}\right)$$

$$R = \text{EHP recombination rate}\left(\text{EHP cm}^{-3}\,\text{s}^{-1}\right)$$

In equilibrium $G_{th} = R$.

It is easy to cause a *nonequilibrium* condition to exist in a semiconductor. For example, we can illuminate the semiconductor with photons whose energy exceeds its energy gap of the semiconductor. We can also cause electric current to flow through the semiconductor by attaching two or more electrodes to the semiconductor and then connecting them across a voltage source.

In an ideal direct gap semiconductor the value of R depends on carrier concentrations. If, for example, the electron concentration n is doubled, R will double since the probability for an electron to reach a hole has doubled. If both n and p are doubled then R will increase by a factor of four since the hole concentration has also doubled. We can state this mathematically as $R \propto np$. We remove subscripts on carrier concentrations when nonequilibrium conditions are present.

Consider a direct gap n-type semiconductor in which $n_0 \gg p_0$. We shall also stipulate that $\delta n(t) \ll n_0$, which states that the excess carrier concentration is small compared to the equilibrium majority carrier concentration. See Example 2.4.

EXAMPLE 2.4

Consider an n-type direct gap semiconductor having equilibrium electron concentration $n_0 = 1 \times 10^{17} \, cm^{-3}$ and equilibrium hole concentration $p_0 = 2.25 \times 10^3 cm^{-3}$ at room temperature.

This semiconductor is now illuminated with light composed of photons with energy larger than the energy gap. The intensity of the light is adjustable. In a first experiment the intensity of the illumination provides excess carrier concentrations Δn and Δp as follows:

Experiment 1) $\Delta n = \Delta p = 1 \times 10^{12} \, cm^{-3}$

In a second experiment the intensity of the illumination is increased and it provides excess carrier concentrations as follows:

Experiment 2) $\Delta n = \Delta p = 1 \times 10^{15} \, cm^{-3}$

Compare the carrier recombination rate R_2 for Experiment 2 to the carrier recombination rate R_1 for Experiment 1.

Solution

Carrier recombination rate is proportional to the product np. In Experiment 1 we have

$$(np)_1 = (n_0 + \Delta n)(p_0 + \Delta p)$$
$$= \left(1 \times 10^{17} \, cm^{-3} + 1 \times 10^{12} \, cm^{-3}\right)\left(2.25 \times 10^3 cm^{-3} + 1 \times 10^{12} \, cm^{-3}\right)$$
$$\cong 1 \times 10^{29} \, cm^{-6}$$

In Experiment 2 we have

$$(np)_2 = (n_0 + \Delta n)(p_0 + \Delta p)$$
$$= \left(1 \times 10^{17} \, cm^{-3} + 1 \times 10^{15} \, cm^{-3}\right)\left(2.25 \times 10^3 cm^{-3} + 1 \times 10^{15} \, cm^{-3}\right)$$
$$\cong \left(1 \times 10^{17} \, cm^{-3}\right)\left(1 \times 10^{15} \, cm^{-3}\right)$$
$$= 1 \times 10^{32} \, cm^{-6}$$

Note the following:

i) The effect of optical carrier generation on n is negligible because $n_0 \gg \Delta n$ and $n_0 \gg \Delta p$.

ii) The equilibrium hole concentration p_0 is negligible compared to Δn and Δp.

We can therefore calculate R_2 / R_1 as follows:

$$\frac{R_2}{R_1} = \frac{(np)_2}{(np)_1} \cong \frac{10^{32} \, cm^{-6}}{10^{29} \, cm^{-6}} = 1 \times 10^3$$

We find that the carrier recombination rate increases by a factor of 10^3 and it varies in proportion to the excess hole concentration Δp for the two experiments. This commonly encountered condition is referred to as a *low-level injection* condition. We can approximate n to be almost constant and essentially independent of the illumination, and p_0 may also be neglected.

In indirect gap semiconductors such as silicon, carrier recombination is more complicated. In practice, the very low recombination rate due to the required participation of a phonon is almost always negligible compared to carrier recombination via defects known as *deep traps*. Nevertheless, this Example *is* valid for silicon because the detailed mechanism of the recombination does not change the result: Provided low level injection conditions are present, *the rate of recombination is proportional to the excess minority carrier concentration*. The n-type semiconductor used in this example can therefore be n-type silicon and the parameters chosen follow from Example 2.3. Deep traps are further discussed in Section 2.20.

We will now look at the time dependence of excess carrier generation. A steady optical generation rate G_{op} is abruptly added at time $t = 0$. This will cause the EHP generation rate to exceed the recombination rate and carrier concentrations will exceed the equilibrium concentrations and will become time dependent. We shall designate $\delta n(t)$ and $\delta p(t)$ to be the time-dependent carrier concentrations in excess of equilibrium concentrations n_0 and p_0.

For an n-type semiconductor in which $n_0 \gg p_0$ and in which $\delta n(t) \ll n_0$, the net rate of increase of p is determined by the net difference between an optical generation rate and a recombination rate. From Example 2.4, the hole recombination rate is shown to be linearly proportional to excess hole concentration $\delta p(t)$ and we can therefore write the simple differential equation

$$\frac{d\delta p(t)}{dt} = G_{op} - \frac{\delta p(t)}{\tau_p} \tag{2.51}$$

where τ_p is a constant. Note that the two terms on the right-hand side of Equation 2.51 represent a generation rate term and a recombination rate term, respectively. The solution to this differential equation is

$$\delta p(t) = \Delta p \left[1 - \exp\left(\frac{-t}{\tau_p}\right) \right] \tag{2.52}$$

This result is sketched in Figure 2.25.

After a time $t \gg \tau_p$, a steady-state value of excess carrier concentration Δp exists. If Equation 2.52 is substituted into Equation 2.51 we obtain

$$\Delta p = G_{op}\tau_p \tag{2.53a}$$

The constant τ_p for this process is called the *carrier recombination time*. Carrier recombination time should not be confused with carrier scattering time of Section 2.15. Scattering times are generally orders of magnitude shorter than recombination times.

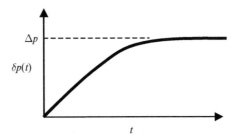

FIGURE 2.25   Plot of excess hole concentration as a function of time. A constant optical generation rate starts at $t = 0$ and continues indefinitely.

If we know the optical generation rate and the characteristic recombination time for a semiconductor, we can calculate the steady-state excess minority carrier concentration using Equation 2.53a). There is also a small steady-state change in the majority carrier concentration because $\Delta n = \Delta p$; however, this is neglected in low-level injection since $\Delta n \ll n_0$.

The same argument can be applied to a p-type semiconductor, and we would obtain the increase in minority carrier concentration as

$$\Delta n = G_{op}\tau_n \qquad\qquad (2.53b)$$

Electron-hole pair recombination may also occur via processes other than direct electron-hole pair recombination events as mentioned in Example 2.4. The concept of a carrier recombination time introduced in this section is still relevant because carrier recombination times may be assigned to a variety of recombination mechanisms. Instead of calling τ_n or τ_p carrier recombination times we often refer to them as *carrier lifetimes* which acknowledges that direct electron-hole pair recombination may not be the dominant mechanism for carrier recombination. This is particularly true in indirect gap semiconductors such as silicon. See Section 2.20.

2.17   CARRIER DIFFUSION AND THE EINSTEIN RELATION

Free carriers that are produced in a spatially localized part of a semiconductor are able to diffuse and thereby move to other parts of the material. The carrier diffusion process is functionally similar to the diffusion of atoms in solids. At sufficiently high temperatures atomic diffusion occurs, which is described as net atomic motion from a region of higher atomic concentration to a region of lower atomic concentration. This occurs due to random movements of atoms in a concentration gradient. There is no preferred direction to the random movement of the atoms; however, provided the average concentration of atoms is not uniform, the result of random movement is for a net flux of atoms to exist flowing from a more concentrated region to a less concentrated region. *Fick's first law* which applies to the diffusion process for atoms as presented in introductory materials science textbooks, is also applicable to electrons.

As with atomic diffusion, the driving force for carrier diffusion is the gradient in electron concentration. For free electrons diffusing along the x-axis, Fick's first law applies and it can be written

$$\phi_n(x) = -D_n \frac{dn(x)}{dx} \tag{2.54}$$

where ϕ_n is the flux of electrons (number of electrons per unit area per second) flowing along the x-axis due to a concentration gradient of electrons. The negative sign in Equation 2.52 indicates that diffusion occurs in the direction of decreasing electron concentration. For holes, Fick's first law becomes

$$\phi_p(x) = -D_p \frac{dp(x)}{dx} \tag{2.55}$$

Since the flow of charged particles constitutes an electric current, we can describe *diffusion currents* due to holes or electrons. These are distinct from drift currents described in Section 2.15 because no electric field is directly involved. Equations 2.54 and 2.55 may be rewritten as currents:

$$J_n(x)_{\text{diffusion}} = qD_n \frac{dn(x)}{dx} \tag{2.56a}$$

$$J_p(x)_{\text{diffsion}} = -qD_p \frac{dp(x)}{dx} \tag{2.56b}$$

Note that there is no negative sign in the case of electrons because electrons carry a negative charge that cancels out the negative sign in Fick's first law.

An interesting situation occurs when both diffusion and drift currents flow. An electric field is present as well as a carrier concentration gradient. The total current densities from Equations 2.50 and 2.56 become

$$J_n(x)_{\text{drift+diffusion}} = q\mu_n n(x)\varepsilon(x) + qD_n \frac{dn(x)}{dx} \tag{2.57a}$$

and

$$J_p(x)_{\text{drift+diffusion}} = q\mu_p p(x)\varepsilon(x) - qD_p \frac{dp(x)}{dx} \tag{2.57b}$$

In semiconductor diodes, both drift and diffusion occur and it is important to become familiar with the situation where drift and diffusion currents coexist in the same part of the semiconductor.

One way to establish an electric field is to have a gradient in doping level by spatially varying the doping concentration in the semiconductor. Consider the example in Figure 2.26. The dopant concentration varies across a semiconductor sample that is in thermal equilibrium. On the left side, the semiconductor is undoped, and an

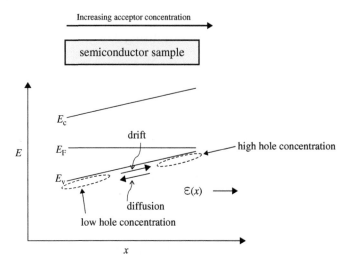

FIGURE 2.26   The energy bands will tilt due to a doping gradient. Acceptor concentration increases from left to right in a semiconductor sample. This causes a built-in electric field, and the hole concentration increases from left to right. The field causes hole drift from left to right, and there is also hole diffusion from right to left due to the concentration gradient.

acceptor dopant gradually increases in concentration from left to right. This causes the Fermi energy to occupy lower positions in the energy gap until it is close to the valence band on the right side of the sample. The Fermi energy does not tilt because it is a thermodynamic quantity and the sample is in equilibrium; however, the valence and conduction bands do tilt as shown.

There will now be a high concentration of holes in the valence band on the right side of the semiconductor, which decreases to a low hole concentration on the left side. Hole diffusion will therefore occur in the negative-*x* direction. At the same time, the tilting of the energy bands means that an electric field is present in the sample. This is known as a *built-in electric field* since it is caused by a spatial concentration variation within the semiconductor material rather than by the application of an applied voltage. The built-in field causes a hole drift current to flow in the positive-*x* direction. Since the semiconductor is in equilibrium, these two hole currents cancel out and *the net hole current flow will be zero*. A similar argument can be made for the electrons in the conduction band and the net electron current will also be zero.

A useful relationship between mobility and diffusivity can now be derived. Since the net current flow illustrated in Figure 2.26 is zero in equilibrium we can write for hole current

$$J_p(x)_{\text{drift+diffusion}} = q\mu_p p(x)\varepsilon(x) - qD_p \frac{dp(x)}{dx} = 0 \tag{2.58}$$

From Equations 2.48 and 2.58 we have

$$q\mu_p p(x)\frac{1}{q}\frac{dE}{dx} - qD_p\frac{dp(x)}{dx} = 0$$

We now calculate $p(x)$. Since the valence band energy E_v is now a function of x we can rewrite Equation 2.37a as

$$p(x) = N_v \exp\left[\frac{-(E_F - E_v(x))}{kT}\right]$$

(2.59)

and we obtain

$$\mu_p N_v \exp\left[\frac{-(E_F - E_v(x))}{kT}\right]\frac{dE_v(x)}{dx} - qD_p\frac{N_v}{kT}\exp\left[\frac{-(E_F - E_v(x))}{kT}\right]\frac{dE_v(x)}{dx} = 0$$

which simplifies to

$$\frac{D_p}{\mu_p} = \frac{kT}{q}$$

(2.60a)

and a similar derivation may be applied to electrons yielding

$$\frac{D_n}{\mu_n} = \frac{kT}{q}$$

(2.60b)

Equation 2.60 is known as the *Einstein relation*. At a given temperature this tells us that mobility and diffusivity are related by a constant factor, which is not unexpected since both quantities express the degree of ease with which carriers move in a semiconductor under a driving force.

2.18   QUASI-FERMI ENERGIES

If a semiconductor is influenced by incident photons or an applied electric current the semiconductor is no longer in equilibrium. This means that we cannot use Equations 2.36a and 2.37 to determine carrier concentrations. In addition Fermi energy E_F is no longer a meaningful quantity since it was defined for a semiconductor in equilibrium in Section 2.9 and the Fermi–Dirac distribution function of Figure 2.10 is also based on equilibrium conditions.

For convenience, two new quantities, F_n and F_p, known as the *quasi-Fermi energy for electrons* and the *quasi-Fermi energy for holes*, respectively are introduced. The quantities may be used even if a semiconductor is not in equilibrium and there are excess carriers. F_n and F_p are defined from the following equations:

$$n = N_c \exp\left[\frac{-(E_c - F_n)}{kT}\right]$$

(2.61a)

and

$$p = N_v \exp\left[\frac{-(F_p - E_v)}{kT}\right]$$

(2.61b)

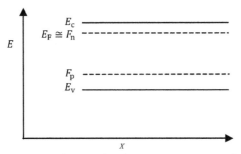

FIGURE 2.27 The quasi-Fermi levels F_n and F_p for an n-type semiconductor with excess carriers generated by illumination. Note the large change in F_p due to illumination and note that F_n is almost the same as the value of E_F before illumination. Low level injection is assumed.

Note the similarity between Equation 2.61 and Equations 2.36 and 2.37. Subscripts for n and p are absent in Equation 2.61 because these carrier concentrations are not necessarily equilibrium values. It follows that if the semiconductor is in equilibrium, the electron and hole quasi-Fermi energies become equal to each other and identical to the Fermi energy. The electron and hole quasi-Fermi levels in an n-type semiconductor will behave very differently upon excess carrier generation, as shown in Figure 2.27. This is examined in Example 2.5.

EXAMPLE 2.5

An n-type silicon sample has a donor concentration of 1×10^{17} cm^{-3}. In Example 2.3 we obtained $n_0 = 1 \times 10^{17}$ cm^{-3}, $p_0 = 2.25 \times 10^3$ cm^{-3} and $E_c - E_F = 0.15$ eV. We now illuminate this sample steadily and introduce a uniform electron-hole pair generation rate of $G_{op} = 5 \times 10^{20}$ cm^{-3} s^{-1}. Assume a carrier lifetime of 2×10^{-6} s.

a) Calculate the resulting electron and hole concentrations.
b) Calculate the quasi-Fermi energy levels.

Solution

(a)

$$\Delta p = G_{op}\tau_p = 5 \times 10^{20}\,\text{cm}^{-3}\text{s}^{-1} \times 2 \times 10^{-6}\,\text{s} = 1 \times 10^{15}\,\text{cm}^{-3}$$

and

$$\Delta n = G_{op}\tau_n = 5 \times 10^{20}\,\text{cm}^{-3}\text{s}^{-1} \times 2 \times 10^{-6}\,\text{s} = 1 \times 10^{15}\,\text{cm}^{-3}$$

Hence

$$p = p_0 + \Delta p = 2.25 \times 10^3\,\text{cm}^{-3} + 1 \times 10^{15}\,\text{cm}^{-3} \cong 1 \times 10^{15}\,\text{cm}^{-3}$$

and

$$n = n_0 + \Delta n = 1 \times 10^{17}\,\mathrm{cm}^{-3} + 1 \times 10^{15}\,\mathrm{cm}^{-3} = 1.01 \times 10^{17}\,\mathrm{cm}^{-3}$$

Therefore carrier concentrations may be strongly affected by the illumination: The hole concentration increases by approximately 12 orders of magnitude from a very small minority carrier concentration to a much larger value dominated by the excess hole concentration.

The illumination only increases the electron concentration slightly (by 1%) since it is a majority carrier. This is therefore an example of low-level injection since the majority carrier concentration is almost unchanged.

b) The quasi-Fermi level for holes may be found from:

$$p = N_v \exp\left(\frac{-(F_p - E_v)}{kT}\right)$$

Solving for $F_p - E_v$ we obtain

$$F_p - E_v = kT \ln\left(\frac{N_v}{p}\right) = 0.026\ \mathrm{eV}\ \ln\left(\frac{1.06 \times 10^{19}\,\mathrm{cm}^{-3}}{1 \times 10^{15}\,\mathrm{cm}^{-3}}\right) = 0.24\,\mathrm{eV}$$

The quasi-Fermi level for electrons may be found from

$$n = N_c \exp\left(\frac{-(E_c - F_n)}{kT}\right)$$

Solving for $E_c - F_n$ we obtain

$$E_c - F_n = kT \ln\left(\frac{N_c}{n}\right) = 0.026\,\mathrm{eV}\ \ln\left(\frac{2.84 \times 10^{19}\,\mathrm{cm}^{-3}}{1.01 \times 10^{17}\,\mathrm{cm}^{-3}}\right) = 0.15\,\mathrm{eV}$$

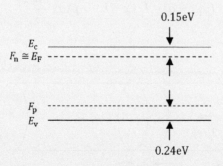

Note that with illumination F_n is almost identical to the original value of E_F but F_p moves significantly lower. This is a consequence of the large excess carrier concentration compared to the equilibrium hole concentration. If a similar problem were solved for a p-type semiconductor then under illumination F_n would move significantly higher and F_p would remain almost identical to the original value of E_F.

The separation between F_n and F_p may be obtained from the product of nonequilibrium carrier concentrations divided by the product of equilibrium carrier concentrations. This can be seen from Equations 2.61 and 2.43 which give us

$$np = N_c N_v \exp\left(\frac{F_n - F_p - E_g}{kT}\right) = n_0 p_0 \exp\left(\frac{F_n - F_p}{kT}\right) \qquad (2.62a)$$

and hence

$$F_n - F_p = kT \ln\left(\frac{np}{n_0 p_0}\right) \qquad (2.62b)$$

2.19   THE DIFFUSION EQUATION

We have introduced carrier recombination as well as carrier diffusion separately; however, carriers in semiconductors routinely undergo both diffusion and recombination simultaneously.

In order to describe this, consider a long semiconductor bar or rod shown in Figure 2.28 in which excess holes are generated at $x = 0$ causing an excess of holes Δp to be maintained at $x = 0$. The excess hole concentration drops off to approach an equilibrium concentration at the other end of the long rod. The excess holes will diffuse to the right. Some of these holes recombine with electrons during this process. We can consider a slice of width dx as shown in Figure 2.28. The hole current $I_p (x = a)$ will be higher than the hole current $I_p (x = b)$ due to the rate of recombination of holes within volume Adx between $x = a$ and $x = b$. From Equation 2.51 $\dfrac{\delta p(x)}{\tau_p}$ is the hole recombination rate per unit volume. In steady-state the net supply and recombination in the slice of width dx may be equated. Expressed mathematically we obtain

$$I_p (x = a) - I_p (x = b) = qAdx \frac{\delta p(x)}{\tau_p} \qquad (2.63)$$

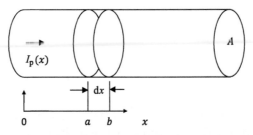

FIGURE 2.28   A solid semiconductor rod of cross-sectional area A has a hole current $I_p (x)$ flowing in the positive x direction. Due to recombination the hole current is dependent on x. At surfaces $x = a$ and $x = b$, $I_p (x)$ changes due to the recombination that occurs within volume Adx.

This may be rewritten as

$$\frac{I_p(x=a) - I_p(x=b)}{dx} = qA\frac{\delta p(x)}{\tau_p}$$

or

$$\frac{dI_p(x)}{dx} = -qA\frac{\delta p(x)}{\tau_p}$$

In terms of current density

$$\frac{dJ_p(x)}{dx} = -q\frac{\delta p(x)}{\tau_p} \qquad (2.64a)$$

and applying the same procedure in the case of electrons

$$\frac{dJ_n(x)}{dx} = q\frac{\delta n(x)}{\tau_p} \qquad (2.64b)$$

If the current is entirely due to the *diffusion* of carriers, the expression for diffusion current from Equation 2.56b for excess carriers $\delta p(x)$ may be rewritten as

$$J_p(x)_{\text{diffusion}} = -qD_p\frac{d\delta p(x)}{dx} \qquad (2.65)$$

Substituting this into Equation 2.64a we obtain

$$\frac{d^2\delta p(x)}{dx^2} = \frac{\delta p(x)}{D_p\tau_p} \qquad (2.66a)$$

This is known as the *steady state diffusion equation for holes*, and the corresponding equation for electrons is

$$\frac{d^2\delta n(x)}{dx^2} = \frac{\delta n(x)}{D_n\tau_n} \qquad (2.66b)$$

The general solution to Equation 2.66a is

$$\delta p(x) = A\exp\left(\frac{-x}{\sqrt{D_p\tau_p}}\right) + B\exp\left(\frac{x}{\sqrt{D_p\tau_p}}\right) \qquad (2.67a)$$

However, considering our boundary conditions, the function must decay to zero for large values of x and therefore $B = 0$ yielding

$$\delta p(x) = \Delta p\exp\left(\frac{-x}{\sqrt{D_p\tau_p}}\right)$$

which may be written

$$\delta p(x) = \Delta p \exp\left(\frac{-x}{L_p}\right) \tag{2.67b}$$

where

$$L_p = \sqrt{D_p \tau_p} \tag{2.68a}$$

is known as the *diffusion length*. The latter determines the position on the x-axis where carrier concentrations are reduced by a factor of e, as shown in Figure 2.29.

The hole current density at any point x may be determined by substituting Equation 2.67b into Equation 2.65, and therefore

$$J_p(x)_{\text{diffusion}} = q\frac{D_p}{L_p}\Delta p \exp\left(\frac{-x}{L_p}\right)$$

This may also be written

$$J_p(x)_{\text{diffusion}} = q\frac{D_p}{L_p}\delta p(x)$$

which shows that both current density and hole concentration have the same exponential form. Figure 2.30 plots current density $J_p(x)_{\text{diffusion}}$ as a function of x. Both diffusion and recombination occur simultaneously, which lowers the diffusion

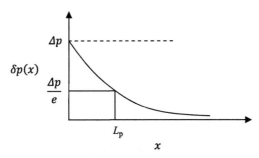

FIGURE 2.29   Plot of excess hole concentration in a semiconductor as a function of x in a semiconductor rod where both diffusion and recombination occur simultaneously. The decay of the concentration is characterized by a diffusion length L_p.

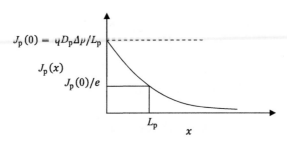

FIGURE 2.30   Hole current density as a function of x for a semiconductor rod with excess carriers generated at $x = 0$.

current exponentially as x increases. This is reasonable since the number of holes that have not recombined drops exponentially with x and therefore L_p also represents the position on the x-axis where current density $J_p(x)$ is reduced by a factor of e.

In the case of excess electron flow, the current flow becomes

$$J_n(x)_{\text{diffusion}} = -q\frac{D_n}{L_n}\delta n(x)$$

where

$$L_n = \sqrt{D_n \tau_n} \qquad\qquad (2.68b)$$

2.20 TRAPS AND CARRIER LIFETIMES

Carrier lifetimes in indirect gap semiconductors such as silicon are, in reality, determined by trapping processes that participate in recombination events because the probability of the needed phonon being available for radiative recombination is very low. *Traps* are impurity atoms or native point defects such as vacancies, dislocations or grain boundaries. There are also *surface traps* due to the defects that inherently occur at semiconductor surfaces, and *interface traps* that form at a boundary between two different material regions in a semiconductor device.

In all cases, traps are physical defects capable of trapping conduction band electrons and/or valence band holes and thereby affecting carrier concentrations, carrier flow and recombination times. Electron-hole pair recombination processes are now *trap-mediated* because a specific defect is involved. After being trapped a carrier may again be released to the band it originated from or it may subsequently recombine at the trap with a carrier of the opposite sign that also gets attracted to the same trap. This is known as *trap-assisted carrier recombination* and is one of the most important phenomena that limits the performance of both direct and indirect gap semiconductors used for solar cells and LEDs. Trap-assisted carrier recombination is often referred to as *Shockley-Read-Hall* (SRH) recombination.

Intentionally introduced n-type and p-type dopants actually are traps, and are referred to as *shallow traps* because they are only separated from either a conduction band or a valence band by a small energy difference that may be overcome by thermal energy. This means that the trap is easy to ionize and the carrier is very likely to be released from the trap. Also since dopants are normally ionized, they do not trap carriers of the opposite sign. For example, phosphorus in silicon is an n-type dopant. Once it becomes a positive ion after donating an electron to the conduction band it has a small but finite probability of recapturing a conduction band electron. Since it spends almost all its time as a positive ion, it has essentially no chance of capturing a hole, which is repelled by the positive charge, and therefore shallow traps generally do not cause trap-assisted carrier recombination.

The traps that we must pay careful attention to are *deep traps*, which exist near the middle of the bandgap. These traps are highly effective at promoting electron-hole pair recombination events. Since their energy levels are well separated from band edges, carriers that are trapped are not easily released. Imagine a deep trap that captures a conduction band electron and is then negatively charged. In this state the negatively charged trap cannot readily release its trapped electron and may therefore attract a positive charge and act as an effective hole trap. Once the hole is trapped it recombines with the trapped electron and the trap is effectively emptied and is again available to trap another conduction band electron. In this manner, traps become a new conduit for electrons and holes to recombine. If the deep trap density is high the average trap-assisted recombination rate is high and the recombination time is low.

Highly purified silicon having excellent crystalline quality exhibits recombination times of hundreds of microseconds or even up to about a millisecond. Often in practical devices, however, the attainable silicon crystal perfection limits carrier recombination times to values below one hundred microseconds.

Understanding deep trap behavior requires that we know the probabilities of the trap being filled or empty. This may be understood in equilibrium conditions by knowing the trap energy level and comparing it to the Fermi energy level. If the trap energy level E_t is above E_F then the trap is more likely to be empty than full. If E_t is below E_F it is more likely to be filled. The terms "filled" and "empty" refer specifically to electrons because the Fermi–Dirac function describes the probability that an electron fills a specific energy level. If we wish to describe the probabilities for a trap to be occupied by holes, we must subtract these probabilities from 1.

We will simplify the treatment of traps by focusing on a very specific situation. Consider a trap *at the Fermi energy and near mid-gap* in a semiconductor. Both E_t and E_F will be at approximately the middle of the energy gap. The ionization energy of the trap for either a trapped electron or a trapped hole is approximately $\dfrac{E_g}{2}$, as illustrated in Figure 2.31. The probability that the trap is empty or filled is 50% since it is at the Fermi energy. The captured electron may be re-released back to the conduction band, or it may be annihilated by a hole at the trap.

There is a simple argument for reasonably assuming the trap is likely to exist at the Fermi level and near mid-gap. At the surface of a semiconductor, electrons that are normally shared between two silicon atoms are only held by a single atom. These *dangling bonds* therefore comprise electrons that are likely to lie at approximately mid-gap because the energy required to excite them into the conduction band is only approximately half as large as the energy E_g required to remove an electron from the complete two-atom covalent bond of the relevant perfect crystal.

Now consider a large number of such dangling bonds at a semiconductor surface. Some of these dangling bonds will have lost electrons and some of them will not.

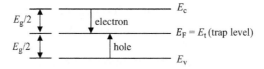

FIGURE 2.31 A trap level at the Fermi energy near mid-gap.

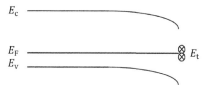

FIGURE 2.32 Surface traps at the surface of a p-type semiconductor comprise electrons held in dangling bonds. The energy needed to release these electrons is approximately $\frac{E_g}{2}$. Since there are large numbers of dangling bond states, some being occupied and some not being occupied by electrons, the Fermi energy becomes pinned at this energy.

Since the Fermi level exists between the highest filled states and the lowest empty state, the Fermi level tends to fall right onto the energy level range of these traps. The Fermi energy gets *pinned* to this trap energy at $\cong \frac{E_g}{2}$. Figure 2.32 shows the pinning of a Fermi level due to surface traps in a p-type semiconductor. These surface traps are often referred to as *surface states*. Notice that at the semiconductor surface the surface traps determine the position of the Fermi energy rather than the doping level. An electric field is established in the semiconductor normal to the surface and *band bending* occurs as shown.

If the semiconductor had been n-type instead of p-type, then the same reasoning would still pin the Fermi energy to mid-gap; however, the band bending would occur in the opposite direction and the resulting electric field would point in the opposite direction, as illustrated in Figure 2.33.

Since traps are often formed from defects other than free surfaces that also involve incomplete bonding like a vacancy, a dislocation line, a grain boundary or an interface between two layers, this simple picture is very useful and will be used in various contexts to explain recombination processes in subsequent sections of this book.

There is a velocity associated with excess minority carriers at a semiconductor surface or at an interface between a semiconductor and another material. Consider diffusion current within a p-type semiconductor having excess holes. Very close to a surface or interface,

$$\phi = -D_n \frac{d\delta n}{dx}$$

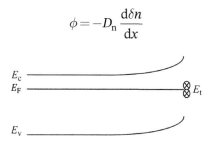

FIGURE 2.33 Surface traps at the surface of an n-type semiconductor causing the Fermi level to be trapped at approximately mid-gap. An electric field opposite in direction to that of Figure 2.32 is formed in the semiconductor.

Flux has units of particles per unit area per unit time. Since we can also define a flux as the product of concentration and velocity, we can write

$$\phi = -D_n \frac{d\delta n}{dx} = S_n \delta n \tag{2.69a}$$

where $|S_n|$ is the *surface recombination velocity* of the electrons, and we evaluate δn and $\frac{d\delta n}{dx}$ at the semiconductor surface. Note that if $S_n = 0$, we can conclude that $\frac{d\delta n}{dx}$ is zero, there is no band bending and the surface does not trap carriers. Conversely if $|S_n| \to \infty$ then $\delta n = 0$ at the surface, which implies that carriers very rapidly recombine at the surface. A similar situation exists at the surface or at an interface of an n-type semiconductor in which holes may recombine and we obtain

$$\phi = -D_p \frac{d\delta p}{dx} = S_p \delta p \tag{2.69b}$$

In both Equations 2.69a and 2.69b it is assumed that the excess carrier concentrations are much larger than the equilibrium minority carrier concentrations. See Example 2.6.

EXAMPLE 2.6

An n-type silicon semiconductor sample having intrinsic electron concentration $n_0 = 1\times10^{17}$ cm^{-3} is uniformly illuminated with light such that excess carrier concentrations are $\delta p = \delta n = 1\times10^{15}$ cm^{-3}. Due to surface recombination, excess hole concentration at the semiconductor surface drops to $\delta p_{surface} = 1\times10^{14}$ cm^{-3}. $\delta p(x)$ is shown in the graph below with surface at $x = 0$:

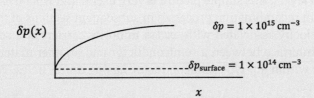

$\delta p(x)$ ⟶ $\delta p = 1 \times 10^{15}$ cm^{-3}

$\delta p_{surface} = 1 \times 10^{14}$ cm^{-3}

x

Find the surface recombination velocity of holes for the illuminated semiconductor. Assume a hole recombination time of 2 microseconds.

Use the diffusion Equation (2.66a) to calculate $\delta p(x)$. Note that the generation rate of carriers must be included in the diffusion equation to offset the carrier recombination rate. The modified form of the diffusion equation will be as follows:

$$D_p \frac{d^2 \delta p(x)}{dx^2} = \frac{\delta p(x)}{\tau_p} - G$$

where G is the generation rate and $\frac{\delta p(x)}{\tau_p}$ is the recombination rate.

Solution

The general solution to the differential equation is $\delta p(x) = A\exp\left(\dfrac{-x}{L_p}\right) + B\exp\left(\dfrac{x}{L_p}\right) + C$ where $L_p = \sqrt{D_p \tau_p}$.

We set $B = 0$ to satisfy the condition that $\delta p(\infty)$ is a constant. Substituting a general solution into the modified diffusion equation and applying boundary conditions

$$\delta p(0) = A + C = 1 \times 10^{14} \text{ cm}^{-3}$$

$$\delta p(\infty) = C = 1 \times 10^{15} \text{ cm}^{-3}$$

Therefore $\delta p(x) = A\exp\left(\dfrac{-x}{L_p}\right) + C$ where $A = -9 \times 10^{14} \text{ cm}^{-3}$ and $C = 1 \times 10^{15} \text{ cm}^{-3}$

Also, $D_p = \dfrac{kT}{q}\mu_p = 12.5 \text{ cm}^2\text{s}^{-1}$ and

$$L_p = \sqrt{D_p \tau_p} = \sqrt{\left(12.5 \text{cm}^2\text{s}^{-1}\right)\left(2 \times 10^{-6}\text{s}\right)}$$

$$= 5 \times 10^{-3} \text{cm}$$

Now, $S_p \delta p\big|_{x=0} = -D_p \dfrac{d\delta p(x)}{dx}\Big|_{x=0} = -D_p \dfrac{A}{L_p}$ and therefore

$$S_p = \frac{-D_p}{\delta p|_{x=0}}\frac{A}{L_p} = \frac{12.5\text{cm}^2\text{s}^{-1}}{1 \times 10^{14}\text{cm}^{-3}} \times \frac{-9 \times 10^{14}\text{cm}^{-3}}{5 \times 10^{-3}\text{cm}} = -22{,}500 \text{ cm/s}$$

A surface recombination velocity is defined by convention as a flux coming *out* of the surface. We therefore remove the negative sign and obtain a surface recombination velocity of $22{,}500 \text{ cm/s}$.

2.21 ALLOY SEMICONDUCTORS

An important variation in semiconductor compositions involves the use of partial substitutions of elements to modify composition. One example is the partial substitution of germanium in silicon that results in a range of new semiconductors of composition $Si_{1-x}Ge_x$, which are known as *alloy semiconductors*. The germanium atoms randomly occupy lattice sites normally occupied by silicon atoms, and the crystal structure of silicon is maintained. Note that Si and Ge are both in the Group IV column of the periodic table and therefore have chemical similarities in terms of valence electrons and types of bonding. This means that provided no additional dopant impurities are introduced into the alloy semiconductor, alloy material with characteristics of an intrinsic semiconductor can be achieved. Of interest in semiconductor

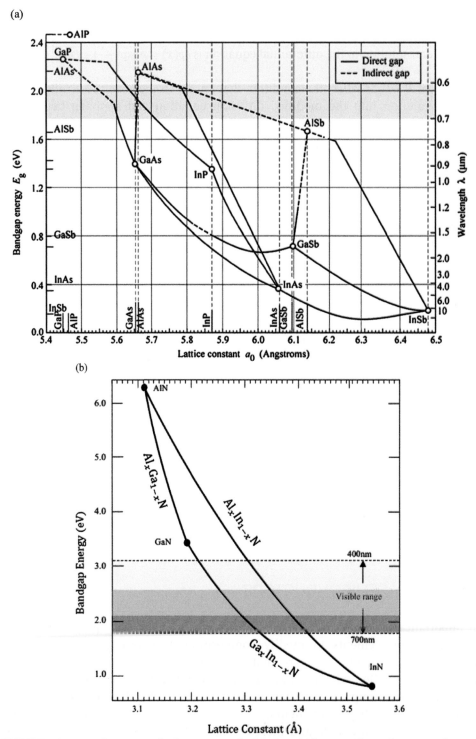

FIGURE 2.34 Bandgap versus lattice constant for (a) phosphide, arsenide, and antimonide III-V semiconductors; (b) nitride-based III-V semiconductors (c) sulfide, selenide, and telluride II-VI semiconductors and phosphide, arsenide, and antimonide III-V semiconductors. (a–b) After E. Fred Schubert, Light-Emitting Diodes, 2e ISBN 978-0-521-86538-82006 (c) After Yong H. Zhang, Arizona State University MBE Optoelectronics.

(c)

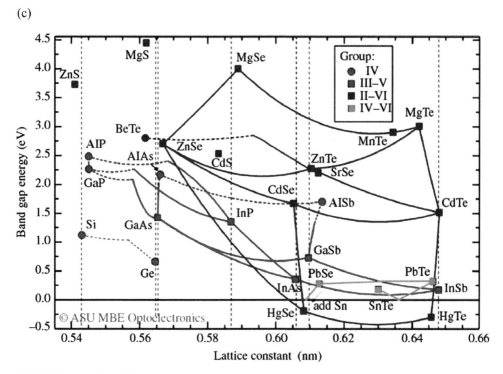

FIGURE 2.34 (Cont'd)

devices is the opportunity to modify the optical and electrical properties of the semi-conductor. Since germanium has a smaller bandgap than silicon, adding germanium decreases bandgap as x increases. In addition, the average lattice constant of the new compound will increase since germanium is a larger atom than silicon. Since both germanium and silicon have the same diamond crystal structure, the available range of x is from 0 to 1 and the indirect bandgaps of the alloy compositions $Si_{1-x}Ge_x$ therefore range between 1.11 eV and 0.67 eV as x varies from 0 to 1 respectively.

Of more relevance to p-n junctions for solar cells and LEDs, alloy semiconductors may also be formed from compound semiconductors. For example $Ga_{1-x}In_xN$ is a *ternary*, or three-component, alloy semiconductor in which a fraction of the gallium atoms in wurtz-ite GaN are replaced by indium atoms. The indium atoms randomly occupy the crystalline sites in GaN that are normally occupied by gallium atoms. Since In and Ga are both Group III elements, the substitution does not act as either an acceptor or a donor. The direct bandgap decreases as x increases. For $x = 0$, $E_g = 3.4$ eV and for $x = 1$, $E_g = 0.77$ eV.

There are many other III-V alloy semiconductors. In $Ga_{1-x}Al_xAs$ alloys the band-gap varies from 1.43 to 2.16 eV as x goes from 0 to 1. In this system, however, the bandgap is direct in the case of GaAs, but indirect in the case of AlAs. There is a transition from direct to indirect bandgap at $x \simeq 0.45$. This transition can be under-stood if we consider the two conduction band minima in GaAs shown in Figure 2.17c. One minimum forms a direct energy gap with the highest energy levels in the valence band; however, the second minimum forms an indirect gap. When $x \simeq 0.45$ these two minima are at the same energy level. For $x < 0.45$ the alloy has a direct gap because the global conduction band minimum forms the direct gap. For $x > 0.45$ the global conduction band minimum is the minimum that forms the indirect gap.

A number of III-V alloy systems are illustrated in Figure 2.34a. An additional set of III-V nitride semiconductors is shown in Figure 2.34b and a set of II-VI semiconductors is included in Figure 2.34c.

To cover ranges of composition in the III-V alloy system, *quaternary* alloys may be formed such as $(Al_xGa_{1-x})_y In_{1-y} P$. The use of ternary and quaternary semiconductor alloys in solar cells and LEDs will be discussed in Chapter 5.

2.23   SUMMARY

2.1 The band theory of solids permits an understanding of electrical and optical properties including electrical conductivity in metals and semiconductors, optical absorption, and luminescence, and properties of junctions and surfaces of semiconductors and metals.

2.2 The underlying form of electron wavefunctions in periodic potentials is found using Fourier series representations. Bloch's theorem is derived in one dimension. The existence of Brillouin zones follows directly from this, and the concept of a wavenumber k within the first Brillouin zone for electrons in a crystal is established.

2.3 In the Kronig–Penney model a periodic potential leads to energy bands and energy gaps by solving Schrödinger's equation for electrons in a periodic potential. The size of the energy gaps increases as the amplitude of the periodic potential increases. As the ionic character of the bonding in the semiconductor increases the energy gap increases. As the size of the atoms decreases the energy gap increases.

2.4 The Bragg model identifies Brillouin zone boundaries as satisfying the Bragg condition for strong reflection. This condition is $2a = n\lambda$. The reduced zone scheme, which shows only the first Brillouin zone, simplifies the representation of energy bands and energy gaps.

2.5 The effective mass m^* is used to quantify electron behavior in response to an applied force. The effective mass depends on the band curvature. Effective mass is constant if the band shape can be approximated as parabolic. The parabolic, isotropic dispersion relation allows for three-dimensional modeling near band edges.

2.6 The number of states in a band n can be determined based on the number of unit cells N in the semiconductor sample. The result $n = 2N$ is obtained for a one-dimensional, two-dimensional, or three-dimensional case.

2.7 The filling of bands in semiconductors and insulators is such that the highest filled band is full and the lowest empty band is empty at low temperatures. In metals the highest filled band is only partly filled. Semiconductors have smaller bandgaps ($E_g = 0$ to 4 eV) than insulators ($E_g > 4$eV).

2.8 The Fermi energy E_F is defined as the energy level at which an electron state has a 50% probability of occupancy at temperatures above 0 K. A hole can be created when an electron from the valence band is excited to the conduction band. The hole can move independently from the electron.

2.9 Carrier concentration in an energy band is determined by (i) finding the probability of occupancy of the states in a band using the Fermi–Dirac distribution function $F(E)$, and (ii) finding the density of states function $D(E)$ for an energy band. Then the integral over the energy range of the band of product $D(E)F(E)$ will determine the number of carriers in the band. In the conduction band the equilibrium electron concentration is n_0 and in the valence band the equilibrium hole concentration is p_0. The product $n_0 p_0$ is a constant that is independent of the Fermi energy.

2.10 A range of semiconductor materials includes Group IV semiconductors, Group III-V semiconductors, and Group II-VI semiconductors as listed in order of increasing ionic character. Bandgap energies decrease for larger atoms that are lower down on the periodic table.

2.11 Most important semiconductor crystals have lattices that are diamond, zinc-blende, or hexagonal. These structures lead to complex band diagrams. The band shapes in E versus k plots depend on crystallographic directions. In addition there are sub-bands in the valence band that correspond to distinct hole effective masses. Bandgaps may be direct or indirect. Conduction bands generally exhibit two minima where one minimum corresponds to a direct gap transition and one minimum corresponds to an indirect gap transition.

2.12 Photon momentum is very small and direct gap transitions in semiconductors are favorable for photon creation and absorption. In indirect gap transitions the involvement of lattice vibrations or phonons is required. The optical absorption coefficient α is higher for direct gap semiconductors and lower for indirect gap semiconductors. Whereas indirect gap silicon has an effective absorption depth of $\cong 100\,\mu m$ for sunlight the corresponding absorption depth in GaAs is only $\cong 1\,\mu m$.

2.13 Pure semiconductors are known as intrinsic semiconductors. The incorporation of low levels of impurity atoms in a semiconductor leads to extrinsic semiconductors, in which the electron concentration n_0 and hole concentration p_0 are controlled by the impurity type and concentration. Donor impurities donate electrons to the conduction band in n-type semiconductors, and acceptor impurities donate holes to the valence band in p-type semiconductors. New shallow energy levels arise within the energy gap, which are called donor and acceptor levels. Carrier concentrations are temperature dependent; however, over a wide intermediate temperature range carrier concentrations are relatively constant as a function of temperature. Minority carriers refer to the carriers having a low concentration in a specific semiconductor region, and majority carriers refer to the carriers having a significantly higher concentration in the same region.

2.14 Carriers move through semiconductors in an electric field ε by a drift process, which is characterized by a drift velocity $\bar{v}$ and mobility μ. Drift current density is given by $J = \sigma \varepsilon$, which is an expression of Ohm's law. The understanding of Ohm's law is based on the concept of a terminal velocity due to scattering events having a characteristic scattering time τ, which depends on impurities, defects, and temperature. At high electric fields drift velocity will eventually saturate.

2.15 Carrier concentrations are not necessarily at equilibrium levels. Photons or applied electric fields can give rise to nonequilibrium excess carrier concentrations, which will return to equilibrium concentrations once equilibrium conditions are restored. Electron-hole pair (EHP) generation and recombination processes G_{th} or G_{op} and R define the resulting rate of generation and recombination. Under equilibrium conditions $G_{th} = R$ where $R \propto np$. The minority carrier lifetime time constants τ_n or τ_p characterize the recombination times of minority carriers.

2.16 Carriers diffuse in semiconductors due to a concentration gradient. The diffusion coefficients D_n and D_p determine the diffusion currents J_n and J_p respectively. The net current flow must include both drift and diffusion current. In equilibrium the net current is zero; however, drift and diffusion currents may be non-zero. The Einstein relation is derived from the requirements for equilibrium conditions and allows D_n and D_p to be derived from μ_n and μ_p.

2.17 In nonequilibrium conditions the Fermi energy is not defined; however, quasi-Fermi energies F_n and F_p may be defined to characterize changes in carrier concentrations due to excess carrier generation.

2.18 Combining the concepts of carrier recombination and carrier diffusion, the diffusion equation leads to the calculation of diffusion length $L_n = \sqrt{D_n \tau_n}$ and $L_p = \sqrt{D_p \tau_p}$ and shows that carrier concentration decays exponentially as a function of distance from a region of excess carrier generation.

2.19 Traps can have a large effect on carrier lifetimes. The most important traps are deep traps that are at or near mid-gap. A high density of such traps occurs at semiconductor interfaces and defects. This is due to dangling bonds at surfaces and defects. Fermi level pinning occurs at or near midgap due to dangling bonds. Surface recombination velocity is a measure of the rate of recombination at semiconductor surfaces.

2.20 The bandgap and direct/indirect nature of a semiconductor can be altered by alloying. Alloying may also change the lattice constant. Industrially important alloy semiconductors exist composed from Group IV elements, Group III-V elements, and Group II-VI elements. Ternary and quaternary compound semiconductors are important for solar cells and LEDs.

PROBLEMS

2.1 Show that Equation 2.13 may be obtained from Equations 2.9, 2.10, 2.11, and 2.12. Note that although this is basic algebra, it does require many steps to complete.

2.2 Show that Equation 2.15 results from Equation 2.14 if we define

$$P = \frac{mU_0ab}{\hbar^2}$$

and take $b \to 0$ and $U_0 \to \infty$ such that bU_0 is constant.

2.3 In Section 2.7 we showed that the number of states in an energy band is $n_b = 2N$ for a one-dimensional semiconductor. Show that the number of states in an energy band in a three-dimensional semiconductor is still $n_b = 2N$ where N is the number of unit cells in the three-dimensional semiconductor.

Hint: Consider a semiconductor in the form of a rectangular box having N_x, N_y, and N_z unit cells along the x, y, and z axes. Assume an infinite walled box with $U = 0$ inside the box. Use the allowed energy values for an electron in Equation 2.30. Include spin.

2.4 A rectangular semiconductor crystal has dimensions $2 \times 2 \times 1$ mm. The unit cell is cubic and has edge length of 2 Å. Find the number of states in one band of this semiconductor.

2.5 A rectangular silicon semiconductor bar of length 12 cm and cross-section 1×5 mm is uniformly doped n-type with concentration $N_d = 5 \times 10^{16}$ cm^{-3}.
   (a) Assuming all donors are ionized, calculate the room temperature current flow if contacts are made on the two ends of the bar and 10 V is applied to the bar.
   (b) Find the electric field in the bar for the conditions of (a).
   (c) What fraction of the current flows in the form of hole current for the conditions of (a)?
   (d) Find the resistivity of the silicon.
   (e) If the silicon were replaced by gallium phosphide and the doping was still $N_d = 5 \times 10^{16}$ cm^{-3} repeat (a), (b), and (c).
   (f) If the silicon temperature was increased to 120° C, repeat (a), (b), and (c). Assume that carrier mobility and bandgap are not affected by the increase in temperature.

2.6 Now, instead of being uniformly doped, the silicon bar of Problem 2.5 is doped with a linearly increasing donor doping concentration, such that the left end of the bar (LHS) is doped with a concentration of 1×10^{16} cm^{-3} and the right end of the bar (RHS) is doped with a concentration of 1×10^{17} cm^{-3}.
   (a) Determine the doping level at three points in the bar:
      (i) at 3 cm from the LHS;
      (ii) at the midpoint;
      (iii) at 9 cm from the LHS.
   (b) Assuming equilibrium conditions (no applied voltage) find the built-in electric field in the bar at each of positions (i), (ii), and (iii). Hint: Find the gradient in the doping about each point.
   (c) Find the electron drift current flowing in the bar at positions (i), (ii), and (iii).
   (d) Explain how the bar can be in equilibrium given the existence of these electric fields and drift currents.
   (e) Find the Fermi level relative to the top of the valence band for each of positions (i), (ii), and (iii).

(f) Sketch the band diagram as a function of position in equilibrium along the length of the bar showing the location of the Fermi energy.

2.7   A square silicon semiconductor wafer 50 cm^2 in area and 0.18 mm in thickness is uniformly doped with both acceptors ($N_a = 5 \times 10^{16}$ cm^{-3}) and donors ($N_d = 2 \times 10^{16}$ cm^{-3}).

(a) Assuming all donors and acceptors are ionized, calculate the room temperature current flow if the silicon is contacted by narrow metal contact strips that run the full length of two opposing edges of the square silicon sheet. 10 V is applied across the contacts.

(b) Repeat (a) but assume that the sheet is 100 cm^2 in area instead of 50 cm^2. Does the current change with area? Explain.

2.8   An undoped square silicon semiconductor wafer 50 cm^2 in area and 0.18 mm thick is illuminated over one entire 50 cm^2 surface and an electron-hole pair generation rate of 10^{21} cm^{-3} s^{-1} is achieved uniformly throughout the material.

(a) Determine the separation of the quasi-Fermi levels. The carrier lifetime is 2×10^{-6} s.

(b) Calculate the room temperature current flow if the silicon is contacted by narrow metal contact strips that run the full length of two opposing edges of the square wafer. 10 V is applied across the contacts under illumination conditions. The carrier lifetime is 2×10^{-6} s.

(c) Explain how a higher/lower recombination time would affect the answer to (b). How does the recombination time of 2×10^{-6} s compare with the transit time of the carriers, which is the time taken by the carriers to traverse the silicon sheet from one side to the other side? This silicon sheet is functioning as a *photoconductive* device since its conductivity depends on illumination. If the transit time is small compared to the recombination time then *gain* can be obtained since more than one carrier can cross the photoconductive sheet before a recombination event takes place on average. Gains of 100 or 1000 may be obtained in practice in photoconductors. How long a recombination time would be required for a gain of 100 to be achieved? Repeat for a gain of 1000.

2.9   A sample of n-type silicon is doped to achieve E_F at 0.3 eV below the conduction band edge at room temperature.

(a) Find the doping concentration.

(b) The n-type silicon sample is in the form of a square cross-section bar at room temperature and it carries a current of 3×10^{-2} A along its length. If the bar is 10 cm long and has a voltage difference of 100 V end-to-end, find the cross-sectional dimensions of the bar.

2.10 A silicon sample is uniformly optically excited such that its quasi-Fermi level for electrons F_n is 0.419 eV above its quasi-Fermi level for holes F_p. The silicon is n-type with donor concentration $N_D = 1 \times 10^{14}$ cm^{-3}. Find the optical generation rate. Assume carrier lifetime of 8×10^{-5} s.

2.11 An n-type silicon wafer is 5.0 mm thick and is illuminated uniformly over its surface with blue light, which is absorbed very close to the silicon surface. Assume that a *surface generation rate* of holes of 3×10^{18} cm^{-2} s^{-1} is obtained over the illuminated surface, and that the excess holes are generated at the silicon surface.

(a) Calculate the hole concentration as a function of depth assuming a hole lifetime of 2×10^{-6} s. Assume that the hole lifetime is independent of depth.

(b) Calculate the hole diffusion current density as a function of depth.

(c) Calculate the recombination rate of holes as a function of depth.

Hint: Note the difference in units between the given surface generation rate of holes (cm^{-2} s^{-1}) which is expressed as a flux density and the recombination rate of holes near the surface (cm^{-3} s^{-1}) which is expressed as a rate per unit volume. You need to think carefully about how to properly use these units.

2.12 Find:

(a) The n-type doping concentration required to cause silicon at room temperature to have electrical conductivity 100 times higher than intrinsic silicon at room temperature.

(b) The p-type doping concentration required to cause silicon at room temperature to have p-type conductivity 100 times higher than intrinsic silicon at room temperature.

2.13 Intrinsic silicon is uniformly illuminated with 10^{14} photons cm^{-2} s^{-1} at its surface. Assume that each photon is absorbed very near the silicon surface, and generates one electron-hole pair. Carrier lifetime is 6×10^{-6} s.

(a) Find the flux density of electrons at a depth of 3 μm. Make and state any necessary assumptions.

(b) Find the total excess electron charge per unit area stored in the silicon, assuming the silicon sample is very thick.

2.14 If the Fermi energy in an n-type silicon semiconductor at 300°C is 0.08 eV below the conduction band, and the donor energy level E_d is 0.02 eV below the conduction band, then find the probability of ionization of the donors.

2.15 For p-type GaAs, the resistance of a rod of the material (measuring 1 mm in diameter and 40 mm in length) from end to end is measured as 4×10^7 ohms at 300 K.

(a) Find the doping concentration.

(b) Find the doping concentration in an n-type, but otherwise identical, GaAs rod having the same resistance at 300 K.

2.16 An intrinsic, planar, room-temperature silicon sample is exposed to a steady flux of light at its surface. The electron concentration as a result of this is measured to be 100 times higher than n_i, the intrinsic equilibrium concentration, at a depth of 100 μm below the silicon surface. Carrier lifetime $\tau_n = 2 \times 10^{-6}$ s.

(a) Assuming that the light is all absorbed very near the silicon surface, and that every incident photon excites one EHP, find the total photon flux density.

(b) Find the quasi-Fermi level for electrons relative to the Fermi level determined without the flux of light at a depth of 100 μm.

2.17 A famous experiment that involves both the drift and the diffusion of carriers in a semiconductor is known as the Haynes–Shockley experiment. Search for the experimental details of this experiment and answer the following:

(a) Make a sketch of the semiconductor sample used in the experiment as well as the location and arrangement of electrodes and the required voltages and currents as well as the connections of electrodes for the appropriate measurements to be made.

(b) Sketch an example of the time dependence of the output of the experiment.

(c) Explain how the Einstein relation can be verified using these data.

2.18 A silicon sample is uniformly optically excited such that its quasi-Fermi level for electrons F_n is 0.419 eV above its quasi-Fermi level for holes F_p. The silicon is n-type with donor concentration $N_D = 1 \times 10^{17}$ cm^{-3}. Find the optical generation rate. Assume a carrier lifetime of 8×10^{-6} s.

2.19 A flash of light at time $t = 0$ is uniformly incident on all regions of a p-type silicon sample with doping of 5×10^{17} cm^{-3}. The resulting EHP concentration is 2×10^{16} EHP cm^{-3}. Find the time-dependence of electron and hole concentrations for time t greater that zero. Assume carrier lifetime of 8×10^{-6} s.

2.20 Carriers are optically generated at an intrinsic silicon surface. The generation rate is 2×10^{19} EHP cm^{-2} s^{-1}. Assume that all the photons are absorbed very close to the silicon surface. Find the diffusion current of electrons just below the surface, and state clearly the assumptions you used to obtain the result. Assume carrier lifetime of 8×10^{-6} s and a surface area of 10 square centimeters.

2.21 A silicon sample is doped with 6×10^{16} donors cm^{-3} and N_a acceptors cm^{-3}. If E_F lies 0.4 eV below E_F in intrinsic silicon at 300 K, find the value of N_a.

2.22 Electric current flows down a silicon rod 1 cm in length and 0.3 mm in diameter. The silicon is n-type with $N_d = 1 \times 10^{17}$ cm^{-3}. A potential difference of 10 V is applied to the rod end-to-end. How many electrons drift through the rod in 60 seconds?

2.23 Find the energy difference between F_n (quasi-Fermi level for electrons under illumination) and E_F (without illumination) for a silicon sample containing 10^{13} donors cm^{-3}. Assume room temperature. The illumination is uniform such that 10^{18} EHP cm^{-3} s^{-1} are generated. Assume carrier lifetime of 8×10^{-5} s.

2.24 (a) A p-type silicon semiconductor is illuminated and excess carrier concentrations $\delta n = \delta p = 2 \times 10^{16}$ cm^{-3} are generated. Find the surface recombination velocity of electrons at the semiconductor surface given the following:

$$p_0 = 5 \times 10^{17} \, \text{cm}^{-3}$$

$$\delta n_{surface} = 2 \times 10^{14} \, \text{cm}^{-3}$$

carrier lifetime $\tau = 100 \, \mu\text{s}$

(b) If the surface recombination velocity is zero for the same sample, calculate $\delta n_{surface}$.

(c) If the surface recombination velocity is infinite for the same sample, calculate $\delta n_{surface}$.

(d) Sketch a band diagram as a function of distance x from the semiconductor surface to a few diffusion lengths away from the surface. What additional carrier transport mechanism besides diffusion will be active near the silicon surface?

(e) Repeat (a) for a GaAs semiconductor with minority carrier lifetime $\tau = 100\,\text{ns}$.

2.25 An n-type silicon sample has a donor concentration of 1×10^{17} cm^{-3}. We now illuminate this sample steadily and introduce a uniform electron-hole pair generation rate of $G_{op} = 3 \times 10^{20}$ cm^{-3} s^{-1}. Assume a carrier lifetime of 2.5×10^{-5} s. Find the surface recombination velocity of holes at the semiconductor surface if the surface hole concentration is 1×10^{14} cm^{-3}.

2.26 Assume a doped (extrinsic) silicon sample at room temperature.

(a) Calculate the separation between E_c and E_F for n-type silicon having a phosphorus impurity concentration of 1×10^{15} cm^{-3}. Find both electron and hole concentrations.

(b) Calculate the separation between E_c and E_F for p-type silicon having an aluminum impurity concentration of 1×10^{15} cm^{-3}. Find both electron and hole concentrations.

(c) Repeat a) and b) for germanium.

2.27 For semiconductor samples of silicon and germanium, assume carrier mobility is temperature independent.

(a) Find the temperature at which the carrier concentration in intrinsic silicon becomes equal to 1×10^{15} cm^{-3}.

(b) Repeat a) for intrinsic germanium.

(c) Find the electrical conductivity σ of each of a) and b) at the temperatures you calculated.

(d) Find the electrical conductivity σ of intrinsic silicon and the electrical conductivity σ of intrinsic germanium at room temperature.

(e) Determine the upper temperature limits for the intermediate temperature ranges of n-type doped silicon and n-type doped germanium samples having doping $N_d = 1 \times 10^{15}$ cm^{-3}. Use the criterion that the conductivity from thermally stimulated carriers is 10% of the conductivity due to carriers supplied by doping.

(f) We have assumed that carrier mobility is temperature independent. A more accurate model includes a temperature-dependent mobility. Use the internet to obtain more details on this dependence and the explanation for it.

2.28 In the Figure 2.21, three temperature regimes are described. Figure 2.11 shows a band diagram and a Fermi-Dirac distribution function that would be relevant to the high temperature case in Figure 2.21.

(a) Modify and redraw Figure 2.11 for the intermediate temperature range of Figure 2.21.

(b) Modify and redraw Figure 2.11 for the low temperature range of Figure 2.21. Hint: Make use of information in the caption of Figure 2.21. Appropriately scale the Fermi-Dirac distribution function for each temperature range.

2.29 The Hall Effect allows information on mobile carriers to be experimentally determined in an extrinsic semiconductor sample of unknown doping. It relies on the Lorentz force (see Section 1.2). The semiconductor sample is subjected to both an electric and a magnetic field causing a Hall Voltage to be established. The magnitude and polarity of this voltage can be used to deduce the carrier type, the carrier mobility and the carrier concentration. Find a reference for this and explain the basis for the Hall Effect in 2 or 3 pages.

SUGGESTIONS FOR FURTHER READING

1. Ashcroft, N.W. and Mermin, N.D. (1976). *Solid State Physics*. Holt, Rinehart and Winston.

2. Kittel, C. (2005). *Introduction to Solid State Physics*, 8e. John Wiley and Sons.

3. Eisberg, R. and Resnick, R. (1985). *Quantum Physics of Atoms, Molecules, Solids, Nuclei and Particles*, 2e. John Wiley and Sons.

4. Neamen, D.A. (2003). *Semiconductor Physics and Devices*, 3e. McGraw Hill.

5. Solymar, L. and Walsh, D. (2004). *Electrical Properties of Materials*, 7e. Oxford University Press.

CHAPTER 3

The p-n Junction Diode

CONTENTS

Objectives

1. Understand the structure of a p-n junction.
2. Obtain a qualitative understanding of diode current flow and the roles of drift and diffusion currents.
3. Derive a quantitative model of diode contact potential based on the band model.

Fundamentals of Semiconductor Materials and Devices, First Edition. Adrian Kitai.
© 2023 John Wiley & Sons Ltd. Published 2023 by John Wiley & Sons Ltd.
Companion Website: www.wiley.com/go/kitai_fundamentals

4. Justify and obtain a quantitative model of the depletion region extending away from the junction.
5. Derive a quantitative model of diode current flow based on carrier drift and diffusion and a dynamic equilibrium in the depletion region.
6. Understand phenomena involved in reverse breakdown of a diode and introduce the Zener diode.
7. Understand the tunnel diode based on a band model and its significance in solar cell applications.
8. Understand generation and recombination currents that exist in the depletion region of a p-n junction.
9. Introduce the physics of metal semiconductor contacts including the Schottky diode and ohmic contacts.
10. Introduce the heterojunction, its band structure, and its relevance to optoelectronic devices such as solar cells and LEDs.
11. Understand p-n junction capacitance and minority charge storage effects relevant to AC and transient behavior.

3.1 INTRODUCTION

A semiconductor device comprising a *p-n junction diode* is illustrated in Figure 3.1. The key components of a crystalline semiconductor p-n junction diode include the following:

(a) a metal *anode* contact applied to a p-type semiconductor forming a metal-semiconductor junction;
(b) a p-type semiconductor;
(c) a p-n semiconductor junction between p-type and n-type semiconductors;
(d) an n-type semiconductor;
(e) a metal *cathode* contact applied to the n-type semiconductor forming another metal-semiconductor junction.

Originally the semiconductor diode was used to provide current flow in one direction and current blocking in the other direction. It found widespread application in early logic circuits as a clamping device and as a logic adder. It is currently widely used as a rectifier in power supplies and as a detector in RF circuits. Two circuit applications are illustrated in Figure 3.1. The applications of specific interest to this book, however, are the more recent derivatives of the semiconductor diode that have become widely used for light emission (LEDs) and for solar power generation (solar cells).

Solar cells and LEDs can be fabricated from a range of both inorganic and organic semiconductor diode materials. Chapter 5 introduces inorganic solar cells and LEDs as well as junction transistors that are based on p-n junctions. The metal oxide semiconductor field effect transistor covered in Chapter 6 and organic semiconductor

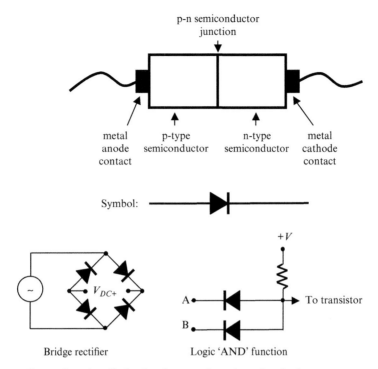

FIGURE 3.1 The p-n junction diode showing metal anode and cathode contacts connected to semiconductor p-type and n-type regions respectively. There are two metal-semiconductor junctions in addition to the p-n semiconductor junction. The diode symbol and two examples of diode applications in circuit design are shown. The diode logic gate was used in early diode-transistor logic solid state computers popular in the 1960s; however, diodes have been replaced by transistor-based designs that consume less power and switch faster.

devices covered in Chapter 8 also make use of p-n junction concepts. A good under-standing of all diode-based semiconductor devices requires a fundamental under-standing of the p-n junction diode including the electrical contacts made to the diode.

We will begin with the *abrupt junction* semiconductor p-n diode in which the transition from n-type material to p-type material occurs abruptly. This is achieved by a step change in the doping type on either side of the semiconductor junction. In practice, it is possible to make such a transition over a distance of just one or two atomic layers of a semiconductor crystal by the use of semiconductor growth methods in which the concentration of dopants can be rapidly and precisely changed during growth. The abrupt junction diode is the easiest diode to model and understand; however, the concepts can be extended to diodes in which the doping transition is gradual rather than abrupt. Chapter 6 describes a range of semiconductor growth and processing techniques.

3.2 DIODE CURRENT

The band model used to describe electron and hole behavior for the p-n junction is shown for equilibrium conditions in Figure 3.2. Since the n-type and p-type

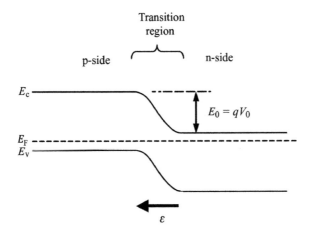

FIGURE 3.2   Band model of p-n junction in equilibrium showing constant Fermi energy and transition region to allow valence band and conduction band to be continuous.

semiconductor regions are in equilibrium, the Fermi energy is constant. Note the difference in position for the valence and conduction band energies on the two sides of the junction. There is a *transition region* in which the energy bands are sloped to provide a continuous conduction band and a continuous valence band extending from the p-side to the n-side of the junction. The transition region is present even though the p-n junction is abrupt.

We saw in Section 2.15 that band slope is caused by an electric field within a semiconductor. This electric field, referred to as a built-in electric field, exists in the transition region of the p-n junction even without an external applied voltage. The origin of this transition region and the associated electric field will be further described in Section 3.3. The direction of the field is shown in Figure 3.2 to be consistent with the direction of the band slope (see Section 2.15). The built-in field gives rise to both electron and hole drift currents across the junction. The change in electron energy across the p-n junction is labeled E_0 provided equilibrium conditions exist and E_0 is referred to as an *energy barrier*.

The energy gained or lost by an electron or a hole flowing across the transition region may be obtained by multiplying the change in potential by the charge. We can write

$$E_0 = qV_0 \tag{3.1}$$

where V_0 is the *contact potential* of the p-n junction.

We will now describe the currents that flow in equilibrium. The built-in electric field ε causes free carriers, both electrons and holes, in the transition region to drift. In addition, there are substantial carrier concentration gradients across the junction. For electrons, the high electron concentration in the n-side falls to a low electron concentration in the p-side. This electron concentration gradient as well as an analogous hole concentration gradient drive diffusion currents across the junction.

Hence, there are four currents to consider:

1. $I_{n, drift}$ Electrons (minority carriers) on the p-side that enter the transition region will drift to the right toward the n-side. This current is driven by the built-in electric field.
2. $I_{p, drift}$ Holes on the n-side (minority carriers) that enter the transition region will drift to the left toward the p-side. This current is driven by the built-in electric field.
3. $I_{n, diffusion}$ Electrons on the n-side (majority carriers) will diffuse to the left. This current is driven by the electron concentration gradient.
4. $I_{p, diffusion}$ Holes on the p-side will diffuse to the right. This current is driven by the hole concentration gradient.

These four currents can be viewed schematically in Figure 3.3. The electron currents are reversed in direction from the directions of electron flow since electrons carry negative charge.

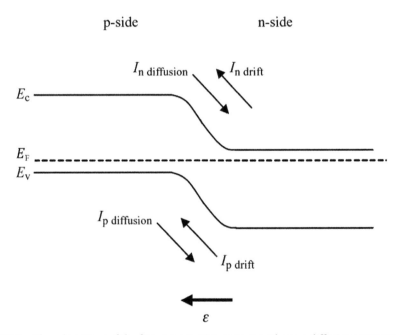

FIGURE 3.3   Flow directions of the four p-n junction currents. The two diffusion currents are driven by concentration gradients of electrons or holes across the junction and the two drift currents are driven by the electric field. Note that the electron currents flow in the direction opposite to the flux or flow of electrons. The electron diffusion flux is to the left and the electron drift flux is to the right.

If the p-n junction is in equilibrium we can conclude that the following equalities apply:

$$I_{p\ drift} + I_{p\ diffusion} = 0 \tag{3.2a}$$

$$I_{n\ drift} + I_{n\ diffusion} = 0 \tag{3.2b}$$

If we apply a voltage to the diode by connecting an external voltage source to the p-n junction, the drift and diffusion currents will no longer completely cancel out and a net diode current now flows. Even small deviations from complete cancellation represent substantial and useful diode currents, and the diode is now no longer in an equilibrium state. This is illustrated in Figure 3.4. This external voltage is called a *bias* voltage.

Let us first consider the application of a *forward bias* with $V > 0$ in which the p-side is connected to the positive output of a voltage source and the n-side to its negative output. The applied voltage V will fall across the transition region of the p-n junction and will *decrease* both the energy barrier height and the electric field ε as shown in Figure 3.5. The decrease in barrier height will result in a net current because the opposing drift current will no longer be sufficient to fully cancel out all the diffusion current. The *net current flow results from a net majority carrier diffusion current* to become

$$I = I_{p\ diffusion} + I_{n\ diffusion} - I_{p\ drift} - I_{n\ drift} > 0 \tag{3.3}$$

If we now consider the application of a *reverse bias* with $V < 0$ the applied voltage will again fall across the transition region of the p-n junction, which will *increase* the magnitude of both the potential barrier height and the electric field ε as shown in Figure 3.6. In spite of the increase in the energy barrier, diffusion current will still effectively oppose drift current. There will, however, be an additional small drift current due to thermally

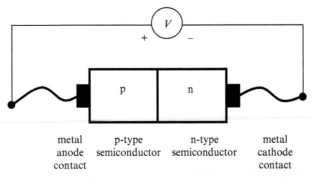

FIGURE 3.4   A p-n junction diode with external voltage source connected. The external bias voltage will modify the built-in electric field. Note that in Figure 3.3 the electron and hole diffusion currents flow in the same direction and may therefore be added together in Equation 3.3 to obtain the total diode diffusion current. The electron and hole drift currents also flow in the same direction and are both negative in Equation 3.3.

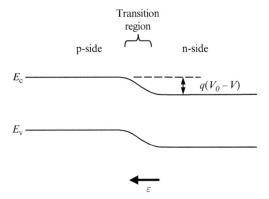

FIGURE 3.5 Diode band model with the application of a forward bias. The energy barrier across the transition region is smaller resulting in net diode current dominated by diffusion currents. In the depletion region ε will be smaller and drift currents no longer fully compensate for diffusion currents. Note that the applied voltage V (in volts) must be multiplied by the electron charge q (in coulombs) to obtain energy (in joules).

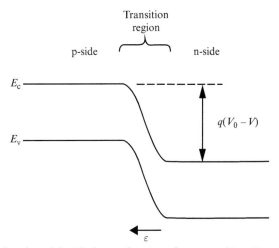

FIGURE 3.6 Diode band model with the application of a reverse bias. Since the applied voltage V is negative, the energy barrier as well as electric field ε become larger across the transition region. A small net minority carrier drift current flows.

generated minority electrons and holes. This results in a small net minority carrier drift current. The *net current flow is dominated by thermally generated minority carrier drift currents.* From Equations 3.2 a and 3.2b we obtain

$$I = I_{p\,\text{diffusion}} + I_{n\,\text{diffusion}} - I_{p\,\text{drift}} - I_{n\,\text{drift}} < 0$$

The total current I is now small and virtually independent of applied voltage V because I is controlled by the supply of thermally generated minority carriers available to drift

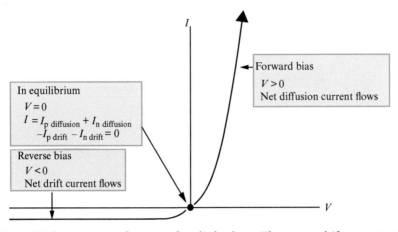

FIGURE 3.7   Diode current as a function of applied voltage. The reverse drift current saturates at a small value called the reverse saturation current due to thermally generated minority carriers. When $V = 0$ the drift and diffusion currents are equal in magnitude and the net current is zero.

and the magnitude of ε is not important. This is analogous to varying the height of a waterfall in a river: The amount of water flowing down the waterfall will depend on the available flow of the water approaching the waterfall and will not be affected by the height of the waterfall. The magnitude of this current is known as the *reverse saturation current*, I_0, and hence I_0 is the net thermally generated drift current supplied by minority carriers.

The diode current may now be plotted as a function of the applied voltage, as shown in Figure 3.7. We will treat diode current quantitatively in Section 3.5 which will further clarify the origin and nature of diode currents.

3.3   CONTACT POTENTIAL

We can calculate the contact potential V_0 using our understanding of energy bands. From Equation 2.37,

$$p_0 = N_v \exp\left(\frac{-(E_F - E_v)}{kT}\right)$$

We can apply Equation 2.37 to both the n-side and the p-side of the junction:

$$(E_F - E_v)_{\text{p-side}} = kT \ln\left(\frac{N_v}{p_0}\right) = kT \ln\left(\frac{N_v}{p_p}\right) \tag{3.4a}$$

where p_p is the equilibrium hole concentration on the p-side. On the n-side,

$$(E_F - E_v)_{\text{n-side}} = kT \ln\left(\frac{N_v}{p_0}\right) = kT \ln\left(\frac{N_v}{p_n}\right) \tag{3.4b}$$

where p_n is the equilibrium hole concentration on the n-side. Subtracting Equation 3.4a from Equation 3.4b we obtain

$$E_0 = (E_v)_{p\text{-side}} - (E_v)_{n\text{-side}} = kT \ln\left(\frac{p_p}{p_n}\right) \tag{3.5}$$

This is illustrated in Figure 3.8.

Now, Equation 3.5 can be expressed in terms of the contact potential V_0 using Equation 3.1, and therefore

$$V_0 = \frac{kT}{q} \ln\left(\frac{p_p}{p_n}\right) \tag{3.6}$$

We can also express this in terms of the doping levels on either side of the junction. From Equations 2.45, 2.47, and 3.6

$$V_0 = \frac{kT}{q} \ln\left(\frac{N_a N_d}{n_i^2}\right) \tag{3.7}$$

The contact potential may also be obtained using electron concentrations. Since $N_a N_d \cong p_p n_n$ and $n_p = \frac{n_i^2}{p_p}$ and $p_n = \frac{n_i^2}{n_n}$ we can also write

$$V_0 = \frac{kT}{q} \ln\left(\frac{n_n}{n_p}\right) = \frac{kT}{q} \ln\left(\frac{p_p}{p_n}\right) \tag{3.8}$$

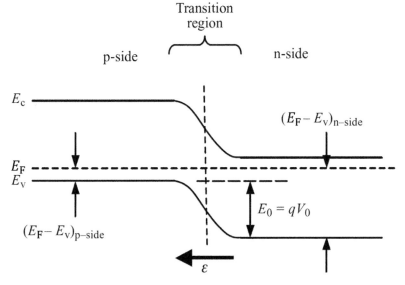

FIGURE 3.8 The equilibrium p-n junction energy barrier height E_0 may be obtained from $(E_F - E_v)_{n\text{-side}} - (E_F - E_v)_{p\text{-side}}$ resulting in Equation 3.5.

It is therefore possible to design a p-n junction having a specific built-in potential V_0 by controlling the doping levels in the p-type and n-type regions.

EXAMPLE 3.1

An abrupt silicon p-n junction diode is doped with $N_a = 1 \times 10^{17}\ \text{cm}^{-3}$ on the p-side and $N_d = 1 \times 10^{17}\ \text{cm}^{-3}$ on the n-side.

Find the built-in potential and sketch the band diagram in equilibrium at room temperature. Include the Fermi level.

Solution

$$V_0 = \frac{kT}{q}\ln\left(\frac{N_a N_d}{n_i^2}\right) = (0.026\ \text{V})\ln\left[\frac{10^{17}\ \text{cm}^{-3} \times 10^{17}\ \text{cm}^{-3}}{(1.5 \times 10^{10}\ \text{cm}^{-3})^2}\right] = 0.817\ \text{V}$$

On the p-side from Example 2.2,

$$N_v = 1.06 \times 10^{19}\ \text{cm}^{-3}$$

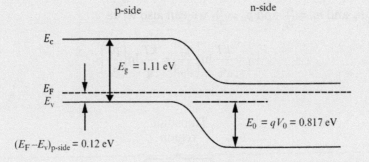

Now,

$$\left(E_F - E_v\right)_{\text{p-side}} = kT\ln\left(\frac{N_v}{p_p}\right) = (0.026\text{eV})\ln\left[\frac{1.06 \times 10^{19}\ \text{cm}^{-3}}{10^{17}\ \text{cm}^{-3}}\right] = 0.12\text{eV}$$

3.4   THE DEPLETION APPROXIMATION

A detailed view of the transition region of Figure 3.2 is redrawn in Figure 3.9. In the central part of the transition region the Fermi energy is close to the middle of the energy gap, which implies that the semiconductor will behave much like intrinsic material even though it is doped either p-type or n-type. This is shown as a *strongly depleted* region in Figure 3.9 because the concentration of charge carriers here is very low as it would be in an intrinsic semiconductor. As we move to the left or right of this

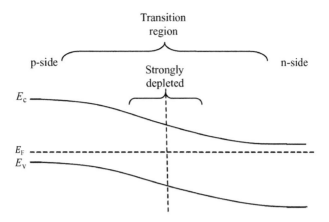

FIGURE 3.9 Depletion occurs near the junction. In order to establish equilibrium conditions, electrons and holes recombine and the Fermi energy lies close to the middle of the bandgap in the strongly depleted region.

region, the carrier concentrations gradually return to their normal p-type or n-type equilibrium values respectively, and the energy bands become horizontal lines.

To understand this depletion it is necessary to consider two separate semiconductors, one n-type and one p-type. If they are brought together, electrons in the n-type material and holes in the p-type material close to the junction will quickly diffuse across the junction and annihilate each other by recombination leaving a deficit of holes and electrons. This diffusion and recombination will be complete after a very short time, and then equilibrium conditions will be established. Electrons and holes further away from the junction will also have a chance to diffuse across the junction; however, the electric field that is built up at the junction opposes this and also causes drift currents and the resulting band diagram is now described by Figure 3.2. The equilibrium current components that continue to flow are described by Equation 3.2.

A simplification called the *depletion approximation* is widely used to model the depletion of charges. We assume complete depletion of charge carriers in a *depletion region* of width W_0 at the junction, and then assume that the carrier concentrations abruptly return to their equilibrium levels on either side of the depletion region as shown in Figure 3.10. Carrier concentrations do not make abrupt concentration changes in real materials or devices and instead gradients in carrier concentrations exist; however, the depletion approximation dramatically simplifies the quantitative model for the p-n junction and the results are still highly relevant to real diodes.

Normally a doped semiconductor consists of both mobile carriers and non-mobile ionized dopants as discussed in Section 2.14. The net charge density in the semiconductor is zero since the dopant ions have a charge density that is equal in magnitude and opposite in sign to the mobile charges that they provide.

The semiconductor material within the depletion region of a p-n junction is depleted of carriers but *it is still doped either p-type or n-type* and dopant ions are

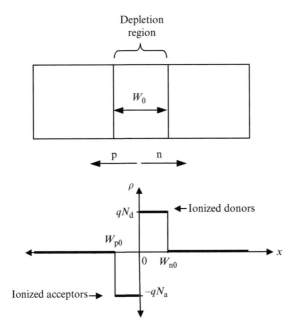

FIGURE 3.10 A depletion region of width W_0 is assumed at the junction. Charge density ρ is zero outside of the depletion region. Inside the depletion region a net charge density due to ionized dopants is established. The origin of the x-axis is placed at the junction for convenience.

therefore present in this region. This means that there is a net fixed (non-mobile) charge density in the depletion region from the ionized dopants.

In the p-side the ions are negative. For example, in silicon with p-type aluminum doping the aluminum ions are negatively charged having accepted an electron to yield a valence band hole. In the n-side the ions are positive. Conversely, in silicon with n-type phosphorus doping, the phosphorus ions would be positively charged having donated an electron to the conduction band. Once these holes and electrons recombine the depletion region will be left with Al^- ions and P^+ ions having concentrations of N_a and N_d respectively.

Charge densities $-qN_a$ and $+qN_d$ (coulombs per cm^3) will reside on the p-side and n-side of the depletion region respectively, as indicated in Figure 3.10. The magnitude of depletion charge on either side of the junction must be the same since one negative and one positive ion results from each recombination of a hole and an electron. Hence the magnitude of the charge present on either side of the junction is

$$Q = qN_aW_{p0}A = qN_dW_{n0}A \tag{3.9}$$

where W_{p0} and W_{n0} are the widths of the depletion regions on the p- and n-sides of the junction in equilibrium respectively, and A is the cross-sectional area of the diode. From Equation 3.9 we can write

$$\frac{W_{p0}}{W_{n0}} = \frac{N_d}{N_a} \tag{3.10}$$

The charged regions in the depletion region give rise to an electric field. This is the same field that causes the energy bands to tilt in Figure 3.2, and it can now be calculated. Using Gauss's law we can enclose the negative charge in the p-side of the depletion region with a rectangular cuboid Gaussian surface as shown in Figure 3.11.

Gauss's law relates the surface integral of electric field ε for a closed surface to the enclosed charge Q or

$$\oint_s \vec{\varepsilon} \cdot \vec{ds} = \frac{Q}{\epsilon} \tag{3.11}$$

If we assume that our Gaussian surface has a sufficiently small dimension W_{p0}, its total surface area will be dominated by cross sectional area A. In this case we can approximate the total surface area as $2A$. Since there is symmetry and the two cross-sectional areas A are equivalent, we can use Equation 3.11 to write

$$2\varepsilon_p A = \frac{Q}{\epsilon_0 \epsilon_r}$$

where ε_p is the magnitude of the electric field caused by the depletion charge on the p-side and ϵ_r is the relative permittivity of the semiconductor, and therefore

$$\varepsilon_p = \frac{Q}{2\epsilon_0 \epsilon_r A} \tag{3.11a}$$

The same reasoning may be applied to a Gaussian surface that encloses depletion charge on the n-side. This will give rise to a field of magnitude

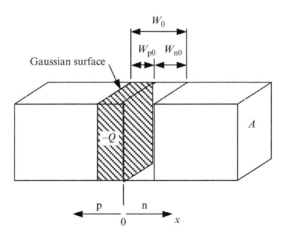

FIGURE 3.11 A Gaussian surface having volume AW_{p0} (shaded) encloses the negative charge of magnitude Q on the p-side of the depletion region.

$$\varepsilon_n = \frac{Q}{2\epsilon_0\epsilon_r A} \qquad (3.11b)$$

Since electric field is a vector quantity the relevant electric field directions are shown in Figure 3.12, and it is clear that the total field in equilibrium at the junction is the vector sum of $\vec{\varepsilon_p}$ and $\vec{\varepsilon_n}$ yielding equilibrium electric field magnitude at the semi-conductor junction

$$\varepsilon_0 = -\frac{Q}{\epsilon_0\epsilon_r A}$$

which may be rewritten using Equation 3.9 as

$$\varepsilon_0 = -\frac{qN_dW_{n0}}{\epsilon_0\epsilon_r} = -\frac{qN_aW_{p0}}{\epsilon_0\epsilon_r} \qquad (3.12)$$

The minus sign indicates that the field points in the negative x direction. The field at other points along the x-axis may also be evaluated. We have assumed that the cross-sectional area of the junction is very large compared to the depletion width. Since the electric field due to an infinite plane of charge is independent of the distance from the plane, the vector quantities in Figure 3.12 will cancel out and the net field will be zero at $x = -W_{p0}$ and at $x = W_{n0}$. At other x values, the electric field will vary linearly as a function of x and may be calculated by appropriately applying Gauss's law (see Problem 3.5). The resulting electric field will also give rise to a potential using Equation 2.48 of

$$V_0(x) = -\int_{-W_{p0}}^{W_{n0}} \varepsilon(x)dx \qquad (3.13)$$

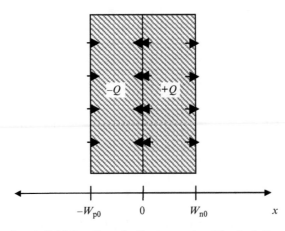

FIGURE 3.12   The electric field directions for the two parts of the depletion region showing that the fields add at the junction.

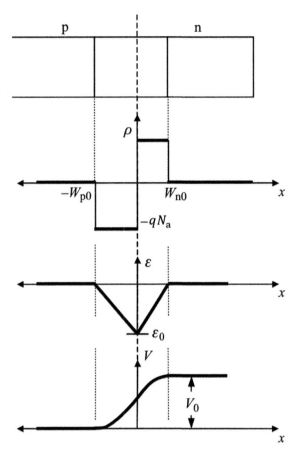

FIGURE 3.13 The equilibrium electric field $\varepsilon(x)$ and potential $V(x)$ for the p-n junction follow from the application of Gauss's law to the fixed depletion charge. Note that V on the n-side is higher compared to V on the p-side, whereas in Figure 3.2 the energy levels on the n-side are lower. This is the case because the energy scale in Figure 3.2 is for electron energy levels; however, the voltage scale in Figure 3.13 is established for a positive charge by convention.

Both $\varepsilon(x)$ and $V(x)$ are shown in Figure 3.13, which also shows that a contact potential V_0 results directly from the depletion model. This contact potential V_0 is the same quantity that we introduced in Section 3.3.

The integral (Equation 3.13) is the area under the ε versus x curve in Figure 3.13, which may be calculated using the area of a triangle, and we obtain

$$V_0 = \frac{1}{2}bh = \frac{1}{2}W_0\varepsilon_0$$

and using Equation 3.12 to express ε_0,

$$V_0 = \frac{qN_dW_{n0}}{2\epsilon_0\epsilon_r}W_0 = \frac{qN_aW_{p0}}{2\epsilon_0\epsilon_r}W_0$$

From this,

$$W_{p0} = \frac{2\epsilon_0\epsilon_rV_0}{qN_aW_0}$$

and

$$W_{n0} = \frac{2\epsilon_0\epsilon_rV_0}{qN_dW_0}$$

and hence

$$W_0 = W_{n0} + W_{p0} = \frac{2\epsilon_0\epsilon_rV_0}{qN_dW_0} + \frac{2\epsilon_0\epsilon_rV_0}{qN_aW_0}$$

which can be rearranged to obtain

$$V_0 = \frac{q}{2\epsilon_0\epsilon_r}\frac{N_aN_d}{N_a+N_d}W_0^2 \tag{3.14}$$

or

$$W_0 = \sqrt{\frac{2\epsilon_0\epsilon_rV_0}{q}\left(\frac{1}{N_a}+\frac{1}{N_d}\right)} \tag{3.15a}$$

with

$$W_{p0} = \frac{W_0N_d}{N_a+N_d} \tag{3.15b}$$

and

$$W_{n0} = \frac{W_0N_a}{N_a+N_d} \tag{3.15c}$$

It is interesting to note that the depletion approximation is consistent with an externally applied voltage falling across the depletion region. Since the depletion region has high resistivity, and the neutral regions on either side of it have high conductivities, we are justified in stating that applied reverse-bias voltages will drop across the depletion region.

EXAMPLE 3.2

a) Find the depletion layer width in both the n-side and the p-side of the abrupt silicon p-n junction diode of Example 3.1 doped with $N_a = 10^{17}\,\text{cm}^{-3}$ on the p-side and $N_d = 10^{17}\,\text{cm}^{-3}$ on the n-side. Find the equilibrium electric field at the semiconductor junction ε_0. Sketch the electric field and the potential as a function of position across the junction.

b) Repeat if the doping in the p-side is increased to $N_a = 10^{18}\,cm^{-3}$ and the doping in the n-side is decreased to $N_d = 10^{16}\,cm^{-3}$. This is called a p^+-n junction to indicate the heavy doping on the p-side.

Solution

(a)

$$W_0 = \sqrt{\frac{2\epsilon_0\epsilon_r V_0}{q}\left(\frac{1}{N_a}+\frac{1}{N_d}\right)}$$

$$= \sqrt{\frac{2\left(8.85\times10^{-14}\times11.8\ F\ cm^{-1}\right)\left(0.817\,V\right)}{1.6\times10^{-19}\,C}\left(\frac{1}{10^{17}\ cm^{-3}}+\frac{1}{10^{17}\ cm^{-3}}\right)}$$

$$= 1.51\times10^{-5}\,cm = 0.15\,\mu m$$

On p-side:

$$W_{p0} = \frac{W_0 N_d}{N_a + N_d} = \frac{0.15\,\mu m \times 10^{17}\ cm^{-3}}{10^{17}\ cm^{-3}+10^{17}\ cm^{-3}} = 0.075\,\mu m$$

On n-side:

$$W_{n0} = \frac{W_0 N_a}{N_a + N_d} = \frac{0.15\,\mu m \times 10^{17}\ cm^{-3}}{10^{17}\ cm^{-3}+10^{17}\ cm^{-3}} = 0.075\,\mu m$$

$$\varepsilon_0 = -\frac{qN_d W_{n0}}{\epsilon_0\epsilon_r} = \frac{1.6\times10^{-19}\,C\times10^{17}\ cm^{-3}\times0.75\times10^{-5}\ cm}{8.85\times10^{-14}\ F\,cm^{-1}\times11.8}$$

$$= 1.15\times10^{5}\,V\,cm^{-1}$$

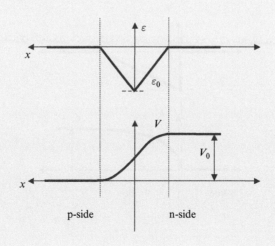

p-side   n-side

(continued)

(continued)

(b)

$$V_0 = kT/q \ln\left(\left(N_a N_d\right)/\left(n_i^2\right)\right)$$

$$= \left(0.026\,\mathrm{V}\right)\ln\left[\left(10^{18}\,\mathrm{cm}^{-3} \times 10^{16}\,\mathrm{cm}^{-3}\right)/\left(1.5 \times 10^{10}\,\mathrm{cm}^{-3}\right)^2\right] = 0.817\,\mathrm{V}$$

$$W_0 = \sqrt{\frac{2\epsilon_0 \epsilon_r V_0}{q}\left(\frac{1}{N_a} + \frac{1}{N_d}\right)}$$

$$= \sqrt{\frac{2\left(8.85 \times 10^{-14} \times 11.8\,\mathrm{F\,cm}^{-1}\right)\left(0.817\,\mathrm{V}\right)}{1.6 \times 10^{-19}\,\mathrm{C}}\left(\frac{1}{10^{18}\,\mathrm{cm}^{-3}} + \frac{1}{10^{16}\,\mathrm{cm}^{-3}}\right)}$$

$$= 3.4 \times 10^{-5}\,\mathrm{cm} = 0.34\,\mu\mathrm{m}$$

On p-side:

$$W_{p0} = \frac{W_0 N_d}{N_a + N_d} = \frac{0.34\,\mu\mathrm{m} \times 10^{16}\,\mathrm{cm}^{-3}}{10^{18}\,\mathrm{cm}^{-3} + 10^{16}\,\mathrm{cm}^{-3}} \cong 0.33 \times 10^{-2}\,\mu\mathrm{m} = 3.3\,\mathrm{nm}$$

On n-side:

$$W_{n0} = \frac{W_0 N_a}{N_a + N_d} = \frac{0.34 \times 10^{18}\,\mathrm{cm}^{-3}}{10^{18}\,\mathrm{cm}^{-3} + 10^{16}\,\mathrm{cm}^{-3}} \cong 0.33\,\mu\mathrm{m}$$

$$\varepsilon_0 = -\frac{q N_d W_{n0}}{\epsilon_0 \epsilon_r} = \frac{1.6 \times 10^{-19}\,\mathrm{C} \times 10^{16}\,\mathrm{cm}^{-3} \times 3.3 \times 10^{-5}\,\mathrm{cm}}{8.85 \times 10^{-14}\,\mathrm{F\,cm}^{-1} \times 11.8}$$

$$= 5.1 \times 10^4\,\mathrm{V\,cm}^{-1}$$

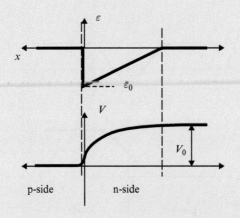

Note that the depletion region is almost entirely in the n-side of the p^+-n junction. Almost all the built-in potential V_0 drops within the n-side. The reverse is true for an n^+-p junction.

3.5 THE DIODE EQUATION

We shall now derive the current–voltage relationship for a p-n junction diode with an externally applied voltage bias, which will result in the *diode equation*. Under equilibrium conditions we saw that diffusion and drift currents cancel out; however, if an external bias is added, then a net current will flow.

In Figures 3.5 and 3.6 we saw that adding an external voltage V creates a new junction potential $V_0 - V$. Since the p-n junction is no longer in equilibrium we cannot continue to use the Fermi energy to describe and derive junction physics; however, we will now show that inside the depletion region of a biased diode, conditions consistent with a *dynamic equilibrium* are maintained even during current flow.

In equilibrium conditions, there are both drift and diffusion currents flowing across the depletion region. These currents cancel out and no net current flows. We will now illustrate, using Example 3.3, that when a diode is biased, very large diffusion and drift currents can flow, and that the range of typical diode currents implies that drift and diffusion currents almost exactly compensate for each other.

The depletion model of a diode is not sufficient to model diode current flow since the depletion model predicts a lack of carriers and hence a lack of current flow within the depletion region. We must therefore remember that gradients in carrier concentrations actually do exist within what we have modeled as the depletion region.

EXAMPLE 3.3

a) For the silicon diode of Example 3.1 assume a junction area of $1\,mm^2$. Estimate the expected equilibrium majority carrier *diffusion* current across the depletion region if drift current is neglected.

b) For the silicon diode of Example 3.1 assume a junction area of $1\,mm^2$. Estimate a typical minority carrier *drift* current if diffusion current is neglected. Assume minority carrier concentrations of only 1% of the majority carrier concentrations.

c) Explain why these current are not measured externally.

Solution

a) From Appendix 6, $\mu_n = 1350\ cm^2\ V^{-1}\ s^{-1}$ and $\mu_p = 480\ cm^2\ V^{-1}\ s^{-1}$ for silicon. Therefore, using the Einstein relation,

$$D_n = \frac{kT}{q}\mu_n = 0.026\ V \times 1350\,cm^2\ V^{-1}\ s^{-1} = 3.51 \times 10^1\ cm^2\ s^{-1}$$

(continued)

(continued)

and

$$D_p = \frac{kT}{q}\mu_p = 0.026\,\text{V} \times 480\,\text{cm}^2\,\text{V}^{-1}\text{s}^{-1} = 1.25 \times 10^1\,\text{cm}^2\,\text{s}^{-1}$$

The depletion model allows us to determine the width of the depletion region. From Example 3.2a, we have majority carrier concentrations on both sides of the depletion region of $1 \times 10^{17}\,\text{cm}^{-3}$ and a depletion region of width $W = 0.15\,\mu\text{m}$. The depletion model predicts physically impossible abrupt changes in majority carrier concentration at the edges of the depletion region. To avoid this difficulty we will assume that the concentration gradient extends all the way across the depletion region. We also restrict our analysis to diffusion currents and assume there is no electric field in the depletion region, which means that drift currents do not flow. The majority diffusion current would therefore flow due to the large concentration gradient of $1 \times 10^{17}\,\text{cm}^{-3}$ across the depletion region of width $0.15\,\mu\text{m}$.

Now,

$$J_n(x)_{\text{diffusion}} = qD_n \frac{dn(x)}{dx} = 1.6 \times 10^{-19}\,\text{C} \times 3.51 \times 10^1\,\text{cm}^2\text{s}^{-1} \times \frac{1 \times 10^{17}\,\text{cm}^{-3}}{1.5 \times 10^{-5}\,\text{cm}}$$

$$= 3.74 \times 10^4\,\text{A}\,\text{cm}^{-2}$$

and

$$J_p(x)_{\text{diffusion}} = -qD_p \frac{dp(x)}{dx}$$

$$= 1.6 \times 10^{-19}\,\text{C} \times 1.25 \times 10^1\,\text{cm}^2\text{s}^{-1} \times \frac{1 \times 10^{17}\,\text{cm}^{-3}}{1.5 \times 10^{-5}\,\text{cm}}$$

$$= 1.33 \times 10^4\,\text{A}\,\text{cm}^{-2}$$

For a $1\,\text{mm}^2$ junction area total diffusion current is $374 + 133 = 507$ A. Typical operating diode current densities are *orders of magnitude* smaller than this. A real diode with a junction area of $1\,\text{mm}^2$ would typically carry currents in the range of microamps or milliamps, with a maximum current on the order of only a few amps.

b. Assume a minority carrier concentration in the depletion region of only 1% of the majority carrier concentration in the diode. Think of these minority carriers as being the result of carrier diffusion. This yields a minority carrier concentration of $1 \times 10^{15}\,\text{cm}^{-3}$. We will now neglect diffusion currents and only consider minority carrier drift currents and an *average* electric field inside the depletion region of $\varepsilon = \dfrac{V_0}{W_0}$. From Example 3.1, $V_0 = 0.817$ V.

Now, $$J_n(x)_{drift} = q\mu_n n\varepsilon$$

$$= 1.6\times10^{-19}\,C\times1350\,cm^2\,V^{-1}s^{-1}\times1\times10^{15}\,cm^{-3}$$

$$\times\frac{0.817\,V}{1.5\times10^{-5}\,cm} = 1.18\times10^4\,A\,cm^{-2}$$

and

$$J_p(x)_{drift} = q\mu_p p\varepsilon$$

$$= 1.6\times10^{-19}\,C\times480\,cm^2\,V^{-1}s^{-1}\times1\times10^{15}\,cm^{-3}$$

$$\times\frac{0.817\,V}{1.5\times10^{-5}\,cm} = 4.2\times10^3\,A\,cm^{-2}$$

For a $1\,mm^2$ junction area total drift current is $118 + 42 = 160$ A. As in part (a), actual operating diode current densities are *orders of magnitude* smaller than this.

c. Measured diode current is the result of a delicate balance between diffusion and drift current. It is clear from parts (a) and (b) that the depletion region in a diode supports both drift and diffusion currents flowing in opposite directions that almost exactly compensate for each other when typical current densities in semi-conductor diode devices flow. This is somewhat analogous to someone running up a high-speed escalator that is heading down. If the runner runs at almost the same speed as the escalator the net speed of the runner is very slow and is much slower than either the running speed or the escalator speed.

Due to the observation from Example 3.3 we are justified in considering the drift and diffusion currents within the depletion region to be virtually equal and opposite *even with the application of a bias voltage V.*

Let us first consider hole current only. Using Equation 2.57 we can write

$$q\mu_p p(x)\varepsilon(x) - qD_p\frac{dp(x)}{dx}\cong 0$$

Solving for $\varepsilon(x)$ we obtain

$$\varepsilon(x) = \frac{D_p}{\mu_p}\frac{1}{p(x)}\frac{dp(x)}{dx}$$

Using the Einstein relation (Equation 2.60) we can write

$$\varepsilon(x) = \frac{kT}{q}\frac{1}{p(x)}\frac{dp(x)}{dx}$$

Using Equation 3.13 and integrating across the depletion region including an applied bias voltage V,

$$V_0 - V = -\int_{-W_p}^{W_n} \varepsilon(x)\,dx = -\frac{kT}{q}\int_{-W_p}^{W_n}\frac{1}{p(x)}\,dp(x) = \frac{kT}{q}\ln\left(\frac{p(-W_p)}{p(W_n)}\right) \qquad (3.16)$$

where $p(-W_p)$ is the majority hole concentration on the p-side of the depletion region and $p(W_n)$ is the minority hole concentration on the n-side of the depletion region.

Substituting for V_0 in Equation 3.16 using Equation 3.6 we obtain

$$\frac{kT}{q}\ln\left(\frac{p_p}{p_n}\right) - V = \frac{kT}{q}\ln\left(\frac{p(-W_p)}{p(W_n)}\right)$$

and rearranging we have

$$-V = \frac{kT}{q}\ln\left(\frac{p(-W_p)}{p(W_n)}\frac{p_n}{p_p}\right)$$

This can be simplified in the case of *low level injection* in which any changes in carrier concentration are small compared to the majority carrier concentrations. This means that $p(-W_p)$ is almost the same as p_p and hence

$$-V = \frac{kT}{q}\ln\left(\frac{p_n}{p(W_n)}\right)$$

Solving for $p(W_n)$, which is the new minority hole concentration at the edge of the depletion region on the n-side, we obtain

$$p(W_n) = p_n \exp\left(\frac{qV}{kT}\right) \qquad (3.17)$$

Note that if $V = 0$ then $p(W_n) = p_n$ as expected. The addition of a bias voltage V therefore multiplies the minority hole concentration at the edge of the depletion region on the n-side by the term $\exp\left(\frac{qV}{kT}\right)$.

The same procedure may be applied to electrons. The drift and diffusion currents for electrons from Equation 2.57 yields

$$q\mu_n n(x)\varepsilon(x) + qD_n\frac{dn(x)}{dx} \cong 0$$

From this, using the Einstein relation, we obtain

$$\varepsilon(x) = \frac{-D_n}{\mu_n}\frac{1}{n(x)}\frac{dn(x)}{dx} = \frac{-kT}{q}\frac{1}{n(x)}\frac{dn(x)}{dx}$$

Now integrating across the depletion region, and using Equation 3.13,

$$V_0 - V = -\int_{-W_p}^{W_n}\varepsilon(x)\,dx = \frac{kT}{q}\int_{-W_p}^{W_n}\frac{1}{n(x)}\,dn(x) = \frac{-kT}{q}\ln\left(\frac{p(-W_p)}{p(W_n)}\right) \qquad (3.18)$$

Substituting Equation 3.8 into Equation 3.18 we obtain

$$\frac{kT}{q}\ln\left(\frac{n_n}{n_p}\right) - V = \frac{-kT}{q}\ln\left(\frac{n(-W_p)}{n(W_n)}\right)$$

and therefore

$$V = \frac{kT}{q}\ln\left(\frac{n(-W_p)\,n_n}{n(W_n)\,n_p}\right)$$

Now for low-level injection, the majority electron concentration at the edge of the depletion region on the n-side $n(W_n)$ is taken to be the same as n_n. Hence,

$$V = \frac{kT}{q}\ln\left(\frac{n(-W_p)}{n_p}\right)$$

Solving for $n(-W_p)$ we obtain

$$n(-W_p) = n_p \exp\left(\frac{qV}{kT}\right) \tag{3.19}$$

The minority electron concentration at the edge of the depletion region on the p-side is now exponentially dependent on the applied voltage V, and if $V = 0$ then $n(-W_p) = n_p$ as expected.

We have now calculated the minority carrier concentrations on either side of the depletion region using the concept of a dynamic equilibrium in the depletion region; however, we have not yet calculated the diode current.

We know that the diode current must be controlled by a process other than the dynamic equilibrium. The rate limiting process in question is the *recombination of the minority carriers* on either side of the depletion region.

Majority carriers are injected across the depletion region and appear on the other side of the junction where they become minority carriers. These minority carriers then diffuse and eventually recombine with majority carriers in order for current to flow through the diode. Recombination is usually a relatively slow process in semiconductors and even more so in indirect gap semiconductors such as silicon, and it is therefore not surprising that this is the rate-limiting step.

In order to analyze the net current, two new coordinates are introduced. The distance to any point in the p-type semiconductor from the depletion region edge will be x_p and the corresponding distance on the n-side will be x_n as shown in Figure 3.14.

It is useful to express Equations 3.17 and 3.19 in terms of *changes* in minority carrier concentrations, and we will use the new coordinate system. From Equation 3.17,

$$\Delta p(W_n) = \Delta p_{n(x_n=0)} = p_n \exp\left(\frac{qV}{kT}\right) - p_n = p_n\left(\exp\left(\frac{qV}{kT}\right) - 1\right) \tag{3.20a}$$

and similarly from Equation 3.19,

$$\Delta n(-W_p) = \Delta n_{p(x_p=0)} = n_p \exp\left(\frac{qV}{kT}\right) - n_p = n_p\left(\exp\left(\frac{qV}{kT}\right) - 1\right) \tag{3.20b}$$

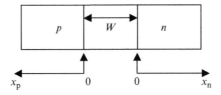

FIGURE 3.14   Coordinates x_p and x_n define distances into the p-type and n-type semiconductor regions starting from the depletion region edges.

From Equations 3.20a and 3.20b for forward bias with positive values of V the minority carrier concentrations increase exponentially with V. In reverse bias for large negative values of V the changes in carrier concentrations Δn_p and Δp_n can be negative and they approach n_p and p_n. This implies that the carrier concentrations of minority carriers decrease to virtually zero at the depletion region edges for reverse bias conditions.

The changes in carrier concentration in Equations 3.20a and 3.20b exist on either side of the depletion region. Far away from the depletion region carrier concentrations will return to their equilibrium levels. This occurs because carriers will be unaffected by the junction if they are several diffusion lengths away (see Figure 2.29), which implies that there are gradients in minority carrier concentration on either side of the depletion region. These concentration gradients will give rise to diffusion currents. The excess minority carrier concentrations exponentially decay with distances x_n and x_p into the neutral regions of the diode.

Another way to view this is using quasi-Fermi levels, which are valid even if there is an applied bias. On the p-side, applying Equation 2.62 and using the low-level injection assumption, we can write

$$F_n - F_p = kT \ln\left(\frac{n_{p(\text{biased})}}{n_p}\right)$$

or

$$\Delta n_{p(x_p=0)} = n_{p(\text{biased})} - n_p = n_p\left[\exp\left(\frac{F_n - F_p}{kT}\right) - 1\right]$$

Comparing this to Equation 3.20b we can see that the separation between the quasi-Fermi levels is equal to the applied voltage V multiplied by q. The same applies to the n-side, and using Equations 2.62 and 3.20a

$$\Delta p_{n(x_n=0)} = p_{n(\text{biased})} - p_n = p_n\left[\exp\left(\frac{F_n - F_p}{kT}\right) - 1\right]$$

The resulting quasi-Fermi levels are shown in Figure 3.15.

The diffusion equation for electrons (Equation 2.66b), which describes both diffusion and recombination written in terms of diffusion length $L_n = \sqrt{D_n \tau_n}$, is

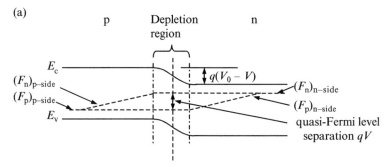

FIGURE 3.15 Quasi-Fermi levels for a forward biased junction. $(F_p)_{p-side}$ and $(F_n)_{n-side}$ are horizontal because for low-level injection the majority carrier concentrations are approximately fixed; however, minority carrier concentrations increase toward the depletion region due to carrier injection and therefore $(F_n)_{p-side}$ and $(F_p)_{n-side}$ are tilted. The separation between quasi-Fermi levels at the depletion region edge is equal to qV and we can write $(F_n)_{n-side} - (F_p)_{p-side} = qV$.

$$\frac{d^2\delta n(x)}{dx^2} = \frac{\delta n(x)}{L_n^2}$$

and may be solved using Equation 3.20 as a boundary condition at $x_p = 0$. For the p-side of the junction the excess electron concentration as a function of x_p becomes

$$\delta n_p\left(x_p\right) = \Delta n_{p(x_p=0)} \exp\left(-\frac{x_p}{L_n}\right) = n_p\left(\exp\left(\frac{qV}{kT}\right) - 1\right)\exp\left(-\frac{x_p}{L_n}\right) \qquad (3.21a)$$

and similarly on the n-side, the excess hole concentration as a function of x_n is

$$\delta p_n\left(x_n\right) = \Delta p_{n(x_n=0)} \exp\left(-\frac{x_n}{L_p}\right) = p_n\left(\exp\left(\frac{qV}{kT}\right) - 1\right)\exp\left(-\frac{x_n}{L_p}\right) \qquad (3.21b)$$

Using Equation 2.56 the electron diffusion current in the p-side will be

$$I_n\left(x_p\right) = qAD_n\frac{d\delta n_p\left(x_p\right)}{dx_p} = -\frac{qAD_n}{L_n}n_p\left(\exp\left(\frac{qV}{kT}\right) - 1\right)\exp\left(-\frac{x_p}{L_n}\right) \qquad (3.22a)$$

and in the n-side,

$$I_p\left(x_n\right) = qAD_p\frac{d\delta p\left(x_n\right)}{dx_n} = -\frac{qAD_p}{L_p}p_n\left(\exp\left(\frac{qV}{kT}\right) - 1\right)\exp\left(-\frac{x_n}{L_p}\right) \qquad (3.22b)$$

The electron and hole minority currents exponentially decay as x_p and x_n increase. This is because the number of minority carriers that have not recombined decreases as we move away from the junction. If we want to know the total injected current *before* any recombination has taken place we must evaluate $I_n(x_p)$ at $x_p = 0$ and $I_p(x_n)$ at $x_n = 0$. This gives

$$I_{\mathrm{n}} = -\frac{qAD_{\mathrm{n}}}{L_{\mathrm{n}}} n_{\mathrm{p}} \left[\exp\left(\frac{qV}{kT}\right) - 1 \right]$$

$$I_{\mathrm{p}} = \frac{qAD_{\mathrm{p}}}{L_{\mathrm{p}}} p_{\mathrm{n}} \left[\exp\left(\frac{qV}{kT}\right) - 1 \right]$$

These results assume that carriers do not recombine while they cross over the depletion region. Carrier recombination or generation in the depletion region will be further discussed in Section 3.8.

Every time a minority carrier recombines, a majority carrier is required for the recombination event. As majority carriers are consumed by the minority carriers that are injected, additional majority carriers must be supplied by the external circuit. The flow of these majority carriers becomes the measured diode current. Since coordinate x_{p} points to the left in Figure 3.14 the total current I will be given by $I_{\mathrm{p}} - I_{\mathrm{n}}$ and we finally obtain the *diode equation*

$$I = qA \left(\frac{D_{\mathrm{n}}}{L_{\mathrm{n}}} n_{\mathrm{p}} + \frac{D_{\mathrm{p}}}{L_{\mathrm{p}}} p_{\mathrm{n}} \right) \left[\exp\left(\frac{qV}{kT}\right) - 1 \right] \tag{3.23a}$$

For large negative values of V, only the reverse saturation current I_0 flows and therefore

$$I_0 = qA \left(\frac{D_{\mathrm{n}}}{L_{\mathrm{n}}} n_{\mathrm{p}} + \frac{D_{\mathrm{p}}}{L_{\mathrm{p}}} p_{\mathrm{n}} \right) \tag{3.23b}$$

Figure 3.16 shows the excess carrier concentrations in the p-n junction for both forward and reverse bias. Note the exponential decay of excess carriers with distance from the depletion region boundaries.

The symmetrical minority hole and minority electron concentration profiles shown in Figure 3.16 imply approximately equal n-type and p-type doping levels N_{d} and N_{a}. In practice, diodes often have $N_{\mathrm{d}} \gg N_{\mathrm{a}}$ or $N_{\mathrm{d}} \ll N_{\mathrm{a}}$, which means that very large ratios between I_{n} and I_{p} can occur.

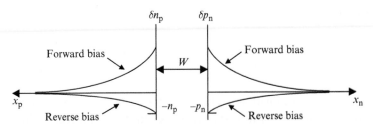

FIGURE 3.16   Excess carrier concentration on either side of the p-n junction depletion region. For forward bias the excess concentration is positive and for reverse bias it is small and negative.

Another way to view the flow of current across a diode is to think in terms of minority carrier recombination. The area under the plot of δn_p versus x_p for the junction in Figure 3.16 is the total minority carrier charge Q_n injected into the p-side. From Equation 3.21a,

$$Q_n = qA \int_0^\infty \delta n_p(x_p) dx_p = qAn_p \left[\exp\left(\frac{qV}{kT}\right) - 1 \right] \int_0^\infty \exp\left(\frac{x_p}{L_n}\right) dx_p$$

$$= qAL_n n_p \left[\exp\left(\frac{qV}{kT}\right) - 1 \right]$$

and similarly,

$$Q_p = qAL_p p_n \left[\exp\left(\frac{qV}{kT}\right) - 1 \right]$$

Now, the total diode current may be determined. The diode current flow due to minority electrons will be $\frac{Q_n}{\tau_n}$ because the charge Q_n will recombine an average of once during recombination time τ_n and therefore charge Q_n must be resupplied by the external circuit. The corresponding current due to minority holes will be $\frac{Q_p}{\tau_p}$. Total diode current is therefore

$$I = \frac{Q_n}{\tau_n} + \frac{Q_p}{\tau_p} = \left(\frac{qAL_n n_p}{\tau_n} + \frac{qAL_p p_n}{\tau_p} \right) \left[\exp\left(\frac{qV}{kT}\right) - 1 \right]$$

which is identical to Equation 3.23 since from Equation 2.68 $\frac{L_n}{\tau_n} = \frac{D_n}{L_n}$ and $\frac{L_p}{\tau_p} = \frac{D_p}{L_p}$. See Example 3.4.

The total diode current makes a transition from being composed of minority and majority carriers at the depletion region boundaries to being composed of majority carriers several diffusion lengths away from the depletion region on either side of the diode. From Equations 3.22b and 3.23 we have

$$I_n(x_n) = I - I_p(x_n) = qA \left[\frac{D_n}{L_n} n_p + \frac{D_p}{L_p} p_n \left(1 - \exp\left(-\frac{x_n}{L_p}\right) \right) \right] \left[\exp\left(\frac{qV}{kT}\right) - 1 \right]$$

At $x_n = 0$, $I_n(x_n)$ does not fall to zero. This is because an additional component of $I_n(x_n)$ must reach the depletion region to be injected to supply the p-side with its minority carriers. A similar expression results for $I_p(x_p)$. The resulting minority and majority currents are plotted as a function of x_p and x_n in Figure 3.17 a). Here, the majority carrier currents flowing toward the junction contribute equally to both injection and recombination. In almost all diodes, however, injection and recombination currents will not be balanced and Figure 3.17 b) shows a more general case.

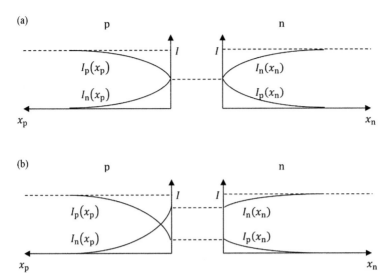

(a)

$I_p(x_p)$

$I_n(x_p)$

$I_n(x_n)$

$I_p(x_n)$

(b)

$I_p(x_p)$

$I_n(x_p)$

$I_n(x_n)$

$I_p(x_n)$

FIGURE 3.17 a) Minority currents $I_n(x_p)$ and $I_p(x_n)$ as well as majority currents $I_p(x_p)$ and $I_n(x_n)$ in a forward biased p-n junction. The sum of the majority and minority currents is always the total current I. Each majority current splits into two parts, one part supplying carriers to recombine with minority carriers and the other part being injected across the junction to supply the other side with minority carriers. Here we assume that each of majority hole and majority electron currents flowing toward the junction contribute equally to injection and recombination. b) In the more general case, injection and recombination current contributions are not equal and an example of current flow for an n^+p junction is shown.

EXAMPLE 3.4

a) The silicon diode of Example 3.1 is forward biased and a current of 1 mA flows. Find the excess minority carrier charge on either side of the junction at room temperature. Assume a carrier recombination time for electrons and holes in silicon of 2 μs and a junction area of $1 \, mm^2$.

b) Repeat for the p^+-n diode of Example 3.2b.

Solution

a) From Appendix 6, for silicon $\mu_n = 1350 \, cm^2 \, V^{-1} \, s^{-1}$ and $\mu_p = 480 \, cm^2 \, V^{-1} \, s^{-1}$. Therefore, using the Einstein relation,

$$D_n = \frac{kT}{q}\mu_n = 0.026 \, V \times 1350 \, cm^2 \, V^{-1} s^{-1} = 3.51 \times 10^1 \, cm^2 \, s^{-1}$$

and

$$D_p = \frac{kT}{q}\mu_p = 0.026 \, V \times 480 \, cm^2 \, V^{-1} s^{-1} = 1.25 \times 10^1 \, cm^2 \, s^{-1}$$

$$L_n = \sqrt{D_n \tau_n} = \sqrt{3.51 \times 10^1 \, cm^2 \, s^{-1} \times 2 \times 10^{-6} \, s} = 8.38 \times 10^{-3} \, cm$$

and

$$L_p = \sqrt{D_p \tau_p} = \sqrt{1.25 \times 10^1 \, cm^2 \, s^{-1} \times 2 \times 10^{-6} \, s} = 5.00 \times 10^{-3} \, cm$$

also,

$$n_p = \frac{n_i^2}{p_p} = \frac{\left(1.5 \times 10^{10}\ cm^{-3}\right)^2}{1 \times 10^{17}\ cm^{-3}} = 2.25 \times 10^3\ cm^{-3}$$

and

$$p_n = \frac{n_i^2}{n_n} = \frac{\left(1.5 \times 10^{10}\ cm^{-3}\right)^2}{1 \times 10^{17}\ cm^{-3}} = 2.25 \times 10^3\ cm^{-3}$$

From Equation 3.23 solving for V we obtain

$$V = \frac{kT}{q} \ln\left[\frac{I}{qA}\left(\frac{D_n}{L_n} n_p + \frac{D_p}{L_p} p_n\right)^{-1} + 1\right]$$

$$\frac{D_n}{L_n} = \frac{3.51 \times 10^1\ cm^2\ s^{-1}}{8.38 \times 10^{-3}\ cm} = 4.19 \times 10^3\ cm\,s^{-1}$$

$$\frac{D_p}{L_p} = \frac{1.25 \times 10^1\ cm^2\ s^{-1}}{5.00 \times 10^{-3}\ cm} = 2.50 \times 10^3\ cm\,s^{-1}$$

Now, $V = 0.026\ V \ln\left| \dfrac{10^{-3}\ A}{1.6 \times 10^{-19}\ C \times 10^{-2}\ cm^2} (4.19 \times 10^3\ cm\,s^{-1} \times 2.25\right.$

$$\left. \times 10^3\ cm^{-3} + 2.50 \times 10^3\ cm\,s^{-1} \times 2.25 \times 10^3\ cm^{-3})^{-1} + 1\right|$$

$$= 0.636\ V$$

On the p-side

$$Q_n = qAL_n n_p \left(\exp\left(\frac{qV}{kT}\right) - 1\right)$$

$$= 1.6 \times 10^{-19}\ C \times 10^{-2}\ cm^2 \times 8.38 \times 10^{-3}\ cm \times 2.25$$

$$\times 10^3\ cm^{-3} \left(\exp\left(\frac{0.636\ V}{0.026\ V}\right) - 1\right) = 1.27 \times 10^{-9}\ C$$

and on the n-side

$$Q_p = qAL_p p_n \left(\exp\left(\frac{qV}{kT}\right) - 1\right)$$

$$= 1.6 \times 10^{-19}\ C \times 10^{-2}\ cm^2 \times 5.00 \times 10^{-3}\ cm \times 2.25$$

$$\times 10^3\ cm^{-3} \left(\exp\left(\frac{0.636\ V}{0.026\ V}\right) - 1\right) = 7.56 \times 10^{-10}\ C$$

(continued)

(*continued*)

(b) For the p^+-n junction

$$n_p = \frac{n_i^2}{p_p} = \frac{\left(1.5 \times 10^{10}\,\text{cm}^{-3}\right)^2}{1 \times 10^{18}\,\text{cm}^{-3}} = 2.25 \times 10^2\,\text{cm}^{-3}$$

and

$$p_n = \frac{n_i^2}{n_n} = \frac{\left(1.5 \times 10^{10}\,\text{cm}^{-3}\right)^2}{1 \times 10^{16}\,\text{cm}^{-3}} = 2.25 \times 10^4\,\text{cm}^{-3}$$

Now using Equation 3.23 to solve for V we obtain $V = 0.601$ V. On the p-side

$$Q_n = qAL_n n_p \left[\exp\left(\frac{qV}{kT}\right) - 1\right]$$

$$= 1.6 \times 10^{-19}\,\text{C} \times 10^{-2}\,\text{cm}^2 \times 8.38 \times 10^{-3}\,\text{cm} \times 2.25$$

$$\times 10^2\,\text{cm}^{-3} \left[\exp\left(\frac{0.601\,\text{V}}{0.026\,\text{V}}\right) - 1\right] = 3.31 \times 10^{-11}\,\text{C}$$

and on the n-side

$$Q_p = qAL_p p_n \left[\exp\left(\frac{qV}{kT}\right) - 1\right]$$

$$= 1.6 \times 10^{-19}\,\text{C} \times 10^{-2}\,\text{cm}^2 \times 5.00 \times 10^{-3}\,\text{cm} \times 2.25$$

$$\times 10^4\,\text{cm}^{-3} \left[\exp\left(\frac{0.601\,\text{V}}{0.026\,\text{V}}\right) - 1\right] = 1.97 \times 10^{-9}\,\text{C}$$

Note that holes injected from the heavily doped p-side into the lightly doped n-side dominate the current flow.

Forward current through a p-n junction diode is a diffusion current. This appears to be an incomplete description of the mechanism for forward current flow because we assume the virtual cancellation of drift and diffusion currents in the depletion region. However we can state that forward current is indeed composed of minority carrier diffusion currents within the n-type and p-type regions of the diode while drift and diffusion currents virtually cancel each other inside the depletion region. The description is therefore valid, but not within the depletion region.

Reverse saturation current is a drift current; however, it is actually calculated from diffusion currents in the neutral parts of the p-n junction. Once again we should not forget the drift and diffusion currents in the depletion region and their virtual cancellation.

If a diode is very strongly forward biased the assumption of low-level injection will no longer be valid. This implies that majority carrier concentrations can no longer be taken as constant. Additional current flows now arise due to gradients in majority

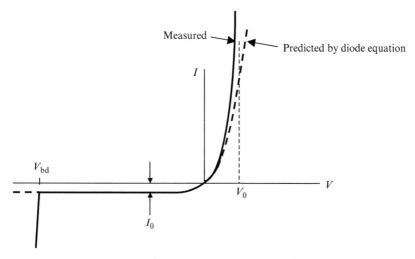

FIGURE 3.18   Measured current–voltage characteristics of a diode as well as predicted characteristics based on the diode equation. A very steep increase in current as applied voltage approaches the built-in potential V_0 is observed in practice as well as an abrupt onset of reverse breakdown current at V_{bd}.

carrier concentrations, and diode currents actually increase more rapidly with applied voltage than predicted by the diode equation. This is illustrated in Figure 3.18 along with the predicted curve from the diode equation. Note that for forward bias the observed current–voltage characteristic rises very steeply as applied voltage V approaches V_0. This is consistent with the band model of Figure 3.5, which shows that the potential barrier height for majority carrier injection will approach zero as V approaches V_0. The reverse bias voltage V_{bd} will be discussed in Section 3.6.

More advanced modelling includes changes in majority carrier concentration to more accurately predict high current characteristics. Additional effects including bulk resistances of the neutral semiconductor regions and contact resistances must also be considered for a more accurate model of measured diode characteristics.

3.6   REVERSE BREAKDOWN AND THE ZENER DIODE

When a large enough reverse bias is applied to a diode there will be an additional current flow, as shown for negative applied voltage in Figure 3.18. This current generally starts abruptly at a well-defined *reverse breakdown voltage* V_{bd} and must be carefully limited due to the large power $P = IV_{bd}$ that must be dissipated by the diode.

There are two possible mechanisms for this current flow. The first is *avalanche breakdown*. This may be understood from Figure 3.6, which shows the increased steepness of the band bending upon application of a reverse bias. This leads to a higher peak value of junction electric field ε_{bias} in the depletion region. If ε_{bias} exceeds the breakdown electric field strength for a given semiconductor then bound electrons that normally do not contribute to current flow may become available for conduction.

This occurs through *field ionization* of semiconductor atoms. Once a small number of normally bound electrons is released these electrons can increase in number by *impact ionization* of other atoms. Since the ionization energy for atoms in a given semiconductor is very specific to that semiconductor material, this process occurs at a specific reverse voltage and then increases rapidly with a further increase in voltage.

If a negative bias voltage V is applied the potential barrier increases to $V_0 - V$. The depletion region width W_{bias} as well as $\varepsilon_{\text{bias}}$ also increase, as shown in Figure 3.19. We can now modify the expressions we have for equilibrium conditions to calculate W_{bias} as well as $\varepsilon_{\text{bias}}$.

The integral of electric field across the depletion region (Equation 3.13) becomes the area under the new ε versus x graph in Figure 3.19 and we obtain

$$V_0 - V = \frac{1}{2}bh = \frac{1}{2}W_{\text{bias}}\varepsilon_{\text{bias}}$$

From Equation 3.12 we can see that $\varepsilon_{\text{bias}}$ increases linearly with an increase in depletion region width on either side of the junction, which means that both W_{bias} and $\varepsilon_{\text{peak}}$ must be proportional to the square root of $(V_0 - V)$. Hence Equation 3.15a may be modified to become

$$W_{\text{bias}} = \sqrt{\frac{2\epsilon_0\epsilon_r\left(V_0 - V\right)}{q}\left(\frac{1}{N_a} + \frac{1}{N_d}\right)} \tag{3.24}$$

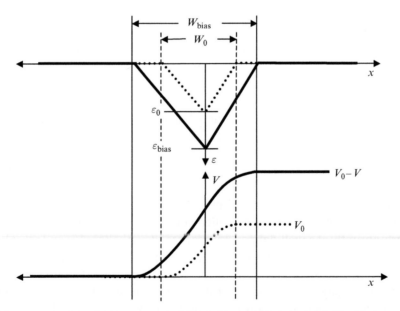

FIGURE 3.19   Increase in depletion region width and increase in junction field with the application of a reverse bias for the p-n junction of Figure 3.13. The equilibrium conditions with depletion width W and peak electric field ε_0 are shown with dotted lines. With the application of reverse bias V (V negative) the depletion width increases to W_{bias} and the peak electric field increases to $\varepsilon_{\text{bias}}$.

and from Equation 3.12 and Equation 3.15b or 3.15c we can write

$$\varepsilon_{\text{bias}} = -\frac{qW_{\text{bias}}}{\epsilon_0 \epsilon_r} \frac{N_a N_d}{N_a + N_d} \tag{3.25}$$

Thus in reverse bias, the depletion region width as well as peak electric field increase with increasingly negative bias voltage until $\varepsilon_{\text{bias}}$ reaches the breakdown field of the semiconductor. At this point, no further increase in $\varepsilon_{\text{bias}}$ can occur; however, electrons are generated at or near the junction, which are swept along by the electric field resulting in the reverse breakdown current.

From Equation 3.25 it is clear that the depletion width increases as the doping levels decrease. Since diodes often have much higher doping levels on one side than the other side we will consider a diode having $N_a \gg N_d$. From Equation 3.24 we can simplify the expression for depletion region width to obtain

$$W_{\text{bias}} = \sqrt{\frac{2\epsilon_0 \epsilon_r (V_0 - V)}{qN_d}} \tag{3.26}$$

We can now use Equations 3.25 and 3.26 and assume $N_a \gg N_d$ to obtain the peak electric field as

$$\varepsilon_{\text{bias}} = -\sqrt{\frac{2qN_d (V_0 - V)}{\epsilon_0 \epsilon_r}} \tag{3.27}$$

Hence the magnitude of the breakdown voltage may be increased by *decreasing* the doping level on at least one side of a diode. This is easy to visualize, since a decreased doping level leads to a wider depletion region with lower electric fields present for a given bias voltage V.

Another mechanism may also cause reverse breakdown current. This occurs in p-n junctions that have simultaneously high values of both n-type and p-type doping. From Equation 3.15a, if both N_a and N_d are large then depletion width W will be small. This is illustrated in Figure 3.20. W may be small enough to allow *tunneling* of electrons directly from the valence band on the p-side into the conduction band on the n-side. This differs from avalanche breakdown since neither field-ionization nor impact-ionization occur.

Tunneling is explained by an electron wavefunction that extends into a potential barrier. In Example 1.9 an electron in a potential well is shown to have wavefunctions that extend into a barrier of height 5 eV. If the barrier is thin enough the electron will have a finite probability of passing through the barrier into an adjacent potential well or an adjacent allowed electron state. Tunneling is further discussed in Chapter 6.

This tunneling type of breakdown is properly referred to as *Zener breakdown* although the *Zener diode* has come to mean a diode used specifically for its reverse breakdown characteristics whether caused by avalanching or tunneling processes. In practice, diodes with small breakdown voltages of a few volts involve tunneling and

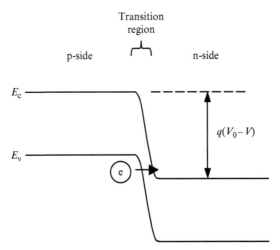

FIGURE 3.20   Tunneling of valence-band electron from valence band on p-side to conduction band on n-side upon application of a small reverse bias voltage. Note that there is a large supply of valence band electrons on the p-side. In comparison there is only a small supply of thermally generated minority carrier electrons that result in current I_0. This explains how the reverse current can be much larger than I_0 as shown in Figure 2.18, when V reaches the breakdown voltage V_{bd}.

diodes with higher breakdown voltages of 5 volts or more involve avalanching. There are often combinations of these two mechanisms occurring simultaneously in diodes with intermediate breakdown voltages.

3.7   TUNNEL DIODES

An important extension of the tunneling mechanism that operates in Zener breakdown occurs in the *tunnel diode*. If the doping levels are further increased to become very high on both the n-side and the p-side of a p-n junction it becomes possible to have tunneling currents flowing in both directions rather than in only one direction. This is illustrated in Figure 3.21. The doping, known as *degenerate* doping, is now high enough to push the Fermi energy into the conduction band on the n-side and into the valence band on the p-side yielding high electron and hole concentrations as well as the alignment of electron energy states. In the n-side, donor states form a sub-band because of their close proximity to each other. This donor sub-band merges with the conduction band. In the p-side an acceptor sub-band forms and merges with the valence band. In practice this occurs in silicon at doping concentrations of over approximately 1×10^{19} cm^{-3}.

As in the Zener diode, electrons in the valence band of the p-type material can tunnel directly into the conduction band of the n-type material. In addition, however, electrons in the conduction band of the n-type material can now tunnel directly into unfilled energy states on the p-side. At the Fermi energy, half the electron states are vacant in the valence band on the p-side. Since current now flows in either direction,

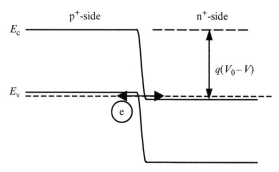

FIGURE 3.21   In a tunnel diode the depletion width is very narrow due to the use of degenerate p^+ and n^+ doping. In addition to the narrow depletion region, the Fermi level enters the bands on either side of the diode resulting in the alignment of electron energy states in the conduction band on the n-side with valence electron states on the p-side. Electron tunneling occurs in either direction as shown.

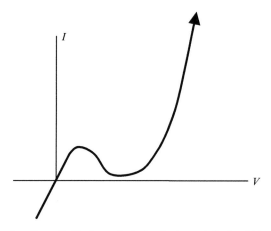

FIGURE 3.22   Current–voltage (I–V) characteristic of a tunnel diode. At low voltages, tunneling currents result in significant current flow in both directions. At higher positive bias voltages, electrons in the conduction band on the n-side will no longer be aligned with the valence band on the p-side. This will prevent tunneling and current flow will therefore decrease. Current flow will eventually rise upon further increase of forward bias since the potential barrier decreases as in a normal p-n junction.

the p-n junction no longer behaves like a normal diode. Even at the lowest applied voltages of either polarity, tunneling current flows. At higher forward bias voltages, however, normal diode behavior is regained because the alignment between electron states on either side of the junction is lost. The resulting I–V characteristic is shown in Figure 3.22.

The tunnel diode is of particular relevance to high performance tandem solar cell devices that require a series connection between two or more p-n junctions. If a pair of p-n junctions, each consisting of an n-type layer and a p-type layer, are to be stacked on top of each other in a series-connected arrangement, an unwanted n-p junction will exist between the two p-n junctions. If the series-connected pair of junctions is illuminated then a forward bias will be produced at each of the pair of junctions. As a result the unwanted n-p junction will be reverse biased, severely limiting the solar cell current that can flow. Provided, however, that the unwanted n-p junction is designed to be a tunnel diode by doping its n- and p-layers heavily, its influence on the overall forward current can be minimized. This will be further discussed in Section 5.6, and the concept can be extended to three or more series-connected p-n junctions.

3.8 GENERATION/RECOMBINATION CURRENTS

We now need to develop a more detailed picture of carrier behavior within the depletion region. Until now, we have neglected carrier recombination or generation in this region; however, for LEDs, solar cells, and other diodes these processes are important.

Let us consider carrier recombination under forward bias conditions in a silicon diode. In forward bias, excess carriers *must* actually be present in the depletion region as they are injected across it. In Figure 3.15 we showed the quasi-Fermi levels for a forward-biased p-n junction. The depletion region is shown again in Figure 3.23. Note that the hole and electron concentrations due to excess carriers are not zero in the depletion region when forward bias is applied.

In the depletion region of a silicon diode, trap-assisted carrier recombination should be considered. This is because carriers may spend enough time traversing this depletion region to make recombination events likely to occur. Section 2.16 developed the concept of low-level injection; however, this is not relevant to the depletion region of a diode where neither carrier concentration dominates. Hole and electron concentrations may be similar.

In Section 2.20 we argued that assuming the existence of traps at approximately mid-gap is justifiable due to defects in the semiconductor crystal. We will therefore assume a trap energy E_t at mid-gap that dominates the recombination process.

We can modify Equation 2.51 to obtain the hole recombination rate. In place of direct electron-hole pair recombination we now consider only the trapping of

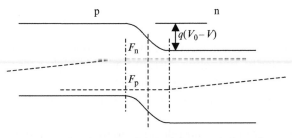

FIGURE 3.23   The quasi-Fermi levels within the depletion region are shown. Although the depletion region is created by the recombination of charges in equilibrium, once injection takes place in forward bias, excess carriers must flow through this region.

carriers. The trap concentration is assumed to be fixed leading to a defined carrier lifetime. Hence, for holes, in the absence of carrier generation,

$$\frac{d\delta p(t)}{dt} = -\frac{\delta p}{\tau_p}. \tag{3.28a}$$

and if the minority carriers were electrons,

$$\frac{d\delta n(t)}{dt} = -\frac{\delta n}{\tau_n}. \tag{3.28b}$$

From Equation 2.62a and Figure 3.15,

$$\delta n \delta p = n_0 p_0 \exp\left(\frac{F_n - F_p}{kT}\right) = n_i^2 \exp\left(\frac{qV}{kT}\right) \tag{3.29}$$

If the quasi-Fermi levels are approximately symmetric about mid-gap then $\delta n = \delta p$. Since trap energy $E_t \cong E_F$, we will consider that half the traps are filled and half are empty. This reduces the recombination rate by a factor of two. From Equations 3.29 and 3.28a and introducing this factor of ½ for the recombination rate, we obtain for holes

$$\frac{d\delta p(t)}{dt} = -\frac{n_i}{2\tau_p} \exp\left(\frac{qV}{2kT}\right)$$

and similarly for electrons,

$$\frac{d\delta n(t)}{dt} = -\frac{n_i}{2\tau_n} \exp\left(\frac{qV}{2kT}\right)$$

A recombination rate R for holes recombining through traps may now be defined as

$$R = -\frac{d\delta p(t)}{dt} = \frac{n_i}{2\tau_p} \exp\left(\frac{qV}{2kT}\right) \tag{3.30a}$$

and for electrons,

$$R = -\frac{d\delta n(t)}{dt} = \frac{n_i}{2\tau_n} \exp\left(\frac{qV}{2kT}\right) \tag{3.30b}$$

Since carriers that recombine must be replaced by current flowing through the external circuit, the portion of total diode current due to depletion region trapping is proportional to R. Since the recombination of minority carriers with majority carriers outside of the depletion region is simultaneously occurring, total diode current will be determined by Equation 3.23, in which diode current is proportional to

$\exp\left(\dfrac{qV}{kT}\right)$ for $qV \gg kT$ as well as by R. The overall diode current including trapping therefore varies with V as

$$I \propto \exp\left(\frac{qV}{nkT}\right)$$

where n is the *ideality factor*, which lies between 1 and 2. The value of n depends on the ratio between the two types of diode current.

Several factors affect ideality factor n including

i) Effective depletion region width: A wider depletion region favors higher values of n since there is more distance for carriers to travel in this region. They spend more time in the region and have a higher chance of trap-induced recombination.

ii) Magnitude of diode current: At high diode currents, n approaches 1 since trap concentration in limited and hence trap availability is limited. Trapping is more significant at low diode currents.

iii) Operating temperature: From the diode equation diode current is controlled by n_p and p_n. See Equation 3.23a. Since $n_p = n_i^2 / N_a$ and $p_n = n_i^2/N_d$ we see that diode current is proportional to n_i^2, whereas from Equations 3.30a and 3.30b diode current due to trapping is proportional to n_i. Hence, due to the exponential temperature dependence of n_i (see Equation 2.39), diode ideality factors approach 1 at higher operating temperatures.

iv) Bandgap: Smaller bandgap semiconductors have larger values of n_i, which means that diode ideality factors approach 1 as bandgap decreases. See iii) above.

The recombination analysis presented here leading to the ideality factor is a special case of a more general analysis of *Shockley-Read-Hall trap-mediated recombination*. See Suggestions for Further Reading.

Carrier generation within the depletion region can also be important in diodes that are reverse biased or weakly forward biased. If trapping sites capture electrons and holes then these sites can also *generate* free carriers by thermal activation. Normally trap generation is in thermal equilibrium with trap recombination. Under reverse bias or weak forward-bias conditions, however, carriers generated by traps may be accelerated away from the depletion region before they can recombine, which will increase the reverse saturation current. This phenomenon can be observed when reverse saturation current I_0 increases with reverse bias rather than remaining fixed. If carriers released from traps in the depletion region dominate I_0 and these traps are uniformly distributed over the depletion region, then I_0 would increase with approximately the square root of the magnitude of reverse-bias voltage: Equation 3.26 shows that depletion width increases as the square root of voltage for $V \gg V_0$. In practical silicon diodes at room temperature, thermal generation current within the depletion region often dominates the reverse saturation current.

If light is incident on the depletion region, carrier generation can occur. This is relevant to solar cells and photodiodes and will be discussed in more detail in Chapter 5.

For direct gap semiconductor p-n junctions, carrier recombination at the junction is discussed in Section 5.13 in the context of LEDs.

3.9   METAL-SEMICONDUCTOR JUNCTIONS

The flow of current between semiconductors and metals that connect semiconductor devices to external circuitry is dependent on a variety of influences. We can model the most important mechanisms at play. The band model will be used to explain current flow.

A metal has a partly filled conduction band, which means that the Fermi energy in the metal lies within this band. If no charges are exchanged between the metal and the semiconductor, we can draw an idealized metal-semiconductor contact in equilibrium by maintaining a constant Fermi level for the metal-semiconductor junction as in Figure 3.24. A small barrier E_b for electrons appears at the junction where $E_b = E_c - E_F$. In this diagram electrons can easily flow from the semiconductor into the metal; however, electrons from the metal need to overcome energy barrier E_b to flow into the semiconductor. Mobile charge can flow across the metal-semiconductor contact, which may either deplete or accumulate near the semiconductor surface and therefore the conditions of Figure 3.24 do not correctly represent equilibrium conditions.

Provided the doping is high enough, $E_c - E_F$ can be made small enough to permit efficient electron flow in both directions because E_F moves closer to the conduction band edge as doping increases, and an *ohmic contact* is formed. In some cases it is of interest to have diode behavior, but in many cases the desire is for ohmic behavior and ohmic contacts are achieved, which allow for current flow in either direction with minimal voltage drop across the junction.

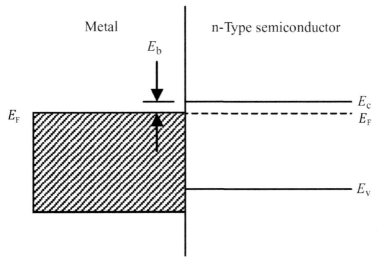

FIGURE 3.24   Metal-semiconductor contact for an n-type semiconductor without any flow of charge between the two sides. The predicted barrier height $E_b = E_c - E_F$ and the flat bands shown are not achieved in real devices due to charge flow and charges at the metal-semiconductor interface that cause band bending and associated electric fields at the interface and in the semiconductor.

Real metal-semiconductor contacts generally result in band-bending due to charges that flow across the metal-semiconductor interface. In addition charges from various sources may accumulate at or near the metal–semiconductor interface. A number of sources of charge at or near the junction exist:

1. Electron flow between the metal and semiconductor. Electron affinities of the semiconductor atoms and the metal atoms do not generally match. This is quantified at a free surface by a workfunction difference between the two materials. Since we are interested in an interface and workfunctions are properly measured at a surface in a vacuum, the contribution of this workfunction difference to the metal-semiconductor junction is understood by using *modified workfunctions* that take the actual interface bonding conditions into account. As a result, electrons will flow between the metal and semiconductor to minimize the potential energy of the system. The result will be a net charge in the semiconductor and a net charge in the metal. This charge will be significant within one diffusion length from the junction in the semiconductor.

2. Dangling bonds at the semiconductor surface. These form since the lattices of the metal and semiconductor do not generally match.

3. Dangling metal bonds at the metal surface. These form when the lattices of the metal and semiconductor do not match. In addition, metals are generally polycrystalline and there will be many crystallographic orientations to consider in one device.

4. Doping effects from metal atoms that diffuse into the semiconductor. If the metal atoms have a different valence number from the semiconductor atom then a charge will be associated with each metal ion in the semiconductor.

5. The *Schottky effect* lowers the effective barrier height due to an electric field at an interface or surface. The flow of electrons in the presence of a potential barrier is discussed later in this section.

6. Oxygen or other impurity incorporation at the metal/semiconductor surfaces near the interface. For example, it is very difficult to completely eliminate oxygen at a metal-semiconductor interface since most metals and common semiconductors such as silicon react strongly with oxygen. Oxide layers tend to trap positive charges due to electron loss from metal/semiconductor dangling bonds within the oxide layer.

The result is that it is virtually impossible to precisely quantify charges and hence barrier heights present near metal-semiconductor interfaces. Charges also control band bending, which has a strong effect on the electrical properties of the metal-semiconductor junction. In practice, to achieve specific electrical properties, there are known recipes for metal-semiconductor junctions that yield good results for a variety of applications.

We will consider two specific cases and then see how these cases affect the behavior of the metal-semiconductor contact using an n-type semiconductor. See Figure 3.25. In Figure 3.25a, a positive charge is trapped at the semiconductor surface. In principle this

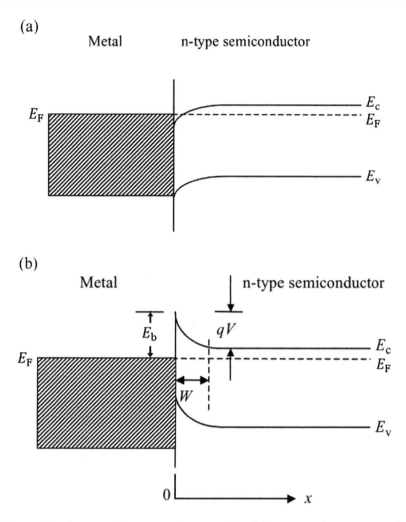

FIGURE 3.25 Metal-semiconductor junction energy band diagrams under various conditions. (a) If the interface is positively charged then the band-bending will be as shown. This forms an ohmic contact. (b) If the interface is negatively charged then band bending will result in a large energy barrier E_b, which blocks electron flow from the metal to the semiconductor, as well as a depletion region in the semiconductor. A Schottky diode is formed.

positive charge could be a result of several of the sources mentioned above. This causes the Fermi energy to move closer to the conduction band in the semiconductor near the interface because the trapped positive charge attracts electrons in the conduction band toward the interface. The Fermi energy may actually enter the conduction band, which means that there is enough positive charge present at the semiconductor surface to attract conduction band electrons to fill the low-lying conduction band states with more than a 50% probability. This results in an ohmic contact.

The formation of a *Schottky diode is* illustrated in Figure 3.25b, in which negative charge is trapped at the junction. The Schottky diode is also referred to as an *MS diode* or *Metal-Semiconductor diode*. This diode will now be discussed and modeled. It has the advantage of being a diode that can be fabricated at low cost since a metal thin film may be deposited on a semiconductor surface by a variety of well-developed vacuum deposition techniques such as evaporation or sputtering. The deposition of an aluminum thin film on the surface of an n-type silicon semiconductor moderately doped n-type to $n \cong 10^{16}$ cm^{-3} is an example of a Schottky diode structure. The diode characteristics will depend on various details of the processing conditions including the crystallographic orientation of the silicon wafer, the degree of trapped oxygen at the interface between the aluminum and the silicon, and the cleaning procedure used to prepare the silicon surface prior to deposition; however, a resulting band diagram similar to that shown in Figure 3.25b is routinely attainable in production.

In Figure 3.25b, the negative charge trapped at the semiconductor surface is a very common situation due to dangling bonds that form at a semiconductor surface. This was discussed in Section 2.20. Conduction band electrons are repelled away from the negative trapped charge leaving a depletion region in the semiconductor near the interface. The Fermi level gets pinned at approximately mid-gap at the semiconductor surface. A large barrier for electron flow from the metal to the semiconductor forms, which blocks current flow even when an external voltage is applied. The barrier for electrons leaving the semiconductor is labeled qV and is formed due to the band bending. If a forward bias voltage is applied across the metal-semiconductor junction such that the metal is positive then this barrier will be reduced or eliminated, allowing current flow. If a reverse bias voltage of magnitude V is applied to the metal relative to the semiconductor the barrier E_b will remain in place and current flow will be blocked. Diode behavior will therefore be obtained.

We can apply the depletion approximation to the Schottky diode to simplify the treatment of the depletion region. If we assume that the depletion region is fully depleted over a width W then the depletion region contains only static charges due to ionized donors. If the semiconductor is n-type with a uniform doping concentration N_d then the total charge in the depletion region in units of coulombs per unit junction area is qN_dW. This charge determines the electric field at the semiconductor surface since we can apply a Gaussian surface to the diode that encloses the entire depletion region. Now, using Gauss's law from Equation 3.11 and assuming a junction area A we have

$$\oint_S \vec{\varepsilon} \cdot \vec{ds} = \frac{Q}{\epsilon} = \frac{qN_dWA}{\epsilon} \tag{3.31}$$

The electric field in the neutral region of the semiconductor is zero. Provided the junction cross-section dimensions are much larger that W we can conclude from Equation 3.31 that

$$\varepsilon_s = \frac{qN_dW}{\epsilon} \tag{3.32}$$

where ε_s is the magnitude of the electric field at the semiconductor surface where $x = 0$. By using Gaussian surfaces that enclose smaller portions of the depletion region, the electric field as a function of depth x may be determined and the magnitude of the electric field as a function of depth $\varepsilon(x)$ over the range $0 \leq x \leq W$ is found to be

$$\varepsilon(x) = \frac{qN_dW}{\epsilon} \frac{(W-x)}{W} \tag{3.33}$$

See Problem 3.21. Note that if $x = 0$ then from Equations 3.32 and 3.33 $\varepsilon(x) = \varepsilon_s$. $\varepsilon(x)$ falls linearly as a function of x until it reaches zero at $x = W$. This is similar to a p^+-n junction. See Example 3.2.

The potential difference arising from this electric field over the depletion region may be obtained by integrating Equation 3.33, and we therefore have

$$V_b = -\int_0^W \varepsilon(x)\,dx = \frac{qN_d}{\epsilon}\int_0^W (W-x)\,dx = \frac{qN_d}{\epsilon}\frac{W^2}{2}$$

Solving for W we obtain

$$W = \sqrt{\frac{2\epsilon V_b}{qN_d}}$$

If a reverse bias voltage V is applied, this voltage will add to the built-in voltage and the depletion width will increase to become

$$W = \sqrt{\frac{2\epsilon\left(V_b + V\right)}{qN_d}}$$

The current flow through a Schottky diode differs from a normal semiconductor p-n junction because majority carriers dominate the process. In the Schottky diode of Figure 3.25b), n-type electron current flow occurs in both the metal and the semiconductor, and we need to consider the flow of electrons across a potential barrier, either from the metal to the semiconductor, or from the semiconductor to the metal. The net current flow will now be the difference between the electron flow in these two directions. We can neglect hole current in the semiconductor, which is insignificant.

Electrons in the metal can overcome a potential barrier by a process called *thermionic emission*. Electrons occupy a range of energy states in the metal and only a fraction of these electrons have enough kinetic energy to overcome the relevant energy barrier at the junction. The fraction of electrons able to overcome the barrier is determined by temperature T as well as the height of the energy barrier E_b.

Consider the Schottky diode in equilibrium in Figure 3.25b. If electron current I_e flows by thermionic emission from the metal into the semiconductor then in order to satisfy equilibrium conditions the *net* current flow must be zero. We can therefore conclude that an equal and opposite electron current flows from the semiconductor into the metal. This latter current flows just as as electron current flows in a p^+-n junction. The electrons in a Schottky diode flowing from the n-type semiconductor toward the metal experience the equivalent depletion region and energy band shape as in the n-type semiconductor of a p^+-n junction and will therefore follow a current–voltage relationship as in a p^+-n junction. Hence

$$I = I_e \left[\exp\left(\frac{qV}{kT} \right) - 1 \right] \tag{3.34}$$

Here we interpret I_e to be the current flowing by thermionic emission from the metal to the semiconductor over energy barrier E_b at temperature T. The form of Equation 3.34 satisfies the condition that for large negative values of V only a net thermionic emission current I_e flows and if $V = 0$ then $I = 0$. Note that the mechanism for the flow of I_e is very different from the origin of saturation current I_0 in a normal p-n junction, the latter being due to minority carrier drift across the depletion region.

The understanding of I_e is based on the behavior of a metal–vacuum interface. Consider a metal–vacuum interface at temperature T where the metal workfunction is Φ_m. Electrons with enough kinetic energy may overcome the workfunction, leaving the metal surface and entering the vacuum. We assume that the electrons in the vacuum cannot flow away unless they return to the metal by again entering the metal surface. This means that an equilibrium between electrons in the metal and electrons in the vacuum will be reached. Figure 3.26 shows an equilibrium system comprising a fixed volume of a vacuum. The vacuum is box-shaped and is completely surrounded by the metal. This geometry ensures that the equilibrium between metal and vacuum is not affected by other interfaces or influences.

The electrons in the metal occupy energy levels that depend upon the details of the band structure in the metal; however, the electrons in the vacuum can be treated more readily since they are free electrons inside a volume that contains the vacuum. We will define the origin of our relative energy scale to be as shown in Figure 3.26b.

The electrons in the vacuum may be modeled in the same way that we modeled electrons in an infinite walled potential box in Chapter 2. A reciprocal space lattice is

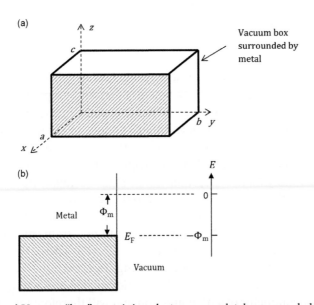

FIGURE 3.26   a) Vacuum "box" containing electrons completely surrounded by metal. Electrons in the vacuum are in thermal equilibrium with electrons in the metal. The vacuum box has dimensions (a,b,c). b) Energy diagram showing the choice of origin for the energy axis.

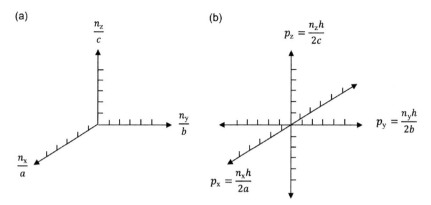

FIGURE 3.27 a) Reciprocal space lattice showing positive standing wave solutions to Schrodinger's equation for electrons in an infinite-walled box. b) Momentum space lattice showing both positive and negative momentum values. Each standing wave solution is composed of pairs of traveling waves proceeding in opposite directions along each axis.

shown in Figure 3.27a. The points in this reciprocal space lattice represent the allowable quantum states for electrons in the vacuum. The boundary conditions for the vacuum effectively confine electron traveling waves just as in Example 2.1 and in Section 2.10 for electrons in an infinite walled potential box. This is not obvious since the potential barriers in Chapter 2 are assumed to be positive whereas the boundaries of the vacuum "box" we are now considering are negative. In Section 1.9 we showed that strong reflections may exist for electrons incident on either positive *or* negative potential steps. The assumption of infinite walls is valid because electron energies in the vacuum are small and much less than Φ_m. See Problem 3.20.

Each point in the reciprocal space in Figure 3.27a represents a standing electron wave that satisfies the boundary conditions of the vacuum. Since we are modeling thermionic emission, we need to examine traveling electrons in the vacuum. Each standing wave corresponding to one reciprocal space lattice point

$$\left(\frac{n_x}{a}, \frac{n_y}{b}, \frac{n_z}{c} \right)$$

is composed of two traveling electron waves along each axis. Along the x-axis, for example, Equation 2.29 describes two electron waves traveling in opposite directions with waves given by $\exp(ikx)$ and $\exp(-ikx)$ where $k = n_x \pi / a$. Equivalent expressions may be written along the y-axis and the z-axis.

The allowed momenta of these traveling waves is shown in Figure 3.27b which includes both positive and negative momentum values given by

$$p_x = \pm \hbar k_x = \pm \frac{n_x h}{2a}$$

$$p_y = \pm \hbar k_y = \pm \frac{n_y h}{2b}$$

$$p_z = \pm \hbar k_z = \pm \frac{n_z h}{2c}$$

We can now determine the number of electrons per unit volume dn that exist within a range of momentum values dp_x, dp_y, and dp_z. Using Figure 3.27b we divide the volume encompassed by this differential range of momentum by the volume associated with each point in momentum space and obtain

$$dn = \frac{1}{abc} F(E) \frac{dp_x dp_y dp_z}{\left(\frac{h}{2a}\right)\left(\frac{h}{2b}\right)\left(\frac{h}{2c}\right)} \times 2 \times \frac{1}{8} = F(E) \frac{2}{h^3} dp_x dp_y dp_z$$

The multiplying factor of 2 accounts for electron spin. Division by 8 is necessary since pairs of traveling waves along each axis represent a single allowed standing wave solution. $F(E)$ is the Fermi Dirac distribution function which defines the probability of occupancy of each allowed electron state.

From Figure 3.26b the electrons in the vacuum are in thermal equilibrium with electrons that overcome the workfunction in the metal. A metal electron that has just enough energy to overcome the workfunction will arrive in the vacuum with virtually zero kinetic energy. We have therefore placed the origin of the energy axis to coincide with the top of the metal workfunction. This means that the relevant Fermi energy is at $E_F = -\Phi_m$. The form of the Fermi–Dirac function of Equation 2.25 should therefore be

$$F(E) = \frac{1}{1 + \exp\left(\dfrac{E + \Phi_m}{kT}\right)}$$

Now if the metal–vacuum interface under consideration lies in a plane normal to the x-axis as illustrated in Figure 3.26a then the vacuum electrons incident on the metal surface will have a momentum in the positive x-direction and the resulting current density magnitude across the interface due to electrons dn will be[1]

$$dJ_x = q v_x dn = \frac{2 q p_x}{m h^3} F(E) dp_x dp_y dp_z$$

since $p_x = m v_x$.

The total current density J_x that can flow across the vacuum–metal interface is now obtained by integrating over all ranges of electron energy. It is possible to simplify the integral since the relevant energy ranges in the Fermi–Dirac function are much larger than kT and therefore the denominator of $F(E)$ is large. We are now justified in replacing the Fermi–Dirac distribution with the Boltzmann function and we have

$$F(E) \cong \frac{1}{\exp\left(\dfrac{E + \Phi_m}{kT}\right)} = \exp\left(-\frac{E + \Phi_m}{kT}\right)$$

where

$$E = \frac{p_x^2 + p_y^2 + p_z^2}{2m}$$

and therefore

$$F(p_x, p_y, p_z) \cong \exp\left(-\frac{p_x^2 + p_y^2 + p_z^2}{2mkT} - \frac{\Phi_m}{kT}\right)$$

Now,

$$J_x = \frac{2q}{mh^3} \exp\left(-\frac{\Phi_m}{kT}\right) \int \int \int p_x \exp\left(-\frac{p_x^2 + p_y^2 + p_z^2}{2mkT}\right) dp_x dp_y dp_z$$

$$= \frac{2q}{mh^3} \exp\left(-\frac{\Phi_m}{kT}\right) \int_0^\infty p_x \exp\left(-\frac{p_x^2}{2mkT}\right) dp_x \int_{-\infty}^\infty \exp\left(-\frac{p_y^2}{2mkT}\right) dp_y$$

$$\times \int_{-\infty}^\infty \exp\left(-\frac{p_z^2}{2mkT}\right) dp_z$$

The limits on the x-axis integral are between 0 and ∞ since only electrons in vacuum traveling in one direction along the x-axis may encounter the metal surface. The limits on the y-axis and z-axis integrals are between $-\infty$ and ∞ since we want to include all momentum components in the y and z directions.

Using the following standard integrals

$$\int_{-\infty}^\infty \exp\left(-ax^2\right) dx = \sqrt{\frac{\pi}{a}}$$

and

$$\int_0^\infty x \exp - (-ax^2) dx = \frac{1}{2a}$$

we obtain

$$J_x = \frac{2q}{mh^3} \pi(2mkT) \exp\left(-\frac{\Phi_m}{kT}\right) \int_0^\infty p_x \exp\left(-\frac{p_x^2}{2mkT}\right) dp_x$$

$$= \frac{qm}{2\pi^2 \hbar^3} (kT)^2 \exp\left(-\frac{\Phi_m}{kT}\right)$$

which gives us a current density from the vacuum to the metal. In thermal equilibrium this is balanced by a current density J_e from the metal to the vacuum of magnitude

$$J_e = \frac{qm}{2\pi^2 \hbar^3} (kT)^2 \exp\left(-\frac{\Phi_m}{kT}\right)$$

and corresponding current I_e of

$$I_e = \frac{qmA}{2\pi^2 \hbar^3} (kT)^2 \exp\left(-\frac{\Phi_m}{kT}\right) \tag{3.35}$$

Equation 3.35 is known as the *Richardson–Dushman equation for thermionic emission*. It is important to remember that this current is an equilibrium current at a temperature T.

If we apply an electric field in the vacuum to draw off the electrons and prevent them from entering the metal we are not truly measuring the equilibrium thermionic emission current; however, the measurement will approach the equilibrium current if the applied electric field is small.

If the vacuum is now replaced by silicon then the work function Φ_m is replaced by E_b and the electron mass m is replaced by the effective electron mass in the semiconductor m_e^*. Now Equation 3.35 becomes

$$I_e = \frac{(kT)^2 m_e^* qA}{2\pi^2 \hbar^3} \exp\left(-\frac{E_b}{kT}\right) \tag{3.36}$$

Finally the Schottky diode current–voltage relationship is obtained from Equations 3.34 and 3.36 as

$$I = \frac{(kT)^2 m_e^* qA}{2\pi^2 \hbar^3} \exp\left(-\frac{E_b}{kT}\right)\left[\exp\left(\frac{qV}{kT}\right) - 1\right]$$

The Schottky diode current increases exponentially as the barrier height E_b decreases. The current increases as a function of increasing temperature. Current dependence on applied voltage approaches an exponential function as forward bias increases.

The Schottky diode is not normally used directly as a solar cell or an LED; however, the concept of thermionic emission is important and is applicable to abrupt interfaces that form abrupt changes in potential energy in devices (see Section 2.10). In addition, organic devices discussed in Chapter 8 often require the application of thermionic emission (see Section 8.3).

We will now return to the formation of ohmic contacts on semiconductors, in which case the current–voltage characteristic of a Schottky diode must be avoided. There is a way to form an ohmic contact even if a negative charge forms at the interface of the n-type semiconductor and metal which would normally form a Schottky diode. Provided the semiconductor is very highly doped near the semiconductor surface, the width of the resulting depletion region can be made small enough to permit electron tunneling directly from the metal to the semiconductor conduction band. This is illustrated in Figure 3.28. The elimination of an energy barrier is therefore not a prerequisite for an effective ohmic contact, and a small barrier width is almost always the key to ohmic behavior.

An analogous set of junctions can form in the case of a metal-semiconductor junction with a p-type semiconductor. We show one example of a p-type semiconductor-metal junction in Figure 3.29, which is equivalent to Figure 3.25a for the n-type case. Here the energy barrier is small and ohmic behavior results. Since current in the p-type semiconductor is predominantly hole current, electrons flowing from the metal into the semiconductor recombine with holes, allowing more holes to flow toward the interface. In practice, however, band bending in the opposite direction

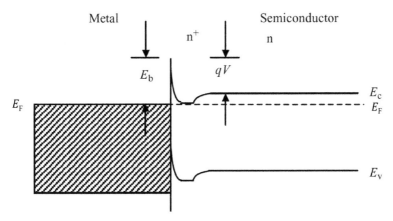

FIGURE 3.28 If the interface is negatively charged and the semiconductor is strongly doped to form an n^+ region near the interface then band bending can result in a very narrow energy barrier of height E_b, which permits electron flow by tunneling through the barrier and an ohmic contact is formed. An additional built-in barrier qV is formed due to band bending in the semiconductor.

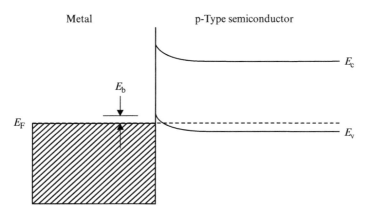

FIGURE 3.29 Example of an ohmic contact between p-type silicon and a metal. Electrons from the metal recombine with a high concentration of holes that accumulate near the surface of the p-type semiconductor. Each hole that recombines allows another hole to take its place, resulting in continuous current flow. In reality it is more likely that an energy barrier forms since the Fermi energy is pinned at mid-gap at the interface.

causing the formation of an energy barrier is much more likely. Tunneling of carriers through a very narrow barrier resulting from highly doped p-type semiconductor at the junction will be the reason for ohmic behavior.

In practice an effective ohmic contact on n-type silicon may be formed by diffusing extra n-type dopant such as phosphorus or antimony into the silicon surface forming an n^+ region before a metal such as aluminum is applied, resulting in a tunneling ohmic contact as in Figure 3.28. For p-type silicon, aluminum may also be used. After depositing aluminum a brief heat treatment can be used to diffuse the aluminum into the silicon forming both a p^+ region and the metal contact. In this way aluminum can serve as metallization for both n-type and p-type ohmic contacts on silicon.

In practice the formation of effective ohmic contacts to any given semiconductor is experimentally determined and optimized and cannot be fully modeled theoretically.

A range of metals, dopants and heat treatments constitute known recipes that form effective ohmic contacts on a wide range of semiconductors.

3.10   HETEROJUNCTIONS

All the p-n junctions we have described until now make use of a single semiconductor material that is doped to form the n and p regions. The dopant concentration is typically in the parts per million range, which means that the semiconductor is really dominated by a single element in the case of elemental semiconductors such as Si, or a single compound in the case of compound semiconductors such as GaAs.

If a p-n junction is formed from two different semiconductors, one on either side of the junction, then a *heterojunction* is formed. An example of this might be p-type GaAs on one side of the junction and n-type $Ga_{0.8}Al_{0.2}As$ on the other side. Since there is only a very small change in lattice constant for this system, as shown in Figure 2.34, a high-quality single-crystal p-n junction may be achieved. A band diagram for this type of junction is shown in Figure 3.30. Note that there is band bending near the junction due to the formation of a depletion region. In addition conduction and valence band offsets ΔE_c and ΔE_v arise. These band offsets imply highly localized electric fields and corresponding charge densities in a very thin sheet at the junction. As in the case for metal-semiconductor junctions, interface charge can arise from several sources. For a single-crystal semiconductor the dominant sources are differing electron affinities between the two semiconductors and anomalies in the bonding at the interface that give rise to localized polarization charges. In addition, since lattice constants are usually not perfectly matched, the inevitable stress and strain induced will further contribute to this local charge. Should dislocations develop, additional charges could exist.

There is no reason for ΔE_c and ΔE_v to be equal. Their relative values are determined by the interface charge; however, in all cases the relationship $\Delta E_c + \Delta E_v = \Delta E_g$ must hold. One obvious consequence of the heterojunction is the difference in the

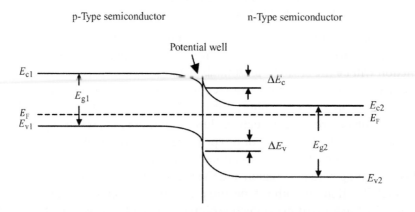

FIGURE 3.30   Example of heterojunction formed between p-type GaAs and n-type $Ga_{1-x}Al_xAs$.

effective potential barrier for electrons and for holes. In Figure 3.30, for example, electrons must overcome an overall potential barrier height of $q(E_{c1} - E_{c2})$, which is smaller than the potential barrier for holes of $q(E_{v1} - E_{v2})$. This favors the injection of electrons rather than holes across the junction. Since the injection of carriers is exponentially dependent on the potential barrier height a large ratio between the hole and electron injection rates may be obtained. The quantitative calculation of the relevant rates requires more detailed knowledge of the shape of the barrier, the effective masses of the carriers in both semiconductor materials and the possibility of tunneling effects when a "spike" exists such as that shown in the conduction band of Figure 3.30. The Richardson–Dushman equation (Equation 3.35) is also often relevant to calculate currents flowing across these abrupt energy barriers.

Of general interest in heterojunctions is the formation of a potential well near the junction. An example of such a potential well is shown in Figure 3.30. Electrons may be trapped in this well, resulting in a thin "sheet" of electrons, which has applications in high-frequency transistors.

In solar cells and LEDs, heterojunctions have important optical and electrical properties that enable high-efficiency device designs. Unwanted photon absorption can be greatly reduced by arranging the wide-gap semiconductor region to be in the path of photons entering or leaving a diode device. Regarding Figure 3.30 as an example of an LED structure, photons emitted by the recombination of electrons and holes across E_{g1} will not be significantly reabsorbed by the semiconductor of energy gap E_{g2}. This effectively reduces light loss from the LED when the light passes through the n-type semiconductor before leaving the device.

In addition, absorption or recombination may be strongly localized by engineering a region of smaller bandgap where absorption or recombination is desired. In a solar cell application, if light of a certain wavelength range reaches a junction made in narrow-gap semiconductor material by passing through a wide-gap semiconductor region then unwanted absorption of light away from the junction can be avoided. In an LED recombination can be confined to a specific layer, which enhances recombination efficiency. These concepts will be further described and developed in the context of solar cells and LEDs in Chapter 5.

3.11   ALTERNATING CURRENT (AC) AND TRANSIENT BEHAVIOR

For solar cells and LED lamp applications, where steady or only slowly varying operating conditions are typical, alternating current (AC) and transient behavior are not important. There are, however, many situations where diodes are used for switching applications and voltages and currents change rapidly with time.

There are two important operating conditions we need to discuss in this section which include *minority carrier recombination time delay* and *junction capacitance*.

In forward bias, a p-n junction diode builds up significant amounts of minority carrier charge on one or both sides of the depletion region. The magnitude of the minority carrier charge depends on the forward diode current; changes in forward current lead to changes in minority carrier concentrations. Since minority carriers do not recombine instantly, there is a time delay associated with the diode reaching steady state as forward bias conditions are changed. If the diode is switched between forward bias conditions and reverse bias conditions there is a time delay involved in eliminating the minority carrier charge. If forward bias is involved we must refer back to the excess minority carrier concentrations shown in Figure 3.16. The maximum speed with which a diode can transition between two forward bias conditions is controlled by how quickly the stored minority charge can be changed. Consider a *reduction* in forward current as a function of time. A reduction in the minority carrier concentrations as a function of time will be required (see Section 3.5).

There are two mechanisms by which the stored charge can be removed. One is by *recombination*, and the rate at which this can occur is controlled by the recombination times τ_n and τ_p. The second mechanism is by charge *extraction*, in which charge flows away from either side of the depletion region by diffusion into the depletion region. This second mechanism is particularly relevant if a diode is rapidly switched from forward bias to reverse bias. The electric field across the depletion region will grow and will assist these minority carriers that diffuse into the depletion region to then drift across to the other side.

The simple exponential shape of the excess minority carrier spatial distribution shown in Figure 3.16 will be made more complex during changes in bias because minority carriers close to the depletion region will have a higher chance of extraction than carriers further from the depletion region edges. In contrast with this the per-carrier rate of recombination is independent of position and is determined by the recombination time.

A very useful quantity that can be used to characterize this process is the *storage delay time* t_{sd}, which is the time needed for the minority carrier concentrations to reach zero after a forward-biased diode is reverse-biased. The storage delay time depends on the initial forward current flowing through the diode, and it can be measured easily since the diode voltage will reach zero when the minority carrier concentrations reach zero. Complex modeling of minority carrier recombination and extraction is required to accurately predict the switching time of a diode, and these models are beyond the scope of this book.

Since recombination time is an important parameter that controls switching speed in diodes, reductions in recombination times are frequently desired and may be achieved by adding traps intentionally (see Section 2.20). A common deep trap applicable to silicon diodes is gold doping added to the silicon in small concentrations in the range of 10^{14} cm^{-3} to 10^{15} cm^{-3}. This can reduce recombination times to several nanoseconds from the microsecond time scale without gold addition. However, adding gold does compromise diode behavior due to additional carrier generation (see Section 3.8).

Schottky diodes do not suffer from minority carrier recombination delays because only majority carriers flow. These diodes can be switched in well under one nanosecond and are therefore useful in high speed switching applications such as in power supplies.

The second topic for discussion here is junction capacitance. Let us consider the case of reverse bias in a diode. Capacitance C is defined by the change in charge Q on either side of an insulator caused by a given change in voltage V across the insulator or

$$C = \left| \frac{dQ}{dV} \right| \tag{3.37}$$

In a reverse-biased p-n junction diode, the insulator is effectively the depletion region of width W. The charge Q on either side of the depletion region is determined by the ionized donor or acceptor concentration. Hence from Equations 3.9 and 3.15,

$$Q = qA \frac{N_d N_a}{N_d + N_a} W$$

Using Equation 3.24 we obtain

$$Q = qA \frac{N_d N_a}{N_d + N_a} \sqrt{\frac{2\epsilon_0 \epsilon_r (V_0 - V)}{q} \left(\frac{1}{N_a} + \frac{1}{N_d} \right)}$$
$$= A \sqrt{\frac{2q\epsilon_0 \epsilon_r (V_0 - V)(N_d N_a)}{N_d + N_a}}$$

Now using Equation 3.37, and recognizing that V_0 is a constant, we have

$$C = \left| \frac{dQ}{dV} \right| = \left| \frac{dQ}{d(V_0 - V)} \right| = \frac{A}{2} \sqrt{\frac{2q\epsilon_0 \epsilon_r}{(V_0 - V)} \frac{N_d N_a}{N_d + N_a}} \tag{3.38}$$

It is clear that C is a function of the reverse bias V. This specific characteristic is exploited in the *varactor diode*, which provides a variable capacitance in some circuits such as tuning circuits. Care must be taken to ensure that the diode remains in reverse bias for this application.

It is interesting to note the similarity between Equation 3.38 and the expression for capacitance of the simple parallel plate capacitor with plate separation d given by

$$C = \frac{\epsilon_0 \epsilon_r A}{d} \tag{3.39}$$

If d in Equation 3.39 is replaced with the expression for W from Equation 3.24 we obtain the identical result to Equation 3.38. This can be understood because in Equation 3.38 a small change in applied voltage dV gives rise to a small change in charge dQ near the edges of the depletion region as the depletion region width changes slightly. This is equivalent to a small change in the charge on the plates on a parallel plate capacitor due to a small change in voltage dV across the plates. See Problem 3.19.

Schottky diodes also exhibit capacitance. The depletion layer acts as a dielectric layer sandwiched between the metal and semiconductor regions.

3.12   SUMMARY

3.1 In a p-n junction an energy barrier is formed, which is characterized by a built-in potential V_0. The Fermi energy is constant across the junction in equilibrium. An electric field is present in the transition region of the junction.

3.2 Four currents flow in a p-n junction. In equilibrium all these currents add to zero. In forward bias the diffusion currents due to majority carriers dominate and high currents can flow. In reverse bias drift currents due to minority carriers flow but these reverse bias currents are limited in magnitude due to the limited number of minority carriers and constitute the reverse saturation current I_0.

3.3 The contact potential V_0 may be calculated from carrier concentrations and the resulting position of the Fermi energy relative to the conduction and valence band edges on either side of the junction. In equilibrium the Fermi energy is constant.

3.4 The transition region can be modeled using the depletion approximation in which a fully depleted space charge layer of width W is assumed. Values for equilibrium charge density $\rho(x)$, electric field $\varepsilon(x)$, and potential $V(x)$ result from the depletion approximation.

3.5 The diode equation determines the net diode current I obtained due to an applied potential V. It may be derived by assuming equilibrium conditions in the depletion region and the consequences of this assumption on minority carrier concentrations at the edges of the depletion region. The diffusion equation can then determine the resulting diffusion currents. Minority carrier concentrations decay exponentially with distance from the depletion region in both n-type and p-type material.

3.6 Reverse breakdown in a diode can arise from carrier avalanching. Field ionization and impact ionization in high electric fields can occur at or near the junction if the reverse bias is large enough. In addition, highly doped p-n junctions may also exhibit electron tunneling with a small reverse bias. Both mechanisms occur in the Zener diode.

3.7 If even higher doping levels are present the Fermi energy can enter the conduction and valence bands and the condition of degenerate doping is established. Donor and acceptor sub-bands form that merge with the conduction and valence bands. Tunneling of electrons in both directions across the junction is enabled, and a tunneling junction results. A tunneling junction allows efficient current flow across a p-n junction, which is important for multi-junction solar cells.

3.8 Carriers crossing over the depletion region may recombine due to deep traps. This will modify the diode equation resulting in a diode ideality factor n with values between 1 and 2. This trapping is most important at low diode currents.

3.9 Diodes require ohmic contacts to allow current to flow between metal contacts and the semiconductor. A metal-semiconductor junction may form a Schottky diode, which can be understood by thermionic emission that occurs at a metal surface. The mechanism for thermionic emission is derived to obtain the Richardson–Dushman equation. By using high doping levels tunneling behavior can be obtained at a metal-semiconductor junction to create a highly conductive or ohmic contact rather than a rectifying contact. Metal-semiconductor junctions are strongly influenced by surface defects, traps, dangling bonds and impurities at the metal-semiconductor interface.

3.10 Heterojunctions may be formed between two semiconductors having different compositions. This permits the bandgap to change through a device. Heterojunctions are important for high performance LEDs and certain solar cells. Undesirable optical absorption may be reduced or eliminated and localization and improved quantum efficiency of electron-hole pair production/recombination may be enabled.

3.11 An understanding of AC and transient diode behavior requires that two mechanisms of charge storage are discussed. In reverse bias, charge is stored in the depletion region in the form of ionized donors and acceptors, which leads to diode capacitance. In forward bias, charges are stored as minority carriers on either side of the depletion region. This leads to switching delays, which can be characterized by a storage delay time. Schottky diodes eliminate this storage delay time.

PROBLEMS

3.1 Draw a diagram showing the excess minority carrier concentrations as a function of position in a p-n junction diode under the following conditions. Assume approximately equal doping levels on both sides of the junction:
(a) Reverse bias
(b) No bias
(c) Weak forward bias
(d) Strong forward bias
What assumption is no longer valid if forward bias current is increased beyond a certain level? Explain carefully.

3.2 An abrupt Si p-n junction has the following properties:
p-side: $N_a = 10^{18} \text{ cm}^{-3}$
n-side: $N_d = 10^{15} \text{ cm}^{-3}$
junction area $A = 10^{-4} \text{ cm}^2$; carrier lifetimes $\tau_n = \tau_p = 2 \text{ μs}$
(a) Sketch a band diagram of the diode under forward bias showing hole and electron quasi-Fermi levels.
(b) Calculate the depletion region width with a reverse voltage of 10 volts.
(c) Find the diode current with the reverse bias of 10 volts. What is this current called?
(d) Find the peak value of depletion region electric field at a reverse bias of 10 volts.
(e) If the silicon exhibits avalanche breakdown at an electric field of 1×10^5 V cm^{-1} find the reverse bias voltage at which breakdown will occur at the junction.

3.3 An abrupt Si p-n junction has the following parameters:
n-side: $N_d = 5 \times 10^{18}$ cm^{-3}
p-side: $N_a = 10^{17}$ cm^{-3}
junction area $A = 10^{-2}$ cm^2; carrier lifetimes $\tau_n = \tau_p = 2$ μs
(a) Find the built-in potential V_0.
(b) Find the reverse saturation current I_0.
(c) Find the ratio of hole injection current to electron injection current at a forward current of 1 mA.
(d) Find the total minority carrier charge on each side of the diode at a forward current of 1 mA.
(e) In the p-side of the diode, at a certain distance away from the depletion region, the hole and electron currents are equal in magnitude and in the same direction. Find this distance.
(f) Find the quasi-Fermi level separations $F_n - F_p$ at a distance of 0.1 μm from the edges of the depletion region on either side of the junction (i.e. at a depth of 0.1 μm into the neutral p-type and n-type regions) at a forward bias of 1 mA.

3.4 An abrupt GaAs p-n junction has the following parameters:
n-side: $N_d = 2 \times 10^{18}$ cm^{-3}
p-side: $N_a = 2 \times 10^{17}$ cm^{-3}
junction area $A = 10^{-4}$ cm^2; carrier lifetimes $\tau_n = \tau_p = 0.2$ μs
(a) Find the built-in potential V_0.
(b) Find the ratio of hole current to electron current crossing the junction at a total forward current of 5 mA.
(c) Find the reverse saturation current I_0.
(d) At a forward current of 5 mA, find the total minority carrier charge on each side of the diode.
(e) Find the distance into the p-side at which the minority electron concentration is half the maximum value. Sketch it as a function of distance into the p-side starting from the edge of the depletion region. Assume the conditions of d).

3.5 Calculate the built-in electric field as a function of position in a p-n junction having a depletion region of width W and constant doping levels of N_d and N_a on the n- and p-sides respectively. To do this, first find the width of the depletion region on either side of the junction as a function of W. Then use Gauss's law to determine the electric field at any point in the depletion region by considering a Gaussian surface that covers only a fraction of the space charge on either side of the junction. Show that the electric field increases linearly and reaches a maximum value at the junction when the Gaussian surface encloses all the space charge on one side of the junction. Sketch the field as a function of position.

3.6 If the diode of Problem 3.4 is forward biased, sketch how the electric field would vary as a function of position throughout the depletion region. Repeat for a reverse bias. Compare these sketches to the sketch without bias.

3.7 A GaAs diode is reverse biased and it exhibits an increasing reverse current as bias increases. Explain how this occurs based on generation or recombination in the depletion region. What changes could be made to the properties of the semiconductor to reduce this effect?

3.8 A tunnel diode is formed using highly doped silicon with a total depletion width under equilibrium conditions of 3 nm. According to the depletion approximation, what doping level would be needed to achieve this? Assume equal doping levels on both sides of the junction.

3.9 The forward current in a silicon diode increases with bias voltage with the following relationship:

$$I \propto \exp\left(\frac{qV}{1.5kT}\right)$$

(a) Explain the physics underlying this dependence. Plot a representative graph of the current versus voltage dependence using linear x- and y-axes and compare its shape to the graph of a diode having a relationship that obeys the diode equation.

(b) At low forward bias voltages a diode is observed to behave according to $I \propto \exp\left(\frac{qV}{1.5kT}\right)$ but at higher voltages its current–voltage dependence approaches the diode equation. Explain carefully.

(c) A silicon diode at a given forward bias voltage range behaves approximately according to $I \propto \exp\left(\frac{qV}{2kT}\right)$ at a junction temperature of $-50°C$ but at $100°C$ its current–voltage dependence for the same bias voltage range approximately follows the diode equation. Explain carefully.

3.10 Figure 3.28 shows the band structure of an effective ohmic contact that works by tunneling applied to n-type silicon.

(a) Sketch the analogous band structure for an ohmic contact applied to p-type silicon that relies on tunneling.

(b) Propose a metal contact material that forms a tunneling type ohmic contact to n-type GaAs. Look in the literature and see if your answer is a good prediction of materials used in practice.

(c) Repeat (b) for p-type GaAs.

3.11 A Schottky diode is composed of a junction between p-type silicon and aluminum. The barrier height E_b is 0.4 eV and junction area is 1000 μm^2.

(a) Calculate I_e at room temperature.

(b) Find and plot the diode current as a function of applied voltage at room temperature for both forward and reverse bias. You may neglect reverse breakdown.

3.12 In solid state textbooks such as those listed under Suggestions for Further Reading, *The Born–von Karman boundary condition* is applied to traveling electron waves to model the density of states. Look up the definition and the application of this boundary condition to electrons in a box having infinite potential barriers. Show that this boundary condition results in the same density of traveling electron waves that we used to derive Equation 3.35. Redraw Figure 3.27 such that is it relevant to the Born–von Karman boundary condition.

3.13 A silicon p-n junction diode has the following parameters: $N_d = 2 \times 10^{18}$ cm^{-3}, $N_a = 2 \times 10^{16}$ cm^{-3}, $\tau_n = \tau_p = 2 \times 10^{-6}$ s, $D_n = 25$ cm^2 s^{-1} and $D_p = 8$ cm^2 s^{-1}. A light source is incident only on the depletion region, producing a generation

current density of $J_{gen} = 50$ mA cm^{-2}. The diode is open circuited. The generation current density forward biases the junction, inducing a forward- bias
current in the opposite direction to the generation current. A steady-state
condition is reached when the generation current density and forward-bias
current density are equal in magnitude. What is the induced forward-bias
voltage at this steady-state condition?

3.14 An abrupt Si p-n junction has the following parameters:
   p-side: $N_a = 10^{17}$ cm^{-3}
   n-side: $N_d = 10^{14}$ cm^{-3}
   junction area $A = 1 \times 10^{-5}$ cm^2; carrier lifetimes $\tau_n = \tau_p = 2$ µs
      Find:
   (a) V_0, the built-in potential.
   (b) I_0, the reverse saturation current.
   (c) The depletion region width at 0 volts.
   (d) The depletion region width at 10 volts reverse bias.
   (e) The peak electric field at 10 volts reverse bias.
   (f) The ratio of hole-to-electron current flow in forward bias.
   (g) The diode capacitance at reverse bias of 5 volts, 10 volts and 15 volts.

3.15 An abrupt Si p-n junction has the following properties:
   p-side: $N_a = 10^{17}$ cm^{-3},
   n-side: $N_d = 10^{15}$ cm^{-3}
   junction area $A = 10^{-4}$ cm^2; carrier lifetimes $\tau_n = \tau_p = 2$ µs
   (a) Sketch an equilibrium band diagram showing E_F and V_0.
   (b) Calculate V_0.
   (c) Calculate the space charge width with zero applied voltage.
   (d) Find the maximum electric field at a reverse bias of 10 volts.
   (e) Find I_0, the diode reverse saturation current.
   (f) Find the breakdown field for silicon if the diode has a reverse breakdown
         voltage of 100 volts. Hint: Use the highest field in the depletion region for
         this calculation.
   (g) Find the depletion region width just before reverse breakdown.
   (h) Find the diode capacitance at a reverse bias of 10 volts.
   (i) Find the voltage across the diode at a forward current at 1 A.

3.16 An abrupt Si p-n junction has the following properties:
   p-side: $N_a = 10^{18}$ cm^{-3}
   n-side: $N_d = 10^{16}$ cm^{-3}
   junction area $A = 10^{-4}$ cm^2; carrier lifetimes $\tau_n = \tau_p = 2$ µs
   (a) Sketch a band diagram of a diode under forward bias showing hole and elec-
         tron quasi-Fermi levels.
   (b) Calculate the space charge width with a reverse voltage of 10 volts.
   (c) Find the diode capacitance at an applied voltage of 10 volts reverse bias.
   (d) Find the diode current with a reverse bias of 10 volts.
   (e) Find the ratio of hole to electron current that crosses over the depletion
         region during forward bias.

(f) Find the peak value of depletion region electric field at a reverse bias of 10 volts.

3.17 In a p^+-n junction at room temperature, the n-doping N_d is doubled. How do the following two parameters change if everything else is unchanged?
(a) breakdown voltage
(b) built-in voltage

3.18 Ohmic contacts are needed on a p-n junction made with silicon. Metal contact pads will be deposited on the p and n regions, and then diffused in for a short time to form ohmic contacts. List some suitable materials for p and n ohmic contacts.

3.19 Show that you can alternatively derive Equation 3.38 using Equation 3.39 as suggested in Section 3.11. Sketch the depletion region of a diode and show the locations of differential charge dQ obtained by a small change in applied voltage dV.

3.20 With reference to Section 1.9, show that vacuum electrons in Figure 3.26b will experience strong reflections at the boundaries of the vacuum "box" and that reflection coefficient $R \cong 1$. Negative potential barriers surrounding the vacuum have step height $U_0 = -\Phi_m$. Assume an electron kinetic energy of kT at room temperature in the vacuum and assume a typical metal workfunction of $\Phi_m = 4eV$.

3.21 Derive Equation 3.33. Make use of the results of Problem 3.5.

NOTE

1. The vacuum box model allows us to understand a density of states for electron energies in a vacuum box. But we must remember that in thermal equilibrium, the flow of electrons in either direction through the box sidewalls between metal and vacuum proceeds freely. In equilibrium there is no net flow of electrons in either direction and we should instead picture *exchanges* of electrons at the box sidewalls in both directions along the x-axis that are not subject to surface reflections. If reflections are considered then the result in Equation 3.35 will overestimate the current.

SUGGESTIONS FOR FURTHER READING

1. Neamen DA. (2003). *Semiconductor Physics and Devices*, 3e. McGraw Hill.
2. Roulston DJ. (1999). *An Introduction to the Physics of Semiconductor Devices*. Oxford University Press.
3. Streetman BG and Banerjee SK. (2015). *Solid State Electronic Devices*, 7e. Pearson.

Photon Emission and Absorption

CONTENTS

Objectives

1. Introduce the basic forms of luminescence.
2. Present the dipole model of luminescence based on radiation from the acceleration of charges.
3. Introduce the quantum mechanical description of acceleration of charges that can be used to calculate radiation rate and radiative power.

Fundamentals of Semiconductor Materials and Devices, First Edition. Adrian Kitai.
© 2023 John Wiley & Sons Ltd. Published 2023 by John Wiley & Sons Ltd.
Companion Website: www.wiley.com/go/kitai_fundamentals

4. Introduce the free exciton and the mechanism by which an exciton emits a photon through dipole radiation.
5. Describe the two-electron atom and the concept of indistinguishable particles.
6. Present the resulting molecular excitons and their classification as singlet excitons or triplet excitons.
7. Describe the luminescent properties of fluorescence and phosphorescence that are observed from singlet and triplet excitons respectively.
8. Describe the band-to-band emission and recombination model that determines absorption or radiation spectra based on band states and their probability of occupancy.
9. Introduce the human visual system and the units of luminescence and color that allow light sources to be described with relevance to human perception.

4.1 INTRODUCTION TO LUMINESCENCE AND ABSORPTION

We have discussed the electronic aspects of a p-n junction in some detail. However, the processes by which light is absorbed or emitted are a crucial aspect of p-n junctions in solar cells and LEDs. The p-n junction is an ideal device to either absorb or emit photons, and these processes occur when an electron-hole pair is generated or annihilated, respectively. The p-n diode efficiently transports both electrons and holes toward a p-n junction for photon production or away from the junction for electric current generation due to photon absorption.

We will discuss in some detail the theory of *luminescence*, in which a photon is created by an electron-hole pair. These concepts can be readily understood in reverse to explain photon absorption also.

Technologically important forms of luminescence may be broken down into several categories as shown in Table 4.1. Although the means by which the luminescence is excited varies, *all luminescence is generated by means of accelerating charges*. The portion of the electromagnetic spectrum visible to the human eye contains wavelengths from 400 to 700 nm. The evolution of the relatively narrow sensitivity range of the human eye is a complex subject, but is intimately related to the solar spectrum, the absorbing behavior of the terrestrial atmosphere, and the reflecting properties of terrestrial materials, green lying near the middle of the useful spectrum. Not surprisingly, the wavelength at which the human eye is most sensitive is also green.

Visible light is the most important wavelength range for both organic and inorganic LEDs since LEDs are heavily used for lighting and display applications. The display

TABLE 4.1 Luminescence types, applications, and typical efficiencies. Efficiency (η) is given in visible light output power as a fraction of input power.

Luminescence type	Examples
Blackbody radiation	Sun
(light generated due to the temperature of a body)	Tungsten filament lamp ($\eta = 5\%$)
Photoluminescence	Fluorescent lamp phosphors ($\eta = 80\%$)
(light emitted by a material that is stimulated by electromagnetic radiation)	
Electroluminescence	Light emitting diode ($\eta = 70\%$)
(light emitted by a material that is directly electrically excited)	

applications typically include red, green, and blue wavelengths in a *trichromatic* scheme that allows humans to perceive a wide range of colors from a set of only three primary colors.

Infrared (IR) and ultraviolet (UV) radiation must also be considered for both solar cells and LEDs. The sun includes IR and UV wavelengths, and solar cells are heavily dependent on IR absorption. Infrared LEDs are well developed and are used for remote control and sensing applications. UV-emitting LEDs are used for industrial processing applications.

4.2   PHYSICS OF LIGHT EMISSION

In order to understand the processes of light emission and light absorption in more detail we will examine the behavior of charges and moving charges. A stationary point charge q results in electric field lines that emanate from the charge in a radial geometry as shown in Figure 4.1. A charge moving with a uniform velocity relative to an observer gives rise to a magnetic field. Figure 4.2 shows the resulting magnetic field when the point charge moves away from an observer.

Since both electric and magnetic fields store energy the total energy density is given by

$$E = \frac{\epsilon_0}{2}\varepsilon^2 + \frac{1}{2\mu_0}B^2$$

It is important to note that this energy field falls off in density as we move away from the charge but it moves along with the charge provided that the charge is either stationary or undergoing uniform motion. There is no flow of energy from the charge.

For a charge undergoing *acceleration*, however, energy continuously travels away from the charge. Consider the charge q in Figure 4.3. Assume it is initially at rest in

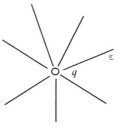

FIGURE 4.1   Lines of electric field ε due to a point charge q.

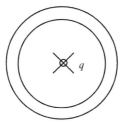

FIGURE 4.2   Closed lines of magnetic field B due to a point charge q moving into the page with uniform velocity.

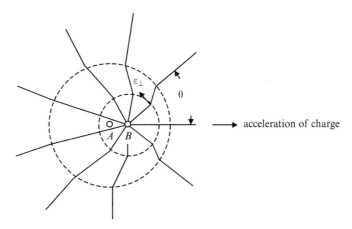

FIGURE 4.3   Lines of electric field emanating from an accelerating charge.

position A, then accelerates to position B, and stops there. The electric field lines now emanate from position B, but further out the lines had emanated from position A. The field lines cannot convey information about the location of the charge at speeds greater than the velocity of light c. This results in kinks in the lines of electric field, which propagate away from q with velocity c. Each time q accelerates, a new series of propagating kinks is generated. Each kink is made up of a component of ε that is transverse to the direction of expansion, which we call $\varepsilon_\perp$. If the velocity of the charge during its acceleration does not exceed a small fraction of c, then for large distances away from charge q,

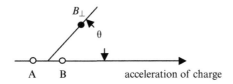

FIGURE 4.4   Direction of magnetic field $B_\perp$ emanating from an accelerating charge. $B_\perp$ is perpendicular to both acceleration and the radial direction.

$$\varepsilon_\perp = \frac{qa}{4\pi\epsilon_0 c^2 r}\sin\theta$$

Here, a is acceleration, r is the radial distance between the charge and the position where the electric field is evaluated, and θ is the angle between the direction of acceleration and the radial direction of the transverse field. The strongest transverse field occurs in directions normal to the direction of acceleration, as shown by Figure 4.3.

Likewise, a transverse magnetic field $B_\perp$ that points in a direction perpendicular to both acceleration and the radial direction is generated during the acceleration of the charge, as shown in Figure 4.4. This is given by

$$B_\perp = \frac{\mu_0 qa}{4\pi cr}\sin\theta$$

In Figure 4.2 the lines of magnetic field are shown as rings perpendicular to the velocity of the charge. In the case of an acceleration, a corresponding change in the magnitude of the magnetic field in the ring occurs but there is no "kink" because the magnetic field lines do not emanate from the charge. Note that this change in magnetic field remains in the form of a ring which is clearly perpendicular to both the acceleration of the charge and the radial direction. Interestingly, $\vec{B}_\perp$ is also perpendicular to $\vec{\varepsilon}_\perp$ which can be understood by comparing Figures 4.2 and 4.3.

The two transverse fields propagate outward with velocity c each time q undergoes acceleration, giving rise to the electromagnetic radiation. The energy density of the radiation is

$$E = \frac{\epsilon_0}{2}\varepsilon_\perp^2 + \frac{1}{2\mu_0}B_\perp^2$$

The Poynting vector, or energy flow per unit area (radiation intensity), is

$$\vec{S} = \frac{1}{\mu_0}\vec{\varepsilon}_\perp \times \vec{B}_\perp$$

$$= \frac{q^2 a^2}{16\pi^2\epsilon_0 c^3 r^2}\sin^2\theta\,\hat{r}$$

since $\vec{B}_\perp$ and $\vec{\varepsilon}_\perp$ are perpendicular to each other. Here, $\hat{r}$ is a unit radial vector.

Maximum energy is emitted in a ring perpendicular to the direction of acceleration, and none is emitted along the direction of acceleration. To obtain the *total* radiated energy per unit time or power P leaving q due to its acceleration, we integrate S over a sphere surrounding q to obtain

$$P = \int_{sphere} S(\theta)\,dA = \int_0^{2\pi}\int_0^{\pi} S(\theta)r^2 \sin\theta\,d\theta\,d\phi$$

or

$$P = \int_0^{\pi} S(\theta)2\pi r^2 \sin\theta\,d\theta$$

Substituting for $S(\theta)$ we obtain

$$P = \frac{1}{16\pi\epsilon_0}\frac{2q^2 a^2}{c^3}\int_0^{\pi}\sin^3\theta\,d\theta$$

which can be integrated (see Problem 4.16) to obtain

$$P = \frac{1}{4\pi\epsilon_0}\frac{2q^2 a^2}{3c^3} \tag{4.1}$$

4.3   SIMPLE HARMONIC RADIATOR

If a charge q oscillates about the origin along the x-axis and its position is given by $x = A\sin\omega t$ then we can calculate the average power radiated away from the oscillating charge. Note that the acceleration a of the charge is given by

$$a = \frac{d^2 x}{dt^2} = -A\omega^2 \sin\omega t$$

and using Equation 4.1

$$P = \frac{2q^2 A^2 \omega^4 \sin^2\omega t}{12\pi\epsilon_0 c^3}$$

which varies with time as $\sin^2\omega t$.
   To obtain *average* power we integrate over one cycle to obtain

$$\bar{P} = \frac{\omega}{2\pi}\frac{2q^2 A^2\omega^4}{12\pi\epsilon_0 c^3}\int_0^{\frac{2\pi}{\omega}}\sin^2\omega t\,dt$$

which yields

$$\bar{P} = \frac{q^2 A^2 \omega^4}{12\pi\epsilon_0 c^3} \tag{4.2}$$

If we now consider that an equal and opposite charge is located at $x = 0$ then we have a *dipole radiator* with electric dipole moment of amplitude $p = qA$. This dipole may consist, for example, of an atom in which the nucleus constitutes an almost stationary charge while the electron oscillates about the nucleus. Now we may write

$$\bar{P} = \frac{p^2 \omega^4}{12\pi\epsilon_0 c^3} \tag{4.3}$$

Radiation that does not rely on dipoles also exists. For example, a synchrotron radiation source is an example of a radiator that relies on the constant centripetal acceleration of an orbiting charge. In a synchrotron the acceleration is in the direction of the radius pointing to the center of the orbit and the radiation is therefore strongest in a direction tangential to the orbit There are also quadrupoles, magnetic dipoles, and other oscillating charge configurations that do not comprise dipoles; however, they have much lower rates of energy release and by far the dominant form of radiation is from dipoles.

Both $\vec{\varepsilon}_\perp$ and $\vec{B}_\perp$ due to the dipole oscillator drop off with distance as $1/r$ whereas the electric field from a static dipole drops much more rapidly as $1/r^3$. This underscores the ability of dipole radiation from an oscillating dipole to travel large distances.

An electron and a hole effectively behave like a dipole radiator when they recombine to create a photon. The photon is not created instantly, and many oscillations of the charges must occur before the photon is fully formed. The description of the photon as a wave-packet is very relevant since the required photon energy is gradually built up as the dipole oscillates to complete the wave-packet. Unless the wave-packet is fully formed no photon exists; the smallest unit of electromagnetic radiation is the photon.

Since we need to describe the positional behavior of electrons and holes using quantum concepts we will now proceed to introduce an important quantum-mechanical expression to allow the calculation of dipole radiation.

4.4 QUANTUM DESCRIPTION

A charge q does not exhibit energy loss or radiation when in a stationary state or eigenstate of a potential energy field. This means that in a stationary state, no net acceleration of the charge occurs, despite its uncertainty in position and momentum within the stationary state. Experience tells us, however, that radiation may be produced when a charge moves from one stationary state to another; it will be the purpose of this section to show that *radiation is produced if an oscillating dipole results from a charge moving from one stationary state to another.*

Consider a charge q initially in normalized stationary state ψ_m and eventually in normalized stationary state $\psi_{m'}$ During the transition, a superposition state is created which we shall call ψ_s:

$$\psi_s = a\psi_m + b\psi_{m'}$$

If

$$|a|^2 + |b|^2 = 1$$

then we have normalized the superposition state. Here a and b are time-dependent coefficients. Initially $a = 1$ and $b = 0$ and finally $a = 0$ and $b = 1$.

Quantum mechanics allows us to calculate the time-dependent expected value of the position $\langle r \rangle (t)$ of a particle in a quantum state. For example, for stationary state ψ_m

$$\langle r \rangle(t) = \langle \psi_m | r | \psi_m \rangle = \int_V \psi_m^* r \psi_m dV = \int_V |\psi_m|^2 r dV$$

where V represents all space. For a stationary state, from Section 1.7

$$\psi(r,t) = \phi(r) \exp \frac{iE}{\hbar} t \tag{4.4}$$

and hence

$$\langle r \rangle(t) = \int_V \left[\phi(r) \exp\left(\frac{iE}{\hbar} t\right) \phi(r) \exp\left(\frac{-iE}{\hbar} t\right) \right] r dV = \int_V r \phi^2(r) dV$$

This expression for $\langle r \rangle$ is therefore not a function of time, which is the fundamental idea underlying the name *stationary* state. A stationary state does not radiate and there is no energy loss associated with the behavior of an electron in such a state. Note that electrons are not truly stationary in a quantum state. It is therefore the quantum state that is described as stationary and not the electron itself. Quantum mechanics sanctions the existence of a charge that has a probability distribution of existing in various places in space but that nevertheless possesses an expected value of acceleration of zero. Classical physics fails to describe or predict this.

If we now calculate $\langle r_s \rangle(t)$, the expectation value of the position of q for the superposition state ψ_s, we obtain, for real values of a and b,

$$\langle r_s \rangle(t) = \langle a\psi_m + b\psi_{m'} | r | a\psi_m + b\psi_{m'} \rangle$$
$$= a^2 \langle \psi_m | r | \psi_m \rangle + b^2 \langle \psi_{m'} | r | \psi_{m'} \rangle + ab \langle \psi_m | r | \psi_{m'} \rangle + ba \langle \psi_m | r | \psi_{m'} \rangle$$

Of the four terms, the first two are stationary but the last two terms are not and therefore the time-dependent part $\langle r_s \rangle(t)$ may be written using Equation 4.4 as

$$\langle r_s \rangle(t) = ab \langle \phi_m | r | \phi_{m'} \rangle \exp\left(\frac{-i(E_m - E_{m'})t}{\hbar}\right) + ba \langle \phi_m | r | \phi_{m'} \rangle \exp\left(\frac{i(E_m - E_{m'})t}{\hbar}\right)$$

Using the Euler formula $e^{ix} + e^{-ix} = 2\cos x$ this may be written as

$$\langle r_s \rangle(t) = 2ab \langle \phi_m | r | \phi_{m'} \rangle \cos\left(\frac{(E_m - E_{m'})t}{\hbar}\right)$$

Defining $\omega_{mm'} = \dfrac{(E_m - E_{m'})}{\hbar}$ we finally obtain

$$\langle r_s \rangle(t) = 2ab \langle \phi_m | r | \phi_{m'} \rangle \cos(\omega_{mm'} t) \tag{4.5}$$

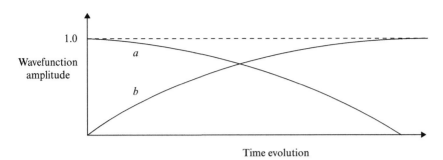

FIGURE 4.5   A time-dependent plot of coefficients a and b is consistent with the time evolution of wavefunctions ϕ_m and $\phi_{m'}$. At $t = 0$, $a = 1$, and $b = 0$. Next a superposition state is formed during the transition such that $a^2 + b^2 = 1$. Finally after the transition is complete $a = 0$ and $b = 1$.

Here, $\langle \phi_m | r | \phi_{m'} \rangle$ is called the *matrix element* for the dipole transition. It is seen that the expectation value of the position of the electron is oscillating with frequency $\omega_{mm'} = \dfrac{E_m - E_{m'}}{\hbar}$ which is the required frequency to produce a photon having energy $E = E_m - E_{m'}$. Product ab also varies with time, but does so very slowly compared with the cosine term. This is illustrated in Figure 4.5.

For $\langle r_s \rangle(t)$, ab is of order unity during most of the transition and the matrix element $\langle \phi_m | r | \phi_{m'} \rangle$ is the important term. Multiplying this matrix element by q yields the *transition dipole moment*

$$q\langle \phi_m | r | \phi_{m'} \rangle = q \int \phi_m^* r \phi_{m'} dV$$

In a *dipole-forbidden transition* $\langle \phi_m | r | \phi_{m'} \rangle$ will be zero and in a *dipole-allowed transition* $\langle \phi_m | r | \phi_{m'} \rangle$ will be greater than zero.

EXAMPLE 4.1

An electron within a dipole oscillates about $x = 0$ with amplitude $A = 1$ Å to produce a photon with $\lambda = 550$ nm.

a) Find the radiation power in watts.
b) Find the photon energy.
c) Find the approximate time taken to release one photon.
d) Find the approximate number of oscillations of the electron required to produce one photon or wave-packet.

Solution

a) From the classical wave equation

$$c = f\lambda = \frac{\omega}{2\pi}\lambda$$

we can solve for ω to obtain

$$\omega = \frac{2\pi c}{\lambda} = \frac{2\pi \times 3.0 \times 10^8 \text{m s}^{-1}}{550 \times 10^{-9} \text{m}} = 3.43 \times 10^{15} \text{rad s}^{-1}$$

now,

$$\bar{P} = \frac{q^2 A^2 \omega^4}{12\pi\epsilon_0 c^3} = \frac{\left(1.6\times10^{-19}\,\text{C}\right)^2 \times \left(10^{-10}\,\text{m}\right)^2 \times \left(3.43\times10^{15}\,\text{rad s}^{-1}\right)^4}{12\pi \times 8.85\times10^{-12}\,\text{Fm}^{-1}\left(3\times10^8\,\text{ms}^{-1}\right)^3}$$

$$= 4\times10^{-12}\,\text{W}$$

b) One photon of this wavelength has energy

$$E_{photon} = \frac{hc}{\lambda} = \frac{6.62\times10^{-34}\,\text{Js} \times 3\times10^8\,\text{m s}^{-1}}{550\times10^{-9}\,\text{m}} = 3\times10^{-19}\,\text{J}$$

c) Hence, the approximate expected length of time taken to release the photon is

$$T = \frac{E_{photon}}{\bar{P}} = \frac{3\times10^{-19}\,\text{J}}{4\times10^{-12}\,\text{W}} = 7.7\times10^{-8}\,\text{s}$$

d) The period of electromagnetic oscillation is

$$T_{oscillation} = \frac{\lambda}{c} = \frac{550\times10^{-9}\,\text{m}}{3\times10^8\,\text{m s}^{-1}} = 1.8\times10^{-15}\,\text{s}$$

Therefore the number of oscillations that take place during the time required to release the photon becomes

$$N = \frac{T}{T_{oscillation}} = \frac{7.7\times10^{-8}\,\text{s}}{1.8\times10^{-15}\,\text{s}} = 4.3\times10^7\,\text{oscillations}$$

Using the quantum description, A may now be defined as the amplitude of $\langle r_s \rangle(t)$. From Equations 4.2 and 4.5, the *photon emission rate* $R_{mm'}$ of a continuously oscillating charge q is

$$R_{mm'} = \frac{\bar{P}}{\hbar\omega} = (ab)^2 \frac{\omega^3}{3\pi\epsilon_0 c^3 \hbar} \left[q\langle \phi_m |r| \phi_{m'}\rangle\right]^2 \frac{\text{photons}}{\text{second}} \tag{4.6}$$

The photon emission rate and hence the *oscillator strength* is proportional to the square of the transition dipole moment. See Problem 4.6.

Note from Section 1.5 that a form of the uncertainty principle relevant to photons is $\Delta E \Delta t \geq \frac{\hbar}{2}$. This would seem to contradict the time calculation for photon release from an oscillating dipole we have just presented. Since photon energy is defined, ΔE is zero and hence we expect time uncertainty Δt for photon release to approach infinity. In fact, there is no contradiction since the time for a photon to be released from an oscillating dipole must be viewed as an expectation value. The release time for any specific photon is not known in advance, but the *expected* release time is known. Once again, classical physics fails to describe how certain photons emitted from an oscillating dipole can be created much more quickly or much more slowly than the expected time.

We are now particularly interested in dipoles formed from a hole-electron pair. A hole-electron pair may produce one photon before it is annihilated, which leads us to examine the hole-electron pair in more detail. Also of relevance is photon absorption, in which a hole-electron pair is created due to photon absorption.

4.5   THE EXCITON

A hole and an electron can exist as a valence band state and a conduction band state. In this model the two particles are not localized and they are both represented using Bloch functions in the periodic potential of the crystal lattice. If the mutual attraction between the two becomes significant then a new description is required for their quantum states that is valid before they recombine but after they experience some mutual attraction.

The hole and electron can exist in quantum states that are actually *within* the energy gap. The band model in Chapter 2 does not consider this situation. Just as a hydrogen atom consists of a series of energy levels associated with the allowed quantum states of a proton and an electron, a series of energy levels associated with the quantum states of a hole and an electron also exists. This hole-electron entity is called an *exciton*, and the exciton behaves in a manner that is similar to a hydrogen atom with one important exception: a hydrogen atom has a lowest energy state or ground state when its quantum number $n = 1$, but a exciton, which also has a ground state at $n = 1$, has an opportunity to be annihilated when the electron and hole eventually recombine.

The energy levels and Bohr radius for a hydrogen atom were derived in Section 1.9. For an exciton we need to modify the electron mass m to become the reduced mass μ of the hole-electron pair which is given by

$$\frac{1}{\mu} = \frac{1}{m_e^*} + \frac{1}{m_h^*}$$

Unlike in the case of the hydrogen atom treated in Section 1.9, the reduced mass is substantially smaller than either the electron effective mass or the hole effective mass. For direct gap semiconductors such as GaAs this turns out to be about one order of magnitude smaller than the free electron mass m. In addition, the exciton exists inside a semiconductor rather than in a vacuum. The relative dielectric constant ϵ_r must be considered, and it is approximately 10 for typical inorganic semiconductors. Using Equation 1.34, the ground state energy for an exciton is

$$E_{excition} = \frac{-\mu q^4}{8\epsilon_0^2 \epsilon_r^2 h^2} \simeq \frac{E_{Rydberg}}{1000}$$

This yields a typical exciton ionization energy or *binding energy* of under 0.1 eV. Using Equation 1.33 the exciton radius in the ground state will be given by

$$a_{exciton} = \frac{4\pi\epsilon_0\epsilon_r\hbar^2}{\mu q^2} \simeq 100 a_0$$

which yields an exciton radius of approximately 50 Å. Since this radius is much larger than the lattice constant of a semiconductor, we are justified in our use of the bulk semiconductor parameters for effective mass and relative dielectric constant.

The electron and hole individually have expected ranges of movement within the exciton that depend on their effective masses. An extreme case of this is illustrated by the hydrogen atom in which the nucleus has a range of movement relative to the center of mass that is approximately three orders of magnitude smaller than that of the electron. In Appendix 6 the effective masses of electrons and holes in a range of semiconductors are listed. If electron effective mass is small compared to hole effective mass, which is true for the listed direct gap semiconductors, expected values of electron cloud radii will be larger than those for holes because the Bohr radius is proportional to the reciprocal of particle mass. See Appendix 4.

Our picture is now of a hydrogen atom-like entity drifting around within the semiconductor crystal and having a series of energy levels analogous to those in a hydrogen atom. Just as a hydrogen atom has energy levels $E_n = \dfrac{-13.6}{n^2} \text{eV}$ where quantum number n is an integer, the exciton has similar energy levels but in a much smaller energy range, and a quantum number $n_{exciton}$ is used.[1]

The exciton must transfer energy to be annihilated. When an electron and a hole form an exciton it is expected that they are initially in a high energy level with a large quantum number $n_{exciton}$. This forms a larger, less tightly bound exciton. Through *thermalization* the exciton loses energy to lattice vibrations and approaches its ground state. Its radius decreases as $n_{exciton}$ approaches 1. Once the exciton is more tightly bound and $n_{exciton}$ is a small integer, the hole and electron can then form an effective dipole and radiation may be produced to account for the remaining energy and to annihilate the exciton through the process of dipole radiation. We can represent the exciton energy levels in a semiconductor as shown in Figure 4.6.

At low temperatures the emission and absorption wavelengths of electron-hole pairs must be understood in the context of excitons in all p-n junctions. The existence of excitons, however, is generally hidden at room temperature and at higher temperatures in *inorganic* semiconductors because of the temperature of operation of the device. The exciton is not stable enough to form from the distributed band states and at room temperature kT may be larger than the exciton binding energy. In this case the spectral features associated with excitons will be absent and direct gap or indirect gap band-to-band transitions occur. Nevertheless photoluminescence or absorption measurements at low temperatures conveniently provided in the laboratory using liquid nitrogen (77 K) or liquid helium (4.2 K) clearly show exciton features, and excitons have become an important tool to study inorganic semiconductor behavior. An example of the transmission as a function of photon energy of a semiconductor at low temperature due to excitons is shown in Figure 4.7.

In a direct-gap semiconductor, dipole radiation can occur. The electron and hole at the band edges in a direct gap semiconductor share the same average linear momentum and may therefore form an oscillating dipole. Electron-hole pair recombination in an indirect gap semiconductor crystal through dipole radiation is forbidden because the electron and hole have different momentum values and dipoles cannot be

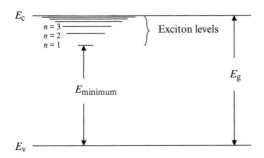

FIGURE 4.6 The exciton forms a series of closely spaced hydrogen-like energy levels that extend inside the energy gap of a semiconductor. If an electron falls into the lowest energy state of the exciton corresponding to $n = 1$ then the remaining energy available for a photon is $E_{minimum}$.

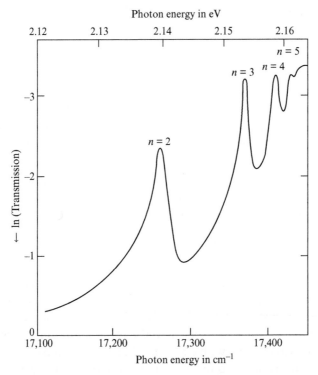

FIGURE 4.7 Low-temperature transmission as a function of photon energy for Cu_2O. The absorption of photons is caused through excitons, which are excited into higher energy levels as the absorption process takes place. Cu_2O is a semiconductor with a bandgap of 2.17 eV. Reprinted from Kittel, C., Introduction to Solid State Physics, 6e, ISBN 0–471-87474–4. Copyright (1986) John Wiley and Sons, Australia.

formed unless a phonon is involved. See Section 2.13. In an indirect gap inorganic semiconductor at room temperature, the hole-electron pair may be fully annihilated through energy loss to phonons rather than through dipole radiation. The requirement of a direct gap for a band-to-band transition that conserves momentum is consistent with the requirements of dipole radiation.

Not all excitons are free to move around in a semiconductor. *Bound excitons* are often formed that associate themselves with defects in a semiconductor crystal such as vacancies and impurities. In *organic* semiconductors *molecular excitons* form, which are very important for an understanding of optical processes that occur in organic semiconductors. This is because molecular excitons typically have binding energies much higher than kT at room temperature. The reason for the higher binding energy is the confinement of the molecular exciton to smaller spatial dimensions imposed by the size of the molecule. This keeps the hole and electron closer and increases the binding energy compared to free excitons. In contrast to the situation in inorganic semiconductors, molecular excitons are thermally stable at room temperature and they generally determine emission and absorption characteristics of organic semiconductors. The molecular exciton will be discussed in Section 4.7. Quantum dots, to be discussed in Chapter 7, also confine and stabilize excitons.

4.6 TWO-ELECTRON ATOMS AND THE EXCHANGE INTERACTION

Until now we have focused on dipole radiators that are composed of two charges, one positive and one negative. In Section 4.3 we introduced an oscillating dipole having one positive charge and one negative charge. In Section 4.5 we discussed the exciton, which also has one positive charge and one negative charge. Many dipoles may exist simultaneously and produce photons in a crystalline semiconductor described with band theory from Chapter 2, but we treat each dipole as a discrete entity.

However, we also need to understand optical absorption and radiation from molecular systems containing two or more electrons per molecule. These electrons cooperate intimately during molecular dipole absorption and radiation events. This cooperation forms the basis of optically active organic semiconductors. Once a system has two or more identical particles (electrons) there is an additional and very fundamental quantum effect that we need to consider. This effect is called the *exchange interaction*.

The best starting point to gain further understanding of this is the helium atom, which has a nucleus with a charge of $+2q$ as well as two electrons each with a charge of $-q$. A straightforward solution to the helium atom using Schrödinger's equation is not possible since this is a *three-body* system; however, we can understand the behavior of such a system by applying the Pauli exclusion principle and by including the spin states of the two electrons.

When two electrons at least partly overlap spatially with one another their wavefunctions must conform to the Pauli exclusion principle; however, there is an additional requirement that must be satisfied. The two electrons must be carefully treated as *indistinguishable* because once they have even a small spatial overlap there is no way to know which electron is which. We can only determine a probability density $|\psi|^2 = \psi^* \psi$ for each wavefunction but we cannot determine the precise location of either electron at any instant in time and therefore there is always a chance that the electrons exchange places. There is no way to label or otherwise identify each electron and the wavefunctions must therefore not be specific about the identity of each electron.

If we start with Schrödinger's equation and write it by adding up the energy terms from the two electrons we obtain

$$-\frac{\hbar^2}{2m}\left(\frac{\partial^2 \psi_T}{\partial x_1^2}+\frac{\partial^2 \psi_T}{\partial y_1^2}+\frac{\partial^2 \psi_T}{\partial z_1^2}\right)-\frac{\hbar^2}{2m}\left(\frac{\partial^2 \psi_T}{\partial x_2^2}+\frac{\partial^2 \psi_T}{\partial y_2^2}+\frac{\partial^2 \psi_T}{\partial z_2^2}\right)+U_T\psi_T=E_T\psi_T \quad (4.7)$$

Here ψ_T $(x_1, y_1, z_1, x_2, y_2, z_2)$ is the wavefunction of the two-electron system, U_T $(x_1, y_1, z_1, x_2, y_2, z_2)$ is the potential energy for the two-electron system and E_T is the total energy of the two-electron system. The spatial coordinates of the two electrons are (x_1, y_1, z_1) and (x_2, y_2, z_2).

To simplify our treatment of the two electrons we will start by assuming that the electrons do not interact with each other. This means that we are neglecting coulomb repulsion between the electrons. The potential energy of the total system is then simply the sum of the potential energy of each electron under the influence of the helium nucleus. Now the potential energy can be expressed as the sum of two identical potential energy functions $U(x, y, z)$ for the two electrons and we can write

$$U_T\left(x_1,y_1,z_1,x_2,y_2,z_2\right)=U\left(x_1,y_1,z_1\right)+U\left(x_2,y_2,z_2\right)$$

Substituting this into Equation 4.7 we obtain

$$-\frac{\hbar^2}{2m}\left(\frac{\partial^2 \psi_T}{\partial x_1^2}+\frac{\partial^2 \psi_T}{\partial y_1^2}+\frac{\partial^2 \psi_T}{\partial z_1^2}\right)-\frac{\hbar^2}{2m}\left(\frac{\partial^2 \psi_T}{\partial x_2^2}+\frac{\partial^2 \psi_T}{\partial y_2^2}+\frac{\partial^2 \psi_T}{\partial z_2^2}\right)$$
$$+U\left(x_1,y_1,z_1\right)\psi_T+U\left(x_2,y_2,z_2\right)\psi_T=E_T\psi_T \quad (4.8)$$

If we look for solutions for ψ_T of the form $\psi_T = \psi(x_1, y_1, z_1)\psi(x_2, y_2, z_2)$ then Equation 4.8 becomes

$$-\frac{\hbar^2}{2m}\psi\left(x_2,y_2,z_2\right)\left(\frac{\partial^2}{\partial x_1^2}+\frac{\partial^2}{\partial y_1^2}+\frac{\partial^2}{\partial z_1^2}\right)\psi\left(x_1,y_1,z_1\right)$$
$$-\frac{\hbar^2}{2m}\psi\left(x_1,y_1,z_1\right)\left(\frac{\partial^2}{\partial x_2^2}+\frac{\partial^2}{\partial y_2^2}+\frac{\partial^2}{\partial z_2^2}\right)\psi\left(x_2,y_2,z_2\right)$$
$$+U\left(x_1,y_1,z_1\right)\psi\left(x_1,y_1,z_1\right)\psi\left(x_2,y_2,z_2\right)$$
$$+U\left(x_2,y_2,z_2\right)\psi\left(x_1,y_1,z_1\right)\psi\left(x_2,y_2,z_2\right)=E_T\psi\left(x_1,y_1,z_1\right)\psi\left(x_2,y_2,z_2\right) \quad (4.9)$$

Dividing Equation 4.9 by $\psi(x_1, y_1, z_1)\psi(x_2, y_2, z_2)$ we obtain

$$-\frac{\hbar^2}{2m}\frac{1}{\psi\left(x_1,y_1,z_1\right)}\left(\frac{\partial^2}{\partial x_1^2}+\frac{\partial^2}{\partial x_1^2}+\frac{\partial^2}{\partial x_1^2}\right)\psi\left(x_1,y_1,z_1\right)-\frac{\hbar^2}{2m}\frac{1}{\psi\left(x_2,y_2,z_2\right)}$$
$$\left(\frac{\partial^2}{\partial x_2^2}+\frac{\partial^2}{\partial x_2^2}+\frac{\partial^2}{\partial x_2^2}\right)\psi\left(x_2,y_2,z_2\right)+U\left(x_1,y_1,z_1\right)+U\left(x_2,y_2,z_2\right)=E_T$$

Since the first and third terms in the LHS are only a function of (x_1, y_1, z_1) and the second and fourth terms are only a function of (x_2, y_2, z_2), and furthermore since the equation must be satisfied for independent choices of (x_1, y_1, z_1) and (x_2, y_2, z_2), it follows that we must independently satisfy two equations namely

$$-\frac{\hbar^2}{2m}\frac{1}{\psi(x_1,y_1,z_1)}\left(\frac{\partial^2}{\partial x_1^2}+\frac{\partial^2}{\partial y_1^2}+\frac{\partial^2}{\partial x_1^2}\right)\psi(x_1,y_1,z_1)+U(x_1,y_1,z_1)=E_1$$

and

$$-\frac{\hbar^2}{2m}\frac{1}{\psi(x_2,y_2,z_2)}\left(\frac{\partial^2}{\partial x_2^2}+\frac{\partial^2}{\partial y_2^2}+\frac{\partial^2}{\partial x_2^2}\right)\psi(x_2,y_2,z_2)+U(x_2,y_2,z_2)=E_2$$

where $E_T = E_1 + E_2$

These are precisely the single-electron Schrödinger equations that we would use if we were to consider each electron separately. We have used the technique of separation of variables.

What has been shown is the justification for forming the total wavefunction ψ_T by the product of the single-electron wavefunctions. We have considered only the spatial parts of the wavefunctions of the electrons; however, electrons also have spin. In order to include spin each wavefunctions must also define the spin direction of the electron.

We will write a complete wavefunction $[\psi(x_1, y_1, z_1)\psi(s)]_a$, which is the wavefunction for one electron where $\psi(x_1, y_1, z_1)$ describes the spatial part and the spin wavefunction $\psi(s)$ describes the spin part, which can be spin up or spin down.[2] There will be four quantum numbers associated with each wavefunction of which the first three arise from the spatial part. A fourth quantum number, which can be $+\frac{1}{2}$ or $-\frac{1}{2}$ for the spin part, defines the direction of the spin part. Rather than writing the full set of quantum numbers for each wavefunction we will use the subscript a to denote the set of four quantum numbers. For the other electron the analogous wavefunction is $[\psi(x_2, y_2, z_2)\psi(s)]_b$, indicating that this electron has its own set of four quantum numbers denoted by subscript b.

Now the wavefunction of the two-electron system including spin becomes

$$\psi_{T_1} = [\psi(x_1,y_1,z_1)\psi(s)]_a[\psi(x_2,y_2,z_2)\psi(s)]_b \qquad (4.10a)$$

The probability distribution function, which describes the spatial probability density function of the two-electron system, is $|\psi_T|^2$ which can be written as

$$|\psi_{T_1}|^2 = \psi_{T_1}^*\psi_{T_1}$$
$$= [\psi(x_1,y_1,z_1)\psi(s)]_a^*[\psi(x_2,y_2,z_2)\psi(s)]_b^*[\psi(x_1,y_1,z_1)\psi(s)]_a \qquad (4.10b)$$
$$\psi(x_2,y_2,z_2)\psi(s)]_b$$

If the electrons are distinguishable then we need also to consider the case where the electrons are in the opposite states, and in this case

$$\psi_{T_2} = \left[\psi(x_1,y_1,z_1)\psi(s)\right]_b \left[x_2,y_2,z_2)\psi(s)\right]_a \tag{4.11a}$$

Now the probability density of the two-electron system would be

$$
\begin{aligned}
\left|\psi_{T_2}\right|^2 &= \psi_{T_2}^* \psi_{T_2} \\
&= \left[\psi(x_1,y_1,z_1)\psi(s)\right]_b^* \left[\psi(x_2,y_2,z_2)\psi(s)\right]_a^* \left[\psi(x_1,y_1,z_1)\psi(s)\right]_b \\
&\quad \psi(x_2,y_2,z_2)\psi(s)]_a
\end{aligned}
\tag{4.11b}
$$

Clearly Equation 4.11b is not the same as Equation 4.10b, because when the subscripts are switched the form of $\left|\psi_T\right|^2$ changes. This specifically contradicts the fundamental requirement that measurable quantities such as the probability density of the two-electron system must remain the same regardless of the interchange of the electrons.

In order to resolve this difficulty it is possible to write wavefunctions of the two-electron system that are linear combinations of the above two electron wavefunctions.

We write a *symmetric* wavefunction ψ_S for the two-electron system as

$$\psi_S = \frac{1}{\sqrt{2}}\left[\psi_{T_1} + \psi_{T_2}\right] \tag{4.12}$$

and an *antisymmetric* wavefunction ψ_A for the two-electron system as

$$\psi_A = \frac{1}{\sqrt{2}}\left[\psi_{T_1} - \psi_{T_2}\right] \tag{4.13}$$

Since ψ_S is used to calculate the probability density function $|\psi_S|^2$, the result will be independent of the choice of the subscripts. In addition since both ψ_{T_1} and ψ_{T_2} are valid solutions to Schrödinger's equation (Equation 4.8) and since ψ_S is a linear combination of these solutions, it follows that ψ_S is also a valid solution. The same argument applies to ψ_A. See Problem 4.17.

The antisymmetric wavefunction ψ_A may be written using Equations 4.13, 4.10a, and 4.11a as

$$
\begin{aligned}
\psi_A &= \frac{1}{\sqrt{2}}[\psi_{T_1} - \psi_{T_2}] \\
&= \frac{1}{\sqrt{2}}\{[\psi(x_1,y_1,z_1)\psi(s)]_a[\psi(x_2,y_2,z_2)\psi(s)]_b \\
&\quad -[\psi(x_1,y_1,z_1)\psi(s)]_b[\psi(x_2,y_2,z_2)\psi(s)]_a\}
\end{aligned}
\tag{4.14}
$$

If, in violation of the Pauli exclusion principle, the two electrons were in the *same* quantum state $\psi_T = \psi_{T_1} = \psi_{T_2}$, which includes both position and spin, then Equation 4.14

immediately yields $\psi_A = 0$, which means that such a situation cannot occur. However, if the symmetric wavefunction ψ_S of Equation 4.12 was used instead of ψ_A, the value of ψ_S would not be zero for two electrons in the same quantum state. See Problem 4.19. For this reason, a more complete statement of the Pauli exclusion principle is that *the total wavefunction of a system of two or more indistinguishable electrons must be antisymmetric.*

We will now examine just the spin parts of the wavefunctions for each electron because there are only a few possible cases. The individual electron spin wavefunctions must be multiplied to obtain the spin part of the wavefunction for the two-electron system as indicated in Equations 4.10a or 4.11a. There are four possibilities, namely $\psi_{\frac{1}{2}}\psi_{-\frac{1}{2}}$ or $\psi_{-\frac{1}{2}}\psi_{\frac{1}{2}}$ or $\psi_{\frac{1}{2}}\psi_{\frac{1}{2}}$ or $\psi_{-\frac{1}{2}}\psi_{-\frac{1}{2}}$.

For the first two possibilities to satisfy the requirement that the spin part of the new two-electron wavefunction does not depend on which electron is which, a symmetric or an antisymmetric spin function is required. In the symmetric case we can use a linear combination of wavefunctions

$$\psi = \frac{1}{\sqrt{2}}\left(\psi_{\frac{1}{2}}\psi_{-\frac{1}{2}} + \psi_{-\frac{1}{2}}\psi_{\frac{1}{2}}\right) \tag{4.15}$$

This is a symmetric spin wavefunction since changing the labels does not affect the result. The total spin for this symmetric system turns out to be one. There is also an antisymmetric spin case for which

$$\psi = \frac{1}{\sqrt{2}}\left(\psi_{\frac{1}{2}}\psi_{-\frac{1}{2}} - \psi_{-\frac{1}{2}}\psi_{\frac{1}{2}}\right) \tag{4.16}$$

Here, changing the sign of the labels changes the sign of the linear combination but does not change any measurable properties and this is therefore also consistent with the requirements for a proper description of indistinguishable particles. In this antisymmetric spin system the total spin turns out to be zero.

The final two possibilities are symmetric cases since switching the labels makes no difference. These cases therefore do not require the use of linear combinations to be consistent with indistinguishability and are simply

$$\psi = \psi_{\frac{1}{2}}\psi_{\frac{1}{2}} \tag{4.17}$$

and

$$\psi = \psi_{-\frac{1}{2}}\psi_{-\frac{1}{2}} \tag{4.18}$$

These symmetric cases both have total spin of one.

In summary there are four cases, three of which, given by Equations 4.15, 4.17, and 4.18, are symmetric spin states and have total spin one, and one of which, given by Equation 4.16, is antisymmetric and has total spin zero. Note that total spin is not always simply the sum of the individual spins of the two electrons, but must take into account the addition rules for quantum spin vectors. See Suggestions for Further

Reading. The three symmetric cases are appropriately called *triplet* states and the one antisymmetric case is called a *singlet* state. Table 4.2 lists the four possible states.

In order to obtain an antisymmetric wavefunction, from Equation 4.10a either the spin part *or* the spatial part of the wavefunction must be antisymmetric. If the spin part is antisymmetric, which is a singlet state, then the Pauli exclusion restriction on the spatial part of the wavefunction is lifted. The two electrons may occupy the same spatial wavefunction or two distinct but symmetric spatial wavefunctions that turn out to have a high probability of being relatively close to each other.

If the spin part is symmetric this is a triplet state and the spatial part of the wave-function must be antisymmetric. The spatial density function of the antisymmetric wavefunction causes the two electrons to have a higher probability of existing further apart. They are in distinct spatial wavefunctions that are antisymmetric.

If we now introduce the coulomb repulsion between the electrons it becomes evident that if the spin state is a singlet state the repulsion will be higher because the electrons spend more time close to each other. If the spin state is a triplet state the repulsion is weaker because the electrons spend more time further apart.

Now let us return to the helium atom as an example of this. Assume one helium electron is in the ground state of helium, which is the 1s state, and the second helium electron is in an excited state. This corresponds to an excited helium atom, and we need to understand this configuration because radiation always involves excited states.

This excited helium atom also qualifies as an example of an exciton because there is a missing electron (hole) in the ground state and there is an electron in the excited state. This exciton is considered a bound exciton because it is confined to the helium atom rather than being free to move around as in a bulk semiconductor.

The two helium exciton electrons can be in a triplet state or in a singlet state. Strong dipole radiation is observed from the singlet state only, and the triplet states do not radiate. We can understand the lack of radiation from the triplet states by examining spin. The total spin of a triplet state is one. The ground state of helium, however, has no net spin because if the two electrons are in the same $n = 1$ energy level the spins must be in opposing directions to satisfy the Pauli exclusion principle. The ground state of helium is therefore always a singlet state and there can be no triplet states in the ground state of the helium atom. See Problem 4.18.

The fundamental quantity of magnetism due to the spin of an electron is the Bohr magneton. If the two helium electrons are in a triplet state there is a net magnetic moment, which can be expressed as two Bohr magnetons since the total spin is one. This means that a magnetic moment exists in the excited triplet state of helium. Photons have no charge and hence no magnetic moment. Because of this a *dipole transition from an excited triplet state to the ground singlet state is forbidden* because the triplet state has a magnetic moment but the singlet state does not, and the net magnetic moment cannot be conserved. In contrast to this the *dipole transition from an excited singlet state to the ground singlet state is allowed* and dipole radiation may be observed provided the matrix element for dipole radiation is not zero.

TABLE 4.2 Possible spin states for a two electron system.

State	Probability	Total spin	Spin arrangement	Spin symmetry	Spatial symmetry	Spatial attributes	Dipole-allowed transition to/from singlet ground state
Singlet	25%	0	$\psi_{\frac{1}{2}}\psi_{-\frac{1}{2}} - \psi_{-\frac{1}{2}}\psi_{\frac{1}{2}}$	Antisymmetric	Symmetric	Electrons close to each other	Yes
Triplet	75%	1	$\psi_{\frac{1}{2}}\psi_{-\frac{1}{2}} + \psi_{-\frac{1}{2}}\psi_{\frac{1}{2}}$ or $\psi_{\frac{1}{2}}\psi_{\frac{1}{2}}$ or $\psi_{-\frac{1}{2}}\psi_{-\frac{1}{2}}$	Symmetric	Anti-symmetric	Electrons far apart	No

The triplet states of helium are slightly lower in energy than the singlet states. The triplet states involve symmetric spin states, which means that the spin parts of the wavefunctions are symmetric. This forces the spatial parts of the wavefunctions to be antisymmetric, as illustrated in Figure 4.8, and the electrons are, on average, more separated. As a result the repulsion between the ground state electron and the excited state electron is weaker. The excited state electron is therefore more strongly bound to the nucleus and it exists in a lower energy state. The observed radiation is consistent with the energy difference between the higher energy singlet state and the ground singlet state. As expected there is normally no observed radiation consistent with the energy difference between the triplet excited state and the ground singlet state. See Figure 4.9.

We have used helium atoms to illustrate the behavior of a two-electron system; however, we now need to apply our understanding of these results to molecular electrons, which are important for organic light-emitting and absorbing materials. Molecules are the basis for organic electronic materials and molecules always contain two or more electrons.

4.7   MOLECULAR EXCITONS

In bulk, inorganic semiconductors electrons and holes normally exist as distributed wavefunctions which prevents the formation of stable excitons at room temperature. In contrast to this, holes and electrons are localized within a given molecule in organic semiconductors. They are also localized in quantum dots to be discussed in Chapter 7. It such materials the exciton is stabilized.

In organic semiconductors which are composed of molecules, excitons are clearly evident at room temperature and even at higher temperatures. An exciton in an organic semiconductor is an excited state of the molecule. A molecule contains a series of electron energy levels associated with a series of *molecular orbitals* that are complicated to calculate directly from Schrödinger's equation due to the complex shapes of molecules. These molecular orbitals may be occupied or unoccupied. When a molecule absorbs a quantum of energy that corresponds to a transition from one molecular orbital to another higher energy molecular orbital, the resulting electronic excited state of the molecule is a molecular exciton comprising an electron and a hole within the molecule. An electron is typically found in the lowest unoccupied molecular orbital and a hole is found in the highest, normally occupied molecular orbital. Since they are both contained within the same molecule the electron-hole state is said to be bound. A bound exciton results, which is spatially localized to a given molecule in an organic semiconductor. Organic molecule energy levels are discussed in more detail in Chapters 8 and 9.

These molecular excitons can be classified as in the case of excited states of the helium atom, and either singlet or triplet excited states in molecules are possible. The results from Section 4.6 are relevant to these molecular excitons and the same concepts involving electron spin, the Pauli exclusion principle, and indistinguishability are relevant because the molecule contains two or more electrons.

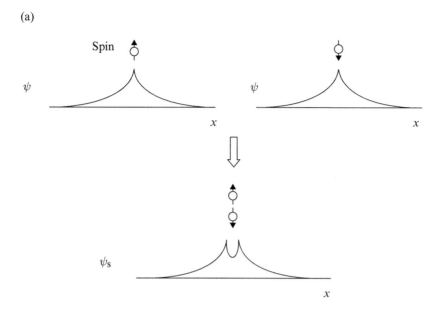

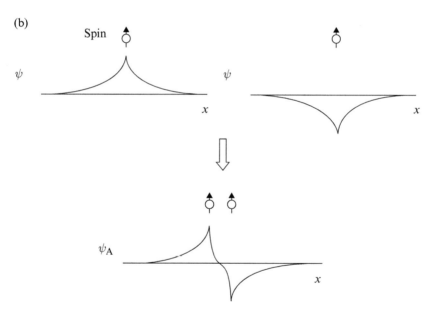

FIGURE 4.8   A depiction of the symmetric and antisymmetric wavefunctions and spatial density functions of a two-electron system. (a) Singlet state with electrons closer to each other on average. (b) Triplet state with electrons further apart on average.

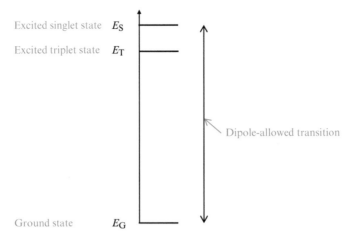

FIGURE 4.9 Energy levels associated with the ground singlet state and both excited singlet and triplet states of a two electron atom. The excited singlet state has the higher energy since the electrons are closer together on average which is less stable than the excited triplet state with electrons further apart on average.

If a molecule in its unexcited state absorbs a photon of light it may be excited forming an exciton in a singlet state with zero spin. These excited molecules typically have characteristic lifetimes on the order of nanoseconds, after which the excitation energy may be released in the form of a photon and the molecule undergoes *fluorescence* by a dipole-allowed process returning to its ground state.

It is also possible for the molecule to be excited to form an exciton by electrical means rather than by the absorption of a photon. This will be described in detail in Chapter 8. Under electrical excitation the exciton may be in a singlet or a triplet state since electrical excitation, unlike photon absorption, does not require the conservation of spin. There is a 75% probability of a triplet exciton and 25% probability of a singlet exciton, as described in Table 4.2. The probability of fluorescence is therefore reduced under electrical excitation to 25% because the decay of triplet excitons is not allowed.

Another process may take place, however. Triplet excitons have a spin state with total spin of one and these spin states can frequently be coupled with the orbital angular momentum of molecular electrons, which influences the effective magnetic moment of a molecular exciton. The restriction on dipole radiation can be partly removed by this coupling, and light emission over relatively long characteristic radiation lifetimes is observed in specific molecules. These longer lifetimes from triplet states are generally on the order of milliseconds and the process is called *phosphorescence*, in contrast with the shorter lifetime fluorescence from singlet states. Since excited triplet states have slightly lower energy levels than excited singlet states, triplet phosphorescence has a longer wavelength than singlet fluorescence in a given molecule.

In addition there are other ways that a molecular exciton can lose energy. There are three possible processes that involve energy transfer from one molecule to another molecule. One important process is known as *Förster resonance energy transfer*. Here a singlet molecular exciton in one molecule is established but a neighboring molecule is not initially excited. The excited molecule will establish an oscillating dipole moment as its exciton starts to decay in energy as a superposition state. The radiation field from this dipole is experienced by the neighboring molecule as an oscillating field and a superposition state in the neighboring molecule is also established. The originally excited molecule loses energy through this resonance energy transfer process to the neighboring molecule and finally energy is conserved since the initial excitation energy is transferred to the neighboring molecule *without the formation of a photon*. This is not the same process as photon generation and absorption since a complete photon is never created; however, only dipole-allowed transitions from excited singlet states can participate in Förster resonance energy transfer.

Förster energy transfer depends strongly on the intermolecular spacing, and the rate of energy transfer falls off as $\dfrac{1}{R^6}$ where R is the distance between the two molecules. An understandingof this can be obtained using the result for the electric field of a static dipole. This field falls off as $\dfrac{1}{R^3}$. Since the energy density in a field is proportional to the square of the field strength it follows that the energy available to the neighboring molecule falls of as $\dfrac{1}{R^6}$. This then determines the rate of energy transfer.

Dexter electron transfer is a second energy transfer mechanism in which an excited electron state transfers from one molecule (the donor molecule) to a second molecule (the acceptor molecule). This requires a wavefunction overlap between the donor and acceptor, which can only occur at extremely short distances typically up to only about 10 Å. Triplet excitons may be transferred which are forbidden to use the Förster resonance energy transfer process because it relies on dipole oscillators. The Dexter process does not rely on dipole interaction.

The Dexter process involves the effective transfer of the electron and hole from molecule to molecule. This will be discussed in Chapter 8 in the context of organic LEDs. The Dexter energy transfer rate is proportional to $e^{-\alpha R}$ where R is the intermolecular spacing. The exponential form is due to the exponential decrease in wavefunction probability density with distance which determines the degree of wavefunction overlap.

Finally. a third process is radiative energy transfer. In this case a photon emitted by the host is absorbcd by the guest molecule. The photon may be formed by dipole radiation from the host molecule and absorbed by dipole absorption in the guest molecule.

4.8 BAND-TO-BAND TRANSITIONS

In inorganic semiconductors the recombination between an electron and a hole occurs to yield a photon, or conversely the absorption of a photon yields a hole-electron pair. The electron is in the conduction band and the hole is in the valence band. It is very useful to analyze these processes in the context of band theory from Chapter 2.

Consider the direct-gap semiconductor having approximately parabolic conduction and valence bands near the bottom and top of these bands respectively, as in Figure 4.10. Parabolic bands were introduced in Section 2.6. Two possible transition energies, E_1 and E_2, are shown, which produce two photons having two different wavelengths. Due to the very small momentum of a photon, the recombination of an electron and a hole occurs almost vertically in this diagram to satisfy conservation of momentum. The x-axis represents the wave-number k, which is proportional to momentum. See Section 2.13.

Conduction band electrons have energy

$$E_e = E_c + \frac{\hbar^2 k^2}{2m_e^*}$$

and for holes we have

$$E_h = E_v - \frac{\hbar^2 k^2}{2m_h^*}$$

In order to determine the emission/absorption spectrum of a direct-gap semiconductor we need to find the probability of a recombination taking place as a function of energy E. This transition probability depends on an appropriate density of states function multiplied by probability functions that describe whether or not the states are occupied.

We will first determine the appropriate density of states function. Any transition in Figure 4.10 takes place at a fixed value of reciprocal space where k is constant. The same set of points located in reciprocal space or k-space gives rise to states both in the valence band and in the conduction band. In the Kronig–Penny model presented in Chapter 2, a given position on the k-axis intersects all the energy bands including the valence and conduction bands. There is therefore a state in the conduction band corresponding to a state in the valence band at a specific value of k.

In order to determine the photon emission rate over a specific range of photon energies we need to find the appropriate density of states function for a *transition* between a group of states in the conduction band and the corresponding group of states in the valence band. This means we need to determine the number of states in reciprocal space or k-space that give rise to the corresponding set of transition energies that can occur over a small radiation energy range ΔE centered at some transition energy in Figure 4.10. For example, the appropriate number of states can be found at E_2 in Figure 4.10b by considering a small range of k-states Δk that correspond to small differential energy ranges ΔE_c and ΔE_v and then finding the total number of band states that fall within the range ΔE. The emission energy from these states will be centered at E_2 and will have an emission energy range $\Delta E = \Delta E_c + \Delta E_v$ producing a portion of the observed emission spectrum. The *density of transitions* is determined by the density of states in the *joint dispersion relation*, which will now be introduced:

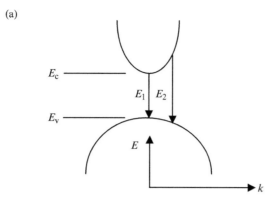

(a)

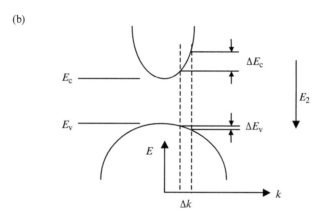

(b)

FIGURE 4.10 (a) Parabolic conduction and valence bands in a direct-gap semiconductor showing two possible transitions. (b) Two ranges of energies ΔE_v in the valence band and ΔE_c in the conduction band determine the photon emission rate in a small energy range about a specific transition energy. Note that the two broken vertical lines in (b) show that the range of transition energies at E_2 is the sum of ΔE_c and ΔE_v.

The energy for any transition is given by

$$E(k) = h\nu = E_e(k) - E_h(k)$$

and upon substitution we can obtain the joint dispersion relation, which adds the dispersion relations from both the valence and conduction bands. We can express this transition energy E and determine the joint dispersion relation from Figure 4.10a as

$$E(k) = h\nu = E_c - E_v + \frac{\hbar^2 k^2}{2m_e^*} + \frac{\hbar^2 k^2}{2m_h^*} = E_g + \frac{\hbar^2 k^2}{2\mu} \qquad (4.19)$$

where

$$\frac{1}{\mu} = \frac{1}{m_e^*} + \frac{1}{m_h^*}$$

Note that a range of k-states Δk will result in an energy range $\Delta E = \Delta E_c + \Delta E_v$ in the joint dispersion relation because the joint dispersion relation sums the relevant ranges of energy in the two bands as required. The smallest possible value of transition energy E in the joint dispersion relation occurs at $k = 0$ where $E = E_g$ from Equation 4.19, which is consistent with Figure 4.10. If we can determine the density of states in the joint dispersion relation we will have the density of possible photon emission transitions available in a certain range of energies.

The density of states function for an energy band was derived in Section 2.10. As originally derived, the form of the density of states function was valid for a box having $U = 0$ inside the box. In an energy band, however, the density of states function was modified. We replaced the free electron mass with the reduced mass relevant to the specific energy band and we replaced m in Equation 2.32 by m_e^* to obtain Equation 2.33a. This is valid because rather than the parabolic E versus k dispersion relation for free electrons in which the electron mass is m, we used the parabolic E versus k dispersion relation for an electron in an energy band as illustrated in Figure 2.8, which may be approximated as parabolic for small values of k with the appropriate effective mass. The slope of the E versus k dispersion relation is controlled by the effective mass, and this slope determines the density of states along the energy axis for a given density of states along the k-axis. We can now use the same method to determine the density of states in the joint dispersion relation of Equation 4.19 by substituting the reduced mass μ into Equation 2.32. Recognizing that the density of states function must be zero for $E < E_g$ we obtain

$$D_{\text{joint}}(E) = \frac{1}{2}\pi\left(\frac{2\mu}{\pi^2\hbar^2}\right)^{\frac{3}{2}}(E - E_g)^{\frac{1}{2}} \tag{4.20}$$

This is known as the *joint density of states* function valid for $E \geq E_g$.

To determine the probability of occupancy of states in the bands, we use Fermi–Dirac statistics, introduced in Chapter 2. The Boltzmann approximation for the probability of occupancy of carriers in a conduction band was obtained in Equation 2.35 as

$$F(E) \cong \exp\left[-\frac{(E_e - E_F)}{kT}\right]$$

and for a valence band the probability of a hole is given by

$$1 - F(E) \cong \exp\left[\frac{(E_h - E_F)}{kT}\right]$$

Since a transition requires both an electron in the conduction band and a hole in the valence band, the probability of a transition will be proportional to

$$F(E)[1 - F(E)] = \exp\left[-\frac{(E_e - E_h)}{kT}\right] = \exp\left[-\frac{E}{kT}\right] \tag{4.21}$$

Including the density of states function, we conclude that the probability $p(E)$ of an electron-hole pair recombination applicable to an LED is proportional to the product of the joint density of states function and the function $F(E)[1 - F(E)]$, which yields

$$p(E) \alpha \ D_{\text{joint}}(E)F(E)\left[1 - F(E)\right] \tag{4.22}$$

Now using Equations 4.20, 4.21, and 4.22 we obtain the photon emission rate $R(E)$ as

$$R(E) \alpha \ (E - E_g)^{1/2} \exp\left[-\frac{E}{kT}\right] \tag{4.23}$$

The result is shown graphically in Figure 4.11.

The result in Equation 4.23 is important for direct gap semiconductors but differs fundamentally from the recombination rate $R \propto np$ previously presented in Section 2.16. This is because in Section 2.16 we assumed that carriers scattered and did not maintain a specific value of k for recombination, which is particularly relevant in an indirect gap semiconductor such as silicon. The recombination events in a semiconductor such as silicon are more likely to be non-radiative than radiative and phonon interactions are involved. In this case the overall carrier concentrations n and p in the bands are important and the consideration of specific values of k used to derive Equation 4.23 are not relevant.

If we differentiate Equation 4.23 with respect to E and set $\dfrac{dR(E)}{dE} = 0$ the maximum is found to occur at $E = E_g + \dfrac{kT}{2}$. From this, we can evaluate the full width at half

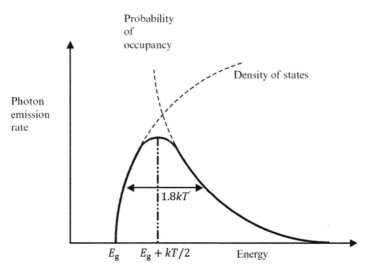

FIGURE 4.11   Photon emission rate as a function of energy for a direct gap transition of an LED. Note that at low energies the emission drops off due to the decrease in the density of states term $(E - E_g)^{\frac{1}{2}}$ and at high energies the emission drops off due to the Boltzmann term $\exp\left[-\dfrac{E}{kT}\right]$.

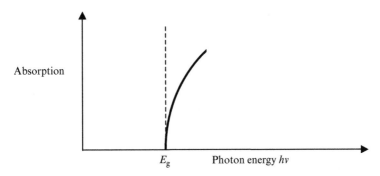

FIGURE 4.12   Absorption edge for direct-gap semiconductor.

maximum to be $1.8kT$. See Problem 4.15. This will be further discussed in Chapter 5 in the context of LEDs.

For a solar cell, the absorption coefficient α provides the rate of absorption of photons which is the converse of the rate of emission R. See Section 2.13. We consider the valence band to be fully occupied by electrons and the conduction band to be empty. In this case the absorption rate depends on the joint density of states function only and is independent of Fermi–Dirac statistics. Using Equation 4.20 we obtain

$$\alpha\left(h\nu\right)\alpha(h\nu - E_{g})^{\frac{1}{2}} \tag{4.24}$$

The absorption edge for a direct-gap semiconductor is illustrated in Figure 4.12.

This absorption edge is only valid for direct-gap semiconductors, and only when parabolic band-shapes are valid. If $h\nu \gg E_{g}$ this will not be the case and measured absorption coefficients will differ from this theory.

In an indirect gap semiconductor the absorption α increases more gradually with photon energy $h\nu$ until a direct-gap transition can occur. This is discussed in more detail in Chapter 5 in the context of solar cells.

4.9   PHOTOMETRIC UNITS

The most important applications of light emitting diodes (LEDs) to be described in Chapter 5 and organic light emitting diodes (OLEDs) to be described in Chapter 8 are visible illumination and displays. This requires the use of units to measure the brightness and color of light output. The power in watts and wavelength of emission are often not adequate descriptors of light emission. The human visual system has a variety of attributes that have given rise to more appropriate units and ways of measuring light output. This human visual system includes the eye, the optic nerve, and the brain, which interpret light in a unique way. Watts, for example, are considered *radiometric* units, and this section introduces *photometric* units and relates them to radiometric units.

Luminous intensity is a photometric quantity that represents the perceived brightness of an optical source by the human eye. The unit of luminous intensity is the *candela* (cd). One cd is the luminous intensity of a source that emits 1/683 watt of light at 555 nm into a solid angle of one steradian. The candle was the inspiration for this unit, and a candle does produce a luminous intensity of approximately one candela.

Luminous flux is another photometric unit that represents the light power of a source. The unit of luminous flux is the *lumen* (lm). A candle that produces a luminous intensity of 1 cd in all directions produces 4π lumens of light power. If the source is spherically symmetrical then there are 4π steradians in a sphere, and a luminous flux of 1 lm is emitted per steradian.

A third quantity, *luminance*, refers to the luminous intensity of a source divided by an area through which the source light is being emitted; it has units of cd m^{-2}. In the case of an LED *die* or semiconductor chip light source the luminance depends on the size of the die. The smaller the die that can achieve a specified luminous intensity, the higher the luminance of this die.

The advantage of these units is that they directly relate to perceived brightnesses, whereas radiation measured in watts may be visible, or invisible depending on the emission spectrum. Photometric units of luminous intensity, luminous flux, and luminance take into account the relative sensitivity of the human vision system to the specific light spectrum associated with a given light source.

The eye sensitivity function is well known for the average human eye. Figure 4.13 shows the perceived brightness for the human visual system of a light source that emits a constant optical power that is independent of wavelength. The left scale has a maximum of 1 and is referenced to the peak of the human eye response at 555 nm. The right scale is in units of *luminous efficacy* (lm W^{-1}), which reaches a maximum of 683 lm W^{-1} at 555 nm. Using Figure 4.13, luminous intensity can now be determined for other wavelengths of light.

An important measure of the overall efficiency of a light source can be obtained using luminous efficacy from Figure 4.13. A hypothetical monochromatic electroluminescent light source emitting at 555 nm that consumes 1 W of electrical power and produces 683 lm has an electrical-to-optical conversion efficiency of 100%. A hypothetical monochromatic light source emitting at 410 nm that consumes 1 W of electrical power and produces approximately 5 lm also has a conversion efficiency of 100%. The luminous flux of a blue LED or a red LED that consumes 1 W of electric power may be lower than for a green LED; however, this does not necessarily mean that the conversion efficiency is lower.

Luminous efficiency values for a number of light sources may be described in units of lm W^{-1}, or light power divided by electrical input power. Table 4.3 adds the luminous efficiency values relevant to Table 4.1. Luminous efficiency can never exceed luminous efficacy for a light source at a given wavelength.

The perceived color of a light source is determined by its spectrum. The human visual system and the brain create our perception of color. For example, we often perceive a mixture of red and green light as yellow even though none of the photons

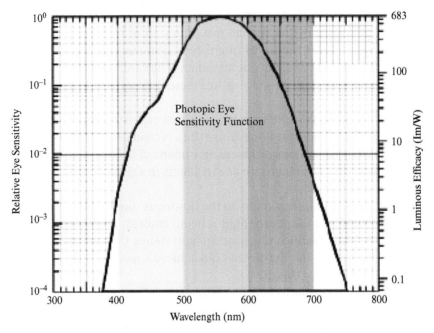

FIGURE 4.13   The eye sensitivity function. The left scale is referenced to the peak of the human eye response at 555 nm. The right scale is in units of luminous efficacy. International Commission on Illumination (Commission Internationale de l'Eclairage, or CIE), 1931.

TABLE 4.3   Luminous efficiency values (in lm W^{-1}) for a variety of light emitters.

Blackbody radiation	Sun
(light generated due to the temperature of a body)	Tungsten filament lamp: $\eta \cong 5\%$ or 15–20 lm W^{-1}
Photoluminescence	Fluorescent lamp phosphors ($\eta \cong 80\%$)
(light emitted by a material that is stimulated by electromagnetic radiation)	The fluorescent lamp achieves 50–80 lm W^{-1} and this includes the generation efficiency of UV radiation in a gas discharge
Electroluminescence	Light emitting diode ($\eta \cong 20 - 50\%$). For a 20% efficient LED this translates into the following approximate values:
(light emitted by a material that is directly electrically excited)	
	red LED at 625 nm 40 lm W^{-1}
	green LED at 530 nm 120 lm W^{-1}
	blue LED at 470 nm 12 lm W^{-1}

arriving at our eyes is yellow. For a LED, both the peak and the full width at half-maximum (FWHM) of the emission spectrum determine the color. More complex spectra due to phosphor-converted LEDs introduced in Chapter 5 can produce perceived colors including white from more than one spectral emission peak.

The human eye contains light receptors on the retina that are sensitive in fairly broad bands centered at the red, the green, and the blue parts of the visible spectrum. Color is determined by the relative stimulation of these receptors. For example, a light source consisting of a combination of red and green light excites the red and green receptors, as does a pure yellow light source, and we therefore perceive both light sources as yellow in color.

Since the colors we observe are perceptions of the human visual system, a color space has been developed and formalized that allows all the colors we recognize to be represented on a two-dimensional graph called the *color space chromaticity diagram* shown in Figure 4.14. The diagram was created by the International Commission on Illumination (Commission Internationale de l'Éclairage, or CIE) in 1931, and is therefore often referred to as the CIE diagram. CIE *x* and *y color coordinates* are shown that can be used to specify the color point of any light source. The outer boundary of this color space represents monochromatic light having single wavelengths. Monochromatic light sources are considered fully *saturated* colors. As we move to the center of the diagram to approach white, the light source becomes increasingly less monochromatic and less saturated. Hence a source having a spectrum of a finite spectral width will be situated some distance inside the boundary of the color space.

If two light sources emit light at two distinct wavelengths anywhere on the CIE diagram and these light sources are combined into a single light beam, the human eye will interpret the color of the light beam as existing on a straight line connecting the

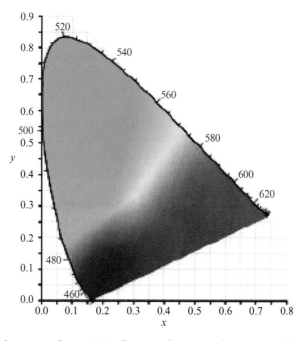

FIGURE 4.14 Color space chromaticity diagram showing colors perceptible to the human eye. Numbers on the boundaries are wavelengths in nanometers. Color saturation or purity is maximum at the boundaries and it decreases toward the center of the diagram, eventually becoming white. International Commission on Illumination (Commission Internationale de l'Eclairage, or CIE), 1931.

locations of the two sources on the CIE diagram. The position on the straight line of this new color will depend on the relative radiation power from each of the two light sources. See Problem 4.20.

If three light sources emit light at three distinct wavelengths that are anywhere on the CIE diagram and these light sources are combined into a single light beam, the human eye will interpret the color of the light beam as existing within a triangular region of the CIE diagram having vertices at each of the three sources. The position within the triangle of this new color will depend on the relative radiation power from each of the three light sources. This ability to produce a large number of colors of light from only three light sources forms the basis for *trichromatic* illumination. Lamps and displays routinely take advantage of this principle. It is clear that the biggest triangle will be available if highly saturated red, green, and blue light sources are selected to define the vertices of the color triangle. This color triangle is often referred to as a *color space* or *color gamut* that is enabled by a specific set of three light emitters.

4.10   SUMMARY

4.1   Luminescence is created by accelerating charges, and examples of luminescence include blackbody radiation, photoluminescence, cathodoluminescence, and electroluminescence.

4.2   Accelerating charges emit energy through the Poynting vector, which carries electric and magnetic field energy. The total power radiated from an accelerating charge is found by integrating the radiated power over a sphere and is found to be proportional to the square of the charge and to the square of the acceleration.

4.3   The simplest mode of charge acceleration is the oscillating dipole radiator in which a charge oscillates sinusoidally. The average power radiated by the dipole is calculated by performing a time average of the instantaneous power.

4.4   Stationary quantum states do not radiate whereas superposition states may radiate through dipole radiation. The expected value of the amplitude of the oscillation of the charge is determined by $\langle \phi_n | r | \phi_{n'} \rangle$, which may be calculated. If $\langle \phi_n | r | \phi_{n'} \rangle = 0$, then radiation will not occur and the transition is forbidden.

4.5   The exciton is formed by a hole and an electron that form a hydrogen-like entity. Excitons in semiconductors give rise to absorption or emission lines that are observable at low temperatures. These lines exist inside the energy gap of the semiconductor. Excitons may either be free to travel through the semiconductor or they might be bound to a defect.

4.6   The two-electron atom involves the consideration of indistinguishable electrons known as the exchange interaction. The wavefunctions for the two-electron atom describe either symmetric or antisymmetric states. The resulting states are known as singlet states in which the electrons are relatively close to each other; dipole radiation is associated with singlet states and not triplet states.

4.7   The molecular exciton comprises an electron and a hole that exist within one molecule. The exciton is bound within the molecule. The molecular exciton can be understood based on the two-electron atom and a singlet or a triplet exciton can be achieved. Singlet molecular excitons involve relatively fast dipole-allowed photon absorption or emission (fluorescence); however, triplet molecular excitons that may be formed by the electrical excitation of molecules decay relatively slowly (phosphorescence).

4.8 Band-to-band transitions in a direct-gap semiconductor produce a range of wavelengths depending on the position in the band. The distribution of electrons and holes in a band as a function of momentum may be determined by the density of states in the bands and the probabilities of occupancy of these states in the bands allowing the radiation spectrum of such a transition to be determined. In addition the absorption spectrum in a direct gap semiconductor can be determined.

4.9 The human eye perceives visible light in conjunction with the human brain, and a set of photometric units has been developed that allows our perception of brightness and color to be quantified. Units of luminance in candelas per meter squared and color coordinates (x,y) in a two-dimensional space form the basis for these units.

PROBLEMS

4.1 Consider an electron that oscillates in a dipole radiator and produces a radiation power of 1×10^{-12} watts. Find the amplitude of the oscillation of the electron if the radiation is monochromatic and at the following wavelengths (i) 470 nm (blue) (ii) 530 nm (green) (iii) 630 nm (red).
(a) Find the time required to produce one photon from the radiator for each wavelength.
(b) Find the number of photons per second required to produce an optical power of 1 watt for each wavelength.
(c) Find the number of photons per second required to produce a luminous flux of one lumen for each wavelength.
(d) Determine the luminous efficacy and the color coordinates of light sources that emit at each wavelength.

4.2 Find the electric field $\varepsilon_\perp$ generated a distance of 100 nm from an electron that accelerates at 10,000 m s^{-2}. Plot your result as a function of the angle between the acceleration vector and the line joining the electron to the point at which the electric field is measured. Repeat for the magnetic field $B_\perp$. Find the magnitude using appropriate units and the direction of the Poynting vector for the resulting traveling wave.

4.3 Find the total power radiated from the charge of Problem 4.2.

4.4 Find the total average power radiated from an electron that oscillates at:
(a) A frequency of 10^{14} Hz and amplitude of oscillation of 0.2 nm.
(b) A frequency of 10^{15} Hz and amplitude of oscillation of 0.2 nm.
(c) A frequency of 10^{15} Hz and amplitude of oscillation of 0.5 nm.

(d) In what part of the electromagnetic spectrum will the radiation be for (a) and (b)? Find the wavelength in free space for the radiation of (a) and (b).

4.5 For each of (a), (b), and (c) in Problem 4.4 find the approximate expected length of time needed to produce one photon. What is the photon energy? What is the photon emission rate in photons per second? How many oscillations of the electron are required to produce one phonon for each case?

4.6 Consider the one-dimensional infinite wall potential well of Example 2.1. Assume $L = 7\text{Å}$. Consider the following three cases:
(i) A single electron occupies a quantum state with $n = 3$ and no other states are occupied. Consider the possibility of a dipole transition for the electron to a state with $n = 1$.
(ii) A single electron occupies a quantum state with $n = 3$ and no other states are occupied. Consider the possibility of a dipole transition for the electron to a state with $n = 2$.
(iii) A single electron occupies a quantum state with $n = 4$ and no other states are occupied. Consider the possibility of a dipole transition for the electron to a state with $n = 1$.

For each of these cases:
(a) Is this a dipole-allowed transition?
   Hint: Calculate the matrix element by performing the relevant integral.
(b) If relevant, find the wavelength of emission and estimate the spontaneous transition time. Assume $(ab)^2 \cong 1$.
(c) Assign an appropriate description to each case. Choices are "Strongly allowed dipole transition," "Weakly allowed dipole transition," and "Dipole-forbidden transition."

4.7 Find the energy of an exciton that is formed from an electron and a hole in gallium arsenide using the appropriate effective masses and relative dielectric constant. Repeat for gallium nitride and cadmium sulphide.

4.8 Find the radius of an exciton in its ground state for each of the semiconductors in Problem 4.7.

4.9 Using a computer, plot the photon emission rate $R(E)$ as a function of energy for a band-to-band recombination event in a direct-gap semiconductor at room temperature. Cover an energy range that extends $10kT$ below and $10kT$ above the bandgap. Assume a bandgap of 2 eV.

4.10 Silicon, being an indirect gap semiconductor, is very inefficient as an emission source for band-to-band radiation. Nevertheless if highly purified and extremely low defect density silicon is prepared then carrier lifetimes can become long enough for radiative recombination to compete effectively with non-radiative emission. Search for information and prepare a short (2–3 page) report on the state of the art on this topic. Some aspects of this will touch on LEDs, which are covered in Chapter 5. Keywords to consider: radiative emission in silicon; the silicon light emitting diode.

4.11 The color space defined by three light emitters to form a trichromatic system in television is an important specification that contributes to the quality of the display. Find the color coordinates of red, green, and blue emitters that are commercially accepted standards for television in both North America and in Europe and plot these on a CIE diagram. Show the correct name for each standard. Show the triangular trichromatic color space on the CIE diagram for each standard Keywords to consider: CIE diagram; RGB color coordinates; Adobe RGB; sRGB, Rec. 2020.

4.12 The color coordinates of displays for portable electronics such as laptop computers generally provide smaller color spaces than for television. Battery power is a critical limitation on the light sources used for the display and maximum display brightness is desired. Explain why a reduction in color space is helpful with reference to the eye sensitivity function and the CIE diagram. See if you can obtain the color spaces used in portable electronics. Keywords to consider: reduced color coordinates; portable electronics; laptop displays.

4.13 Luminance of light sources and displays varies according to the application. For the following, find, using on-line resources, the luminance levels in units of cd m^{-2} that have become standard in the industry:
(a) Movie screen in movie theater
(b) Laptop screen display
(c) Desktop computer monitor
(d) Cell-phone display
(e) Television
(f) Drive-through restaurant menu display
(g) Large outdoor electronic billboard

4.14 The light output from small area sources that approximate a directed point source, such as an inorganic light emitting diode, is specified in terms of a plot of luminous intensity as a function of angle of emission θ, zero degrees corresponding to the central axis of the emission cone. The luminous intensity of the device is generally quoted along the central axis of the emission cone.

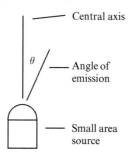

(a) Find a manufacturer's data sheet of the light spread from a high-efficiency red LED light source specified as a 30 degree device and plot the luminous intensity as a function of angle of emission using units of candelas. How is the 30 degree angle range defined? Your plot should cover the angle range from $\theta = -50$ to $\theta = +50°$. The value on your graph at $\theta = 0$ should correspond to the LED's quoted luminous intensity.

(b) Use the plot from (a) and integrate the total light output from the red LED to obtain the luminous flux in units of lumens. This luminous flux represents the total light output from the LED.

Hint: The emission pattern is circular and you must use a spherical-polar coordinate system to perform this integral correctly. The area under the plot of luminous intensity as a function of angle is NOT the correct answer.

(c) Refer to the test conditions used to obtain the quoted luminous intensity for the LED of (a). Using typical values of voltage and current quoted by the manufacturer, calculate the electrical power in watts used by the device. Now divide the result of (b) by the electrical power in watts to obtain the LED efficiency. This efficiency is quoted in lumens per watt.

(d) Obtain the wavelength of emission of the red LED. Refer to Figure 4.13 and determine the luminous efficacy of the LED. Note that luminous efficacy is not the same as luminous efficiency. Divide the luminous efficiency of (c) by the luminous efficacy. This unit-less quantity is the power efficiency of the LED and is a measure of the fraction of electrical input power that gets converted to light. By way of reference, high-efficiency red LEDs can achieve a power efficiency of approximately 50%. See Chapter 5.

(e) Repeat (a) to (d) for high-efficiency green and blue LEDs.

4.15 Differentiate Equation 4.23 with respect to E and set $\dfrac{dR(E)}{dE} = 0$. Show that the maximum will occur at $E = E_g + \dfrac{kT}{2}$. From this, show that the full width at half maximum (FWHM) of the LED emission spectrum is $\cong 1.8kT$.

4.16 Integrate

$$P = \frac{1}{16\pi\epsilon_0} \frac{2q^2 a^2}{c^3} \int_0^\pi \sin^3\theta\, d\theta$$

to obtain Equation 4.1.

4.17 Show that both the symmetric wavefunction ψ_s and the antisymmetric wavefunction ψ_a of Equations 4.12 and 4.13, respectively, will yield probability density functions that are not affected by the labeling of the two electrons.

4.18 Explain why the ground ($n = 1$) state of an atom containing two electrons can only exist as a singlet state and the triplet state cannot occur.

Hint: Consider the options available for both the spatial and the spin portions of the wavefunction to ensure that labeling of the two electrons is consistent with the requirement that the electrons are indistinguishable.

4.19 Find an expression for ψ_s using Equations 4.12, 4.10a, and 4.11a. Show that for two electrons in the same quantum state, ψ_s does not become zero as required by the Pauli Exclusion Principle.

4.20 Consider a red light source having color coordinates $x = 0.65$, $y = 0.35$ and a green source having color coordinates $x = 0.4$, $y = 0.6$.

(a) If the two light sources are combined on a screen such that the screen luminance due to the red source is 50 cd m^{-2} and the screen luminance due to the green source is 50 cd m^{-2}, find the resulting color coordinates of the combined light at the screen.

Hint: Use Figure 4.13 to determine the ratio r of the radiation power from each source. Plot the two color coordinates on the color space chromaticity diagram, and determine the color coordinates of the resulting color by dividing the line into two parts with lengths of this ratio r.

(b) Repeat (a), but now add a third blue light source having color coordinates $x = 0.15$, $y = 0.1$. This blue source provides screen luminance 20 cd m^{-2}.

NOTES

1. Section 1.9 derived the ground state of a hydrogen atom and mentioned, but did not derive, higher excited state eigenfunctions and eigenenergies. A principal quantum number n is defined as $n = 1$ for the ground state and excited states are assigned quantum numbers $n = 2,3,4....$ The associated eigenenergies are $\dfrac{-13.6}{n^2}$ eV. The Rydberg constant of 13.6 eV is obtained if $n = 1$.

2. Proper expression of spin requires the *spinor* which is a two-element column matrix. See Griffiths in Suggestions for Further Reading. We will continue to use simplified notation.

SUGGESTIONS FOR FURTHER READING

1. Ashcroft NW and Mermin ND. (1976). *Solid State Physics*. Holt: Rinehart and Winston.
2. Eisberg R and Resnick R. (1985). *Quantum Physics of Atoms, Molecules, Solids, Nuclei and Particles*, 2e. Wiley.
3. Kittel C. (2005). *Introduction to Solid State Physics*, 8e. Wiley.
4. Schubert EF. (2006). *Light Emitting Diodes*, 2e. Cambridge University Press.
5. Griffiths D and Schroeter D. (2018). *Introduction to Quantum Mechanics*. 3e.

CHAPTER 5

Semiconductor Devices Based on the p-n Junction

CONTENTS

Fundamentals of Semiconductor Materials and Devices, First Edition. Adrian Kitai.
© 2023 John Wiley & Sons Ltd. Published 2023 by John Wiley & Sons Ltd.
Companion Website: www.wiley.com/go/kitai_fundamentals

Objectives

1. Understand basic solar cell p-n junction design and its application to the PV industry.
2. Understand the solar spectrum as well as light absorption in semiconductors.
3. Introduce the model for the p-n junction when functioning as a solar cell.
4. Introduce parameters that define the optimum solar cell operating point.
5. Understand basic LED device operation and LED device structures based on the p-n junction.
6. Understand the LED emission spectrum and emission linewidth based on direct gap recombination.
7. Review the sources of non-radiative recombination that must be recognized and minimized in effective LED design.
8. Introduce and model optical outcoupling concepts that determine the efficiency with which light can emerge from LED devices.
9. Describe GaAsP LED technology and performance levels.
10. Describe AlGaAs LED technology and introduce the double heterostructure design and performance levels.
11. Describe AlGaInP LED technology and performance levels.
12. Describe GaInN LED technology and performance levels.
13. Introduce the principle of down-conversion of LED light to create white-emitting LEDs.
14. Motivate the BJT concept using the photodiode and describe the rational for two closely spaced p-n junctions.
15. Model base diffusion to derive a current gain model for the BJT.
16. Describe JFET operation using the p-n junction depletion concept in both a linear region and a saturation region.
17. Model the linear region of JFET operation.
18. Describe uses of the BJT and the JFET.

5.1 INTRODUCTION

The p-n junction forms the basis of semiconductor devices that have had dramatic impacts on society. The bipolar junction transistor was the first commercialized transistor that emerged in the middle of the twentieth century to replace vacuum tubes. This ushered in an electronics revolution that impacted communications, computing, digital entertainment and many other sectors.

More recently the optical properties of the p-n junction are playing an ever-increasing role in the mitigation of global warming. Electric lighting has transitioned over the past century from tungsten filament lamps to fluorescent lamps and now to light emitting diodes. This progression in technology has improved the conversion efficiency from electrical energy to visible light by an order of magnitude.

Our most abundant form of sustainable energy, sunlight, is now being harvested by p-n junction solar cells that currently offer a dramatically lower cost per installed kilowatt generation capacity than generators using hydrocarbon fuels or nuclear fuels. Annual worldwide solar cell capacity is increasing steadily. For example in the year 2022 alone, additional solar cell power capacity installed was 175 GW which is equivalent to 175 traditional 1 GW coal-fired power stations.

A range of key applications of the p-n junction will be introduced in this chapter.

5.2   THE P-N JUNCTION SOLAR CELL

We will first study the basic operation of inorganic p-n junctions specifically designed and optimized for solar cells. The focus will be on silicon which is the most important PV material. The physics of the p-n junction solar cell is common to a wide range of semiconductor materials, however. For example, organic solar cells are described in Chapter 8.

The solar cell functions as a forward-biased p-n junction; however, current flow occurs in a direction opposite to a normal forward-biased p-n diode. This is illustrated in Figure 5.1. Light that enters the p-n junction and reaches the depletion region of the solar cell generates electron-hole pairs. The generated carriers will *drift* across the depletion region and enter the n- and p-regions as majority carriers as shown. It is also possible for electron-hole pairs to be generated within about one diffusion length on either side of the depletion regions and through diffusion, to reach the depletion region where drift will again allow these carriers to cross to the opposite side. It is crucial to minimize carrier recombination, allowing carriers to deliver as much as

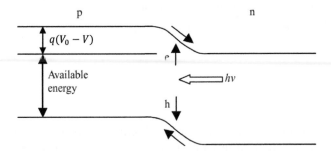

FIGURE 5.1   Band diagram of a solar cell showing the directions of carrier flow. Generated electron-hole pairs drift across the depletion region. After becoming majority carriers, current will flow through an external circuit producing power.

possible of the available energy to the external circuit. This means that the carriers must cross the depletion region and become majority carriers on the opposite side of the junction. If they are generated but then recombine before they are collected, they will not contribute to the flow of current.

The available energy may also be optimized by minimizing the potential barrier $q(V_0 - V)$ that is required to facilitate carrier drift across the depletion region. The magnitude of $q(V_0 - V)$ is subtracted from the semiconductor bandgap, which reduces the available energy difference between electrons and holes traveling in the n-type and p-type semiconductors respectively. This causes a reduction in the operating voltage of the solar cell. If $q(V_0 - V)$ is too small, however, carriers will not be effectively swept across the depletion region making them more susceptible to recombination.

A p-n junction exhibits current–voltage behavior as in Figure 3.7. If the p-n junction is illuminated in the junction region then in reverse bias the reverse current increases substantially due to the electron-hole pairs that are optically generated. Without optical generation, the available electrons and holes that comprise reverse saturation current are thermally generated minority carriers, which are low in concentration.

In forward bias the reverse drift current still flows but it is usually *smaller* than diffusion current. In a solar cell, however, the optically generated current is *larger* than diffusion current and it continues to dominate current flow until stronger forward bias conditions are present. The *I–V* characteristic of a solar cell is shown in Figure 5.2. The appropriate operating point for a solar cell is shown in which current flows out the positive terminal of the p-n junction (p-side), through the external

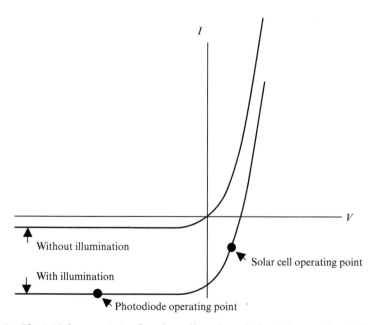

FIGURE 5.2 The *I–V* characteristic of a solar cell or photodiode without and with illumination. The increase in reverse current occurs due to optically generated electron-hole pairs that are swept across the depletion region to become majority carriers on either side of the diode.

circuit, and then into the negative (n-side) terminal. At this operating point the current flow is dominated by optically generated depletion region carrier drift and not by majority carrier diffusion.

A photodiode is a light detector that operates in reverse bias, as shown in Figure 5.2. In this case, current flow is in the same direction as for solar cells, but energy is consumed rather than generated because they are reverse-biased. Photodiodes are closely related to solar cells despite their energy-consuming mode of operation and are important as optical detectors used in applications such as infrared remote controls and optical communications.

5.3 LIGHT ABSORPTION

In order to efficiently generate electron-hole pairs in a direct gap solar cell, light must reach the junction area and be absorbed effectively. Total energy will be conserved during the absorption process. Photons have energy $E = h\nu$, which must be at least as large as the semiconductor bandgap.

Total momentum will also be conserved. Photon momentum $p_{\text{photon}} = \dfrac{h}{\lambda}$ is very small compared to electron and hole momentum values in typical semiconductors. From Figure 2.8, for example, the band states range in momentum $p = \hbar k$ from $p = 0$ to $p = \pm\dfrac{\hbar\pi}{a}$. Since the lattice constant a is generally in the range of a few angstroms whereas visible light has wavelengths λ in the range of 5000 angstroms, it is clear that p_{photon} is much smaller than the possible electron and hole momentum values in a band. The absorption of a given photon energy therefore proceeds as an almost vertical transition in k-space, as illustrated in Figure 5.3.

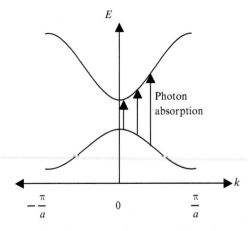

FIGURE 5.3 The absorption of a photon in a direct-gap semiconductor proceeds in an almost vertical line since the photon momentum is very small on the scale of the band diagram. There are many possible vertical lines that may represent electron-hole generation by photon absorption as shown. These may exceed the bandgap energy.

Photon absorption for an ideal direct-gap semiconductor was derived in Equation 4.24, and therefore

$$\alpha(hv) \approx A(hv - E_g)^{1/2}$$

where A is a constant that depends on the material. Examples of direct-gap semiconductors used in solar cells are shown in Figure 2.14. These include the III-V semiconductor gallium arsenide and the II-VI semiconductor cadmium telluride.

In indirect gap semiconductors, the absorption of a photon having energy $hv \approx E_g$ would appear to be forbidden due to the requirement of momentum conservation, illustrated in Figure 5.4 and discussed in Section 2.13. Absorption is possible, however, if phonons are available to supply the necessary momentum. Typical phonons in crystalline materials can transfer large values of instantaneous momentum to an electron since atomic mass is much higher than electron mass. Phonon energies are small, however. At temperature T the average phonon energy will be of the order of kT, or only 0.026 eV at room temperature. The absorption process involving a phonon is a two-step process, as shown in Figure 5.4, along with a single-step absorption for higher energy photons. The result is a low but steadily increasing absorption coefficient as photon energy increases above E_g followed by a much steeper increase in absorption once photon energies are high enough for direct absorption.

The wavelength dependence of absorption coefficients for a number of important semiconductors is shown in Figure 5.5. Note the long absorption tails for silicon and germanium which are indirect gap materials. The other semiconductors are direct-gap materials that exhibit much sharper absorption edges with higher absorption coefficients compared to silicon.

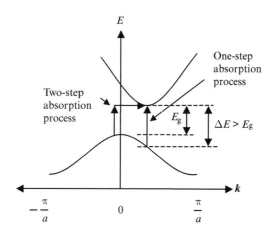

FIGURE 5.4 Indirect gap semiconductor showing that absorption near the energy gap is only possible if a process involving phonon momentum is available. The indirect, two-step absorption process involves a phonon to supply the momentum shift that is necessary to absorb photons. For higher energy photons a direct one-step absorption process is possible.

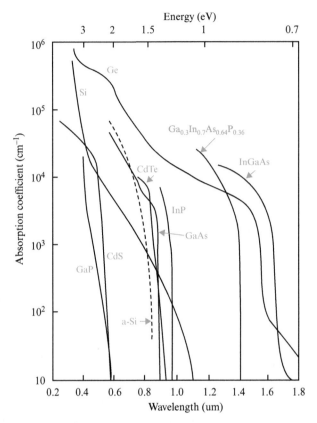

FIGURE 5.5   Absorption coefficients covering the solar spectral range for a range of semiconductors. Note the absorption tails in silicon and germanium arising from two-step absorption processes. Amorphous silicon is a non-crystalline thin film that has different electron states and hence different absorption coefficients compared with single-crystal silicon. After M. Shur, Physics of semiconductor devices. Prentice Hall 1990.

Although silicon is an indirect gap semiconductor, it is much lower in cost than other semiconductors and it therefore dominates the solar cell industry. Silicon solar cells are approximately two orders of magnitude thicker ($100 - 200\,\mu$m) than the semiconductor thickness in direct-gap semiconductor solar cells ($1 - 2\,\mu$m) to compensate for silicon's lower absorption coefficient.

5.4   SOLAR RADIATION

Sunlight is caused by blackbody radiation from the outer layer of the sun. At a temperature of approximately 5250°C, this layer emits a spectrum as shown in Figure 5.6, which represents the solar spectrum in space and is relevant to solar cells used on satellites and space stations. Terrestrial solar cells, however, rely on the terrestrial solar spectrum, which suffers substantial attenuation at certain wavelengths. In particular, water molecules absorb strongly in four infrared bands between 800 and 2000 nm as shown.

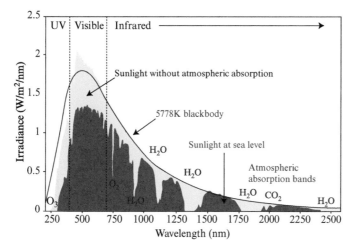

FIGURE 5.6   Solar radiation spectrum for a 5250°C blackbody, which approximates the space spectrum of the sun, as well as a spectrum at the earth's surface that survives the absorption of molecules such as H_2O and CO_2 in the earth's atmosphere. Note also the substantial ozone (O_3) absorption in the UV part of the spectrum. Robert A. Rohde / Wikimedia / CC BY 3.0.

5.5   SOLAR CELL DESIGN AND ANALYSIS

The design of a practical silicon solar cell can now be considered. In order for light to reach the junction area of the p-n junction, the junction should be close to the surface of the semiconductor. The junction area must be large enough to capture the desired radiation. This dictates a thin n- or p-region on the illuminated side of the solar cell. A significant challenge is to enable the thin region to be sufficiently uniform in potential to allow the junction to function over its entire area. If a contact material is applied to the surface of the cell, sunlight will be partly absorbed in the contact material. The common solution to this is to make the thin region of the silicon as conductive as possible by doping it heavily. In this way, the highly doped thin region simultaneously serves as a front electrode with high lateral conductivity (conductivity in the plane of the junction) *and* as one side of the p-n junction. Since n-type silicon has higher electron mobility and therefore higher conductivity than is achieved by the lower mobility of holes in p-type material, the thin top layer in silicon solar cells is frequently n-type in practice.

A crystalline silicon solar cell is shown in Figure 5.7. It consists of a thin n^+ front layer. A metal grid (usually silver) is deposited on this layer and it forms an ohmic contact to the n^+ material. The areas on the n^+ front layer that are exposed to sunlight are coated with an antireflection layer. The simplest such layer is a quarter wavelength in thickness such that incident light waves reflecting off the front and back surfaces of this layer can substantially cancel each other (see Problem 5.2). The metal

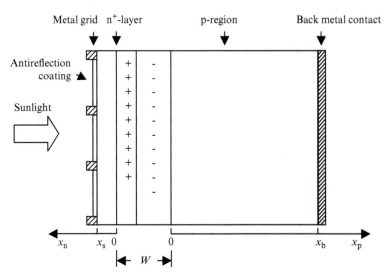

FIGURE 5.7   Cross-section of a silicon solar cell showing the front contact metal grid that forms an ohmic contact to the n^+-layer. The depletion region at the junction has width W.

grid does block some sunlight; however, in practice the grid lines are narrow enough to prevent excessive light loss. A thick p-type region absorbs virtually all the remaining sunlight and is contacted by a rear metal ohmic contact.

Because most of the photons are absorbed in the thick p-type layer, the minority carriers that need to be collected are predominantly electrons. The goal is to have these electrons reach the front contact. There will also be a small contribution from minority holes generated in the thin n^+ region that reach the p-region.

Sunlight entering the solar cell will be absorbed according to the relationship introduced in Section 2.13

$$I(x)=I_0 e^{-\alpha x}$$

In order to simplify the treatment of the solar cell we will firstly assume that the optical generation rate G is uniform throughout the p-n junction. This implies that the absorption constant α is very small. Real silicon solar cells are approximately consistent with this assumption only for photons of longer wavelengths of sunlight very close in energy to the bandgap. Shorter wavelengths, however, should really be modeled as providing an exponentially decaying generation rate with depth.

Secondly, we will assume that the relevant diffusion lengths of minority carriers in both the n-type and p-type regions are much shorter than the thicknesses of these regions. This means that the p-n junction may be regarded as possessing semi-infinite thickness as far as excess minority carrier distributions are concerned, and in Figure 5.7 the front surface and back surface at $x_n = x_s$ and $x_p = x_b$ respectively are far away from regions influenced by the junction. The surface recombination velocity of Section 2.20 is included in more complete solar cell modeling.

For the n-side, the diffusion equation (Equation 2.66a) for holes may be rewritten as

$$D_p \frac{d^2 \delta p_n (x_n)}{dx_n^2} = \frac{\delta p_n (x_n)}{\tau_p} - G \tag{5.1}$$

The term G must be subtracted from the hole recombination rate $\frac{\delta p_n (x_n)}{\tau_p}$ because G is a hole generation rate. The solution to this equation where

$$L_p = \sqrt{D_p \tau_p}$$

is

$$\delta p_n (x_n) = G\tau_p + C \exp\left(\frac{x_n}{L_p}\right) + D \exp\left(\frac{-x_n}{L_p}\right) \tag{5.2}$$

Note that the solution is the same as Equation 2.67a except for the added term $G\tau_p$. See Problem 5.3.

The boundary conditions we shall satisfy are:

$$\delta p_n (0) = p_n \left(\exp\frac{qV}{kT} - 1\right)$$

and

$$\delta p_n (x_n \rightarrow \infty) = G\tau_p$$

The first boundary condition is as discussed in Section 3.5. The dynamic equilibrium in the depletion region still determines the carrier concentrations at the depletion region boundaries. For the second boundary condition $G\tau_p$ is the excess carrier concentration optically generated far away from the depletion region (see Equation 2.53).

Substituting these two boundary conditions into Equation 5.2 we obtain (see Problem 5.3)

$$\delta p_n (x_n) = G\tau_p + \left[p_n \left(\exp\frac{qV}{kT} - 1\right) - G\tau_p\right] \exp\left(\frac{-x_n}{L_p}\right) \tag{5.3a}$$

The analogous equation for the p-side is

$$\delta n_p (x_p) = G\tau_n + \left[n_p \left(\exp\frac{qV}{kT} - 1\right) - G\tau_n\right] \exp\left(\frac{-x_p}{L_n}\right) \tag{5.3b}$$

Note that if $V = 0$, Equations 5.3a and 5.3b yield $\delta p_n (x_n) = G\tau_p$ and $\delta n_p (x_p) = G\tau_n$ for large values of x_n and x_p respectively. In addition at $V = 0$ these equations give zero for both $\delta p_n(x_n = 0)$ and $\delta n_p(x_p = 0)$. We can show this more clearly in Figure 5.8 for an illuminated p-n junction under short circuit conditions ($V = 0$). We use Equations

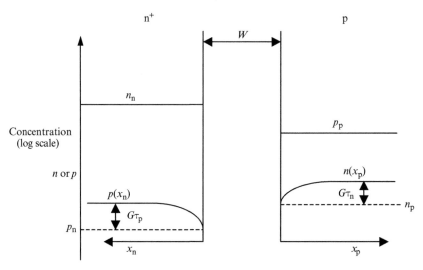

FIGURE 5.8 Concentrations are plotted on a log scale to allow details of both the minority carrier concentrations and the majority carrier concentrations to be shown on the same plot. Note that the n⁺-side has higher majority carrier concentration and lower equilibrium minority carrier concentration than the more lightly doped p-side corresponding to Figure 5.7. Diffusion lengths are assumed to be much smaller than device dimensions. The p-n junction is shown in a short circuit condition with $V = 0$.

5.3a and 5.3b to plot $p_n(x_n) = \delta p_n(x_n) + p_n$ and $n_p(x_p) = \delta n_p(x_p) + n_p$, where p_n and n_p are the equilibrium minority carrier concentrations.

Having determined the minority carrier concentrations, we can now determine diffusion currents $I_n(x)$ and $I_p(x)$ using the equations for diffusion currents (Equations 2.56). By substitution of Equations 5.3a and 5.3b into Equations 2.56 we obtain

$$I_p(x_n) = \frac{qAD_p}{L_p} p_n \left[\exp\left(\frac{qV}{kT}\right) - 1\right] \exp\left(-\frac{x_n}{L_p}\right) - qAGL_p \exp\left(-\frac{x_n}{L_p}\right) \tag{5.4a}$$

for holes diffusing in the n-side and

$$I_n(x_p) = -\frac{qAD_n}{L_n} n_p \left[\exp\left(\frac{qV}{kT}\right) - 1\right] \exp\left(-\frac{x_p}{L_n}\right) + qAGL_n \exp\left(-\frac{x_p}{L_n}\right) \tag{5.4b}$$

for electrons diffusing in the p-side (see Problem 5.4). Note that the first term on the RHS in these equations are identical to Equations 3.22b and 3.22a for an unilluminated diode.

Evaluating Equations 5.4 at the edges of the depletion region, the two currents that enter the depletion region and contribute to the total solar cell current are given by

$$I_p(0) = \frac{qAD_p}{L_p} p_n \left[\exp\left(\frac{qV}{kT}\right) - 1\right] - qAGL_p \tag{5.5a}$$

and

$$I_n(0) = -\frac{qAD_n}{L_n} n_p \left[\exp\left(\frac{qV}{kT}\right) - 1 \right] + qAGL_n \tag{5.5b}$$

Since there is uniform illumination, we need also to consider generation within the depletion region. We shall neglect recombination of electron-hole pairs here and assume that drift is sufficient to transport all carriers to either edge of the depletion region. This means that every electron and every hole created in the depletion region contributes to diode current. The generation rate G must be multiplied by depletion region volume WA to obtain the total number of carriers generated per unit time in the depletion region. Carrier current optically generated from inside the depletion region therefore becomes the total charge generated per unit time or

$$I_{(\text{depletion})} = qGWA \tag{5.6}$$

Although both an electron and a hole are generated by each absorbed photon, each charge pair is only counted once: one generated electron drifts to the n-side metal contact, flows through the external circuit, and returns to the p-side. In the meantime, one hole drifts to the p-side metal contact and is available there to recombine with the returning electron. Solar cell junction area is A.

It now follows that using Equations 5.5a, 5.5b, and 5.6, the total solar cell current I, defined by convention as a current flowing from the p-side to the n-side, is given by the sum of three currents

$$I = I_p(0) - I_n(0) - I_{\text{depletion}}$$

$$= \frac{qAD_p}{L_n} p_n \left[\exp\left(\frac{qV}{kT}\right) - 1 \right] - qAGL_p + \frac{qAD_n}{L_n} n_p \left[\exp\left(\frac{qV}{kT}\right) - 1 \right] - qAGL_n - qAGW$$

No negative sign is required in front of $I_p(0)$ because the x_n axis is already reversed. Using Equation 3.23b the total diode current I may be written

$$I = I_0 \left[\exp\left(\frac{qV}{kT}\right) - 1 \right] - I_L \tag{5.7a}$$

where I_L, the current optically generated by sunlight, is

$$I_L = qAG(L_n + W + L_p) \tag{5.7b}$$

This confirms that Figure 5.2 is valid and the $I-V$ characteristic based on the diode equation is shifted vertically by amount I_L upon illumination. Of the three terms in Equation 5.7b, the second term is generally smallest due to small values of W compared to carrier diffusion lengths in silicon solar cells. It is reasonable that diffusion lengths L_n and L_p appear in Equation 5.7b: minority carriers must cross over the depletion region to contribute to solar cell output current. They have an opportunity to diffuse to the depletion region before they drift across it, and the diffusion lengths are the appropriate length scales over which this is likely to occur.

The solar cell *short circuit current* I_{SC} can now be seen to be the same as I_L by setting $V = 0$ in Equation 5.7a. Hence

$$I_{SC} = qAG(L_n + W + L_p) \tag{5.8}$$

In addition, the solar cell *open circuit voltage* V_{OC} can be found by setting $I = 0$ in Equation 5.7a and solving for V to obtain

$$V_{OC} = \frac{kT}{q}\ln\left(\frac{I_{SC}}{I_0} + 1\right) \tag{5.9}$$

These quantities are plotted in Figure 5.9 together with the solar cell operating point. The fill factor *FF* is defined as

$$FF = \frac{I_{MP}V_{MP}}{I_{SC}V_{OC}} \tag{5.10}$$

In crystalline silicon solar cells *FF* may be in the range of 0.7–0.85.

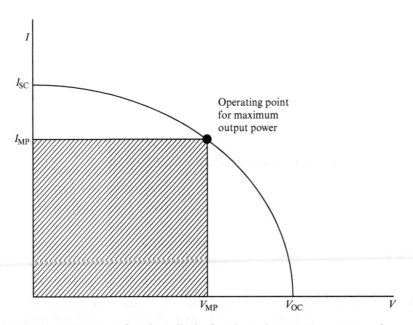

FIGURE 5.9 Operating point of a solar cell. The fourth quadrant in Figure 5.2 is redrawn as a first quadrant for convenience. Open circuit voltage V_{oc} and short circuit current I_{oc} as well as current I_{MP} and voltage V_{MP} for maximum power are shown. Maximum power is obtained when the area of the shaded rectangle is maximized.

EXAMPLE 5.1

An abrupt silicon p-n junction solar cell at room temperature is exposed to sunlight. Assume that the sunlight is uniformly intense throughout the silicon yielding an optical generation rate of 5×10^{19} cm^{-3} s^{-1}. The solar cell has a junction area of 100 cm^2, a depletion region width of 3 μm and reverse saturation current density of $J_0 = 1 \times 10^{-11}$ A cm^{-2}. The silicon has a carrier lifetime of 2×10^{-6} s.

a) Find the optically generated current that is generated inside the depletion region.
b) Find the total optically generated current.
c) Find the short-circuit current.
d) Find the open-circuit voltage.
e) If the solar cell fill factor is 0.75, find the maximum power available and discuss this in terms of the expected total available sunlight.

Solution

a)
$$I_{(\text{depletion})} = qGWA = 1.6 \times 10^{-19} \text{C} \times 5 \times 10^{19} \text{cm}^{-3}\text{s}^{-1} \times 3 \times 10^{-4} \text{cm}$$
$$\times 100 \, \text{cm}^2 = 0.24 \, \text{A}$$

b) From Example 3.3,

$$L_n = \sqrt{D_n \tau_n} = \sqrt{3.51 \times 10^1 \, \text{cm}^2 \, \text{s}^{-1} \times 2 \times 10^{-6} \, \text{s}} = 8.38 \times 10^{-3} \, \text{cm}$$

and

$$L_p = \sqrt{D_p \tau_p} = \sqrt{1.25 \times 10^1 \, \text{cm}^2 \, \text{s}^{-1} \times 2 \times 10^{-6} \, \text{s}} = 5.00 \times 10^{-3} \, \text{cm}$$

and therefore

$$I_L = qAG(L_n + W + L_p)$$
$$= 1.6 \times 10^{-19} \text{C} \times 100 \, \text{cm}^2 \times 5 \times 10^{19} \, \text{cm}^{-3} \, \text{s}^{-1} (8.38 \times 10^{-3} + 3 \times 10^{-4}$$
$$+ 5.00 \times 10^{-3} \text{cm}) - 10.9 \, \text{A}$$

c) The short circuit current I_{SC} is the same as I_L. Therefore $I_{SC} = 10.9$ A.
d) Open-circuit voltage:

$$I_0 = J_0 A = 1 \times 10^{-11} \text{A cm}^{-2} \times 100 \, \text{cm}^2 = 1 \times 10^{-9} \text{A}$$

and now

$$V_{OC} = \frac{kT}{q} \ln\left(\frac{I_L}{I_0} + 1\right) = 0.026 \, \text{V} \ln\left(\frac{10.9 \, \text{A}}{1 \times 10^{-9} \text{A}} + 1\right) = 0.601 \, \text{V}$$

(continued)

(continued)

e) The maximum output power is obtained at the operating point V_{MP} and I_{MP}. Therefore

$$I_{MP}V_{MP} = FF \times I_{SC}V_{OC} = 0.75 \times 10.9\,\text{A} \times 0.601\,\text{V} = 4.91\,\text{W}$$

The available sunlight per square meter for full sun conditions on the earth's surface is approximately 1000 W, which yields 10 W over an area of 100 cm^2. The best silicon solar cell, however, is not more than about 25% efficient and therefore the most electrical power that we could expect to be available from the solar cell should be closer to 2.5 W.

In practice, achieving an electrical output power of 4.91 W from a silicon solar cell of 100 cm^2 would require a concentration of sunlight by a factor of approximately two using reflective or refractive optical concentrators. Concentrating sunlight onto solar cells can generate more power per cell, however this approach is not widely employed. One key disadvantage is the need to re-align the cell optics in real time to track the changing directions of the sun.

5.6   SOLAR CELL EFFICIENCY LIMITS AND TANDEM CELLS

Single junction solar cells are limited to under 30% conversion efficiency given the range of photon energies according to the solar spectrum. The primary considerations that give rise to this limit are the following:

(i) A p-n junction cannot harvest photons having energies below the bandgap of the semiconductor. This means that a portion of the solar spectrum will not be absorbed. To minimize this issue, the bandgap of the semiconductor should be small.

(ii) When photons having energy larger than the semiconductor bandgap are absorbed, this excess energy is lost to thermalization. Generated electrons and holes initially populate energetic band states in the conduction and valence bands, respectively. These carriers rapidly lose energy to approach the lowest available energy levels in their respective bands that are near the band edges. This energy loss produces heat and is therefore called thermalization. To minimize thermalization losses the bandgap of the semiconductor should be large.

Given the opposing bandgap requirements of i) and ii), the bandgap of single junction solar cell semiconductors is a compromise and it lies between 1 eV and 1.5 eV. Production silicon cells are up to 25% efficient. The physical limit in efficiency is captured by a more detailed analysis called the *Shockley-Queisser limit*. See Problem 5.7.

Other solar cell types exist that serve specific market needs. In space applications, III-V semiconductor-based solar cells are justifiable in spite of their high costs.

They offer a wide range of available band gaps. Thickess of the light absorbing layer is minimized in the case of direct gap materials. Improvements in conversion efficiency can be achieved by layering III-V p-n junctions such that sunlight enters a first p-n junction with a high bandgap and only short wavelength photons are harvested producing a higher output voltage for these photons. Then lower energy photons that are not absorbed pass through into an underlying, second p-n junction formed in a lower bandgap semiconductor that produces a lower output voltage. Additional junctions with progressively lower bandgaps may follow. These *multijunction* or *tandem* solar cells can provide conversion efficiencies approaching 50%. See Figure 5.10 and Problem 5.8.

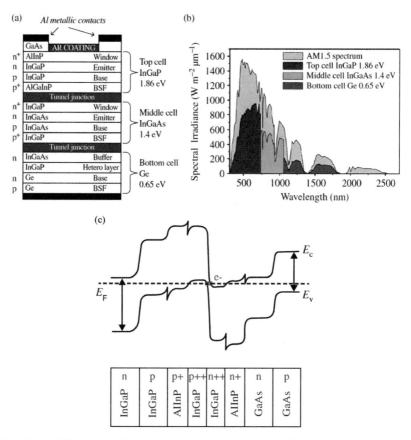

FIGURE 5.10 Multijunction solar cell (a) showing the portions of the solar spectrum that are covered by each material (b) Wavelengths up to 1800 nm are absorbed. In this example, germanium is used as both the supporting substrate and the third semiconductor. The III-V junctions are epitaxially grown on the germanium substrate. Tunnel junctions are formed between the cells to ensure that current produced by each cell can flow to the adjacent cell. The three cell voltages add together; however, the overall current is controlled by the minimum current that is supplied by each of the cells. This means that current matching between multiple cells in a multijunction cell is required. In (c) an example of the energy band diagram for an InGaP tunnel junction between an InGaP cell and a GaAs cell is shown. (a,b) after N.V.Yastrebova, High-efficiency multijunction solar cells: current status and future potential (2007). (c) Wikimedia Commons.

5.7   THE LIGHT EMITTING DIODE

The inorganic light emitting diode (LED) has shown remarkable development since the early 1960s when the first practical red emitting devices were introduced to the market. This early start gave the LED a place in portable electronics such as calculator and watch displays. The advantage of a very long life enabled market penetration.

Following this start, the development of new LED semiconductor materials and new device structures resulted in a wider range of colors as well as increased efficiency. The earliest generations of LEDs were red and infrared only. During the decades following the 1960s, bandgaps gradually increased in stages allowing intense red, orange and yellow LEDs to be mass produced by the 1990s. During this decade, the breakthrough discovery of a material for bright blue and saturated green LEDs occurred. By combining a blue LED with a yellow-emitting phosphor material, bright white-emitting LEDs for illumination purposes paved the way for LEDs to significantly outperform incandescent and fluorescent lamps.

Today, efficient LED emission, not only in the visible wavelength range, but also in the UV and IR ranges is available. It should be noted that UV and IR emitting devices are still commonly called LEDs. These devices find application in IR remote controls and as UV light sources for industrial processes such as for promoting chemical reactions that are triggered by a UV light source.

Inorganic p-n junction LEDs and p-n junction solar cells are very different in terms of junction area. LEDs are typically diced into individual chips or *die* that have junction areas of under 10 mm^2 whereas silicon solar cell wafers have junction areas of approximately 10,000 mm^2. The selling price per unit junction area for production LEDs is about 1000 times higher than the selling price per unit junction area of widely used solar cells. This means that more costly materials and processing steps are commonly employed for LED production than for solar cell production.

Figure 5.11 shows a historical plot of the efficiencies and colors achieved by LEDs over the past decades. Of significance is the achievement of LED efficiency values that exceed other traditional lamp technologies such as incandescent lamps and fluorescent lamps.

The simplest LED p-n junction is illustrated in Figure 5.12. Light is emitted when electrons and holes that flow as majority carriers are injected and recombine at or near the junction causing radiative recombination to take place. This is a diffusion-driven process at the junction as opposed to the drift mechanism in solar cells. Radiative recombination is favored in direct-gap semiconductors and as a result, the materials chosen for most practical LEDs are direct gap. There are exceptions to this provided that some mechanism exists to overcome the lack of radiative recombination in indirect gap material. See Section 5.12.

LED device structures vary; however, a common structure of a small, lensed LED showing both the packaging as well as the semiconductor die is illustrated in Figure 5.13. Since light is emitted from the sides as well as the front of the die, a reflective cup is used to allow the light to propagate forward and through a molded polymer lens.

Figure 5.14 shows the internal structure of three examples of LED die. A single crystal substrate is used to support the active layers. A thick layer above the active layers spreads the current from the top contact over the junction area. The top contact is made

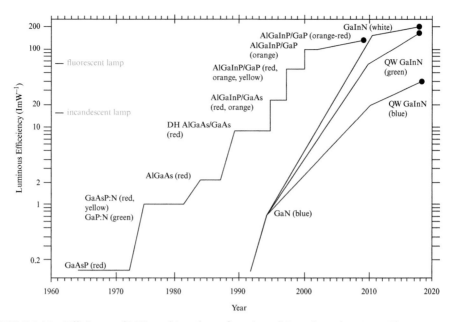

FIGURE 5.11 Efficiency of LEDs achieved as a function of time since the 1960s. The photometric quantity lmW^{-1} (lumens per watt) was introduced in Chapter 4.

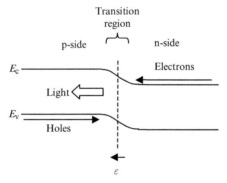

FIGURE 5.12 Forward-biased LED p-n junction. When one electron and one hole recombine near the junction one photon of light may be emitted. The achievement of high efficiency requires that there is a good chance that recombination events are radiative and that the generated photons are not reabsorbed or trapped in the device.

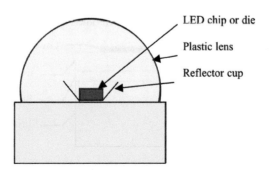

FIGURE 5.13 LED packaging may include a transparent lens which is made from an epoxy or a silicone material, and a reflector cup into which the LED die is mounted. The radiation pattern is determined by the combination of the die emission pattern, the reflector cup design, and the shape and refractive index of the lens.

(a)

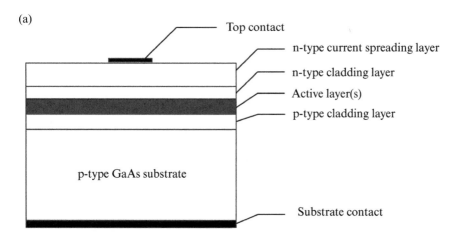

Top contact

n-type current spreading layer

n-type cladding layer

Active layer(s)

p-type cladding layer

p-type GaAs substrate

Substrate contact

(b)

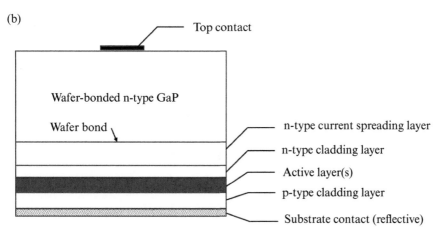

Top contact

Wafer-bonded n-type GaP

Wafer bond

n-type current spreading layer

n-type cladding layer

Active layer(s)

p-type cladding layer

Substrate contact (reflective)

(c)

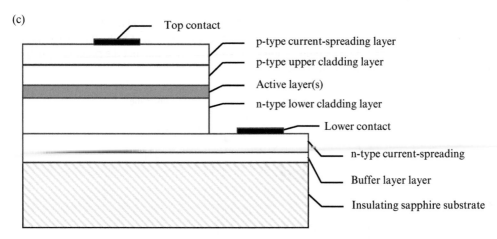

Top contact

p-type current-spreading layer

p-type upper cladding layer

Active layer(s)

n-type lower cladding layer

Lower contact

n-type current-spreading

Buffer layer layer

Insulating sapphire substrate

FIGURE 5.14 Examples of LED die are shown. (a) A series of layers are epitaxially grown on a p-type GaAs substrate. (b) The light absorbing GaAs substrate of (a) is removed and the active layers are wafer-bonded to a transparent n-type GaP substrate to prevent light absorption and improve outcoupling efficiency. (c) For GaInN LEDs, an insulating sapphire (Al_2O_3) substrate is widely used. This requires the use of a current spreading layer for the lower contact.

as small as possible to avoid blocking generated light leaving through the top of the die. Additional information on these structures will be presented in the following sections.

5.8 EMISSION SPECTRUM

Consider a direct-gap semiconductor having approximately parabolic conduction and valence bands near the bottom and top of the bands respectively. The photon emission rate $R(E)$ relevant to an LED was derived in Chapter 4, and the result from Equation 4.23 is

$$R(E) = R(hv) \propto (E - E_g)^{\frac{1}{2}} \exp\left[-\frac{E}{kT}\right] \qquad (5.11)$$

From this, the intensity profile of an LED may be plotted as a function of wavelength λ where $E = \frac{hc}{\lambda}$. The result is shown graphically as a function of energy in Figure 4.11. The same data can be replotted on an intensity versus wavelength graph. Figure 5.15 shows measured emission spectra from commercially available red and amber LEDs.

EXAMPLE 5.2

Obtain the full width at half maximum (FWHM) for the amber LED of Figure 5.15 and compare with the theoretical value. Assume room temperature.

Solution

From Section 4.8, the full width at half maximum (FWHM) for an LED is found to be $\Delta E = 1.8kT$ (see Figure 4.11). For the amber LED of Figure 5.15 the graph gives us $\Delta \lambda = 13.5$ nm.

At the lower FWHM point, reading the graph, $\lambda_{lower} = 584$ nm.
At the upper FWHM point, reading the graph, $\lambda_{upper} = 597$ nm.
Now,

$$E_{lower} = \frac{hc}{\lambda_{lower}} = \frac{(6.63 \times 10^{-34}\,\text{Js})(3 \times 10^8\,\text{ms}^{-1})}{584 \times 10^{-9}\,\text{m}}$$

$$= (3.40 \times 10^{-19}\,\text{J})\left(\frac{1}{1.6 \times 10^{-19}}\frac{\text{eV}}{\text{J}}\right) = 2.13\,\text{eV}$$

and

$$E_{upper} = \frac{hc}{\lambda_{upper}} = \frac{(6.63 \times 10^{-34}\,\text{Js})(3 \times 10^8\,\text{ms}^{-1})}{597 \times 10^{-9}\,\text{m}}$$

$$= (3.33 \times 10^{-19}\,\text{J})\left(\frac{1}{1.6 \times 10^{-19}}\frac{\text{eV}}{\text{J}}\right) = 2.08\,\text{eV}$$

Hence, $\Delta E = 2.13 - 2.08 = 0.05$ eV.
    From Figure 4.11 at 300 K:
    Emission linewidth $= 1.8\,kT = 1.8 \times 0.026 = 0.047$ eV in reasonable agreement with the measured result.

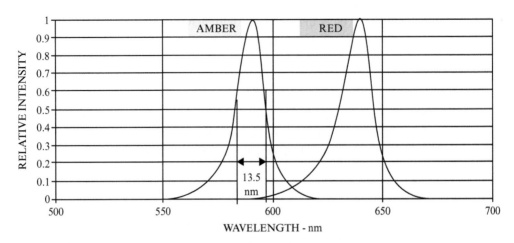

FIGURE 5.15 Emission spectra of AlGaInP LEDs. The linewidth of the amber LED is measured as the full width at half maximum as shown and is 13.5 nm. Reproduced after data from Avago Technologies.

Other broadening mechanisms exist for LED emission including inhomogeneous broadening in which non-uniformities in LED material modify the semiconductor locally. This is particularly prevalent for alloy semiconductors where small deviations in composition are possible. Emission intensity, peak emission wavelength, and line-width may vary from die to die even if they are from the same semiconductor wafer. Improvements in as-grown uniformity are a key goal of the LED industry.

For this reason, LEDs intended for applications where device-to-device uniformity is important are *binned*, or sorted into batches based on measured values of emission color and luminous intensity. Units used to specify visible light emitting LEDs include luminous intensity and color coordinates which were introduced in Section 4.9.

Ultimate LED efficiency is limited by both *non-radiative recombination* and *optical outcoupling* losses. The next two sections discuss these in more detail.

5.9 NON-RADIATIVE RECOMBINATION

Non-radiative electron-hole pair recombination can occur through a variety of mechanisms. These include traps and surface states, which were discussed in Section 2.20. Due to the high electron mobility and high diffusivity in materials such as GaAs, surface recombination velocities are high. One solution is to ensure that carrier recombination occurs at distances well over one diffusion length from surfaces. This can be achieved by making contacts to the p-n junction that are smaller in area than the semiconductor die dimensions. Within a few carrier diffusion lengths of a surface the probability of radiative recombination events falls off rapidly.

In small LEDs, however, another approach is needed. Surface passivation coatings can be used to minimize surface recombination. Dangling bonds are terminated by the coating which can reduce or eliminate surface traps. Surface passivation is further discussed in Chapter 7.

Additional traps can occur due to native defects such as interstitials and vacancies, and in compound semiconductors *anti-sites* can result in energy levels within the forbidden gap. Anti-sites refer to the substitution of the cation in the place of the anion or the anion in the place of the cation in a compound III-V or II-VI semiconductor. Particularly if the levels are near mid-gap they become deep traps and cause non-radiative recombination.

In addition, as the operating temperature of the LED rises, thermal energy promotes electrons to higher energy states in the conduction band, which has the effect of giving a nominally direct-gap semiconductor some indirect-gap character. This can occur since electrons can be thermally promoted to occupy a secondary valley in the conduction band of direct gap semiconductors. These electrons are then forced to recombine indirectly. See band diagrams in Figure 2.17 for a variety of semiconductors, including GaAs, that have a primary direct-gap conduction band valley as well as a secondary indirect gap valley. The smaller the energy barrier between the primary and secondary valleys, the easier it becomes to thermally excite electrons between them.

Another important mechanism for non-radiative recombination is *Auger recombination*. An electron near the bottom of the conduction band can be promoted to a high-energy state within the conduction band due to the recombination of a nearby electron with a hole. Rather than the recombination event producing a photon, the EHP recombines and conserves both energy and momentum by kicking this additional electron to the high energy state followed by lattice heating as the promoted electron subsequently thermalizes. An Auger process involving a hole is also possible.

Auger processes become important especially at high rates of recombination. This is because either two electrons or two holes that are closely spaced are required for this process and the chance of having the carrier pair available to take part in the Auger process is proportional to the square of the carrier concentration. Auger processes are further discussed in the context of quantum dots in Chapter 7.

Achieving a balance between electron and hole currents is also important for high-efficiency LEDs. The total current flow in a p-n junction diode can arise from various ratios of holes and electrons crossing the junction; however, a similar number of electrons and holes arriving near the junction per second is needed to maximize recombination efficiency in an LED. This requires that the doping levels on either side of the LED junction be carefully controlled.

5.10 OPTICAL OUTCOUPLING

The issue of optical outcoupling plays a key role in LED performance. Generated photons can be reabsorbed in LEDs, and they may be reflected back into the device by surface or interface reflections.

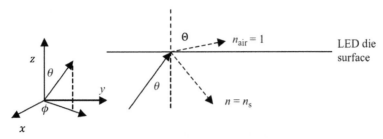

FIGURE 5.16   Light generated in the semiconductor will reach the surface and either reflect or be able to exit depending on the angle of incidence. The critical angle is θ_c.

Due to total internal reflection, only a fraction of the light generated in an LED die will be emitted. Light crossing an interface between two regions of refractive indices n_1 and n_2 must obey Snell's law,

$$\frac{\sin \theta_1}{\sin \theta_2} = \frac{n_2}{n_1}$$

In Figure 5.16 the critical angle for light generated in the semiconductor and reaching an air interface occurs when the angle in the air is $\frac{\pi}{2}$. If θ_c is the critical angle in the semiconductor then $n_s \sin\theta_c = 1$ where n_s is the index of refraction of the semiconductor.

Since θ_c is small for practical semiconductors we can often use the approximation $\sin\theta_c \cong \theta_c$, and now the critical angle occurs when

$$\theta_c \cong \frac{1}{n_s} \tag{5.12}$$

Using the spherical polar coordinate system in Figure 5.16, an *escape cone* for the light reaching the interface results because all values of ϕ need to be considered. The portion of the surface area of a sphere of radius r that corresponds to radial directions that lie within the escape cone is given by

$$A = \int dA = \int_0^{2\pi} \int_0^{\theta_c} r^2 \sin\theta \, d\theta d\phi = 2\pi r^2 (1 - \cos\theta_c) \tag{5.13}$$

The resulting outcoupling efficiency η_{out} is now the fraction of the sphere through which light may escape:

$$\eta_{out} = \frac{2\pi r^2 (1 - \cos\theta_c)}{4\pi r^2} = \frac{1 - \cos\theta_c}{2} \tag{5.14}$$

EXAMPLE 5.3

Calculate and compare the optical outcoupling at a GaAs die surface given the following parameters: index of refraction of GaAs $n_s = 3.4$; index of refraction of epoxy $n_e = 1.5$.

a) in air
b) in an epoxy encapsulation.

Solution

(a) In air:

critical angle in air

$$\theta_c = \frac{1}{n_s} = \frac{1}{3.4} = 0.29 \text{ radians or } 16.9°$$

$$\eta_{out} = \frac{1 - \cos\theta_c}{2} = \frac{1 - \cos(16.9°)}{2} = 0.021 \text{ or } 2.1\%$$

(b) In epoxy resin:

for critical angle θ_c in epoxy:

$$\frac{\sin 90°}{\sin\theta_c} = \frac{3.4}{1.5}$$

Solving, $\theta_c = 26.2°$ and

$$\eta_{out} = \frac{1 - \cos\theta_c}{2} = \frac{1 - \cos(26.2°)}{2} = 0.051 \text{ or } 5.1\%$$

It is interesting to examine the far-field dependence of LED intensity on exit angle of the light that does outcouple. Consider light emission from a single LED surface. If $\theta < \theta_c$ then light will be emitted at an angle Θ, as shown in Figure 5.16.

For light within angle range $d\theta$ the output angle range can be determined using Snell's law:

$$n_s \sin\theta = n_a \sin\Theta \cong n_s\theta$$

since θ is often a small angle and hence

$$\theta \cong \frac{n_a \sin\Theta}{n_s}$$

Now, differentiating with respect to Θ we obtain

$$\frac{d\theta}{d\Theta} = \frac{n_a}{n_s}\cos\Theta \tag{5.15}$$

or

$$d\Theta = \frac{n_s}{n_a}\frac{1}{\cos\Theta}d\theta$$

Normally light emission inside the semiconductor is angle independent. This means that light reaching the surface of the semiconductor has an intensity independent of arrival angle θ. If Θ was proportional to θ then the emission distribution of an LED die surface would also provide the same intensity independent of Θ. However due to the above equation, the light intensity being emitted into higher angle ranges decreases. This is because a specific light flux arriving over an incoming angle range $d\theta$ now emerges from the LED over a larger angle range $d\Theta \propto d\theta / \cos \Theta$ yielding an intensity versus angle dependence of $\cos \Theta$. Light intensity reaches zero as Θ approaches $90°$. This is referred to as a *lambertian* source in which luminous intensity varies as $\cos \Theta$. Note that since the area of the LED die surface subtended by a viewer at Θ also varies as $\cos \Theta$, the luminance of the die surface is independent of Θ. Paper reflecting ambient light is also a lambertian source.

In practice, the output characteristics of an LED are modified by refractive and reflective optics to suit particular applications. Figure 5.13 shows a transparent epoxy polymer lens injected around an LED chip, which helps to increase outcoupling. The polymer has a refractive index of approximately 1.5, which increases the critical angle compared to a die operating in air with refractive index of 1. The outside shape of the epoxy lens can be tailored to achieve a specific beam profile. In many LEDs the amount of light emitted from the sidewalls is greater than the light emitted from the top surface of the LED die. The reflector cup is used to reflect light emitted through the sides of the die toward the lens. Lensed LEDs with narrow emission cones are available.

The primary outcoupling efficiency from an LED die may be improved dramatically. For example if the side cuts of the die are angled then outcoupling can be improved. Light that internally reflects from the top face may reach the angled facets at an angle smaller than the critical angle and outcoupling can be enhanced. Surface texturing can also be used to reduce internal reflection losses. In addition, high refractive index encapsulants reduce losses. State-of-the-art commercialized LEDs achieve approximately 90% outcoupling efficiency.

5.11 GaAs

Understanding LED development from a historical perspective is a useful way to gain a comprehensive understanding of LED technology. The first type of LED made in the 1960s used GaAs semiconductor material. In Chapter 2 we introduced GaAs as a direct-gap semiconductor with an energy gap of 1.43 eV, which is in the infrared wavelength range. In order to create n-type and p-type GaAs, it is necessary to dope the GaAs with an impurity. Silicon, being a group IV element, is capable of doping GaAs n-type if it substitutes for a Ga atom and p-type if it substitutes for an As atom.

Liquid phase epitaxy (LPE) is a growth technique that has been widely used to grow GaAs and related semiconductors. In LPE, a liquid phase melt containing a solution of Ga and As atoms is prepared. The Ga concentration in the melt is much higher than the As concentration. In this way the melt consists of Ga solvent and As solute atoms. If such a melt is slowly cooled in contact with a single crystal GaAs substrate, a layer of GaAs will nucleate and grow on the substrate as the temperature drops causing the As to become supersaturated in the solution. The grown layer will be single crystal and will follow the crystallographic structure of the substrate.

Si atoms can also be incorporated into the melt of Ga and As. If the epitaxial layer grows at 900°C then the silicon atoms cause the GaAs to be n-type because they preferentially substitute for Ga atoms. If the epitaxial layer grows at 850°C the silicon atoms result in p-type GaAs because they preferentially substitute for As atoms. The resulting p-n junction behaves as illustrated in Figure 5.12 and a typical structure is shown in Figure 5.14a).

Since GaAs has a small energy gap the photons are infrared with a wavelength of 900–980 nm. GaAs LEDs continue to be manufactured in high volume. These devices are useful for applications such as remote controls and for short distance optical fibre communications.

5.12   GaP:N LEDS

Interest in visible light emission led to the incorporation of phosphorus into LEDs grown on GaAs substrates in the 1960s. From Figure 2.34a it is seen that as x increases in $GaAs_{1-x}P_x$ the bandgap increases. Visible emission starts near 700 nm, which corresponds to a bandgap of approximately 1.75 eV. Therefore x may be as low as about 0.25 for visible light emission.

The visible emission from these $GaAs_{1-x}P_x$ LEDs is not efficient, as shown in Figure 5.11, for a number of reasons:

(a) As the amount of phosphorus increases, the lattice constant also increases (see Figure 2.34a). The result is that high-quality epitaxial growth of $GaAs_{1-x}P_x$ on GaAs substrates cannot be achieved unless x is very small. At x values of 0.25 or more, lattice defects and dislocations are present in high concentrations, which lead to non-radiative carrier recombination.

(b) The human eye sensitivity function (luminous efficacy) is weak at long and short wavelengths near the edges of the visible spectrum (see Chapter 4). An efficient red wavelength for human vision is approximately 620 nm, which requires phosphorus concentrations that are high enough to produce a substantial dislocation density.

(c) $GaAs_{1-x}P_x$ undergoes a direct–indirect transition near $x = 0.5$, which places a further constraint on bandgap. Even for x values below 0.5, electrons have a chance of being thermally excited into the secondary valley of the conduction band and recombining non-radiatively through an indirect gap transition. This process becomes more serious as the LED junction temperature increases.

(d) A significant fraction of light generated at the junction will be reabsorbed by the GaAs substrate, which has a smaller bandgap than the $GaAs_{1-x}P_x$ light emitting material, as well as by $GaAs_{1-x}P_x$ material surrounding the junction.

An interesting and unexpected way to improve the performance of GaP LEDs is through nitrogen doping of the transition region of the p-n junction only. Nitrogen replaces phosphorus atoms in the lattice. Since both nitrogen and phosphorus are in the same column of the periodic table, the nitrogen doping is called an *isoelectronic* defect as it does not act as n-type or p-type doping, but instead it provides a recombination center inside the bandgap of the GaP material about 100–200 meV below the conduction band minimum. It does this by effectively trapping electrons before recombination occurs. The observed photons therefore have longer wavelengths than from material without nitrogen doping.

The performance improvement from nitrogen doping is explained by the Heisenberg uncertainty principle. This states that both momentum and position of a particle cannot be measured with certainty at the same time, and the relevant uncertainties obey the relationship $\Delta p \Delta x \geq \frac{\hbar}{2}$. See Chapter 1. The localization of the electron at the nitrogen impurity substantially reduces Δx compared to a conduction band electron. This results in a high degree of uncertainty in momentum and hence uncertainty in wavevector $\Delta k = \frac{\Delta p}{\hbar}$. The probability of a direct transition is therefore enhanced. In addition, nitrogen traps effectively compete with other non-radiative crystal defect traps in the lattice-mismatched material, which increases the probability of radiative transitions. At the same time, the decreased energy of the photons generated compared to the host material leads to less reabsorption of light in the host. By restricting nitrogen doping to the junction region, unnecessary traps levels that would contribute to reabsorption in surrounding GaP are avoided.

As a result, GaP:N LEDs have been in production since the 1970s and continue to be widely used for low-cost LEDs in indicator lamps, numerical readouts and display applications. Offsetting the relatively low efficiency of these LEDs (1–2 lm/W) is the low production cost and high yield. Yellowish-green-emitting GaP:N LEDs are popular since their emission near 530 nm is close in wavelength to the peak of the human eye sensitivity curve at 555 nm. See Figure 4.13.

GaP substrate material which is now routinely used for low defect density green-emitting GaP:N devices constitutes an ideal lattice-matched substrate in place of the light absorbing GaAs substrate shown in Figure 5.14a).

5.13   DOUBLE HETEROJUNCTION Al$_x$Ga$_{1-x}$As LEDS

Another way to increase the bandgap of GaAs is to partly substitute aluminum for gallium in GaAs as discussed in Section 2.21. Figure 2.34a shows the dependence of bandgap on x in Al$_x$Ga$_{1-x}$As. Of particular interest is the virtually invariant lattice constant for all values of x which allows high-quality epitaxial growth on GaAs substrates. The semiconductors are direct gap when $0 < x < 0.45$ and indirect for $0.45 < x < 1.0$. In practice, high efficiency requires that the chosen bandgap be several kT smaller than the critical bandgap of approximately 2 eV at the direct/indirect transition point of $x = 0.45$ to avoid populating the indirect gap valley. The method used for high-quality epitaxial growth is discussed in Section 5.14.

In practice, the highest *external quantum efficiency* may be obtained in this system for active layers with $x = 0.35$ and emission near 650 nm. Quantum efficiency refers to the probability of a photon being created per EHP recombination event and external

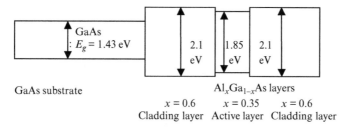

FIGURE 5.17   Band gaps of double heterojunction using Al$_x$Ga$_{1-x}$As layers grown epitaxially on a GaAs substrate. Carriers recombine in the active layer of width W. The cladding layers are doped such that one layer is n-type and one layer is p-type. Similar structures are used in GaInN LEDs. See Section 5.15.

quantum efficiency includes outcoupling efficiency, which further specifies that only the photons that outcouple are counted. 650 nm emission constitutes a deep red color which is not favorable for the eye sensitivity function shown in Figure 4.13. This limits the luminous efficiency achieved to be approximately 2 lm/W.

A further breakthrough in the performance of these devices in the 1980s was due to the *double heterojunction* device design. A diagram of the device structure is shown in Figure 5.17. Here, the aluminum content is lowered at the junction compared to the remainder of the n- and p-regions. This forms an active layer with a reduced band-gap in the form of an energy well that serves two important purposes:

(a) The active layer energy well captures and confines both electrons and holes in the same spatial region thereby improving the radiative recombination efficiency. Electrons and holes become more likely to encounter one another rather than recombining non-radiatively due to defects, traps, and surface states.

(b) Photons generated in the active layer will have a lower energy compared to the surrounding or *cladding* semiconductor material and will therefore not be reabsorbed. Since this material does not need to be direct gap, x values well over 0.45 are possible and $x > 0.6$ is common in cladding material.

The double heterojunction can be approximated as an infinite-walled potential box which was used in Section 2.10. The dimensions of the box include the cross-section area A and the width W shown in Figure 5.18. Although a double heterojunction does not have infinitely high walls, the infinite-walled model may be applied as an approximation. Conduction band and valence band wells of height ΔE_c and ΔE_v respectively, are illustrated.

In an operating double heterojunction LED we need to consider the flow of current and the rate at which carriers arrive at the conduction and valence band wells.

If an electron current density J flows under a steady state condition into the conduction band well as shown in Figure 5.18, the current adds to the electron population of the well. In one second the number of electrons added to the well will

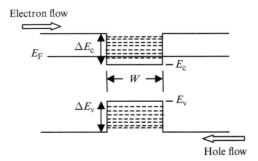

FIGURE 5.18   Electron and hole energy levels within the double heterojunction wells of height ΔE_c in the conduction band and ΔE_v in the valence band.

be $\frac{JA}{q}$. Since the volume of the well is AW the number of electrons added per second per unit volume in the well is $\frac{J}{Wq}$. This can be expressed in differential form as

$$\frac{dn}{dt} = \frac{J}{Wq} - R \tag{5.16}$$

where R is the recombination rate of the carriers in the well.

As discussed in Section 2.16, the recombination rate R in a direct gap semiconductor is proportional to product np. Silicon recombination times are primarily determined by trap density because silicon is an indirect gap semiconductor and generally a fixed recombination time is valid particularly for minority carriers where a large supply of majority carriers in available and low level injection is assumed.

For a forward-biased LED we cannot assume a fixed recombination time because both n and p can vary over orders of magnitude in the recombination region depending on drive current. High forward current densities of $100\ \mathrm{A\ cm^{-2}}$ or more are possible. The recombination rate relationship $R \propto np$ is expected and the recombination time is not fixed during injection; it decreases as the carrier concentration increases. In high-quality direct gap LED materials, trapping does not control recombination rates.

To model this we introduce a *recombination coefficient B* such that

$$R = Bnp \tag{5.17}$$

Values of B for direct-gap semiconductors are approximately $10^{-10}\ \mathrm{cm^3\,s^{-1}}$, whereas in indirect gap semiconductors such as GaP, Si, and Ge the values of B are three to four orders of magnitude smaller than this since the participation of a phonon is required for momentum conservation. Auger recombination is not included in this model.

Substituting Equation 5.17 into Equation 5.16, and assuming that the supply of holes and electrons is balanced ($n = p$), we obtain

$$n = \sqrt{\frac{J}{qBW}} \tag{5.18}$$

The number of available states in an infinite-walled well was derived in Chapter 2 and can be used to estimate how many electron states the well can accommodate. If we regard the bottom of the upper well as zero in energy and assume complete filling of all available states up to the top of the well, then we can identify ΔE_c as the highest energy level E in the well. Integrating Equation 2.33a over E with upper limit $E = \Delta E_c$ and using effective mass m_e^* we obtain

$$n = n(E) = \frac{\pi}{3}\left(\frac{2m_e^*\Delta E_c}{\hbar^2\pi^2}\right)^{\frac{3}{2}} \tag{5.19}$$

Equating 5.18 and 5.19, and solving for J, which now represents the maximum current density that can be accommodated before the well overflows, we obtain

$$J_{max} = qBW\left(\frac{\pi}{3}\right)^2\left(\frac{2m_e^*\Delta E_c}{\hbar^2\pi^2}\right)^3 \tag{5.20a}$$

and using a similar argument for the lower well we obtain

$$J_{max} = qBW\left(\frac{\pi}{3}\right)^2\left(\frac{2m_h^*\Delta E_v}{\hbar^2\pi^2}\right)^3 \tag{5.20b}$$

The well that overflows first effectively limits the current density.

This shows that the maximum current flow that can be accommodated in the well is proportional to the well width and the cube of the well height. For high-current-density LEDs a wide well is required. Typically several thousand amps per square centimeter can be accommodated in a double heterostructure LED.

EXAMPLE 5.4

Calculate the current density level at 300 K at which the electron well overflows in an Al$_x$Ga$_{1-x}$As double heterostructure with the following parameters:

Barrier height $\Delta E_c = 200$ meV
Well width $W = 200$ Å
Effective density of states $N_c = 5 \times 10^{17}$ cm^{-3}
Recombination coefficient $B = 2 \times 10^{-10}$ cm^3 s^{-1}

Solution

From Equation 2.36b:

$$N_c = 2\left(\frac{2\pi m_e^* kT}{h^2}\right)^{\frac{3}{2}}$$

Solving for m_e^* we obtain

(continued)

(continued)

$$m_e^* = \frac{h^2}{2\pi kT}\left(\frac{N_c}{2}\right)^{\frac{2}{3}} = \frac{(6.62\times10^{-34}\,\text{Js})^2}{2\pi\left(0.026\times1.6\times10^{-19}\,\text{J}\right)}\left(\frac{5\times10^{17}\,\text{cm}^{-3}\times10^6\,\text{cm}^3\text{m}^{-3}}{2}\right)^{\frac{2}{3}}$$

$$= 6.63\times10^{-32}\,\text{kg}$$

Now,

$$J_{max} = qBW\left(\frac{\pi}{3}\right)^2\left(\frac{2m_e^*\Delta E_c}{\hbar^2\pi^2}\right)^3$$

$$= 1.6\times10^{-19}\,\text{C}\times2\times10^{-10}\,\text{cm}^3\,\text{s}^{-1}\times10^{-6}\,\text{m}^3\text{cm}^{-3}\times200\times10^{-10}\,\text{m}$$

$$\times\left(\frac{\pi}{3}\right)^2\left[\frac{2\times6.63\times10^{-32}\,\text{kg}\times0.2\,\text{eV}\times1.6\times10^{-19}\,\text{J}\,\text{eV}^{-1}}{\left(1.05\times10^{-34}\,\text{Js}\right)^2\times\pi^2}\right]^3$$

$$= 4162\,\text{A}\,\text{cm}^{-2}$$

This shows that this double heterojunction functions well at high current densities. Auger processes may limit performance at high currents.

Since J_{max} decreases as W decreases, a very narrow double heterojunction well of width W under about 100 Å, which is referred to as a *quantum well*, will overflow at considerably lower current densities. The bonding evidence for this overflow is an observed saturation in quantum well LED brightness at a current density that is lower than that of the same LED but having a wider well. In a quantum well LED the number of available states in the well is small; however, more than one quantum well may be used in an LED. There are advantages to the use of quantum wells and this is discussed further in Section 5.15.

Still higher efficiencies are possible if the light absorbing GaAs substrate shown in Figure 5.14a) can be removed. A wide-bandgap foreign wafer such as GaP may be bonded to the device whereupon the original GaAs substrate may chemically or mechanically removed.

A technique known as *wafer bonding* is employed to achieve this. The two materials are bonded together by applying heat and pressure. Chemical bonds are formed between the top layer of the GaAs-based LED and a GaP wafer even though they are lattice-mismatched materials. Extreme care to maintain cleanliness, flatness and smoothness is essential for wafer bonding to be successful. Finally, the original GaAs substrate may be removed by polishing and/or etching. The transparent GaP substrate provides a stable support for the original epitaxial layers forming the double heterojunction before the GaAs substrate is removed. It also provides the electrical connection to the new top contact. See Figure 5.14b). Efficiencies close to 10 lmW^{-1} are achieved in practice. Wafer bonding is widely used in LED manufacturing.

One challenge in high-aluminum-content $Al_xGa_{1-x}As$ material is the tendency for aluminum to react with oxygen and moisture in the surrounding atmosphere. For this reason the lifetimes of these devices can be limited unless they are properly protected.

5.14 AlGaInP LEDS

The leading LED material system for high-brightness red, orange, and amber/yellow LEDs is the quaternary compound AlGaInP. Group III elements Al, Ga, and In are collectively combined with P in a 1:1 ratio. The compound family may be written as $(Al_xGa_{1-x})_yIn_{1-y}P$. This system was developed in the late 1980s and permits an additional degree of freedom in controlling bandgap and lattice constant compared to ternary alloys. By varying the composition, both the bandgap and the lattice constant may be varied independently. The lattice constant is maintained to match that of GaAs while the bandgap may be direct gap up to 2.33 eV. This has yielded orange LEDs with efficiencies of 100 lmW^{-1} by incorporating a double heterostructure. A more detailed picture of the composition range is given in Figure 5.19.

The preferred growth method for these quaternary materials is *metal-organic vapor phase epitaxy* (MOVPE), also referred to as *metal-organic chemical vapor*

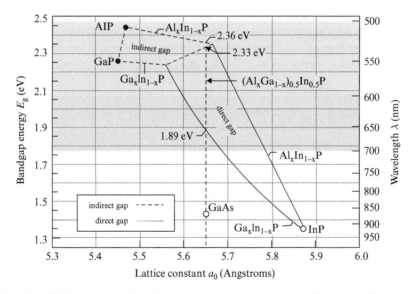

FIGURE 5.19 $(Al_xGa_{1-x})_yIn_{1-y}P$ bandgap versus lattice constant graph showing the composition ranges in this quaternary system. By adjusting the two available parameters, x and y, a field of compositions is possible represented by the labelled areas. A range of energy gaps from 1.89 eV to 2.33 eV is available in the direct bandgap region while matching the GaAs lattice constant. After C.H. Chen, S.A. Stockman, M.J. Peanasky, C.P. Kuo, OMVPE Growth of AlGaInP for High-Efficiency Visible Light-Emitting Diodes, Semiconductors and Semimetals, Volume 48, 1997.

deposition (MOCVD). Unlike the use of the liquid melt used in LPE growth, MOVPE growth is a gas-phase growth technique in which molecules containing the required semiconductor atoms are allowed to pass over a heated substrate in a gas stream usually combined with an inert gas such as argon. The molecules contain an organic group composed of elements hydrogen and carbon in combination with one of the desired Al, Ga, In, or P atoms (or the required dopant atom). The gas stream is composed of a well-controlled mixture of the desired molecules. By carefully establishing the ratios of the respective molecules in the gas stream, the desired semiconductor material is grown on the substrate because the substrate temperature is high enough to crack the molecules allowing the semiconductor atoms to deposit on the substrate while the remaining organic species flow away with the gas stream.

Advantages of MOCVD growth include the ability to grow ternary and quaternary semiconductors covering a wide composition range while simultaneously achieving a high degree of control of composition. During growth, composition changes may be made by changing the flow rates of specific molecular species.

Although the highest achievable bandgap of 2.33 eV in the AlGaInP system corresponds to green emission at 532 nm, efficient green emission is not available from this system because of the need to reduce bandgap several kT below the direct–indirect transition point. In addition, the bandgaps of this quaternary system are sensitive to ordering of the atoms. Figure 5.20 is for unordered material in which the Al, Ga, and In atoms are randomly positioned in the lattice. Unordered material is not always obtainable since there is a thermodynamic drive for atoms to order during growth to minimize energy. Atomic ordering can lower the band energy of the crystal and the degree of ordering is dependent on composition. It is experimentally found that when ordering occurs the effective bandgap drops by as much as

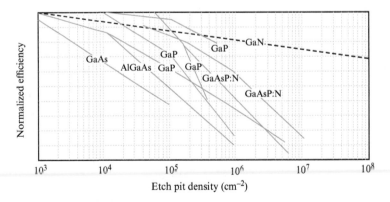

FIGURE 5.20   Radiative efficiency as a function of dislocation (etch pit) density for a variety of III-V semiconductors. Dislocation density is determined by etching the crystal surface and then counting the number of resulting etch pits per unit area. Etch pits form at the dislocations. After E. Fred Schubert, Light-Emitting Diodes, 2e, ISBN 978-0-521-86538-8, 2006.

190 meV. For these reasons yellow is the shortest practical emission wavelength for this system.

The efficiency peaks in the orange wavelength range since this represents the best compromise between the human eye response (luminous efficacy), which falls off as wavelength increases above 555 nm, and the recombination efficiency, which falls off as wavelength decreases and gets closer to the direct–indirect transition. The ability of the quaternary AlGaInP system to reach shorter wavelengths with high efficiency results in an order of magnitude improvement in luminous efficiency compared to the Al$_x$Ga$_{1-x}$As system.

As with AlGaAs LEDs the double heterostructure is used to improve the probability of radiative electron-hole recombination but in addition, larger values of ΔE_c and ΔE_v in the heterostructure can be achieved due to the bigger bandgap range available, which further improves carrier confinement even at higher operating temperatures.

For the highest efficiencies the removal of the GaAs growth substrate is required since GaAs absorbs a portion of the emitted light. Wafer bonding shown in Figure 5.14b) is employed as well as further outcoupling improvements as discussed in Section 5.10.

5.15 Ga$_{1-x}$In$_x$N LEDS

Since the early 1990s the ternary nitride materials system Ga$_{1-x}$In$_x$N capable of yielding high-efficiency LEDs covering the green and blue parts of the visible spectrum has been developed and widely commercialized using nitride semiconductors.

Prior to this breakthrough using nitrides, II-VI semiconductors were thought to be ideal candidates for green and blue LEDs due to their direct bandgaps. These materials, including direct-gap semiconductors ZnO, CdS and CdSe, did not demonstrate sufficiently stable performance partly due to the difficulty in incorporating p-type dopants. For the substitution of a group II metal (Zn or Cd) by a p-type dopant the choices consist of group I metals such as sodium or other singly ionized transition metals such as copper. Unfortunately, singly ionized atoms are only weakly bonded in II-VI crystals and they are therefore mobile and hence unstable.

The development of GaN-based LEDs was limited prior to the 1990s. The perceived challenges included the following:

(a) GaN is not lattice matched with readily available single crystal substrates.
(b) The growth conditions for crystalline GaN by MOCVD require high-temperatures of 900°C or more whereas other non-nitride III-V LEDs are grown at temperatures between 400 and 600°C. This further limits the range of suitable substrates.
(c) Doping to achieve p-type material is required and had not been successfully demonstrated and understood.

GaN is a direct-gap semiconductor closely related to the family of other III-V semiconductors used for LEDs, although it has the wurtzite crystal structure (see Sections 2.11

and 2.12). GaInN has now proved to be the most viable approach to large, direct band-gap materials for LEDs. The challenges listed above have been overcome, allowing dramatic new results to be obtained, and $Ga_{1-x}In_xN$ alloy semiconductors are now established as being of primary importance to LED technology. See Figure 2.34b.

Using an MOCVD growth method, smooth, high-quality films of GaN may be grown on substrates that are not lattice matched with GaN. The most important substrates are sapphire and SiC, which are mismatched by approximately 12% and 3% respectively. The achievement of high-performance GaN material despite a poor substrate lattice match has been the subject of much investigation. It is now understood to be possible due to the formation of dislocations in the GaN that form near the surface of the substrates but that reduce in density as the GaN film grows thicker. The dislocation densities achieved of 10^7 to $10^9 \, cm^{-2}$ are still high compared to the dislocation densities of well below $10^6 \, cm^{-2}$ achievable in non-nitride III-V materials; however, they are not as detrimental to LED performance as was expected. A comparison is shown in Figure 5.20 which plots the observed efficiency drop as a function of dislocation density for a variety of semiconductors. The nitrides show a significantly higher tolerance to dislocation density. Figure 5.21 illustrates the density of dislocations on a sapphire substrate and how this dislocation density drops with further growth. By growing the highly faulted layer at lower temperatures the dislocation density in the subsequently grown material is greatly reduced. Even this reduced dislocation density would be too high for all but the nitride materials. The structure of a typical GaInN LED on a sapphire substrate is shown in Figure 5.14c).

The issue of doping also had to be resolved. GaN is readily doped n-type using the group IV atoms Si or Ge to substitute for Ga. p-Type doping is achieved with Mg, although in other III-V semiconductors such as GaAs, both Zn and Mg are used as

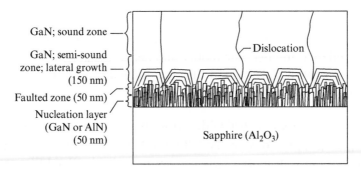

FIGURE 5.21 Dislocations in GaN epitaxial layer grown on sapphire. In addition to these factors, the 12% lattice mismatch of GaN with respect to sapphire is effectively much less apparent since it turns out that a rotation about the c-axis of GaN relative to the sapphire substrate allows a far better lattice match of the GaN system relative to the sapphire in the plane normal to the c-axis. See Problems 5.17 and 5.20. Both sapphire and SiC are very stable substrate materials that may be heated to over 1000°C during GaN growth, and both substrates are used in the high-volume production of GaInN LEDs. After The Blue Laser Diode, S. Nakamura, S. Pearton, G. Fasol, Springer Science & Business Media, 2000.

p-type dopants. The Mg^{2+} ion is very stable and it can substitute for Ga in spite of the fact that Mg^{2+} is usually incorporated into crystals with six nearest neighbors (octahedral symmetry) rather than the four nearest neighbors (tetrahedral symmetry) found in GaN. The effective p-type doping of GaN with Mg was not initially successful. Eventually a high-temperature anneal after dopant incorporation at temperatures near 900°C was discovered to activate the dopant allowing it to act as a stable shallow acceptor. The ability to dope GaN p-type became the final key to the success of the GaInN system.

GaN has a bandgap that is higher than needed for visible light emission, and indium is added to form the alloy semiconductor Ga$_{1-x}$In$_x$N. The incorporation of In decreases the bandgap and it may be added in the appropriate proportion for blue or green light emission using the MOCVD growth process. In addition, In plays a role in reducing the requirement for crystal perfection. It is believed that In incorporation in GaN effectively traps electrons locally and assists in the recombination process, competing effectively with traps due to dislocations. Finally, the double heterojunction structures that were used for red and yellow LEDs are directly applicable to GaInN LEDs and the bandgap engineering required to form these heterojunctions are achieved by controlling indium/gallium ratios.

GaN has a hexagonal crystal structure and is usually grown in the c-axis direction such that close-packed planes alternating between Ga and N atoms lie perpendicular to the growth direction (see Figure 2.16c). Since these atomic planes contain alternating charge densities, with positively charged Ga planes and negatively charged N planes, there is a natural polarization due to charge distribution along the c-axis of the crystal. If these opposing charge densities are equal and opposite there will be no *average* polarization along the c-axis; however, in practice there is a significant average polarization, which depends strongly on compressive strain in the crystal. Consider the active layer in the double heterostructure of Figure 5.22a. The In added to the active layer induces lateral compressive strain in the active layer, which significantly contributes to a net polarization field along the c-axis. This means that the energy band diagram must be modified as shown in Figure 5.22b. When an electron and a hole are injected into the active layer they are naturally pulled to opposite sides of the layer and a reduced radiation efficiency results. This difficulty can be overcome by making the active layer thinner, and layer thicknesses in the range of 30 Å are suitable. Due to their small dimension the energy wells are referred to as *quantum wells*.

These quantum well LEDs suffer from low maximum current densities due to the narrow well width (see Equation 5.20). This difficulty can be addressed by increasing the number of quantum wells to form a *multiple quantum well* LED. The standard high efficiency GaInN LED has multiple quantum wells as shown in Figure 5.22c. The MOCVD growth technique is capable of producing accurate and reproducible multiple quantum well structures.

The desire for white emission from LEDs has motivated the development of materials that convert blue light into yellow light. Since blue-emitting In$_x$Ga$_{1-x}$N LEDs are highly efficient, high-efficiency white LEDs may be achieved by *down-conversion* in which a phosphor material situated in close proximity to the blue-emitting GaInN die

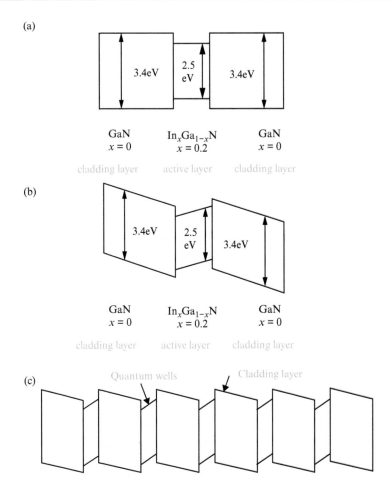

FIGURE 5.22 (a) Band diagram of a double heterostructure using In$_x$Ga$_{1-x}$N active layer grown epitaxially. Carriers recombine in the active layer. The cladding layers are doped such that one layer is n-type and one layer is p-type. (b) Band diagram including the effect of polarization. (c) band diagram of multiple quantum well GaInN LED. Between 4 and 8 quantum wells are commonly employed for high efficiency nitride-based LEDs.

absorbs blue LED emission and then re-emits light by fluorescence or phosphorescence at longer wavelengths. The material most widely used to achieve this is YAG:Ce which is abbreviated notation for cerium-doped Y$_3$Al$_5$O$_{12}$ phosphor material having the garnet crystal structure. This phosphor has strong absorption in a narrow band at 460 nm and broadband yellow emission centered at 550 nm. A fine powder of this phosphor can be included in the epoxy-encapsulated LED package such that a portion of the blue light emitted from the LED die is converted to yellow emission and the remaining blue emission mixes with the yellow to create white light. Figure 5.23 shows the emission spectrum of an LED of this type. The CIE diagram (Figure 4.14) can be used to understand the achievement of white light emission based on an appropriate combination of blue and yellow light.

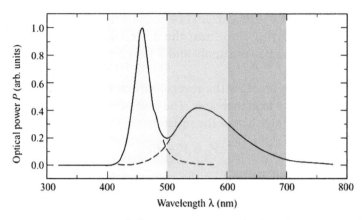

FIGURE 5.23 Emission spectrum of white-emitting LED. Blue light from the LED die is down-converted using YAG:Ce to produce broadband yellow fluorescence emission. The yellow fluorescence contains both green and red wavelengths. When combined with the blue LED emission white light results. Data from Nichia Corp.

The ideal white light source for many lighting applications is the solar spectrum shown in Figure 5.6, and a deficiency in wavelengths around 500 nm is clearly evident in the spectrum of Figure 5.23. A quantity called the *color rendering index* (CRI) specifies the ability of a white light source to illuminate color samples having a range of test colors. Standard, carefully formulated color samples are illuminated with the LED light source under test and the color coordinates of the reflected light from each sample are measured and recorded. The same color samples are then illuminated using a reference black-body light source that closely matches the terrestrial solar spectrum. The distance on the CIE diagram between the measured color coordinates from the LED and reference light sources are compared for each sample. These differences are suitably combined and subtracted from 100 to obtain the CRI index. For most practical white light sources the resulting CRI values are in the range between 50 (poor) and 100 (ideal). LEDs achieving CRI values between 90 and 100 are considered high CRI lamps and may be used to view art, color samples, and pigments with a good degree of accuracy. For more general illumination tasks a CRI between 70 and 90 is often considered adequate.

There are a variety of phosphor materials that have specific emission spectra that favor either green or red emission. These phosphors are excited with blue LED light and may be blended with YAG:Ce to improve the overall emission spectrum. See Suggestions for Further Reading.

5.16 BIPOLAR JUNCTION TRANSISTOR

It is now appropriate to describe and model the first commercialized transistor, the *bipolar junction transistor (BJT)* developed in the 1950s at Bell Labs. The p-n junction diode already discussed in detail in Chapter 3 provides an excellent background for this.

In Section 5.2 the IV curves are shown for a p-n junction illuminated by a light source. This lays the foundation for the BJT. Figure 5.2 as described by Equation 5.7a shows how the reverse saturation current of a p-n junction is increased by the optical generation of minority carriers at or near the depletion region. Consider a reverse-biased p-n junction with a power supply and a load resistor connected as shown in Figure 5.24.

If light shines on this junction the reverse saturation current increases and the power flowing through the load increases. The reverse voltage across the diode is the supply voltage less the voltage across the resistor:

$$-V_D = V_S - IR$$

A graph showing a load line based on the resistor R is shown in Figure 5.25. The voltage across the resistor can be varied by the intensity of the light and can range from almost zero if only the reverse saturation current flows through the diode (no illumination) to almost V_S if illumination is strong enough. If we regard the resistor as

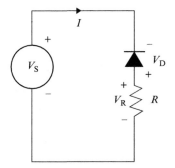

FIGURE 5.24 Reverse-biased p-n junction in series with a resistive load.

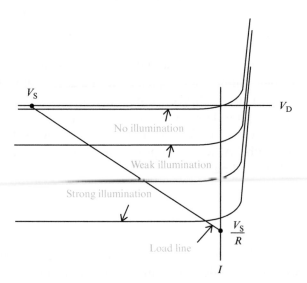

FIGURE 5.25 Characteristics of a reverse-biased photodiode connected to a resistive load as in Figure 5.24.

a load, the photodiode becomes a way to control a current I flowing through the load and a resulting voltage $V_R = IR$ across the load.

A transistor is a device that controls a current and voltage for a load in response to an electrical input rather than by using light intensity as an input. In order to achieve this we might consider how to vary the reverse saturation current in a reverse-biased diode in response to an electrical input voltage or current rather than by the use of light. Reverse saturation current is the result of minority carriers that drift across a reverse biased p-n junction.

We will now show that a concentration of minority holes can be electrically pro-duced and controlled in the n-type material of a reverse-biased p-n junction. Since only one minority carrier type is required for reverse diode current to flow, it is accept-able to only generate holes. Later we will show that minority electrons could also be electrically generated in the p-type material of a reverse-biased p-n junction to achieve the same goal.

In Chapter 3 we concluded that the p-n junction in forward bias causes minority carriers to be injected into both the p-side and the n-side of the junction. If one side of the junction has much higher doping than the other side then most of the minority carrier injection and hence most of the overall current flow is attributable to only this one minority carrier type.

We will now consider a p^+-n junction. Since the p-side is heavily doped, the equilibrium minority electron concentration n_p in the p-side will be small compared to the equilibrium minority carrier concentration p_n in the n-side. From Section 3.5 we see that minority carriers exist as excess holes in the n-side of a p-n junction and from Equation 3.21b) we obtain

$$\delta p(x_n) = \Delta p_{n(x_n=0)} \exp\left(-\frac{x_n}{L_p}\right) = p_n \left(\exp\left(\frac{qV}{kT}\right) - 1\right)\exp\left(-\frac{x_n}{L_p}\right) \qquad (5.21)$$

The excess hole concentration $\delta p(x_n)$ is exponentially related to applied voltage V and therefore only a small change of V in forward bias across this p-n junction can lead to orders of magnitude of change in $\delta p(x_n)$. If these minority holes could somehow be made available to the reverse-biased p-n junction of Figures 5.24 then we would have the ability to control the current flow through the reverse biased p-n junction by varying the voltage V across a forward-biased p-n junction.

Since the excess minority holes in the n-side of a forward-biased p-n junction drop off exponentially with distance away from the edge of the depletion region, we could consider positioning the reverse-biased p-n junction as closely as possible to the forward-biased junction, and preferably well within a diffusion length of the forward-biased junction. The resulting device is shown in Figure 5.26 with a forward bias connected to a first junction and a reverse bias connected to a second junction. This is the structure of a bipolar junction transistor.

Since we produce excess holes in the n-type region due to the p^+-n junction this p^+ region is called the *emitter*. Since we remove the minority holes from the reverse-biased junction using the p-type material on the right side we call this the *collector*. The n-type region is called the base for historical reasons since the first BJT was mechanically supported by this n-type layer.

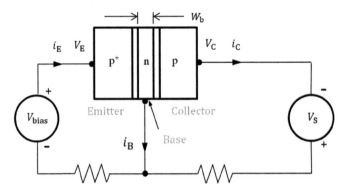

FIGURE 5.26 Voltages and currents in a PNP BJT.

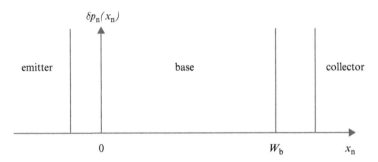

FIGURE 5.27 Base region of width W_b is located between two depletion regions.

The most important region to understand is the base of the transistor because holes injected from the emitter-base junction diffuse through the base and then drift across the base-collector junction into the collector.

We can establish a coordinate system as shown in Figure 5.27. Here the excess minority electron concentration in the base $\delta p_n(x_n)$ closest to the base-emitter depletion region is $\Delta p_{n(x_n=0)}$ and in the base closest to the base-collector depletion region it is $\Delta p_{n(x_n=W_b)}$.

We can now determine two boundary conditions relevant to the base region at $x_n = 0$ and at $x_n = W_b$.

$\Delta p_{n(x_n=0)}$ is obtained from the excess carrier concentration that exists on the p-side of the base-emitter p-n junction for a given base-emitter bias voltage. Using Equation 3.21b) we obtain

$$\Delta p_{n(x_n=0)} = p_n \left[\exp\left(\frac{qV_{BE}}{kT} \right) - 1 \right]$$

where V_{BE} is the base-emitter voltage. If $\dfrac{qV_{BE}}{kT} \gg 1$ for a forward biased base-emitter junction then we can write

$$\Delta p_{n(x_n=0)} \approx p_n \exp\left(\frac{qV_{BE}}{kT} \right)$$

We can also determine $\Delta p_{n(x_n = W_b)}$ by considering the base-collector p-n junction:

$$\Delta p_{n(x_n = W_b)} = p_n \left[\exp\left(\frac{qV_{CB}}{kT} \right) - 1 \right]$$

where V_{CB} is the collector-base voltage. If $\dfrac{qV_{CB}}{kT} << 0$ for a reverse-biased base-collector junction then we can write

$$\Delta p_{n(x_n = W_b)} \approx -p_n$$

Since p_n is a very small number we will approximate this to zero and hence

$$\Delta p_{n(x_n = W_b)} \approx 0$$

The excess minority carrier concentration in the base can now be obtained by solving the diffusion Equation 2.66a). The general solution is

$$\delta p_n (x_n) = C_1 \exp\left(-\frac{x_n}{L_p} \right) + C_2 \exp\left(\frac{x_n}{L_p} \right) \tag{5.22}$$

where C_1 and C_2 are constants. Since the base is finite in extent we cannot argue the elimination of one term and both terms must be used. Applying our two boundary conditions to Equation 5.22 we obtain

$$\Delta p_{n(x_n = 0)} = C_1 + C_2$$

and

$$C_1 \exp\left(-\frac{W_b}{L_p} \right) + C_2 \exp\left(\frac{W_b}{L_p} \right) = 0$$

Solving for C_1 and C_2 we obtain

$$C_1 = \Delta p_{n(x_n = 0)} \frac{\exp\left(\dfrac{W_b}{L_p} \right)}{\exp\left(\dfrac{W_b}{L_p} \right) - \exp\left(-\dfrac{W_b}{L_p} \right)} \tag{5.23}$$

and

$$C_2 = \Delta p_{n(x_n = 0)} \frac{-\exp\left(-\dfrac{W_b}{L_p} \right)}{\exp\left(\dfrac{W_b}{L_p} \right) - \exp\left(-\dfrac{W_b}{L_p} \right)} \tag{5.24}$$

Substituting C_1 and C_2 into Equation 5.22 we have

$$\delta p_n(x_n) = \Delta p_{n(x_n=0)} \frac{\exp\left(\dfrac{W_b}{L_p}\right)\exp\left(-\dfrac{x_n}{L_p}\right) - \exp\left(-\dfrac{W_b}{L_p}\right)\exp\left(\dfrac{x_n}{L_p}\right)}{\exp\left(\dfrac{W_b}{L_p}\right) - \exp\left(-\dfrac{W_b}{L_p}\right)}$$

This can be plotted as shown in Figure 5.28.

In order to calculate the current flow through the base we consider minority hole diffusion current density (Equation 2.65) and assume a BJT junction area A to obtain

$$I_p(x_n) = -qAD_p\frac{d\delta p(x_n)}{dx_n}$$

Substituting $\delta p(x_n)$ from Equation 5.22 and evaluating $I_p(x_n)$ at $x_n = 0$ we obtain

$$I_p(0) = qA\frac{D_p}{L_p}(C_1 - C_2)$$

and evaluating $I_p(x_n)$ at $x_n = W_b$ we obtain

$$I_p(W_b) = qA\frac{D_p}{L_p}\left(C_1\exp\left(-\frac{W_b}{L_p}\right) - C_2\exp\left(\frac{W_b}{L_p}\right)\right)$$

Using Equations 8.3 and 8.4 we can write $I_p(0)$ and $I_p(W_b)$ as

$$I_p(0) = qA\Delta p_{n(x_n=0)}\left(\frac{D_p}{L_p}\right)\frac{\exp\left(\dfrac{W_b}{L_p}\right) + \exp\left(-\dfrac{W_b}{L_p}\right)}{\exp\left(\dfrac{W_b}{L_p}\right) - \exp\left(-\dfrac{W_b}{L_p}\right)} \tag{5.25}$$

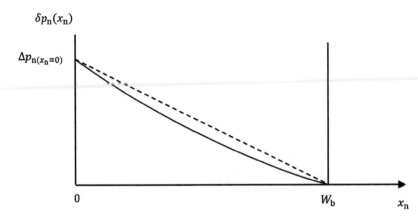

FIGURE 5.28   Plot of excess carrier concentration in the base under normal operating conditions for a PNP BJT.

and

$$I_p\left(W_b\right) = qA\Delta p_{n(x_n=0)}\left(\frac{D_p}{L_p}\right)\frac{2}{\exp\left(\dfrac{W_b}{L_p}\right) - \exp\left(-\dfrac{W_b}{L_p}\right)} \tag{5.26}$$

Note that $I_p(0)$ and $I_p(x_n)$ are determined by the slopes of the curve in Figure 5.28 at $x_n = 0$ and at $x_n = W_b$ respectively. We require that $\frac{W_b}{L_p} \ll 1$ and only a small fraction of the holes entering the base from the emitter junction recombine before they reach the base-collector junction. This therefore implies that the two slopes are similar. As a BJT becomes more close to ideal in performance the shape of the curve approaches a straight line shown as a dotted line in Figure 5.28.

The holes that do recombine in the base cause a base current to flow. This is because for every hole that recombines in the base, an electron must be supplied to the base to maintain base change neutrality. These electrons are supplied to the base from the external circuit resulting in base current flowing *out* of the base. Another contribution to base current exists due to electrons being injected from the base into the emitter. Since we have intentionally specified base doping much lower than emitter doping in the p$^+$-n emitter-base junction this electron injection current will be small. See Example 3.4 in which holes injected from a heavily doped p-side into a lightly doped n-side dominate current flow. The goal in BJT transistor design is to minimize total base current.

We will define the total base current as $i_B = I_B + I_n$ where I_B represents electrons required for hole recombination in the base and I_n represents electrons injected from the base to the emitter. We can quantify I_B as the difference between the hole current entering the base and the hole current leaving the base and therefore

$$I_B = I_p(0) - I_p(W_b) \tag{5.27}$$

Using Equations 5.25 and 5.26 we obtain

$$I_B = qA\Delta p_{n(x_n=0)}\left(\frac{D_p}{L_p}\right)\frac{\exp\left(\dfrac{W_b}{L_p}\right) + \exp\left(-\dfrac{W_b}{L_p}\right) - 2}{\exp\left(\dfrac{W_b}{L_p}\right) - \exp\left(-\dfrac{W_b}{L_p}\right)}$$

Since $\frac{W_b}{L_p} \ll 1$ a Taylor series expansion may be used for the exponential functions. To the second order,

$$e^x \approx 1 + x + \frac{1}{2}x^2$$

Upon substitution we obtain the approximation

$$I_{\mathrm{B}} \approx qA\Delta p_{\mathrm{n}(x_{\mathrm{n}}=0)} \left(\frac{D_{\mathrm{p}}}{L_{\mathrm{p}}}\right) \frac{1+\dfrac{W_{\mathrm{b}}}{L_{\mathrm{p}}}+\dfrac{1}{2}\left(\dfrac{W_{\mathrm{b}}}{L_{\mathrm{p}}}\right)^{2}+1-\dfrac{W_{\mathrm{b}}}{L_{\mathrm{p}}}+\dfrac{1}{2}\left(\dfrac{W_{\mathrm{b}}}{L_{\mathrm{p}}}\right)^{2}-2}{1+\dfrac{W_{\mathrm{b}}}{L_{\mathrm{p}}}+\dfrac{1}{2}\left(\dfrac{W_{\mathrm{b}}}{L_{\mathrm{p}}}\right)^{2}-\left(1-\dfrac{W_{\mathrm{b}}}{L_{\mathrm{p}}}+\dfrac{1}{2}\left(\dfrac{W_{\mathrm{b}}}{L_{\mathrm{p}}}\right)^{2}\right)}$$

$$= qAW_{\mathrm{b}}\Delta p_{\mathrm{n}(x_{\mathrm{n}}=0)} \left(\frac{D_{\mathrm{p}}}{2L_{\mathrm{p}}^{2}}\right)$$

This can be rewritten in terms of minority hole lifetime using $L_{\mathrm{p}}^{2}=D_{\mathrm{p}}\tau_{\mathrm{p}}$ to yield

$$I_{\mathrm{B}} = \frac{qAW_{\mathrm{b}}\Delta p_{\mathrm{n}(x_{\mathrm{n}}=0)}}{2\tau_{\mathrm{p}}} \tag{5.28}$$

As expected, if the minority hole lifetime increases, less recombination will occur and base current decreases.

There is another way to approximate the base current. Using Figure 5.28 we can obtain the total charge in the base due to minority holes using the area under the curve. To simplify this we will assume the dotted line and a triangular profile allows us to use the area of a triangle $\frac{1}{2}bh$ to obtain

$$Q_{\mathrm{p(base)}} = \frac{1}{2}qAW_{\mathrm{b}}\Delta p_{\mathrm{n}(x_{\mathrm{n}}=0)}$$

Since this charge recombines on average in time τ_{p} the recombination current or base current becomes

$$I_{\mathrm{B}} = \frac{Q_{\mathrm{p(base)}}}{\tau_{\mathrm{p}}} = \frac{1}{2\tau_{\mathrm{p}}}qAW_{\mathrm{b}}\Delta p_{\mathrm{n}(x_{\mathrm{n}}=0)}$$

which is identical to Equation 5.28.

We can now obtain a more complete view of the BJT and its performance and express this using a few key parameters.

The *base transport factor B* is the fraction of holes injected by the base-emitter junction that avoid recombination and cross the base to the base-collector junction and we therefore define B as

$$B = \frac{I_{\mathrm{p}}(W_{\mathrm{b}})}{I_{\mathrm{p}}(0)}$$

Typical values of base transport factor in a good BJT transistor approach unity. The value is maximized by minimizing the base width and by doping the base lightly since increased doping causes undesirable carrier scattering and lowers minority carrier lifetimes and diffusion lengths in semiconductors. Note also that we can express I_B in terms of B from Equation 5.27 as

$$I_B = I_p(0) - I_p(W_b) = (1 - B)I_p(0)$$

Since not all the current flowing through the base-emitter junction is hole current we define the *emitter injection efficiency* γ as

$$\gamma = \frac{I_p(0)}{I_p(0) + I_n}$$

where I_n is due to electron injection into the emitter from the emitter-base junction. The value of emitter injection efficiency also approaches unity and is maximized by having a high emitter doping level and a low base doping level.

The ratio between collector current and emitter current is called the *current transfer ratio* α and is given by

$$\alpha = \frac{i_C}{i_E} = B\gamma$$

which also approaches unity for a good BJT transistor. The terminal currents flowing in or out of the BJT terminals are labeled using lower case symbols as shown in Figure 5.26.

The total base current i_B is now attributable to both minority hole recombination in the base and to electron injection from base to emitter. Hence,

$$i_B = I_B + I_n = (1 - B)I_p(0) + I_n$$

An important parameter that determines the current amplification available in a transistor is the *collector current amplification factor* β defined as

$$\beta = \frac{i_C}{i_B} = \frac{BI_p(0)}{(1 - B)I_p(0) + I_n}$$

Dividing both numerator and denominator on the right-hand side by $I_p(0) + I_n$ we obtain

$$\beta = \frac{i_C}{i_B} = \frac{B\gamma}{1 - B\gamma} = \frac{\alpha}{1 - \alpha}$$

Since α is almost unity, typical values of β are large and can be on the order of 100. In this case a base current of 1 mA for a BJT would result in a collector current of 100 mA.

The resulting electrical behavior of a PNP transistor may now be presented by a family of characteristic curves shown in Figure 5.29. Each curve shows the collector current for a given value of base current.

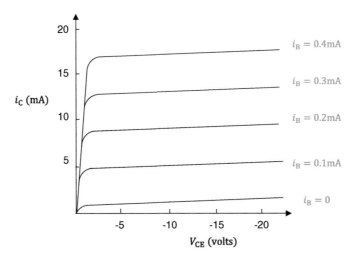

FIGURE 5.29   Plot of collector current i_C versus collector-emitter voltage $V_{CE} = V_C - V_E$ for a typical PNP BJT. V_{CE} is negative since the base-collector junction is reverse biased. See Figure 5.26. The collector current amplification factor $\beta = \frac{i_C}{i_B}$ for this transistor can be read from the graph to be approximately 45. Note that collector current slightly increases with increasing collector voltage: As the reverse bias on the collector-emitter junction increases the effective base width W_b narrows which slightly increases base transport factor B.

Although we have presented a PNP transistor, the converse transistor type, the NPN transistor, operates in a completely analogous manner. In the NPN transistor, the minority carriers crossing the p-type base are electrons rather than holes. Since electrons have higher diffusion lengths than holes in silicon, NPN transistors generally outperform PNP transistors, although both transistor types are used.

5.17   JUNCTION FIELD EFFECT TRANSISTOR

A commercially successful *junction field effect transistor (JFET)* was developed about 10 years after the BJT. Both of these p-n junction-based transistors are still in use today. The JFET is also built on the basis of the p-n junction but its operating principle is quite different from a BJT. The JFET has the advantage over the BJT of virtually no input current.

Instead of using minority carriers, the JFET relies on current flow due to majority carriers. Control of current flow is provided by changing the cross-sectional area of a *channel* through which majority carriers flow. By reducing the channel width to virtually zero the flow of current can be minimized and by maximizing the channel width a maximum current can flow.

In Chapter 3 we saw that the depletion width of a p-n junction depends on a reverse bias. If reverse-biased junctions are arranged on either side of a channel, the effective width of the channel can be controlled by means of the reverse bias voltage applied to the p-n junctions.

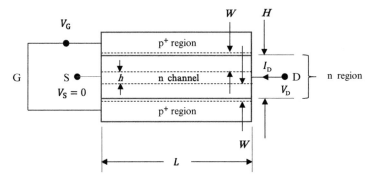

FIGURE 5.30   Diagram of n-channel JFET. Channel width h is defined as the portion of the n-region of thickness H that remains un-depleted between two depletion regions each having width W. The depletion region boundaries are shown as dotted lines. For convenience we define source voltage $V_s = 0$.

Figure 5.30 shows an n-type channel of width h located between two p$^+$ regions. The equilibrium depletion regions of width W for the two resulting p$^+$n junctions are shown using dashed lines. They extend further into the n-type region than into the p$^+$ regions as expected. See Example 3.2.

The n-type material has ohmic contacts at the beginning and end of the channel. These are labeled Source (S) and Drain (D) and the channel carries the load current. The two p-type regions are connected together and they are connected to a Gate (G) electrode using two ohmic contacts.

The resulting JFET characteristic curves are shown in Figure 5.31.

To understand and model this graph we will start by confining our analysis to the behavior closer to the origin. Consider the case in which $V_S = 0$ and V_D is small. As the gate voltage is made negative relative to the source region and the drain region, the two p-n junctions become reverse biased and the depletion regions of the two p-n junctions increase in width. See Figure 3.19. This decreases the effective channel width h. As gate voltage continues to become more negative the channel width reaches zero.

An expanded version of Figure 5.31 is shown in Figure 5.32 in which $V_{DS} = V_D - V_S$ is small. In this regime we can consider the width h of the channel to be constant along the channel. For convenience we shall again set $V_S = 0$ and therefore we can state that $V_D \cong V_S = 0$. The width W of a depletion region with bias was derived in Chapter 3 and using Equation 3.26 we obtain

$$W \cong \sqrt{\frac{2\epsilon_0\epsilon_r(-V_G)}{qN_d}}$$

If the doping in the p$^+$ regions is much higher than the doping in the n-type channel we can assume the depletion regions exist almost entirely within the n-type region. Hence from Figure 5.30 we can express channel width h as

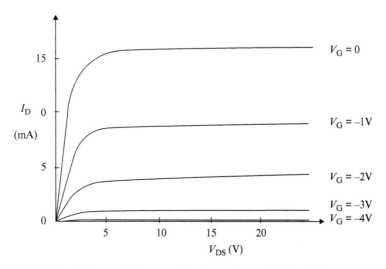

FIGURE 5.31　Family of characteristic curves for the n-channel JFET of Figure 5.30.

$$h(V_G) = H - 2W = H - 2\sqrt{\frac{2\epsilon_0\epsilon_r(-V_G)}{qN_d}}$$

Note that the channel width h reaches zero when $H = 2W$. We define this condition as *pinch-off* which is achieved if $H - 2W = 0$. Solving for the corresponding gate voltage

$$V_p = -V_{G \text{ at pinch off}} = \frac{qN_dH^2}{8\epsilon_0\epsilon_r}$$

where *pinch-off voltage* V_p is defined as the magnitude of gate voltage at pinch-off.

The electrical resistance R between source and drain of a channel of length L is therefore

$$R(V_G) = \rho\frac{L}{A} = \rho\frac{L}{hZ} = \rho\frac{L}{Z}\left[H - 2\sqrt{\frac{2\epsilon_0\epsilon_r(-V_G)}{qN_d}}\right]^{-1}$$

where A is the cross-sectional area of the channel and Z is the channel depth. For the JFET of Figure 5.32 the curve having the smallest slope is for the case of a channel very close to the pinch-off condition and hence $V_p \cong 4V$ for this JFET. Since the slope of each line in Figure 5.32 is equal to $1/R$ we can express it as

$$\text{slope} = \frac{dI_D}{dV_{DS}} = \frac{1}{R} = \frac{Z}{\rho L}\left[H - 2\sqrt{\frac{2\epsilon_0\epsilon_r(-V_G)}{qN_d}}\right] \tag{5.29}$$

Equation 5.29 and corresponding Figure 5.32 represent widely applicable *voltage-controlled resistor* behavior. Depending on gate voltage the resistance between drain and

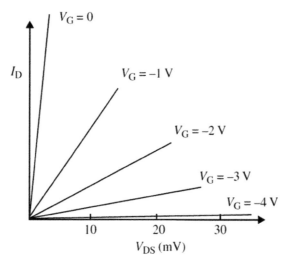

FIGURE 5.32   Detail of Figure 5.31 for small values of V_{DS}. The approximately straight lines have slopes that depend on V_G according to Equation 5.29. This is often referred to as the linear region of JFET operation.

source varies. If the FET is connected such that drain current I_D flows through a load then the FET can control load current. A particular advantage of a JFET is that gate current is negligible since the p^+n junctions are never forward-biased during normal operation.

We can also consider larger values of V_{DS} which are included in Figure 5.31. Modeling a JFET over a wide range of voltages is more complicated since the channel width is no longer uniform from source to drain. As the drain voltage shown in Figure 5.30 becomes more positive it adds a further reverse bias to the p^+n junctions near the drain contact thereby narrowing the channel width at the drain end of the JFET. Figure 5.33 shows the tapered shape of a channel with $V_G = 0$ under these circumstances that reaches pinch-off at the drain. Further increasing V_{DS} will no longer substantially increase drain current and the JFET is said to be in saturation. Figure 5.34 shows the point on each characteristic curve of the JFET at the onset of saturation.

More complete JFET modeling requires two-dimensional treatment of the tapered channel to obtain the conductivity of the channel of varying cross-sectional area. It is made still more complex due to carrier velocity effects since at pinch-off, drain current I_D is crowded into a very narrow channel. Electric fields in this portion of the channel

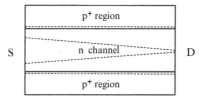

FIGURE 5.33   The tapered shape of a channel with V_{DS} sufficient to cause the onset of pinch-off at the drain. $V_G = 0$ and $V_{DS} \cong 4$ volts for the JFET of Figure 5.31. Further increasing V_{DS} will no longer substantially increase drain current and the JFET is said to be in saturation.

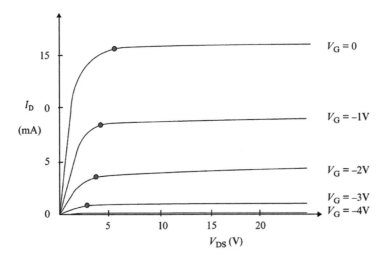

FIGURE 5.34   Circles show the onset of saturation for each characteristic curve of Figure 5.31. Note that when $V_G = 0$ pinch-off occurs at $V_{DS} \cong 6$ volts. For $V_{DS} > 4$ volts there is saturation and little increase in drain current. Corresponding points in the case of $V_G = -1$ volt, -2 volts, and -3 volts occur at $V_{DS} \cong 4$ volts, 3 volts, and 2.5 volts, respectively. When $V_G = -4$ volts the channel is almost completely pinched off for all values of V_{DS}.

can be sufficiently high that carrier velocity saturation occurs, especially in short channel JFETs. See Figure 2.24.

The converse JFET device has a p-type channel between two n^+p junctions and its modes of operation are completely analogous to the n-channel JFET. Both n-channel and p-channel JFETs are in use. Variations on the JFET design we have described also exist. For example JFETs may be fabricated in which the channel is formed between only one p-n junction and an insulating layer on the opposing side of the channel. The same operating principle still applies.

5.18   BJT AND JFET SYMBOLS AND APPLICATIONS

Figure 5.35 shows both PNP and NPN BJT transistor symbols as well as n-channel and p-channel JFET symbols.

Today's electronics industry is dominated by digital electronics and the MOSFET (metal oxide field effect transistor) serves almost all of this market. The MOSFET is a variation of the JFET and is covered in detail in Chapter 6.

There are many other more specialized applications of transistors in linear amplifiers, temperature sensors, low-noise amplifiers, analog switches, and power electronics for which BJTs and JFETs can outperform MOSFETs. The p-n junctions common to both the BJT and the JFET are very robust and these devices are more rugged and less prone to damage than are MOSFETs. JFETs have inherently lower

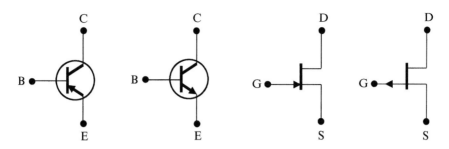

FIGURE 5.35 Symbols for BJT and JFET transistors.

noise operation than MOSFETs. The BJT is widely used in discrete applications due to the very broad selection of individually packaged BJT types available covering a wide spectrum of applications from small signal amplifiers to power electronics. The BJT is also used in very high frequency applications, such as radio frequency circuits for wireless communications systems. As a temperature sensor, the BJT, due to its well understood temperature dependence, is widely used. The JFET offers virtually infinite input resistance and is ideal for high performance amplifier input stages.

5.19 SUMMARY

5.1 A solar cell consists of a p-n junction in which sunlight generates electron-hole pairs that are collected by the external circuit and provide electrical power. Carrier flow occurs by drift across the depletion region in the solar cell and operation is in the fourth quadrant of an I–V diagram. The potential barrier in a solar cell p-n junction at the operating point should be as small as possible to maximize the output voltage of a solar cell, but must be large enough to maintain a suitable reverse current.

5.2 Light absorption in a solar cell is sensitive to the type of bandgap. Indirect gap semiconductors have weaker absorption than direct-gap semiconductors over a range of photon energies. The absorption in indirect gap material requires the participation of phonons and thicker cells are required.

5.3 The solar spectrum is based on the blackbody radiation spectrum at a temperature corresponding to the surface of the sun at 5250°C. This spectrum is affected by atmospheric absorption due principally to oxygen, water vapor, and carbon dioxide.

5.4 A crystalline silicon solar cell consists of a thin n-type front layer and a thick p-type layer. A front electrode metal grid, back contact metallization, and an antireflection layer complete the device. Sunlight that is absorbed within the depletion region or within a diffusion length on either side of the depletion region has the best chance of generating power in an external circuit. Short circuit current, open circuit voltage, and fill factor are useful characterization parameters.

5.5 The limiting solar cell efficiency is dependent on the bandgap of the solar cell. Limitations in efficiency arise from photons that are too low in energy to be absorbed or photons with too much energy that must thermalize before they can be collected. In addition the operating voltage is lower than the bandgap energy, which represents a further energy loss. Silicon has achieved a conversion efficiency of approximately 25%. Maximum efficiencies are predicted by the Shockley-Queisser limit.

5.6 Multijunction solar cells have achieved well over 30% conversion efficiency. The best-known devices are grown by epitaxial layer growth of III-V semiconductors lattice matched to a germanium substrate. The solar spectrum is more efficiently utilized than in single-junction solar cells.

5.7 Operation of LEDs is by means of electron-hole pair recombination. Photons generated must be outcoupled from the device and outcoupling losses should be minimized. Further losses occur due to non-radiative recombination.

5.8 The measured emission spectra of LEDs confirm the model presented in Chapter 4 for direct-gap electron-hole pair recombination.

5.9 Non-radiative recombination can take place at surfaces, interstitials, vacancies, and anti-sites. Auger recombination is another possible recombination mechanism and is also non-radiative.

5.10 Optical outcoupling of an LED can be modeled by using Snell's law for light paths intersecting the semiconductor surfaces. Total internal reflection gives rise to an escape cone for light emerging through a planar surface of the LED die, and a lambertian emission pattern is typically present from the top surface of the die.

5.11 The interest in visible light emission and the incorporation of phosphorus into LEDs in the 1960s resulted in $GaAs_{1-x}P_x$ LED technology. As x increases in $GaAs_{1-x}P_x$ the bandgap increases. Visible emission starts at 750 nm, which corresponds to a bandgap of approximately 1.65 eV. Doping with nitrogen helps to promote radiative recombination and allows indirect gap materials such as GaP to exhibit sufficiently efficient luminescence for medium brightness green LEDs.

5.12 In the 1980s the $Al_xGa_{1-x}As$ system was developed in which the lattice constant is virtually independent of the aluminum content. Bandgap engineering opens the door to the double heterostructure, which provides carrier confinement and higher radiative recombination rates. In addition optical outcoupling is significantly improved. The range of current density that a well can accommodate may be determined using a calculation based on a well density of states and a recombination coefficient.

5.13 AlGaInP LEDs developed in the late 1980s achieve improved efficiency and a wider range of colors. The use of a quaternary system provides independent control of lattice constant and bandgap and shorter wavelength direct-gap emission is enabled. Luminous efficiency improves because of increased eye sensitivity as wavelength decreases. A transparent substrate may be wafer bonded to further increase efficiency. The MOCVD growth technique is versatile and allows a wide range of ternary and quaternary LED materials to be grown.

5.14 Ga$_{1-x}$In$_x$N developed by the early 1990s has allowed high-efficiency high-brightness LEDs to cover the entire visible and the near-UV wavelength ranges. Development of nitride LEDs required new high-temperature growth conditions and the use of lattice-mismatched substrates and p-type doping with magnesium. Polarization effects favor the use of quantum wells, which provide better carrier confinement. White emission may be achieved by down-converting blue emission using YAG:Ce phosphor.

5.15 LED optical outcoupling can be improved by fabricating die with tilted side-walls to reduce total internal reflections. Surface texturing can also improve outcoupling as can high index plastic encapsulation. Emission color and color rendering index are used to judge the white emission quality of an LED designed for general illumination purposes.

5.16 Two well-known transistor types include the Bipolar Junction Transistor (BJT) and the Junction Field Effect Transistor (JFET). These devices both take advantage of the properties of p-n junctions.

5.17 The BJT uses a reverse-biased p-n junction and controls the reverse current through this diode by means of another forward-biased p-n junction in close proximity to the reverse-biased p-n junction.

5.18 Modeling the BJT requires examining the base region of the device. The base region is common to both p-n junctions in the device. The minority carrier concentration must be calculated within this base region in order to calculate the current flow through the device.

5.19 Current flow between the emitter and the collector of the BJT is shown to be controlled using the base electrode. Base current can be calculated and by suitably designing the BJT, base current can be minimized. Collector current is related to base current by the collector current amplification factor β.

5.20 The JFET operates on a very different principle from the BJT. A channel formed laterally between two p-n junctions carries current between a source electrode and a drain electrode. A gate electrode allows reverse bias to be simultaneously applied to both p-n junctions simultaneously.

5.21 Increasing the reverse bias on the two p-n junctions causes their depletion widths to increase. This causes the channel to be narrowed thereby restricting the current flow between source and drain. In this manner, the JFET operates as a voltage-controlled resistor provided drain voltages are small.

5.22 For higher drain voltages the JFET channel becomes tapered and current flow between source and drain is more complicated to calculate. Above a pinch-off voltage the device goes into saturation. Beyond saturation the channel remains pinched off and drain current does not significantly increase further with increasing drain voltage.

5.23 Both PNP and NPN versions of the BJT exist and both n-channel and p-channel JFETs exist. Although not as widely used as MOSFETS, BJTs and JFETs remain popular in a variety of specialized electronics applications such as low noise amplifiers and very high frequency circuits. The BJT, in particular, is widely used as a discrete transistor component.

PROBLEMS

5.1 Solar radiation is fundamentally determined by *blackbody radiation*. Blackbody radiation produces a radiation spectrum described by *Planck's law*. Planck's law describes the *radiance* as a function of frequency v of electromagnetic radiation emitted from a non-reflective surface of a body in thermodynamic equilibrium at temperature T. As a function of v in cycles per second and T in kelvins, Planck's law is

$$I(v,T) = \frac{2hv^3}{c^2} \frac{1}{\exp\left(\dfrac{hv}{kT}\right) - 1}$$

and represents the emitted power in watts per unit area of an emitting surface at temperature T per unit solid angle per unit frequency. It provides what is referred to as a *specific radiative intensity*.

(a) Using a computer, plot $I(v, T)$ as a function of v on a single set of coordinates for the following temperatures: $T = 1500$ K, $T = 2500$ K, $T = 3500$ K, $T = 4500$ K, and $T = 5500$ K. For each plot estimate the perceived color of the observed emission and find the wavelength at which $I(v)$ is a maximum. At what wavelength is sunlight a maximum?

(b) For each temperature, find the fraction of total radiated power that falls in the visible light range between 400 and 700 nm. *Hint*: Use the area under the curve of $I(v,T)$.

(c) Find a CIE diagram similar to Figure 4.14 that shows the locus of color coordinates x and y due to blackbody radiation. Show the point on this locus corresponding to a tungsten filament lamp operating at 3500 K. How much below the melting point of tungsten is this? Also show the point on the locus corresponding to the light emitted by the sun.

5.2 Antireflection layers applied to silicon solar cells function in a manner similar to those used on camera lenses and eyeglass lenses. If light is incident normal to a first interface between two transparent optical media with different indices of refraction, reflection will take place. If the reflection at a second interface is arranged to cancel out the first reflection by ensuring that both the magnitude and the phase of the reflections will cancel out the first reflection, then overall reflection is reduced or eliminated. The reflection coefficient R at a single interface is

$$R = \left(\frac{n_2 - n_1}{n_2 + n_1}\right)^2$$

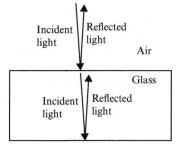

A simple example of this is the interface between air and glass shown above. Air has index of refraction $n_1 \cong 1.0$ and glass has index of refraction $n_2 \cong 1.5$. Therefore,

$$R = \left(\frac{n_2 - n_1}{n_2 + n_1}\right)^2 = \left(\frac{1.5 - 1.0}{1.5 + 1.0}\right)^2 = 0.04 \text{ or } 4\%$$

Note that the same reflectivity is obtained from the air–glass interface whether the light is incident from air onto the air–glass interface or from the glass onto the glass–air interface.

An antireflection layer having index of refraction n_2 as shown below creates two interfaces, one between media 1 and 2, and one between media 2 and 3:

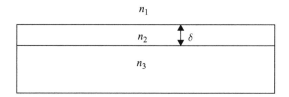

(a) Regard n_1 and n_3 as given, but n_2 as an adjustable parameter. Find the value of n_2 in terms of n_1 and n_3 that will cause the reflection from the interface between 1 and 2 to be the same in magnitude as the reflection from the interface between 2 and 3.

(b) If medium 1 is air and medium 3 is glass, find the numerical value for n_2 that satisfies the criterion of (a).

(c) In order for the reflections to cancel out, the phases of the light waves leaving the interfaces must also be arranged to be 180 degrees out of phase with each other. Show that the smallest possible thickness δ of medium 2 is $\frac{\lambda}{4}$ where λ is the wavelength of the light in medium 2.

(d) For the conditions in (b) find the smallest possible thickness of medium 2 to allow for cancellation of reflection for light having wavelength 600 nm. *Hint:* Wavelength in a medium having index of refraction n is reduced relative to its wavelength in vacuum by a factor of n.

(c) Assuming that silicon is a transparent medium, and using information in the appendices, find both the thickness and index of refraction of an antireflection layer that could be deposited on silicon if it is situated in air.

(f) Tantalum oxide is a popular antireflection material for silicon solar cells. Look up the refractive index of tantalum oxide and compare with your answer to (e).

(g) Since silicon solar cells are generally bonded to a glass front sheet with $n \cong 1.5$, repeat steps (e) and (f) if the air is replaced with glass. In which case (air or glass) does tantalum oxide best match the calculated refractive index? You may neglect the front reflection off the glass sheet at the air interface.

(h) Compare the expected efficiency of a silicon solar cell with an antireflection layer to that without an antireflection layer. *Hint*: Light that is not reflected will be transmitted into the silicon. Assume a wavelength of 550 nm.

(i) Silicon solar cells with an antireflection coating generally have a deep-blue color. Explain this based on your understanding of antireflection layers.

5.3 Show that Equation 5.2 is the solution to Equation 5.1 by substituting it into the differential equation. Also show that Equation 5.3 is obtained from Equation 5.2 by applying the appropriate boundary conditions.

5.4 Using the expressions for diffusion current,

$$J_n(x)_{\text{diffusion}} = qD_n \frac{dn(x)}{dx}$$

and

$$J_p(x)_{\text{diffusion}} = -qD_p \frac{dp(x)}{dx}$$

show that Equations 5.4a and 5.4b are obtained from Equations 5.3a and 5.3b.

5.5 The optical generation rate inside a silicon solar cell at room temperature in a certain flux of sunlight is assumed to be a constant value $G = 5 \times 10^{19}$ EHP cm^{-3} s^{-1}. If $W = 1.0$ μm, the electron and hole lifetimes are 2×10^{-6} s, $I_0 = 1 \times 10^{-9}$ A, and the cell area is 100 cm^2, calculate

(a) The optically generated current I_L.
(b) The short circuit current I_{SC}.
(c) The open circuit voltage V_{OC}.
(d) The maximum output power if the fill factor is 0.8.

5.6 A silicon solar cell of area 100 cm^2 has doping of $N_a = 5 \times 10^{18}$ cm^{-3} and $N_d = 5 \times 10^{16}$ cm^{-3}. If the carrier lifetimes are 5×10^{-6} s and the optical generation rate inside the silicon solar cell is assumed to be a constant value of $G = 7 \times 10^{19}$ EHP cm^{-3} s^{-1}, calculate

(a) The optically generated current I_L. State and justify any assumptions you used to obtain your solution.
(b) The short circuit current I_{SC}.
(c) The open circuit voltage V_{OC} at room temperature.
(d) The open circuit voltage at 100°C and at −50°C.
(e) The maximum output power if the fill factor is 0.8 at both room temperature and at 100°C.
(f) If G is increased by a factor of 100 by a solar concentration system to 7×10^{23} EHP cm^{-3} s^{-1} calculate the maximum power output. Confirm that it rises by more than a factor of 100, and state the assumptions you are making in this calculation.

5.7 Obtain a copy of the original paper presenting the Shockley-Queisser limit: William Shockley, Hans J. Queisser, Detailed Balance Limit of Efficiency of *p-n* Junction Solar Cells, Journal of Applied Physics 32, 510 (1961).

(a) Read the paper carefully.

(b) Write a well-constructed summary of the physics used to obtain the results and the efficiency limiting aspects for each of the following

(i) spectral absorption losses

(ii) thermalization losses

(iii) potential losses

(iv) thermodynamic limit

5.8 A tandem solar cell is formed as shown in the figure below.

(a) Find the lattice constant of germanium and that of GaAs. Compare these and look at literature showing the epitaxial growth of GaAs thin films on Ge substrates. How is the lattice mismatch accommodated?

(b) How thick should the GaAs layer be to absorb 30% of sunlight at a wavelength of 800 nm? Use Figure 5.5 for data on the absorption coefficient of GaAs.

(c) Explain how GaInP can be lattice matched to GaAs and have an energy gap of 1.95 eV. Use Figure 2.34a) to answer this question.

(d) Explain in your own words how this tandem cell could outperform a single junction cell in terms of efficiency.

(e) Look up information about triple junction tandem solar cells. What triple junction cell currently has the highest conversion efficiency? What is this efficiency value?

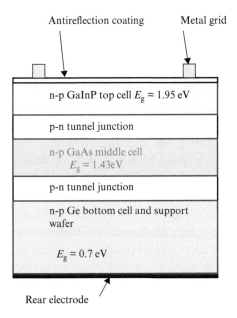

5.9 Consider the LED emission spectrum and answer the following:

(a) Obtain the full width at half maximum (FWHM) for the emission spectrum of the red LED of Figure 5.15 at room temperature.

(b) Find the bandgap of the semiconductor used for the red LED of Figure 5.15. *Hint*: Treating E_g as a parameter, plot $R(E)$ vs E using Equation 5.11 and find the value of E_g that allows $R(E)$ to best match the spectrum of (a).

(c) Using the value of E_g obtained from (b), plot Equation 5.11 at $-50°C$, $+20°C$, and $+100°C$ on the same graph and measure the FWHM for each plot.

(d) Calculate the FWHM for the three temperatures of (c) using FWHM $= 1.8kT$ and compare with the answers of (c).

(e) Is the photon emission rate $R(E)$ really equivalent to luminous intensity as a function of E? There are two sources of discrepancy, one based on photon properties and one based on photometric concepts. Explain both. Now explain why we are justified in neglecting these for the answers in (a) to (d).

5.10 The emission spectrum broadening discussed in Problem 5.9 is not the only possible broadening mechanism in an LED. List a few other additional contributions to linewidth broadening for the spectrum of an LED that may occur.

5.11 Look up more detailed information on liquid phase epitaxy (LPE). Explain the principles of the technique using a phase diagram and discuss which semiconductor types have been successfully grown using LPE. Why is LPE suitable for GaAs LEDs but not for GaN devices?

5.12 LED pioneer Nick Holonyak is credited with being the "father of the LED." Find and summarize a short biography of his life and discuss his contributions to LEDs.

5.13 Shuji Nakamura is a famous pioneer in the field of GaN LEDs. He, along with two others, won the Nobel Prize for the development of the blue LED. Find and summarize a short biography of his life and explain his and the other two winners' contributions to the development of GaInN LEDs.

5.14 Look up more detailed information on molecular beam epitaxy (MBE). Explain the principles of the technique using a diagram and discuss which semiconductor types have been successfully grown using MBE. What are the important advantages and disadvantages of MBE?

5.15 Research more detailed information on the semiconductor growth technique metal-organic chemical vapor deposition (MOCVD). Explain the principles of the technique using a diagram and discuss which semiconductor types have been successfully grown using MOCVD. Why is MOCVD of particular importance in growing the more complex quaternary AlGaInP LED devices? Also show how it is used to grow GaInN LEDs.

5.16 The substrate materials used for GaN-based LEDs do not have ideal lattice constants or matched values of thermal coefficient of expansion to GaN. Find both lattice mismatch and thermal coefficient of expansion (TCE) mismatch for the leading substrate types Al_2O_3 and SiC relative to GaN.

5.17 The desire for affordable and high-quality single-crystal GaN wafers is an ongoing issue. Look into the following issues:

(a) What is the fundamental difficulty in growing large single-crystal boules of GaN?

(b) What methods have been developed to grow bulk GaN material?

(c) For what applications is single-crystal GaN seen as important? Does this include LEDs?

5.18 The growth of GaN is generally based on *c*-axis growth on silicon carbide and sapphire substrates; however, the growth of GaN has been achieved in other crystallographic directions yielding non-polar M-planes or A-planes as well as semi-polar planes. What is the potential benefit of using these growth directions? Find more information about work done on these alternative growth directions and the measured LED performance that is achieved. What specific challenges are limiting the widespread application of these alternative growth directions?

5.19 At 300 K at the electron well overflows in an $Al_xGa_{1-x}As$ double heterostructure with the following parameters:
Barrier height $\Delta E_c = 180$ meV
Well width $W = 150$ Å
Effective density of states $N_c = 5 \times 10^{17}$ cm^{-3}
Recombination coefficient $B = 1.9 \times 10^{-10}$ cm^{-3} s^{-1}
Calculate the current level at which the well overflows.

5.20 Look up information on the bandgap of the wide gap semiconductor SiC. It was previously used as the semiconductor for making blue LEDs but these LEDs were no longer viable after GaInN LEDs were developed. You will find several distinct bandgaps for this material. Explain this. What is the reason for the very low efficiency of SiC p-n junction LEDs? SiC subsequently became an important substrate material for GaN-based LEDs.

5.21 The growth of GaN on sapphire appears to have a very large lattice mismatch of 12%; however, a rotation of the axes in the growth plane of the GaN relative to the sapphire occurs that reduces the effect of this mismatch substantially. Look up more information on this topic. *Hint*: Use keywords in-plane rotation; GaN on sapphire growth.
(a) Construct a diagram showing the planar arrangement of atoms in the sapphire substrate as well as the rotated GaN crystal plane that provides an optimum fit to the sapphire in the same diagram using two different colors to distinguish the layers.
(b) What rotation angles are optimum?
(c) Review recent journal papers covering details of dislocation formation and dislocation accommodation during the growth of GaN on sapphire. Write a short review of this literature.

5.22 Look up the light emission patterns of a few lens-free commercially available white-emitting LEDs that use powder phosphors. Note that they are approximately lambertian. Explain why a powder phosphor source of light is lambertian. Hint: White paper forms a very good approximation to a lambertian light source when it reflects ambient light. The explanation for this is similar to that of the powder phosphor layer in an LED.

5.23 A PNP BJT has emitter doping 100 times larger than base doping. The base width is much smaller than the carrier diffusion length.
(a) Assuming we increase base doping by a factor of 10. Qualitatively, what effect will this have on the collector current if any? Why?

(b) Assume we decrease base width by a factor of 2. Qualitatively, what effect will this have on collector current?

5.24 Calculate and plot the excess hole concentration in the base of a PNP BJT as a function of position assuming $W_b / L_p = 0.4$. Use spreadsheet software. You can assume a value of $\Delta p_E = 1 \times 10^{15}$ cm^{-3}. Repeat for $W_b / L_p = 2.0$.

5.25 A PNP BJT has the following properties at room temperature:

Carrier lifetimes of 0.1 μs

Emitter doping 10^{19} cm^{-3}

Collector doping 10^{17} cm^{-3}

Base doping 10^{17} cm^{-3}

Metallurgical base width (junction to junction based on a step doping profile): 1μm

Cross-sectional area 10^{-5} cm^2

(a) Calculate the depletion width at the collector-base junction with $V_{CB} = 50$V
(b) Calculate B for the transistor with $V_{CB} = 50$V
(c) Calculate γ for the transistor with $V_{CB} = 50$V
(d) Calculate β for the transistor with $V_{CB} = 50$V
(e) Plot the excess hole concentration in the base region for the following conditions:

(i) $I_C = 1\mu A$

(ii) $I_C = 1mA$

(iii) Calculate the hole charge in the base for each of i) and ii) above. Use these values to estimate β and compare this to the answer you got in part d) above. Explain any discrepancy carefully.

(f) Sketch the ideal collector characteristics for the transistor (i_C versus $-V_{CE}$). Let i_B vary from zero to 0.2 mA in increments of 0.02 mA. Let $-V_{CE}$ vary from 0 up to –50 volts.

(g) Explain how it is possible for the average time an injected hole spends in transit across the base to be shorter than the hole lifetime in the base.

(h) Repeat parts a), b), c), d) for an NPN transistor with the same doping levels.

(i) Why are NPN transistors preferred if there is a choice?

5.26 Look up further information as needed in Suggestions for Further Reading or using the internet to answer the following questions:

(a) Find the meanings of "saturation" and "oversaturation" as applied to the BJT.

(b) Explain why the turn-on time of a BJT is faster when the transistor is driven into oversaturation.

(c) Explain why the turn-off time of a BJT is slower after it had been driven into oversaturation.

(d) Explain why PNP transistors are frequently needed in circuits that also contain NPN transistors.

5.27 A symmetrical P $^+$NP$^+$ transistor is connected in four ways to a DC voltage source V as shown in the table. Assume that $V \ll kT/q$.

Configuration	Collector	Base	Emitter
1	−V	No connection	+V
2	−V	−V	+V
3	+V	−V	No connection
4	+V	−V	+V

(a) Sketch a diagram for each connection scheme using the appropriate BJT symbol in Figure 5.35.

(b) Sketch (x_n) in the base region for each case.

(c) Explain how you can justify each sketch in part b).

(d) Which situation is most likely to exist in a real transistor application? Why?

5.28 An n-channel JFET is described as follows:

Semiconductor is silicon at room temperature

p^+ doping: $5 \times 10^{18} \, \text{cm}^{-3}$

n doping: $2 \times 10^{16} \, \text{cm}^{-3} \, L = 100 \, \mu\text{m}$

The JFET is biased such that $V_S = V_D = 0$

(a) Find the value of H in Figure 5.30 that would cause the channel to reach pinch-off when $V_G = -4$ volts.

(b) Using the value of H obtained in a) calculate the resistance of the channel from source to drain for each of the following voltages:

   (i) $V_G = -3$ volts

   (ii) $V_G = -2$ volts

   (iii) $V_G = -1$ volt

   (iv) $V_G = 0$

(c) Calculate the maximum drain voltage V_D that could be applied to the FET with $V_G = -2$ volts such that the channel width h remains at 90% of its width at the source-end of the JFET. Use the value of H obtained in a) and continue to assume that $V_S = 0$.

5.29 Look into Suggestions for Further Reading or similar resources to find a full derivation of JFET channel resistance for a tapered channel. Obtain a good understanding of the model and its limitations and understand the mathematical justification for the range of validity of the model.

5.30 Look up further information as needed in Suggestions for Further Reading or using the internet and answer the following questions:

(a) The JFET is often referred to a *depletion mode* transistor. Explain the origin of this terminology.

(b) Explain why the JFET has a much higher *input impedance* than the BJT. What values of input impedance are typical for JFETs? Why is this important?

FURTHER READING

1. Kitai, A.H. (2019). *Principles of Solar Cells LEDs and Related Devices*, 2e. Wiley.
2. Green, M.A. (1992). *Solar Cells, Operating Principles, Technology and System Applications*. University of New South Wales.
3. Luque, A. and Hegedus, S. (eds.) (2003). *Handbook of Photovoltaic Science and Engineering*. Wiley.
4. Nakamura, S., Pearton, S., and Fasol, G. (2000). *The Blue Laser Diode, The Complete Story*, 2e. Springer.
5. Schubert, E.F. (2006). *Light Emitting Diodes*, 2e. Cambridge University Press.
6. Kitai, A.H. (ed.) (2017). *Materials for Solid State Lighting and Displays*. Wiley.
7. Neamen, D.A. (2003). *Semiconductor Physics and Devices*, 3e. McGraw Hill.
8. Roulston, D.J. (1999). *An Introduction to the Physics of Semiconductor Devices*. Oxford University Press.
9. Streetman, B.G. and Banerjee, S.K. (2015). *Solid State Electronic Devices*, 7e. Pearson.

CHAPTER 6

The Metal Oxide Semiconductor Field Effect Transistor

CONTENTS

Fundamentals of Semiconductor Materials and Devices, First Edition. Adrian Kitai.
© 2023 John Wiley & Sons Ltd. Published 2023 by John Wiley & Sons Ltd.
Companion Website: www.wiley.com/go/kitai_fundamentals

Objectives

1. Introduce the structure, basic operation, and advantages of the MOSFET.
2. Analyze channel conductivity as a function of gate voltage.
3. Introduce semiconductor–oxide interface effects including charge trapping.
4. Model and predict inversion layer and accumulation layer thicknesses.
5. Summarize advanced nanoscale lithography techniques.
6. Introduce ion beam implantation and MOSFET fabrication.
7. Motivate and explain CMOS structures.
8. Introduce and explain flash memory.
9. Describe nanoscale effects and the two-dimensional electron gas.
10. Derive tunneling and apply it to the gate oxide layer in a MOSFET.

6.1 INTRODUCTION TO THE MOSFET

A transistor is a three terminal semiconductor device as distinct from a diode which is a two terminal device. In operation, applying voltages to a third terminal enables the control of electric current passing between the remaining two terminals. Transistors are the building blocks of virtually all *integrated circuits* (ICs) including advanced nanoscale ICs.

The Metal-Insulator-Semiconductor transistor is by far the most widely used type of transistor. A conducting *channel* is controlled by a voltage applied to a *gate* that is isolated from the channel by an insulator layer such that an electric field can be produced and controlled in the semiconductor. Since this insulator layer is typically an oxide, the device name has become the Metal-Oxide-Semiconductor Field Effect Transistor (MOSFET).

The structure of a basic MOSFET is shown in Figure 6.1. The MOSFET shown here is an *n-channel* MOSFET and *p-channel* MOSFETs are basically identical but have the n-and p-type doping regions reversed.

A suitable positive voltage applied to the gate will cause the formation of an n-type conducting channel within the p-type semiconductor, allowing the flow of electrons between source and drain. This channel forms within approximately 5 nm of the surface of the silicon and just below the oxide insulating layer starting from the edge of the source region and terminating at the edge of the drain region.

Unlike other types of transistor devices such as the *bipolar junction transistor* (BJT) or the *junction field effect transistor* (JFET) introduced in Chapter 5 that require rather awkward structures, the MOSFET is formed at, or very close to, the surface of

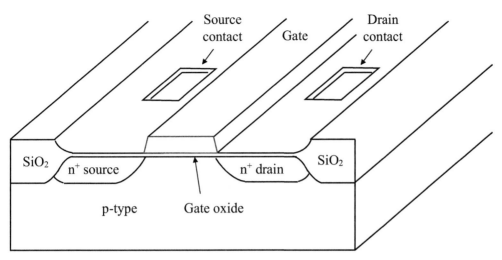

FIGURE 6.1 Cross section of a basic n-channel MOSFET structure. Electrons (n-type carriers) are blocked from passing from source to drain because the substrate is p-type. Depending on the voltage applied to the gate electrode, a thin n-type channel can be induced near the substrate surface and electrons can then flow between source and drain.

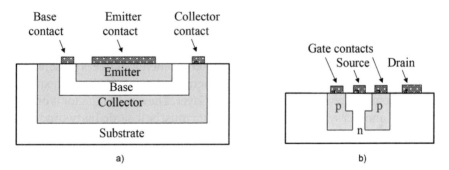

FIGURE 6.2 Cross sections of a) Bipolar Junction Transistor (BJT) and b) Junction Field Effect Transistor (JFET). These devices require buried structures when implemented using planar processing on a silicon wafer. The buried structures make size reduction challenging and costly. Transistor density for integrated circuits using these transistors is limited compared to MOSFET devices which are formed very near the silicon surface.

the silicon. This important attribute has been a key enabler of MOSFET size reduction. See Figure 6.2.

We firstly examine the portion of the MOSFET that lies between the source and drain in order to understand how the gate voltage controls the formation of a conducting channel. Following this in Section 6.9 the remarkable series of size reductions of nanoscale MOSFETs is discussed in the context of Moore's Law. Fabrication steps along with the relevant underlying physics and chemistry of photolithography, ion implantation, etching, oxidation, and deposition that result in nanoscale MOSFETs are covered in Sections 6.10–6.12.

6.2   MOSFET PHYSICS

The MOS capacitor is shown in Figure 6.3. This is the important portion of a MOSFET that exists between the source and drain regions of an n-channel MOSFET. We arbitrarily choose to focus on the n-channel MOSFET. For a p-channel MOSFET the semiconductor would be n-type.

Depending on the gate voltage relative to the silicon potential, an n-type channel may, or may not, be formed near the upper surface of the p-type semiconductor. Figure 6.4 shows four important cases that can result from four different gate voltages applied to a MOS capacitor with a p-type semiconductor.

At the semiconductor–oxide interface we define the modified work function based on the Fermi energy in the semiconductor. This is done to be consistent with the definition at the metal–oxide interface even though no electrons normally exist at the semiconductor Fermi level.

We also define $q\phi_F$ as the difference between the Fermi energy of the doped semiconductor and a reference energy level E_i located approximately mid-gap that follows the band bending of the conduction and valence bands. E_i will be more precisely defined later.

To simplify our initial treatment of the MOS capacitor, we assume that $q\Phi_m = q\Phi_s$ in which case the work function difference $\Phi_{ms} = \Phi_m - \Phi_s$ is zero. In Section 6.6 a more general case will be discussed.

Each example of biasing in Figure 6.3 may be viewed as a capacitor holding a specific static charge. There is no flow of current between metal gate and semiconductor because of the insulating layer. No external current or external energy source is present to maintain the state. Each of these conditions may therefore be regarded as an equilibrium state of the system *on a given side* of the insulator layer. The semiconductor is one system and the metal is another system. The MOS capacitor must be regarded as two independent systems because charge cannot flow between the two systems. This is important as it gives us license to plot the Fermi energy as a straight line within each system.

With no net charge on either the metal or the semiconductor as shown in Figure 6.4a) the *flat-band* condition exists. The energy bands in the semiconductor are horizontal and no electric fields exist.

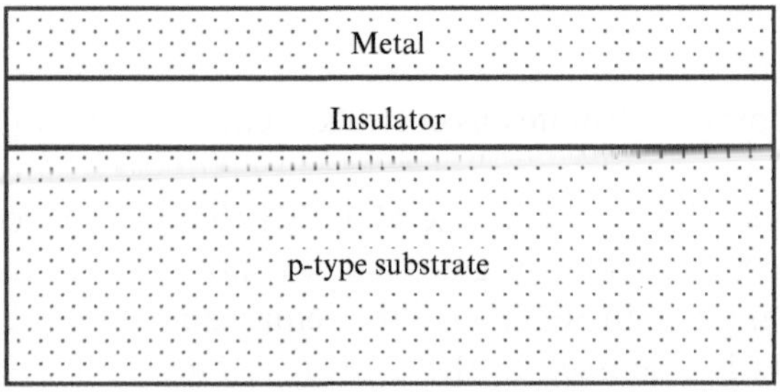

FIGURE 6.3   The MOS capacitor is composed of a p-type semiconductor substrate, an oxide insulator, and a metal. This structure extends from the source to the drain of a completed n-channel MOSFET.

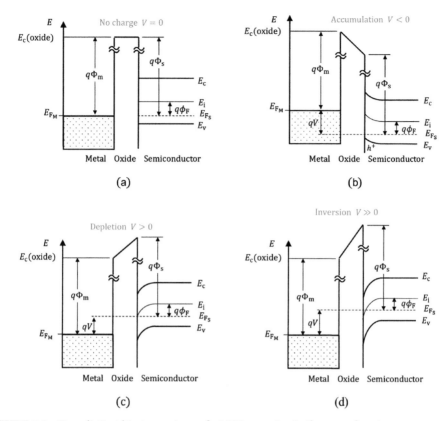

FIGURE 6.4 Four distinct biasing regimes of a MOS capacitor in the case of a p-type semiconductor. Each diagram plots electron energy versus position. a) flat-band: No charge is stored in the capacitor and the energy bands are horizontal. b) Accumulation: The metal is negatively charged with an equivalent positive charge in the semiconductor due to excess holes. c) Depletion: A small positive charge on the metal induces an equivalent negative charge in the semiconductor due to a depletion layer. d) Inversion: A larger positive charge on the metal induces an equivalent negative charge in the semiconductor. Electrons are attracted to near the silicon surface forming a channel.

Note that *modified work functions* are defined as $q\Phi_m$ at the metal surface and as $q\Phi_s$ at the semiconductor surface. Normally, a work function is defined as the energy required to remove an electron from the Fermi energy of a metal into the lowest vacuum energy level. In place of a vacuum level we need to consider the lowest available energy state in the oxide insulator layer. An insulator can be viewed as a semiconductor with a very wide bandgap. The location of the conduction band in the oxide layer defines this lowest available energy state. The energy difference between the Fermi energy of the metal and the conduction band in the oxide is the energy required for an electron to escape from the semiconductor surface. The modified work function is this energy difference. Note that the oxide layer affects the surface condition of both the metal and the semiconductor compared to standard surface conditions in a vacuum environment. This influences the effective work function. Many oxide materials are amorphous. In such cases the conduction band energy is not well defined as for crystalline materials and a less precisely defined band edge of the conduction band may exist.

When a negative charge is deposited on the gate by the application of a negative voltage to the gate relative to the semiconductor, the MOS capacitor goes into accumulation indicating that holes are attracted toward the semiconductor surface so as to balance the negative gate charge. The semiconductor surface is therefore induced to act more strongly p-type than in the bulk. This is called *accumulation* as shown in Figure 6.3b). Note that the valence band in the semiconductor moves closer to the Fermi energy which is consistent with a higher hole concentration. Bands are no longer horizontal in both the oxide and in the semiconductor due to the presence of electric fields.

Note that the vertical axis in Figure 6.3 is opposite in direction to the polarity of the gate bias. This is because energy band diagrams are oriented to show electron energy on the vertical axis whereas a voltage axis is, by convention, defined in the direction of energy for a positive charge.

Making the gate somewhat positive causes *depletion* as shown in Figure 6.3c). A negative charge in the capacitor is now achieved due to negatively charged acceptor ions in the semiconductor near its surface which balance the positive charge on the gate. Here, the semiconductor is depleted of holes near the surface which is consistent with the Fermi energy lying near mid-gap as shown.

Further increasing the positive gate bias will lead to *inversion* as shown in Figure 6.3d). Electrons in the conduction band of the semiconductor build up near the semi-conductor surface. The existence of these electrons is consistent with the Fermi energy at the semiconductor surface lying above mid-gap as shown. The increased positive gate charge remains balanced by the total negative charge in the semiconductor.

6.3   MOS CAPACITOR ANALYSIS

A more precise measure of the degree of accumulation, depletion, or inversion is available if we define E_i such that when $E_F - E_i > 0$ the semiconductor acts n-type and when $E_i - E_F > 0$ the semiconductor acts p-type. Since from Equation 2.36 the equilibrium electron concentration is

$$n_0 = N_c \exp\left(-\frac{(E_c - E_F)}{kT}\right)$$

we define E_i from

$$n_i = N_c \exp\left(-\frac{(E_c - E_i)}{kT}\right)$$

where n_i is the intrinsic electron concentration. Calculating the ratio of these equations

$$\frac{n_0}{n_i} = \exp\left(\frac{E_F - E_i}{kT}\right)$$

and therefore

$$n_0 = n_i \exp\left(\frac{E_F - E_i}{kT}\right) \tag{6.1}$$

Note that the semiconductor has intrinsic free electron concentration n_i if $E_F = E_i$.

An analogous result exists for holes. From Equation 2.37

$$p_0 = N_v \exp\left(-\frac{(E_F - E_v)}{kT}\right)$$

We again define E_i using

$$p_i = N_v \exp\left(-\frac{(E_i - E_v)}{kT}\right)$$

where p_i is the intrinsic hole concentration. The ratio of these equations is

$$\frac{p_0}{p_i} = \exp\left(\frac{E_i - E_F}{kT}\right)$$

and therefore

$$p_0 = p_i \exp\left(\frac{E_i - E_F}{kT}\right) \tag{6.2}$$

This again confirms that the semiconductor acts as intrinsic material if $E_F = E_i$. The way we have defined E_i clearly satisfies the condition that when $E_F - E_i > 0$ the semiconductor provides n-type behavior and when $E_i - E_F > 0$ the semiconductor provides p-type behavior.

The most interesting part of Figure 6.4 is the case of inversion shown in Figure 6.4d). If this MOS capacitor were part of a MOSFET device with n-type source and n-type drain, then the inversion layer just beneath the semiconductor surface would form a channel enabling the flow of electrons between source and drain. In Figure 6.5 the

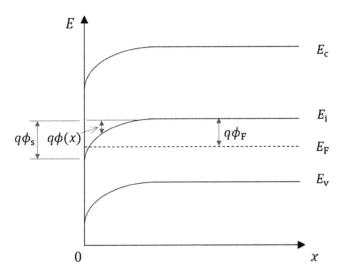

FIGURE 6.5 The semiconductor band diagram for the case of inversion in a MOS capacitor. The origin of the x-axis is defined at the semiconductor–oxide interface. Position-dependent band bending is defined as $q\phi(x)$. Band bending is $q\phi_s$ at the surface and $q\phi_F$ depends on the p-type doping level in the semiconductor.

inversion condition is shown in more detail. The amount of band bending at the semi-conductor surface is defined as $q\phi_s$. Here, ϕ_s is a quantity of electric potential.

From Equation 6.1

$$n_0 = n_i \exp\left(\frac{E_F - E_i}{kT}\right) = n_i \exp\left(\frac{-q\phi_F}{kT}\right)$$

If we want to evaluate electron concentration as a function of x we can write

$$n(x) = n_i \exp\left(-\frac{q\phi_F - q\phi(x)}{kT}\right) = n_0 \exp\left(\frac{q\phi(x)}{kT}\right) \tag{6.3a}$$

From Equation 6.2 for holes

$$p_0 = p_i \exp\left(\frac{E_i - E_F}{kT}\right) = p_i \exp\left(\frac{q\phi_F}{kT}\right)$$

and hole concentration as a function of x is

$$p(x) = p_i \exp\left(\frac{q\phi_F - q\phi(x)}{kT}\right) = p_0 \exp\left(\frac{-q\phi(x)}{kT}\right) \tag{6.3b}$$

We can now write down all the sources of charge that exist as a function of x. Expressing this as charge density $\rho(x)$ and including charges due to both free carriers and ionized dopants

$$\rho(x) = q\left(N_d^+ - N_a^- + p(x) - n(x)\right)$$

or

$$\rho(x) = q\left[p_0\left(\exp\left(\frac{-q\phi(x)}{kT}\right) - 1\right) - n_0\left(\exp\left(\frac{q\phi(x)}{kT}\right) - 1\right)\right] \tag{6.4}$$

All dopants are assumed to be ionized as in the intermediate temperature approxima-tion. Equilibrium hole concentration p_0 is therefore equal to N_a^-. Even though the current example is for a p-type semiconductor, this expression includes term N_d^+ which would be relevant in the case of an n-type semiconductor.

We will now make use of Gauss's law in order to obtain an expression for the electric field $\varepsilon(x)$ inside the semiconductor. Gauss's law was used previously to deter-mine the electric field in the p-n junction depletion layer in Chapter 3.

Imagine a thin, closed Gaussian box containing a portion of the semiconductor as shown in Figure 6.6. The Gaussian box is shown in red. The box faces perpendicular to the x-axis each have area A and these faces are assumed to be large relative to the sidewalls of the box such that infinite planes of charge can be assumed.

The Gaussian box encloses the region of the MOS capacitor from depth x_1 to depth x_2. Our goal is to find the electric field in the semiconductor layer at depth x_1. Depth x_2 is chosen such that the electric field there is zero because at this depth there is no band bending as shown in Figure 6.6b).

a)

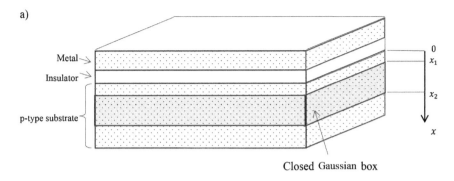

b)

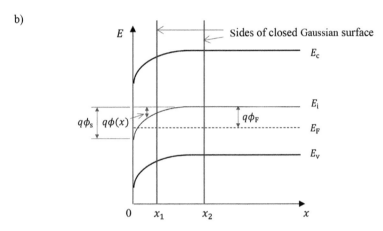

FIGURE 6.6 a) Gaussian box starting at depth x_1 below the semiconductor surface and extending to depth x_2. b) Band diagram of p-type semiconductor showing box faces, each of area A, at depth x_1 and at depth x_2. The origin of the x-axis is at the semiconductor surface.

Not all the charge in the MOS capacitor is inside this Gaussian box. Both the charge at the surface of the metal gate and the charge in the semiconductor in the depth range $0 \leq x \leq x_1$ lie outside the box. We have already defined the charge per unit area at the metal gate as Q_m. In addition, we define the charge per unit area in the semiconductor over depth range $0 \leq x \leq x_1$ to be Q_1 and the charge per unit area inside the box over depth range $x_1 \leq x \leq x_2$ to be Q.

The positive and negative MOS capacitor charge per unit area must remain balanced and therefore

$$Q_m = -Q_1 - Q$$

The total charge per unit area outside the Gaussian box is now $Q_m + Q_1 = -Q$. Note that this is equal in magnitude to the charge per unit area inside the Gaussian box but opposite in sign.

Now we will apply Gauss's law

$$\oint \varepsilon ds = \frac{QA}{\epsilon}$$

to the Gaussian box yielding

$$2\varepsilon_{box} A = \frac{QA}{\epsilon}$$

Since the charge inside the Gaussian box is negative, field lines considering only the charge inside the box point into the box and thread both the box face at x_1 and the box face at x_2. The field at x_1 points in the positive x direction and the field at x_2 points in the negative x-direction.

Electric field magnitude at each box face is therefore

$$\varepsilon_{box} = \left| \frac{Q}{2\epsilon} \right|$$

We know, however, that the net electric field at x_2 must be zero because there is no band bending there. This can be explained by the additional charge outside the Gaussian surface that gives rise to an additional electric field component pointing in the positive direction along the x-axis. We can regard this charge to be infinite in area in which case the electric field is independent of the distance from the charge. We determined that this charge per unit area is equal in magnitude but opposite in sign to the charge inside the Gaussian box, and the additional electric field therefore cancels the field due to the charge inside the Gaussian box at x_2.

At position x_1 the net electric field $\varepsilon(x_1)$ is the sum of the fields contributed by charge inside the box and by charge outside the box. These fields both point in the positive x-direction. Hence

$$\varepsilon(x_1) = \left| \frac{Q}{2\epsilon} \right| + \left| \frac{Q}{2\epsilon} \right| = \left| \frac{Q}{\epsilon} \right| \tag{6.5}$$

It is reassuring to know that at the metal gate, the net electric field contributed by both charge inside the box and charge outside the box is zero. See Problem 6.1.

In order to obtain Q we need to perform an integral of the charge density inside the Gaussian box. Using Equation 6.4 we obtain

$$Q = \int_{x_1}^{x_2} \rho(x)\,dx = q \int_{x_1}^{x_2} \left[p_0 \left(\exp\left(\frac{-q\phi(x)}{kT} \right) - 1 \right) - n_0 \left(\exp\left(\frac{q\phi(x)}{kT} \right) - 1 \right) \right] dx$$

Clearly, Q is a function of $\phi(x)$ as defined in Figure 6.5. We cannot directly integrate charge density as a function of x because we do not have an expression for $\phi(x)$.

However, both Q and ε depend on x_1. The electric field ε is now understood to decrease from $\varepsilon(x_1) = \left| \frac{Q}{\epsilon} \right|$ until it reaches $\varepsilon(x_2) = 0$. We could express this as follows[1]

$$\varepsilon(x_1) = -\int_{x_1}^{x_2} \frac{d\varepsilon}{dx}\,dx$$

Using Equation 6.5, and remembering that Q is negative, we can write

$$Q = -\epsilon\varepsilon(x_1) = \epsilon \int_{x_1}^{x_2} \frac{d\varepsilon}{dx}\,dx = q \int_{x_1}^{x_2} \left[p_0 \left(\exp\left(\frac{-q\phi(x)}{kT} \right) - 1 \right) - n_0 \left(\exp\left(\frac{q\phi(x)}{kT} \right) - 1 \right) \right] dx$$

We must therefore conclude that

$$\frac{d\varepsilon}{dx} = \frac{q}{\epsilon}\left[p_0\left(\exp\left(\frac{-q\phi(x)}{kT}\right)-1\right)-n_0\left(\exp\left(\frac{q\phi(x)}{kT}\right)-1\right)\right]$$

Also, by carefully examining Equation 2.48 and Figure 6.6

$$\varepsilon(x) = -\frac{d\phi(x)}{dx}$$

and we can write

$$\frac{d^2\phi(x)}{dx^2} = -\frac{q}{\epsilon}\left[p_0\left(\exp\left(\frac{-q\phi(x)}{kT}\right)-1\right)-n_0\left(\exp\left(\frac{q\phi(x)}{kT}\right)-1\right)\right] \tag{6.6}$$

Note that this equation may have been obtained directly using Poisson's equation. See Problem 6.2.

In order to obtain ε we will integrate both sides over the variable ϕ rather than over the variable x. It is not immediately obvious that this will be helpful, but the approach is motivated by the fact that the RHS can be readily integrated over ϕ.

Now,

$$\int_0^{\phi} \frac{d^2\phi}{dx^2}d\phi = -\frac{q}{\epsilon}\int_0^{\phi}\left[p_0\left(\exp\left(\frac{-q\phi}{kT}\right)-1\right)-n_0\left(\exp\left(\frac{-q\phi}{kT}\right)-1\right)\right]d\phi$$

The knot is unraveled by the following ingenious rewrite of the LHS:

$$\int_0^{\phi}\frac{d^2\phi(x)}{dx^2}d\phi = \int_0^{\phi}\frac{d^2\phi(x)}{dx^2}\frac{dx}{dx}d\phi$$

$$= \int_0^{\phi}\frac{d^2\phi(x)}{dx^2}\frac{d\phi(x)}{dx}dx = \int_0^{\phi}\frac{d\varepsilon(x)}{dx}\varepsilon(x)dx = \int_0^{\phi}\varepsilon(x)d\varepsilon$$

which is easy to integrate. We define the semiconductor dielectric constant as ϵ_s. We also define ε_s to be the electric field at the semiconductor surface. Since ϕ_s is the band bending at the semiconductor surface and because we are particularly interested in ε_s we now set $x_1 = 0$. This locates one face of the Gaussian box at the semiconductor surface meaning that

$$\int_0^{\varepsilon_s}\varepsilon(x)d\varepsilon = -\frac{q}{\epsilon_s}\int_0^{\phi_s}\left[p_0\left(\exp\left(\frac{-q\phi}{kT}\right)-1\right)-n_0\left(\exp\left(\frac{q\phi}{kT}\right)-1\right)\right]d\phi$$

and therefore

$$\frac{1}{2}\varepsilon_s^2 = \left(\frac{kTp_0}{\epsilon_s}\right)\left[\left(e^{-\frac{q\phi}{kT}}+\frac{q\phi}{kT}-1\right)+\frac{n_0}{p_0}\left(e^{\frac{q\phi}{kT}}-\frac{q\phi}{kT}-1\right)\right] \tag{6.7}$$

$$\varepsilon_s = \sqrt{\frac{2kTp_0}{\epsilon_s}}\left[\left(e^{-\frac{q\phi_s}{kT}}+\frac{q\phi_s}{kT}-1\right)+\frac{n_0}{p_0}\left(e^{\frac{q\phi_s}{kT}}-\frac{q\phi_s}{kT}-1\right)\right]^{\frac{1}{2}}$$

We can convert the semiconductor surface electric field ε_s to the total charge density. From Equation 6.5,

$$Q_s = -\sqrt{2kT\epsilon_s p_0} \left[\left(e^{-\frac{q\phi_s}{kT}} + \frac{q\phi_s}{kT} - 1 \right) + \frac{n_0}{p_0} \left(e^{\frac{q\phi_s}{kT}} - \frac{q\phi_s}{kT} - 1 \right) \right]^{\frac{1}{2}}$$

where Q_s is the total charge per unit area in the semiconductor. This equation is plotted in Figure 6.7 for silicon with $N_a = 4 \times 10^{15}\,\text{cm}^{-3}$ at room temperature. When the surface potential is zero there is no space charge because there is no band bending. For negative values of surface potential, the term $e^{-\frac{q\phi_s}{2kT}}$ will dominate and an exponential rise of positive charge density occurs which corresponds to accumulation. Holes accumulate very near the semiconductor surface. Due to the log scale in Figure 6.7, the plot of this exponential term approaches a straight line.

As the surface potential becomes positive, the term $\sqrt{\frac{q\phi_s}{2kT}}$ dominates at first and a slowly rising negative charge builds up in the semiconductor principally due to

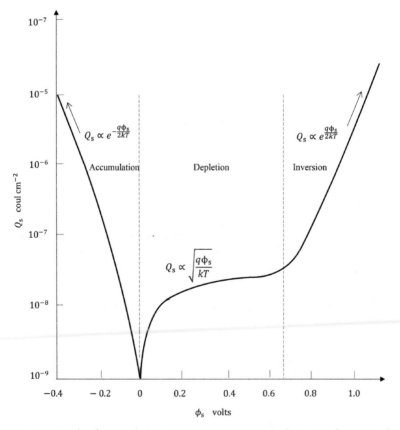

FIGURE 6.7 Magnitude of space charge per unit area in p-type silicon as a function of surface potential ϕ_s plotted for $N_a = 4 \times 10^{15}\,\text{cm}^{-3}$ at room temperature.

negatively ionized dopant ions. This is shown as positive in Figure 6.7 since charge density magnitude is shown. The semiconductor moves into depletion up to a maximum depth of hundreds of nanometers typically.

For larger values of positive surface potential, the term $e^{\frac{q\phi_s}{2kT}}$ will eventually dominate in spite of the fact that it is multiplied by $\sqrt{\frac{n_0}{p_0}}$ which is a very small quantity. Now the semiconductor moves into inversion, the depletion width having reached its maximum value. Negative charge builds up exponentially in the form of mobile electrons that form the inversion layer very close to the semiconductor surface and the curve again approaches a straight line. The magnitude of this negative charge density is plotted.

In Figure 6.8 we plot the charge density, the electric field, and the potential for a MIS capacitor in inversion. The metal has charge density Q_m which is sufficiently positive to cause inversion as shown in the charge distribution plot. This positive charge is balanced by a negative charge comprising two negative components. The first is due to negatively ionized acceptors in the substantially depleted layer of the semiconductor. The second component is the very thin inversion layer in which electrons concentrate. This may be expressed mathematically if we reintroduce the concept of a depletion approximation as was done in Chapter 3. If we define depletion region width as W_d and the inversion layer charge per unit area as Q_n (a negative quantity in this case) we can write

$$Q_m = qN_aW_d - Q_n \tag{6.8}$$

The applied voltage V is the sum of the voltage drop V_i across the insulator layer and ϕ_s across the semiconductor depletion region and therefore

$$V = V_i + \phi_s$$

We can use the depletion approximation and observe the analogy between the depleted semiconductor layer in this MOS capacitor and a depletion region in an n^+p junction. See Equation 2.15. The MOS capacitor depletion region width is therefore

$$W_d = \sqrt{\frac{2\epsilon_s\phi_s}{qN_a}} \tag{6.9}$$

As applied voltage V becomes more positive, band bending will increase and the depletion layer will tend toward a maximum width W_m. After this the additional negative charge required to balance Equation 6.8 will be in the form of inversion layer charge and band bending will no longer increase appreciably. Figure 6.6 shows the semiconductor just after it enters the inversion range. At the condition corresponding to the onset of inversion, the term $\sqrt{\frac{n_0}{p_0}}e^{\frac{q\phi_s}{kT}}$ becomes significant (i.e., ≥ 1). A definition of the threshold of "strong" inversion is therefore

$$\frac{n_0}{p_0}e^{\frac{q\phi_s}{kT}} = \frac{n_i^2}{N_a^2}e^{\frac{q\phi_s}{kT}} = 1$$

or

$$\phi_s \left(\text{threshold of strong inversion} \right) = \phi_{\text{strong}} = \dfrac{2kT \ln \left(\dfrac{N_a}{n_i} \right)}{q} \tag{6.10}$$

and from Equation 6.9

$$W_m = 2 \sqrt{\dfrac{\epsilon_s kT \ln \left(\dfrac{N_a}{n_i} \right)}{q^2 N_a}}$$

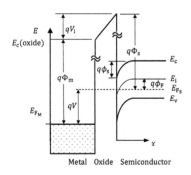

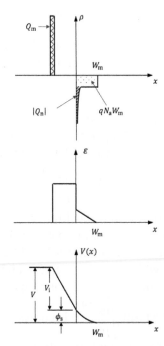

FIGURE 6.8 Band diagram, charge density, electric field, and potential plotted as a function of position in a MOS capacitor biased in inversion. The shaded areas in the charge density plot correspond to the labeled quantities.

Equation 6.10 may be simplified using the definition of ϕ_F from Figure 6.5. Using Equation 6.2 we see that

$$\phi_F = \frac{kT}{q}\ln\left(\frac{p_0}{p_i}\right) = \frac{kT}{q}\ln\left(\frac{N_a}{n_i}\right)$$

Therefore

$$\phi_{strong} = 2\phi_F \tag{6.11}$$

and

$$W_m = 2\sqrt{\frac{\epsilon_s\phi_F}{qN_a}}$$

We can now define V_T as the *threshold gate voltage* at the threshold of strong inversion. Using the depletion approximation, the depletion region charge per unit area is

$$Q_d = -qN_aW_m = -qN_a 2\sqrt{\frac{\epsilon_s\phi_F}{qN_a}}$$

The voltage drop across the dielectric layer is

$$V_i = \frac{Q_d}{C_i}$$

where C_i is the capacitance of the insulator layer per unit area.

The surface potential at threshold is given by Equation 6.11 and therefore at the onset of inversion,

$$V_T = -\frac{Q_d}{C_i} + 2\phi_F \tag{6.12}$$

6.4 ACCUMULATION LAYER AND INVERSION LAYER THICKNESSES

In the above analysis of the MOSFET we have not obtained information on the thicknesses of accumulation and inversion layers in which mobile charges exist.

We firstly need to introduce a new concept called the *Debye length*. This length is determined by a screening consideration since charges gathering near the semiconductor surface screen applied electric field at depths beyond the Debye length.

Consider a p-type silicon MOS capacitor in accumulation as shown in Figure 6.9. A negative charge is introduced on the gate electrode resulting in accumulation of holes near the semiconductor surface.

The electric field produced by the negatively charged gate electrode causes holes in the originally neutral semiconductor to move toward the surface. However the holes will be prevented from crowding right at the semiconductor surface due to hole diffusion away from the surface.

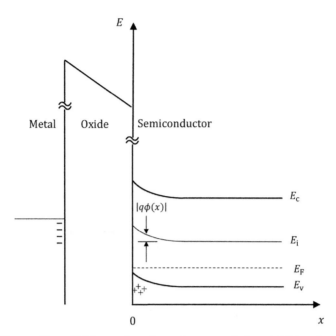

FIGURE 6.9 P-type silicon MOS capacitor shown in accumulation. Additional positive charge accumulates at the p-type semiconductor surface to balance negative charges on the metal. Note that $q\phi(x)$ is negative in this figure due to the way it was defined previously. See Figure 6.5.

We will define $\delta p(x)$ as the change in hole concentration relative to the equilibrium concentration far away from the semiconductor surface.

Therefore

$$p(x) = p_0 + \delta p(x)$$

Using Equation (6.3b)

$$\frac{p(x)}{p_0} = \exp\left(\frac{-q\phi(x)}{kT}\right) = 1 + \frac{\delta p(x)}{p_0}$$

If $\delta p(x)$ is small compared to p_0 then with a Taylor series approximation to first order

$$\exp\left(\frac{-q\phi(x)}{kT}\right) = 1 + \frac{\delta p(x)}{p_0} \cong 1 - \frac{q\phi(x)}{kT}$$

and therefore

$$\frac{\delta p(x)}{p_0} \cong -\frac{q\phi(x)}{kT}$$

or

$$\phi(x) = -\frac{kT}{q}\frac{\delta p(x)}{p_0}$$

Using Equation 6.6, and neglecting electron concentration which is much smaller than hole concentration,

$$\frac{d^2\phi(x)}{dx^2} = -\frac{q}{\epsilon}\left[p_0\left(\exp\left(\frac{-q\phi(x)}{kT}\right)-1\right)\right] = -\frac{q}{\epsilon}\left[p(x)-p_0\right] = -\frac{q}{\epsilon}\delta p(x)$$

and therefore

$$\frac{kT}{qp_0}\frac{d^2\delta p(x)}{dx^2} = \frac{q}{\epsilon}\delta p(x)$$

For this second-order differential equation the general solution is

$$\delta p(x) = A\exp\left(\frac{-x}{d}\right) + B\exp\left(\frac{x}{d}\right) + C$$

where

$$d = \sqrt{\frac{kT\epsilon}{q^2 p_0}}$$

From the band diagram in Figure 6.4b) we can see that as depth x increases, the excess hole concentration should approach zero. This means that $B = C = 0$ and we obtain solution

$$\delta p(x) = A\exp\left(\frac{-x}{d}\right)$$

The Debye length L_D is defined as characteristic layer depth

$$L_D = d = \sqrt{\frac{kT\epsilon}{q^2 p_0}}$$

and

$$\delta p(x) = A\exp\left(\frac{-x}{L_D}\right)$$

Debye length L_D depends on p_0 which, in turn, depends on semiconductor doping level N_a. At depths greater than the Debye length, the semiconductor is substantially shielded from the gate charge.

EXAMPLE 6.1

Find the Debye length of holes in a MOS capacitor in accumulation at room temperature if the semiconductor is silicon with a p-type doping level of $N_a = 2\times10^{16}\,\text{cm}^{-3}$.

Solution

$$L_D = \sqrt{\frac{kT\epsilon}{q^2 p_0}} = \sqrt{\frac{0.026\text{eV}\times1.6\times10^{-19}\text{coul}\times11.8\times8.85\times10^{-14}\text{Fcm}^{-1}}{\left(1.6\times10^{-19}\text{coul}\right)^2\times2\times10^{16}\text{cm}^{-3}}} = 29\,\text{nm}$$

Therefore the characteristic thickness of the accumulation layer in this MOS capacitor is 29 nm. L_D decreases if N_a is increased.

The above Debye length derivation assumes weak electric fields which are consistent with only slight band bending in accumulation as shown in Figure 6.4b). However in the case of a p-type semiconductor and an n-type inversion layer, the surface electric field causing electrons to drift toward the semiconductor surface is stronger as shown in Figure 6.4d). It can be evaluated at the onset of strong inversion by considering the charge in the depletion layer.

$$Q_d = -qN_aW_m$$

From Equation 6.5 this electric field is

$$\varepsilon = \left|\frac{Q_d}{\epsilon}\right| = \frac{qN_aW_m}{\epsilon}$$

The inversion layer is subject to diffusion which, in equilibrium, balances the electron drift. Therefore we can write

$$q\mu_n n(x)\varepsilon + qD_n \frac{dn(x)}{dx} = 0$$

or

$$\frac{dn(x)}{dx} = -\frac{\mu_n \varepsilon}{D_n} n(x)$$

which has solution

$$n(x) = Ae^{-\frac{x}{\delta}}$$

where

$$\delta = \frac{D_n}{\mu_n \varepsilon} = \frac{kT}{q\varepsilon}$$

becomes the characteristic thickness of the inversion layer.

EXAMPLE 6.2

Estimate the thickness of the inversion layer at the onset of strong inversion for the MOS capacitor of Example 6.1.

Solution

The width of the depletion region at the onset of strong inversion is

$$W_m = 2\sqrt{\frac{\epsilon_s \phi_F}{qN_a}}$$

where

$$N_a = 2 \times 10^{16} \, cm^{-3}$$

$$\phi_F = \frac{kT}{q}\ln\left(\frac{N_a}{n_i}\right) = 0.026 \text{ eV } \ln\left(\frac{2\times10^{16} \text{ cm}^{-3}}{1.5\times10^{10} \text{ cm}^{-3}}\right) = 0.367 \text{ V}$$

$$W_m = 2\sqrt{\frac{\epsilon_s\phi_F}{qN_a}} = 2\sqrt{\frac{8.85\times10^{-14} \text{ Fcm}^{-1}\times11.8\times0.367 \text{ V}}{1.6\times10^{-19} \text{ coul}\times2\times10^{16} \text{ cm}^{-3}}} = 2.19\times10^{-5} \text{ cm}$$

$$\varepsilon = \frac{qN_aW_m}{\epsilon} = \frac{1.6\times10^{-19} \text{ coul}\times2\times10^{16} \text{ cm}^{-3}\times2.19\times10^{-5} \text{ cm}}{8.85\times10^{-14} \text{ Fcm}^{-1}\times11.8}$$

$$= 6.7\times10^4 \text{ Vcm}^{-1}$$

$$\delta = \frac{D_n}{\mu_n\varepsilon} = \frac{kT}{q\varepsilon} = \frac{0.026 \text{ V}}{6.7\times10^4 \text{ Vcm}^{-1}} = 3.9\times10^{-7}\text{cm} = 3.9 \text{ nm}$$

Therefore the characteristic thickness of the inversion layer at the onset of inversion is 3.9 nm. Note that the inversion layer is thinner than the accumulation layer of Example 6.1 as expected, and that δ decreases if N_a is increased.

6.5 CAPACITANCE OF MOS CAPACITOR

The MOS capacitor capacitance C per unit area may be modeled as two capacitors in series. The first capacitance C_i represents the oxide layer capacitance per unit area and the second capacitance C_d represents the depletion region capacitance per unit area.

While C_i is a fixed capacitance, C_d depends on the bias conditions.

In accumulation there is no depletion layer and the net capacitance is simply C_i per unit area. The p-type silicon acts as a conductive electrode due to the mobile hole charge. In Section 3.11 it was shown that capacitance is defined as $\left|\frac{dQ}{dV}\right|$ which, in the context of our MOS semiconductor layer, is $C_d = \left|\frac{dQ_s}{d\phi_s}\right|$

Figure 6.7 plots Q_s versus ϕ_s. The steep portion of this figure where accumulation exists yields values of $\left|\frac{dQ_s}{d\phi_s}\right|$ that are sufficiently large such that $C_d \gg C_i$ and therefore

$$C = \frac{C_iC_d}{C_i+C_d} \cong C_i$$

As the bias conditions change to depletion, a depletion layer forms and then grows. $C_d = \frac{\epsilon_s}{W}$ will decrease to a minimum value as depletion layer thickness reaches its maximum value W_m and now the series capacitance reaches a minimum value

$$C_{min} = \frac{C_iC_d}{C_i+C_d} = \frac{C_i\dfrac{\epsilon_s}{W_m}}{C_i+\dfrac{\epsilon_s}{W_m}} = \frac{C_i\epsilon_s}{C_iW_m+\epsilon_s}$$

This is plotted in Figure 6.10. C_{min} exists at the bias voltage where the slope in Figure 6.7 is smallest which is just below the threshold voltage for strong inversion V_T.

Once bias reaches the condition of inversion, a new source of charge must be considered. This is inversion layer electron charge which is produced in the silicon.

From Figure 6.7, the steep portion of this figure where inversion exists yields values of $\left|\dfrac{dQ_s}{d\phi_s}\right|$ that are again sufficiently large that $C_d \gg C_i$ and therefore $C \cong C_i$. The fact that this inversion charge can vary in magnitude near the silicon surface in response to bias conditions means that the depletion layer capacitance is effectively bypassed by the inversion charge and is no longer relevant. The inversion charge is supplied and depleted by generation and recombination processes, respectively, in order to attain the equilibrium conditions of Figure 6.7. The black curve in Figure 6.10 is valid.

If an applied AC voltage of a high enough frequency is applied, however, the generation and recombination processes do not occur fast enough and the red curve will apply. Inversion layer charge will not have sufficient time to achieve equilibrium levels in response to the applied voltage. At sufficiently high applied frequencies the inversion layer charge will remain fixed. Carrier recombination times in the silicon used for MOSFETs are typically in the 100 µs time scale which means that the frequency range should be in the MHz range to be considered "high." The capacitance of the depletion layer must therefore be included.

Detailed time-dependent modeling of inversion layer charge on gate voltage in a MOS capacitor is a complex topic because it involves charge generation or depletion in both the bulk and the depletion regions, changes in depletion layer thickness, and drift current that flows due to the lack of equilibrium conditions. This is studied by

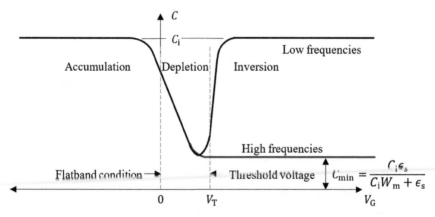

FIGURE 6.10 MOS capacitor series capacitance per unit area versus gate voltage for p-type silicon. This capacitance is the series capacitance of the oxide layer capacitance C_i and the depletion layer capacitance C_d, the latter being both voltage and frequency dependent. Just below V_T the series capacitance reaches a minimum value which is consistent with the position of minimum slope in Figure 6.7. The flat-band condition is shown at zero gate voltage; however, in real devices the gate voltage will not be zero. See Section 6.6.

analyzing the impulse response of the MOS capacitor to a step change in gate voltage and will not be discussed in detail here.

In a completed MOS transistor, carriers are readily available near the silicon surface from the source and drain regions to flow in and out of the inversion layer very rapidly. The black curve in Figure 6.10 will therefore be observed even at high frequencies.

6.6 WORK FUNCTIONS, TRAPPED CHARGES, AND ION BEAM IMPLANTATION

In a practical MOSFET we cannot assume that $\Phi_m = \Phi_s$. In reality Φ_s depends on both the semiconductor and the semiconductor doping condition. Also the type of conductor used for the gate will affect Φ_m.

Due to the high temperatures required for MOSFET fabrication, the gate material of choice was historically polycrystalline, highly doped silicon, which can be deposited on top of the gate oxide, although refactory metal gates have now largely replaced polycrystalline silicon.

The correction term Φ_{ms} must be added to Equation 6.12. The flat-band condition in Figure 6.3a) is now only reached if an applied bias of Φ_{ms} is applied to the gate. Specific values of Φ_{ms} depend on the gate material, the semiconductor used for the channel, and the doping concentration of the semiconductor. For a p-channel MOS capacitor,

$$V_T = \Phi_{ms} - \frac{Q_d}{C_i} + 2\phi_F$$

Trapped charge in a MOS capacitor is yet another consideration that affects threshold voltage. There are several types of trapped charge:

1. Bulk oxide layer trapped charge can be due to impurity ions such as Na^+ or native defects in the oxide insulator layer associated with incomplete or dangling bonds. The location of these charges within the insulator layer affects the influence of these charges on the threshold voltage. Charges closer to the semiconductor interface will influence threshold voltage the most.

   Mobile ions such as Na^+ must be carefully avoided because they can migrate during the operation of a MOSFET due to the operational electric fields present. The existence of such ions can be checked by applying and maintaining a strongly positive gate voltage at elevated temperature for a few minutes. Then threshold voltage can be measured. Following this, a strongly negative gate voltage may be applied for a similar period of time at elevated temperatures. If threshold voltage shifts it can be explained by the movement of impurity ions in the applied electric field.

   Until about 2007 the oxide layer in MOSFETs was almost exclusively SiO_2 however, this layer has now been largely replaced by HfO_2 (hafnium oxide). HfO_2 has a relative dielectric constant of approximately 25 which somewhat depends on the deposition method used, whereas SiO_2 has a relative dielectric constant of 3.9. Hf^{4+} is one of only a few cations that has a higher affinity for oxygen than silicon. This is critical since the goal is to prevent the formation of SiO_2 to maintain a high relative dielectric constant. Nevertheless the interface of HfO_2 with the silicon surface includes at least one or two atomic layers of SiO_2. HfO_2 is commonly referred to as a *high K* dielectric material and is deposited as an amorphous layer.

2. Trapped interface charges at the SiO_2/Si interface include charges due to incomplete oxidation of SiO_2 at the silicon surface, as well as charges due to dangling silicon bonds. Interface charge depends on the growth method and growth rate used for the SiO_2 as well as on the crystalline orientation of the silicon surface. A [111] surface typically has a positive charge density of 10^{11} cm^{-2} and a [100] surface has an order of magnitude lower charge density. Therefore silicon wafers for MOS devices are generally cut as [100] wafers.

An *effective positive interface charge density* denoted Q_i just inside the oxide layer at the silicon interface can be defined to represent all sources of trapped charges. This positive charge reduces threshold voltage since it reduces the voltage drop across the oxide by an amount $\dfrac{Q_i}{C_i}$. Therefore

$$V_T = \Phi_{ms} - \frac{Q_i}{C_i} - \frac{Q_d}{C_i} + 2\phi_F$$

The flat-band condition shown in Figure 6.4a) and again in Figure 6.10 will no longer occur at gate voltage $V_G = 0$. It will instead occur at

$$V_G = \Phi_{ms} - \frac{Q_i}{C_i}$$

In some cases, trapped charges as well as work function differences can lead to *depletion mode* MOSFETs in which an inversion layer can form even with a gate voltage of zero, whereas usually the more desirable situation is to require a positive gate voltage (for p-type silicon) or a negative gate voltage (for n-type silicon) for inversion to occur. This is known as *enhancement mode* operation.

It is possible to set the threshold voltage by *ion beam implantation* on a given silicon wafer. By embedding a carefully controlled dopant dose very near the silicon surface, the effective value of Q_d can be altered to meet specifications during production. Since dopants are ionized, their charge contributes to the term Q_i which therefore influences V_T. The technique is particularly important and effective since the dose can be customized for each silicon wafer in production to account for unavoidable variations in other parameters such as substrate doping concentration, interface charge density, and oxide thickness. See Section 6.11 for details of this technique.

6.7   SURFACE MOBILITY

Surface mobility in silicon is reduced relative to bulk mobility. In Chapter 2 a series of bulk scattering processes are discussed including phonon, impurity, and defect scattering. In the case of the channel in a MOSFET, however, additional scattering types must be considered when considering lateral current flow. These include

(i) surface scattering in which the abrupt termination of the silicon surface causes a new scattering mechanism
(ii) surface roughness scattering. Scattering from atomic-scale surface roughness of the silicon results from deviations from a planar surface
(iii) coulombic scattering due to trapped charges in the gate oxide close to the silicon surface
(iv) surface phonon scattering that results from surface phonon modes which are distinct from bulk phonon modes. Since phonons are modeled as the vibrations of masses (atoms) coupled by springs (bonds), surface phonons in the silicon will be strongly influenced by the incomplete bonding at the silicon–SiO$_2$ interface

All the scattering rates need to be summed together to obtain a total scattering rate. The reciprocal of the total scattering rate is the effective scattering time τ_s which then determines surface mobility $\mu_s = \dfrac{q\tau_s}{m^*}$.

The degree to which surface mobility drops below bulk mobility depends on the proximity of the charge carriers to the silicon surface and therefore on the strength of the transverse electric field provided by the gate bias that draws carriers toward the silicon surface. Consider the charge per unit area for an n-channel MOSFET device shown in Figure 6.11.

We can calculate an "effective" transverse electric field $\varepsilon_{\text{effective}}$ halfway through inversion layer charge density Q_n assuming the gate area A is very large compared to

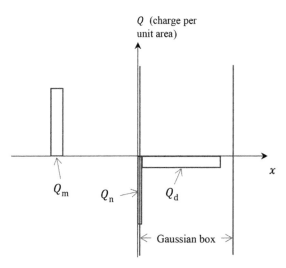

FIGURE 6.11   Charge per unit area versus depth in an n-channel MOSFET. Note that a Gaussian box has been introduced that cuts through the center of the inversion layer.

device thickness. We will assume that a uniform charge density exists through the channel thickness. Q_n represents charge per unit gate area.

From Equation 6.5, assuming we define x_1 to be halfway through the inversion layer,

$$\varepsilon_{\text{effective}} = \left| \frac{Q}{\epsilon} \right|$$

where Q is the charge per unit area inside the Gaussian box shown in Figure 6.11.

Therefore

$$\varepsilon_{\text{effective}} = \left| \frac{Q}{\epsilon} \right| = \frac{1}{\epsilon_s}\left(\frac{Q_n}{2} + Q_d \right)$$

It is experimentally found that, for a given silicon substrate orientation, a relationship between μ_s for electrons and $\varepsilon_{\text{effective}}$ exists. The measured result is shown in Figure 6.12. Note that the curve is temperature dependent due to temperature dependence of the phonon scattering rate. The electron mobility values are well below the pure bulk silicon value of $1350 \, \text{cm}^2\text{V}^{-1}\text{s}^{-1}$. Surface effects on mobility are clearly evident. For p-channel MOSFETs, the hole mobility can also be predicted based on a suitably calculated effective electric field.

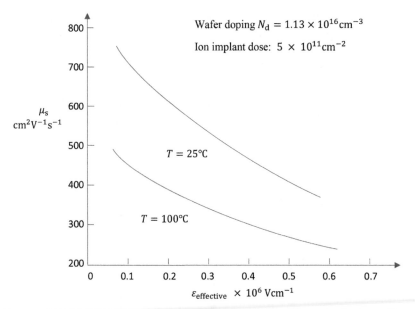

FIGURE 6.12 Based on numerous experimental results this plot shows the "universal" relationship between surface mobility and effective transverse electric field for the $\langle 100 \rangle$ silicon surface at temperatures of 25°C and 100°C. The relationship is found to be insensitive to doping levels provided doping is in a reasonable range that is normally used in MOSFET production. A typical bulk wafer doping level is in the range of $10^{16} \, \text{cm}^{-3}$ followed by ion implantation of P and/or B dopants. Specific parameters are shown in the graph, but the *shape* of the curves is universal and will apply to a range of other doping and implant values. Adapted from Sabnis and Clemens, 1979.

6.8   MOSFET TRANSISTOR CHARACTERISTICS

As detailed in Figure 6.13, the gate voltage V_G is equal to the sum of insulator layer voltage V_i and the total band bending in the semiconductor ϕ_s.

We know that at the onset of strong inversion, gate threshold voltage is

$$V_T = \Phi_{ms} - \frac{Q_i}{C_i} - \frac{Q_d}{C_i} + 2\phi_F \qquad (6.13)$$

For simplicity, we will make the approximation that the depletion width does not increase beyond the threshold voltage. An increase in gate voltage beyond threshold voltage which further increases positive gate charge will induce negative charge density Q_n in the inversion layer. This approximation implies that the MOS capacitance at high frequencies reaches its minimum value at V_T. The approximation can be understood from Figure 6.10.

Therefore

$$V_G - V_T = \frac{|Q_n|}{C_i}$$

and

$$|Q_n| = C_i (V_G - V_T)$$

Consider the MOSFET shown in Figure 6.14. A voltage drop from drain to source will cause lateral drift current I_D to flow in the inversion layer which constitutes the channel current of the MOSFET. Until now we have defined the x-axis to be normal to the semiconductor surface but we now redefine it along the channel length with its origin at the source end of the inversion layer as shown.

The voltage at any point x along the channel length is V_x. This voltage falls to zero at the source since the source voltage V_s is set to zero, and it rises to drain voltage V_D at the drain. Channel current I_D flows from drain to source along the channel of length L and width W. Since V_x reduces the voltage across the insulator we obtain

$$|Q_n(x)| = C_i (V_G - V_T - V_x) \qquad (6.14)$$

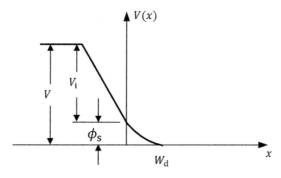

FIGURE 6.13   Electrostatic potential in MOSFET as a function of position.

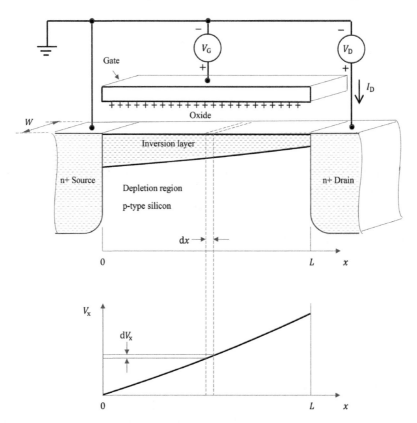

FIGURE 6.14 Transverse potential along the channel of length L and width W of an n-channel MOSFET. The x-axis lies along the direction of channel length.

where $Q_n(x)$ is now a position-dependent quantity. Since $Q_n(x)$ is composed of electrons it is negative. We will make use of the *gradual channel approximation* in which we assume that the voltages vary gradually along the channel from the drain to the source. This approximation allows the use of one-dimensional modeling in the x-axis and one-dimensional modeling perpendicular to the x-axis rather than more complex two-dimensional modelling.

Now consider the lateral flow of current along the channel. Consider a charge dQ contained in differential channel length dx that moves a distance dx in time interval dt. Also, carrier mobility relates carrier velocity to electric field.

Current from drain to source in the channel is therefore given by

$$I_D = \frac{dQ}{dt} = \frac{W dx |Q_n(x)|}{dt} = W \mu_s \varepsilon(x) |Q_n(x)| = W \mu_s \frac{dV_x}{dx} |Q_n(x)|$$

where μ_s is surface mobility. For simplicity, μ_s is assumed to be independent of position x.

Substituting from Equation 6.14 and integrating along the length of the channel from source to drain we obtain

$$\int_0^L I_D dx = W\mu_s C_i \int_0^{V_D} (V_G - V_T - V_x) dV_x$$

and therefore

$$I_D = \frac{W\mu_s C_i}{L}\left[(V_G - V_T)V_D - \frac{V_D^2}{2}\right] \tag{6.15}$$

This equation assumes that the drain voltage is not high enough to substantially change the depletion layer thickness and the resulting depletion layer charge density Q_d. More accurate modeling that will not be considered here should include a dependence upon x of depletion layer charge $Q_d(x)$.

For low drain voltages, neglecting the quadratic term in Equation 6.15, we obtain

$$g = \frac{dI_D}{dV_D} = \frac{W\mu_s C_i}{L}(V_G - V_T)$$

where we define g to be the *conductance* of the MOSFET at low drain voltages. As shown in Figure 6.15, g is equal to the slope of each curve near the origin at a given gate voltage. The MOSFET drain and source terminals therefore act as a resistor at low drain voltages and the resistance can be set by V_G.

Due to the decrease of effective gate voltage toward the drain side of the channel, each curve in Figure 6.15 saturates. The channel is now pinched off. In saturation, the inversion layer starts from the source but only exists for a portion of the distance

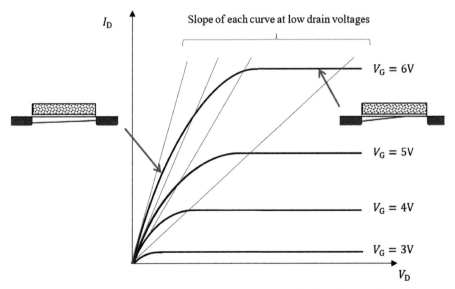

FIGURE 6.15   Drain current for an n-channel MOSFET. Threshold voltage in this case is just below 3 volts. The slopes of the curves representing conductance values g at low drain voltages are shown in blue. The inset on the left illustrates the continuous inversion layer below saturation. In the saturation regime the inversion layer is pinched off and does not reach the drain. See inset on the right.

between source and drain from an equilibrium viewpoint. See insets in Figure 6.15. However electrons in the portion of the channel starting from the source will nevertheless stream across the pinched-off region very near the silicon surface until they reach the drain. Drain current is independent of drain voltage in the saturation regime.

The onset of saturation can be understood from Equation 6.15. At the drain end of the channel

$$|Q_n(L)| = C_i(V_G - V_T - V_D)$$

The channel enters the pinch-off regime when $|Q_n(L)| = 0$ and therefore the drain voltage at which saturation starts is

$$V_{D \text{ onset of saturation}} = V_G - V_T$$

Saturation current can be found by substituting this drain voltage into Equation 6.15 to obtain

$$I_{D \text{ sat}} = \frac{W \mu_s C_i}{2L}(V_{D \text{ onset of saturation}})^2 = \frac{W \mu_s C_i}{2L}(V_G - V_T)^2$$

We also define saturation *transconductance* $g_{m(\text{sat})}$ as

$$g_{m(\text{sat})} = \frac{dI_{D \text{ sat}}}{dV_G} = \frac{W \mu_s C_i}{L}(V_G - V_T)$$

Saturation transconductance is the change in saturation current with respect to a change in gate voltage. Note that g_m increases as gate voltage increases, which explains the increasing spacings between saturation currents in Figure 6.15 as gate voltage increases.

The above expressions for conductance g and saturation transconductance $g_{m(\text{sat})}$ contain the insulator capacitance C_i which, in turn, depends on the relative dielectric constant of the oxide layer. This explains the current use of HfO_2, its dielectric constant being approximately six times higher than SiO_2. Higher values of g and g_m allow MOSFET operation with higher switching speeds, lower operating voltages, and lower total power consumption.

In the pinch-off region, carrier velocity is high and, particularly in nanoscale MOSFETs the carrier velocity can reach saturation velocity. This has consequences. See Section 6.16.

EXAMPLE 6.3

An n-channel MOSFET is based on the MOS capacitor of Example 6.1. Assume that the insulator layer is hafnium oxide with relative dielectric constant 25 and thickness 10 nm. The gate metal is aluminum with modified work function 4.1 eV.

The modified work function of silicon depends on doping and is 4.05 eV for $N_a = 2 \times 10^{16} \, \text{cm}^{-3}$. If the effective positive trapped charge density at the silicon surface is $2 \times 10^{13} \, \text{cm}^{-2}$, find

a) Threshold voltage
b) Conductance if V_G is 0.5 volts above threshold
c) Saturation transconductance if V_G is 0.5 volts above threshold

You may neglect the very thin SiO_2 layer at the silicon surface. Assume the surface mobility is $800 \, \text{cm}^2\text{V}^{-1}\text{s}^{-1}$. Channel length is 100 nm and channel width is 100 nm.

Solution

a)

$$C_i = \frac{\epsilon_0 \epsilon_r}{d} = \frac{8.85 \times 10^{-14} \, \text{Fcm}^{-1} \times 25}{10 \times 10^{-7} \, \text{cm}} = 2.21 \times 10^{-6} \, \text{Fcm}^{-2}$$

From Example 6.2, $W_m = 2.19 \times 10^{-5}$ cm, $N_a = 2 \times 10^{16} \, \text{cm}^{-3}$, $\phi_F = 0.367$ V
Therefore

$$Q_d = -q N_a W_m = -1.6 \times 10^{-19} \, \text{C} \times 2 \times 10^{16} \, \text{cm}^{-3} \times 2.19 \times 10^{-5} \, \text{cm}$$
$$= -7.01 \times 10^{-8} \, \text{C cm}^{-2}$$

and

$$V_T = \Phi_{ms} - \frac{Q_i}{C_i} - \frac{Q_d}{C_i} + 2\phi_F = \Phi_m - \Phi_s - \frac{Q_i}{C_i} - \frac{Q_d}{C_i} + 2\phi_F$$

Therefore

$$V_T = 4.1 \, \text{V} - 4.05 \, \text{V} - \frac{2 \times 10^{13} \times 1.6 \times 10^{-19} \, \text{C cm}^{-2}}{2.21 \times 10^{-6} \, \text{F cm}^{-2}}$$
$$- \frac{-7.01 \times 10^{-8} \, \text{C cm}^{-2}}{2.21 \times 10^{-6} \, \text{Fcm}^{-2}} + 2 \times 0.367 \, \text{V}$$

$$= 0.05 \, \text{V} - 1.45 \, \text{V} + 3.17 \times 10^{-2} \, \text{V} + 0.73 \, \text{V} = -0.64 \, \text{V}$$

This is therefore a depletion mode n-channel MOSFET.

b)

$$g = \frac{W \mu_s C_i}{L}(V_G - V_T) = \frac{100 \times 10^{-7} \, \text{cm} \times 800 \, \text{cm}^2\text{V}^{-1}\text{s}^{-1} \times 2.21 \times 10^{-6} \, \text{Fcm}^{-2}}{100 \times 10^{-7} \, \text{cm}}$$
$$\times 0.5 \text{V} = 8.84 \times 10^{-3} \, \text{AV}^{-1}$$

or $g = 8.84 \times 10^{-3}$ S (siemens)

c) $g_{m(sat)} = g = 8.84 \times 10^{-3}$ S

6.9   MOSFET SCALING

Gate length (distance from source to drain) is in the nanometer regime in virtually all of today's high density integrated circuits that rely on the MOSFET as the building block for complex functionality in memory chips, graphic processors, and microprocessors. Gate length is an important measure of the generation or *technology node* of MOSFET integrated circuits. Gate lengths have been decreasing since the inception of the MOSFET in the 1970s and are migrating to well below 10 nm in state-of-the-art integrated circuits. A graph of MOSFET gate length versus year is shown in Figure 6.16. This graph leads directly to Moore's Law shown in Figure 6.17 which correctly predicted, in 1975, that the number of devices that can be integrated onto a single microchip would double every two years.

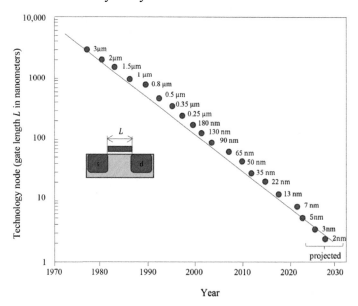

FIGURE 6.16   MOSFET gate length versus year. In addition to gate length reduction over time, MOSFET designs deviated from planar structures after 2010. See Chapter 9.

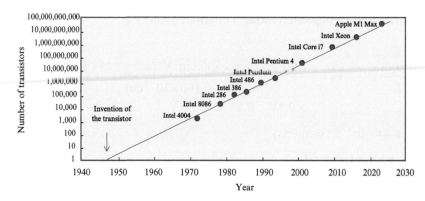

FIGURE 6.17   Moore's law showing that the number of MOSFET transistors on a silicon chip doubles every 2 years.

6.10   NANOSCALE PHOTOLITHOGRAPHY

Since its inception in the 1960s, integrated circuit manufacturing has relied on the technique of *photolithography*. This versatile fabrication method has evolved through successive generations from the microscale and into the nanoscale.

Cost reductions per MOSFET are associated with each technology node. In 1990, memory device MOSFET costs were on the order of $100,000 for 10^9 MOSFETs, enough for a gigabit of memory storage. In 2020 the cost had dropped by 6 orders of magnitude to $0.10 for 10^9 MOSFETs. This dramatic cost reduction could not have been achieved without photolithography.

The photolithography process allows for the patterning of various layers in nanoscale electronic devices. To perform a photolithography step, a polymer *photoresist* material is first applied as a thin film coating to a surface supported by a silicon wafer. The photoresist is then patterned by *exposing* only selected areas of the photoresist to photon irradiation by means of a *photomask*. The photomask causes light that illuminates the photoresist to be spatially patterned. A very basic photomask comprises an optically transparent glass plate with an opaque metal film such as chromium deposited on it in regions where light is to be blocked.

The light affects the ability of the photoresist to be chemically dissolved and the photoresist is subsequently selectively removed by a liquid *developer*. After this step, the remaining photoresist is baked at a low temperature to be strengthened and hardened. It now forms a patterned barrier layer on the silicon wafer. See Figure 6.18.

An example of a simple photoresist is polymethyl methacrylate (PMMA). PMMA is available in liquid form as a long chain polymer (typical chain length may be on the order of 10^5 mers) dissolved in an organic solvent such as methyloxybenzene (anisole). This solvent evaporates and is released during the baking step soon after the PMMA is deposited on the silicon wafer.

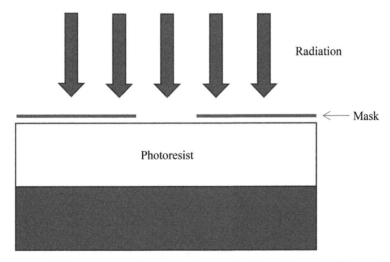

FIGURE 6.18   Radiation exposes photoresist after passing through a mask. The mask spatially defines those regions of the photoresist that are exposed and those that are not.

Because UV light exposure positively affects the ability for a liquid solvent to dissolve and remove the exposed areas of the resist, PMMA functions as a *positive* photoresist. The total flux of light arriving per unit area of the photoresist is called the *dose*. UV photons with wavelengths below 250 nm are able to cut the polymer chains in the PMMA. This process is referred to as chain *scission*. This dramatically increases the solubility of the PMMA by a factor of 1000 or more compared to the unexposed PMMA in the appropriate solvent. Solvents suited to this purpose include methyl isobutyl ketone or isopropyl alcohol (colorless organic liquids). A solvent in the context of this role is referred to as a *developer* because it allows the desired pattern to develop in the photoresist. The *development rate* controls the length of time it takes the developer to remove the photoresist for a given dose. If the dose is too small then the photoresist may be thinned but not fully removed by the developer.

Several photoresist types other than PMMA are available including *negative* photoresists in which exposure to radiation negatively affects their ability to dissolve in the relevant developer due to photo-initiated polymer chain *crosslinking*. In this case the developer removes only the unexposed photoresist.

The smallest feature size that can be formed, and hence the smallest possible gate length of a device such as a MOSFET, is associated with the wavelength λ of the photons used to selectively irradiate the photoresist through the photomask. As MOSFETs have become smaller, light used in integrated circuit production equipment has decreased through a range of UV wavelengths from 435 nm in the 1980s down to 13.5 nm (extreme UV) by 2020.

The dependence of achievable feature size on wavelength is understood by considering single slit diffraction. The actual patterns produced in the photoresist include these diffraction effects. See Figure 6.19. A basic optical mask can be thought of as a pattern of openings acting like slits.

Single slit diffraction occurs as light in the form of a plane wave of wavelength λ passing through a slit of width d spreads out beyond the slit forming a larger feature in the photoresist layer. Each point on the wavefront arriving at the slit must be considered an effective point source emerging from its specific position in the slit. At one specific angle θ_{min} the phase shift between any pair of point sources spaced apart by $d/2$ is 180 degrees. Summing all possible such pairs accounts for all the possible point sources in the slit and therefore complete cancellation of the light sources occurs at this angle.

The path length from each point source of a given such pair of sources to the photoresist therefore must differ by

$$\delta = \frac{\lambda}{2}$$

and based on the triangle in the detailed slit drawing on the right side of Figure 6.19 we see that

$$\sin \theta_{min} = \frac{\dfrac{\lambda}{2}}{\dfrac{d}{2}} = \frac{\lambda}{d} \tag{6.16}$$

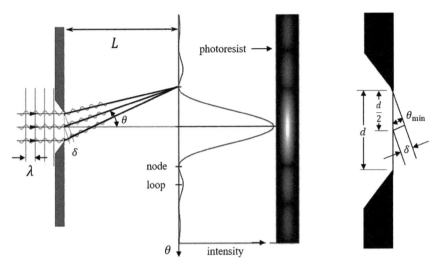

FIGURE 6.19 Single slit diffraction. Light in the slit is modeled as an array of point sources of waves. If a phase shift of 180 degrees exists between waves from two point sources traveling toward the photoresist layer at angle θ_{min} then destructive interference occurs and the photoresist reached at this angle by these two light sources is not exposed. Waves traveling at $\theta = 0$ are in-phase and constructive interference occurs resulting in a maximum of intensity. It is assumed that $L \gg d$.

This destructive interference forms a node in the diffraction pattern at angle θ_{min}. Additional photoresist exposure takes place due to small lobes in the diffraction pattern that exist at higher angles, but the primary lobe within angle range $-\theta_{min} < \theta < \theta_{min}$ determines the spatial resolution of the photoresist exposure pattern. This is referred to as the *diffraction limit*. Although reducing mask-to-photoresist distance L will reduce the size of the primary lobe, reduction of L is restricted because the mask cannot physically touch the photoresist layer as damage will occur. In addition the surface of a silicon wafer is never perfectly flat due to wafer warp and/or pre-existing features on the wafer.

From Equation 6.16 it can be seen that the primary lobe will decrease in width if wavelength λ is reduced. In real two-dimensional masks these diffraction effects degrade pattern resolution in both dimensions. Minimizing wavelength is again the most straightforward way to improve pattern resolution in the photoresist layer.

Figure 6.20 shows photolithography wavelength versus year used in industry. Wavelengths have decreased since the mercury discharge lines at 435 nm and 365 nm were in use. The introduction of the *excimer laser* light source revolutionized lithography due to its combination of high beam power and short wavelength. An excimer laser uses a combination of a noble gas (argon, krypton, or xenon) and a reactive gas (fluorine or chlorine). Under the appropriate conditions of electrical stimulation forming a plasma discharge at high pressure, a pseudo-molecule is created which can only exist in an energized state and can give rise to laser light in the ultraviolet range.

Laser action in an excimer molecule occurs because it has a bound excited state, but a repulsive ground state. Noble gases such as xenon and krypton are highly inert

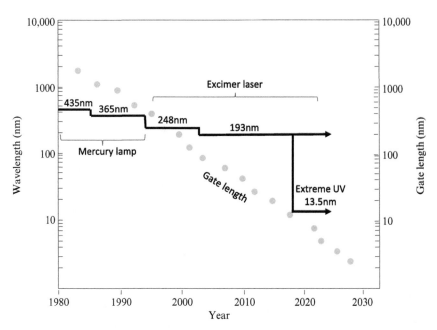

FIGURE 6.20 Photolithography light source wavelength used in semiconductor production plotted from 1980. Note how gate length continued to decrease in spite of the reliance on the 193 nm excimer laser light source since 2002. Currently both 193 nm and 13.5 nm light sources are used.

and do not usually form chemical compounds. However, when in an excited state induced by electrical discharge, noble gas molecules can form a temporarily bound molecule with halogens such as fluorine and chlorine known as an *exciplex*. The exciplex can release its excess energy by undergoing spontaneous or stimulated emission, resulting in a strongly repulsive ground state molecule which very quickly (on the order of a picosecond) dissociates back into two unbound atoms. This forms a population inversion and hence lasing occurs. ArF forms the exciplex producing 193 nm laser light whereas the earlier generation of excimer lasers using the KF exciplex lased at 248nm.

193 nm light produced by excimer lasers is able to produce feature sizes having a smallest Critical Dimension (CD) or gate length in the case of MOSFETs of below 25 nm. A series of innovations has allowed this remarkable outcome in which critical dimensions are much smaller than the wavelength of the light.

To achieve still smaller feature sizes in the range 2–15 nm, Extreme UV (EUV) lithography at $\lambda = 13.5$ nm was introduced in 2020 and is now entering commercial use for the most advanced MOSFET integrated circuits. Challenges presented by EUV light include its inability to propagate through air, liquids, and solids. Only reflective optics in a vacuum environment may be used. EUV light sources use laser-excited plasmas and light production efficiency is currently very low ($\cong 0.02\%$) in this wavelength range. Refinements and improvements to this newest generation of lithography equipment are being introduced to the market as they become available. Eximer laser sources at 193 nm are still widely used.

The following innovations can enable critical dimensions smaller than the wavelength of light:

(i) The use of water coupling between the light source and the photoresist known as *immersion lithography* takes advantage of the refractive index of the water. The effective wavelength $\lambda_{\text{effective}}$ of light is given by $\lambda_{\text{effective}} = \dfrac{\lambda}{n}$ where λ is the vacuum wavelength and n is the refractive index of the medium through which the light propagates. A higher value of n therefore results in smaller feature sizes. At the excimer wavelength of 193 nm, the refractive index of water is $n = 1.44$ compared to air with $n = 1$ resulting in $\lambda_{\text{effective}} = 134\text{nm}$. See Figure 6.21.

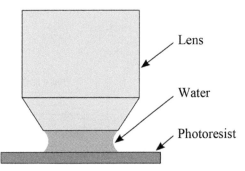

FIGURE 6.21   Water-coupled lens.

(ii) *Optical Proximity Correction* makes intentional but small alterations to the detailed shape of the mask to reduce or negate certain resulting pattern distortions due to the rather complex diffraction effects that occur in two dimensions. The corrections are complex since the diffraction explained in Figure 6.19 is a *far-field* analysis with $L \gg d$ in contrast with the *near-field* case. Computer simulations are used in the industry to optimize the optical proximity correction mask patterns. See Figure 6.22.

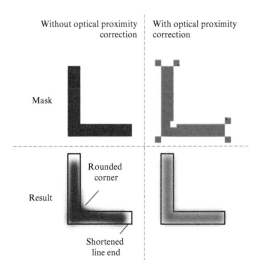

FIGURE 6.22   A simple example of optical proximity correction. By modifying the mask pattern, a more accurate photoresist exposure takes place.

(iii) Instead of relying solely on a patterned mask for blocking or passing light from the excimer laser, the technique of *Phase Shift Lithography* employs masks that can spatially vary the phase of the light reaching the photoresist. To understand the principle, consider that selected portions of such a mask are coated with a transparent phase shift layer causing a phase shift of $180°$. Now, light emerging from the locations in the mask between the non-phase-shifted and the phase-shifted portions suffers cancellation interference yielding an unexposed region of photoresist which is highly spatially resolved. See Figure 6.23. The monochromatic nature of excimer laser light makes this interference technique effective. A related lithography technique involves angling the light reaching the mask. This technique is called *off-axis illumination* which takes further advantage of light phase cancellation effects.

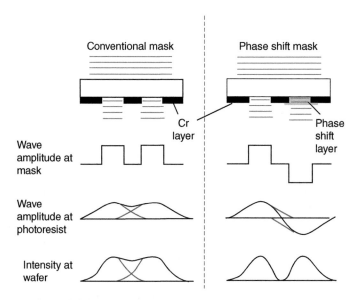

FIGURE 6.23   A phase shift layer selectively added to the mask causes wave cancellation and results in improved feature resolution.

(iv) *Multiple patterning lithography* makes use of two or more masks in stages to complete the exposure of the photoresist. For example, two exposure steps using two distinct masks, each having more masked area than the desired pattern, can complete one photoresist exposure and avoid unwanted adjacent-slit diffraction effects which are based on the well-known physics of multi-slit diffraction. There are many variations on uses of multiple patterning to reduce effective feature size and improve critical dimensions. See Figure 6.24.

(v) To help manufacturers optimize results, *grid design rules* are used. These organize high resolution patterns into simplified and standardized grid patterns for which advanced lithography techniques such as those already described have

FIGURE 6.24 Double patterning by pitch splitting. Two different masks are used to expose the final pattern. A first mask with the pattern shown in red is used for a first exposure. Then a second mask with the features shown in blue is used to further expose the photoresist. Variations on this exist. For example it may be advantageous to develop and expose the photoresist following the first exposure with the red features and to complete the etch or deposition process based on this mask. Then another layer of photoresist could be deposited and the process could be repeated using the mask with blue features to yield the final result.

been carefully optimized through extensive computer modeling. Grids are scaled and designed to take advantage of spatial phase interference effects to optimize pattern integrity. Benefits of grid design rules include tighter pitches, less distorted features, less variability over the design area, and higher manufacturing yield. See Figure 6.25.

(vi) Development rate versus exposure dose for the photoresist is an important contributor to the available resolution of photolithography. Due to diffraction effects, the exposure dose in the photoresist at the edges of a mask feature does not abruptly drop to zero. If a linear dependence between development rate and dose exists, then upon the completion of the developer step, the photoresist will taper

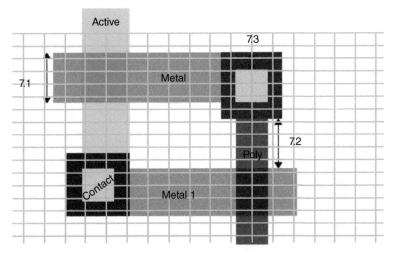

FIGURE 6.25 Grid design rules simplify the design process.

off at feature edges, whereas a steeper photoresist sidewall is preferred. In order to modify photoresist response to exposure, the chemistry of the chain scission can be designed to achieve multiple chain scissions per photon. This is classified as a *chemically amplified* resist. In one such resist an acid (H^+ ion) is liberated during chain scission. The H^+ can attack remaining chain linkages causing further scission events and liberating more H^+ ions. Advanced chemically amplified resists can undergo on the order of 10 chain scissions per photon absorbed. See Figure 6.26.

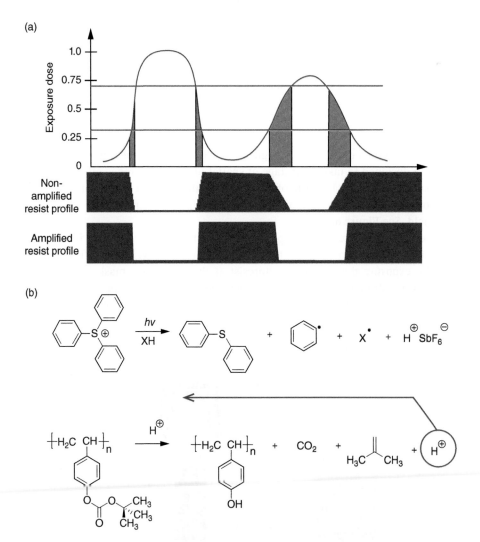

FIGURE 6.26   a) Development rate versus exposure dose for non-amplified and amplified photoresists. b) The steeper photoresist sidewall is achieved with a chemically amplified photoresist in which an acid (H^+ ion) is liberated during chain scission. The H^+ can attack remaining chain linkages causing further scission events and liberating more H^+ ions.

(vii) Standing waves in photoresist can modify resist profiles. See Figure 6.27. These standing waves are set up during the exposure step in which excimer laser light reflects back and forth between the upper and lower surfaces of the photoresist forming a series of electromagnetic standing wave loops and nodes. This can be detrimental to the accuracy of photoresist edges. A light absorbing primer layer deposited under the photoresist can be used to reduce reflections of laser light at the wafer–photoresist interface. This therefore minimizes or eliminates these standing waves.

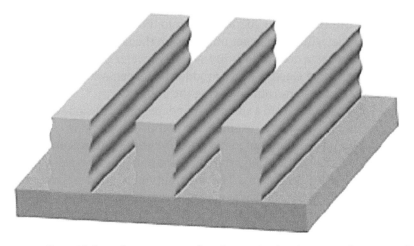

FIGURE 6.27 Sinusoidal roughness pattern after photoresist development where a standing wave of light intensity existed during exposure. The unwanted sidewall roughness can be eliminated using an light absorbing layer applied under the photoresist.

As a result of the resolution-enhancing photolithography techniques just discussed 193 nm excimer laser photolithography is widely used for critical dimensions as small as approximately 20 nm. The technology is both mature and cost-effective. For many microscale and nanoscale semiconductor devices the feature size capability is sufficient.

13.5 nm extreme UV lithography is now an emerging technology and similar techniques to those just described are also being applied to this wavelength. As a result, MOSFETs with gate lengths as low as 1 nm are being actively developed for future ultra-high density integrated circuits.

6.11   ION BEAM IMPLANTATION

Ion Beam Implantation is a technique that embeds low concentrations of atomic or ionic impurities below the surface of a target material. This occurs by accelerating the desired species to high energies after which they impinge on the material surface. A high voltage power supply provides the acceleration potential. Collisions within the atomic structure of the target material slow down and stop the species at a statistically controlled depth according to the *stopping power* of the material. Ion implantation is suitable for doping a semiconductor since only a low concentration of impurities is required.

In an ion beam implanter, a vapor of ions of the desired dopant species is produced in a plasma chamber. This forms an ion source. These ions undergo mass selection of the desired ion to optimize purity. Mass selection is achieved in a curved beamline by using a magnetic field to correspondingly bend the ion beam. The transverse Lorentz force on the ions is proportional to ion charge, longitudinal ion velocity and magnetic field strength. The magnetic field is therefore selected such that those ions having the appropriate charge and mass will follow the curved beamline allowing them to exit through an aperture and into the acceleration tube while other ions are trapped. The exiting ion beam impinges on the target comprising the silicon wafer. To ensure ion beam dose uniformity across the wafer surface, the beam may be scanned using transverse electric fields produced between scan plates as shown in Figure 6.28. The wafer may be rotated during the implant.

The depth of the implant into a silicon wafer depends on the ion beam energy and species. Figure 6.29 shows an example of depth distribution specifically for phosphorus and boron ions implanted at various energies into silicon . The *range* of the implant defines the depth of highest concentration and *straggle* is the standard deviation of the depth distribution. The distribution is often approximated by a Gaussian function due to the statistical nature of the collisions of the implanted ions in the silicon lattice as they give up their initial kinetic energy and come to rest. In practice, a desired depth distribution of an implant can be achieved using multiple implants having distinct implant energies.

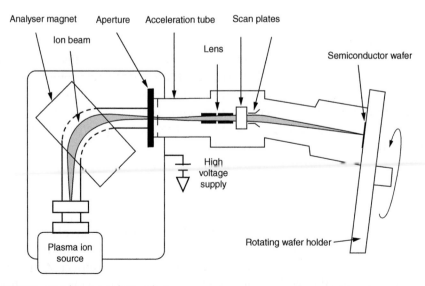

FIGURE 6.28   Ion beam implanter design.

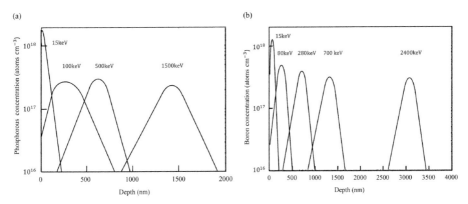

FIGURE 6.29   Distribution of a) implanted phosphorous and b) implanted boron in silicon. In both cases the implant dose is 10^{13} atoms cm^{-2} and the implant is into a (100) silicon surface. Note that boron, being slightly smaller than phosphorous, reaches a greater implant depth for a given implant energy. Straggle for phosphorous is greater than for boron due to increased collisions. Adapted from Hernández-Mangas et al., 2002.

For other dopant species such as indium, arsenic, or antimony that are larger than boron or phosphorous the implant range values will be significantly smaller for a given implant ion energy. The larger electron shells of these atoms increase collisions during the implant.

6.12   MOSFET FABRICATION

Fabrication steps for a basic n-channel MOSFET will now be described. Figure 6.30 illustrates the key steps.

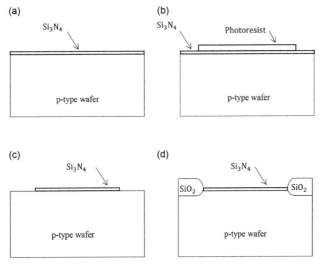

FIGURE 6.30   MOSFET fabrication steps.

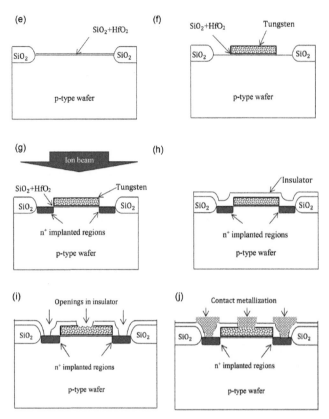

FIGURE 6.30 (Cont'd)

a) Nitride growth, Figure 6.30a).

A polished p-type silicon wafer with doping of approximately $10^{16}\,\mathrm{cm}^{-3}$ is chemically cleaned and residual SiO_2 is removed by HF acid etching.

By the process of *low pressure chemical vapor deposition* (LPCVD), a thin ($\cong 20\mathrm{nm}$) layer of Si_3N_4 is deposited on the entire wafer surface. The gas-phase reaction of silane (SiH_4), a colorless, flammable, and toxic gas, with ammonia (NH_3), a colorless gas, is carried out in a tube furnace in the presence of the clean silicon wafer surface. Waste gases are removed and a Si_3N_4 layer is deposited. A few percent of hydrogen may be incorporated into the layer. Growth reaction

$$3\ SiH_4 + 4\ NH_3 \rightarrow Si_3N_4 + 12\ H_2$$

occurs at $\cong 700°\mathrm{C}$.

b) Nitride patterning, Figures 6.30b) and 6.30c).

Using a first photolithographic step, Si_3N_4 is removed outside the transistor areas as shown A layer of photoresist is applied over the Si_3N_4 and is selectively exposed using a suitable mask. The photoresist is then chemically developed to remove it in areas where the Si_3N_4 is not wanted. See Section 6.10.

The wafer is now subjected to an acid etch to selectively remove Si_3N_4. The acid will only come in contact with those portions of the Si_3N_4 layer that are not covered by photoresist. An aqueous solution of orthophosphoric acid (H_3PO_4) may be used for this purpose. Dangling bonds of surface silicon atoms become terminated with OH^- groups (hydrolysis reaction). After further similar reaction steps, $Si(OH)_4$ acid forms which is soluble, thereby depleting silicon from the nitride layer. The $Si(OH)_4$ rapidly converts to end product SiO_2 which is eventually removed by rinsing in water. At the same time, nitrogen forms soluble NH_4 cations which are also removed after rinsing. This etch is selective to the Si_3N_4 layer and does not attack the silicon beneath.

Finally, the remaining photoresist is dissolved in a solvent such as acetone. Acetone dissolves all the photoresist regardless of light exposure.

c) Oxide growth, Figure 6.30d)
Next, a SiO_2 layer of about 0.3 μm in thickness is grown in the regions not coated by Si_3N_4. This is known as a selective or local oxidation process. Silicon oxide is an excellent insulator and has specific and useful electrical properties which will be discussed in the next sections. Very thin oxides (10Å to 1000 Å) are grown using the dry oxidation techniques in which O_2 gas reacts at the silicon surface to form a SiO_2 layer. O_2 diffusion through the already-grown SiO_2 layer allows further reaction at the Si–SiO_2 interface and the oxide layer grows in thickness at a rate predominantly controlled by this diffusion process. Thicker oxides (>1000 Å) are grown using faster wet oxidation techniques in which H_2O is used in place of O_2. The H_2O molecule is smaller than the O_2 molecule and therefore H_2O diffusion through the silicon oxide is faster. Oxidation generally takes place in a tube furnace at temperatures in the range from 700°C to 1200°C. The remaining Si_3N_4 layer prevents the oxidation process from taking place underneath it since oxygen cannot easily diffuse through this layer to reach the silicon substrate. Each region of the remaining Si_3N_4 film now defines the location of one MOSFET.

Note that the oxide layer rises above the original silicon surface. This in due to the volume expansion of the original silicon as it combines with oxygen to form SiO_2. It turns out that 44% of the SiO_2 lies below the original silicon surface and 56% forms above the original surface.

The remaining Si_3N_4 layer is now removed by an etch that is *selective* to this material but that has a much lower etch rate for SiO_2. For example phosphoric acid H_3PO_4 etches Si_3N_4 at a rate 10 times faster than SiO_2. Other etchants are available with even higher selectivities. In place of wet etching, dry plasma and beam type etching is now commonly used. In this case energetic and reactive molecules chemically attack the Si_3N_4 while not substantially attacking the SiO_2 regions. Examples of this dry etch approach include mixtures of energetic molecules such as $SF_6/CH_4/N_2/O_2$ in the form of a gas plasma.

d) *Gate oxide growth*, Figure 6.30e)
An ultra-thin ($\cong 0.3$nm) layer of electrically insulating SiO_2 is firstly grown in the region of the exposed silicon surface by an oxidation process as discussed in step c).A lithography step is avoided since it is not an issue for the already-oxidized regions to thicken very slightly. A dry oxidation process is used. This SiO_2 will become the first portion of the gate oxide layer producing a highly stable and well-understood inter-face to the silicon beneath. This SiO_2 layer is then followed by the growth of 5–10 nm of HfO_2 to form the complete insulating oxide layer.

A type of chemical vapor deposition known as *atomic layer deposition* (ALD) is favored for HfO_2 growth. A source of hafnium is tetrakis(ethylmethylamino)hafnium (TEMAH). This *organometallic* molecule with chemical formula $[(CH_3)(C_2H_5)N]_4Hf$ flows through a tube furnace and over a heated substrate. The TEMAH molecule is cracked in the furnace to release hafnium onto the heated substrate. The remaining organic components of the molecule are removed by the flowing gas stream. After a monolayer of Hf is formed, the TEMAH source is turned off and a pulse of oxygen enters the gas flow. This allows one layer of oxygen atoms to bond with the available hafnium radicals on the substrate. Following this the TEMAH source is reintroduced and the sequence repeats allowing atomic layer-by-layer growth. Thickness control is precise since it depends on the number of steps in the ALD growth sequence.

e) *Metal growth and patterning*, Figure 6.30f)
The entire wafer is now coated with a layer of a metal such as tungsten. In practice a variety of other metals including titanium or cobalt may be used. Highly conductive ceramics such as titanium nitride may also be used. The choice of material depends on whether the MOSFET is p-channel or n-channel and it also depends on the desired threshold voltage because the work function of the metal affects threshold voltage. See Example 6.3.

A second lithographic step is now required to pattern the metal to form the gate electrode region. Details of this second step are very similar to the first lithographic step already described. The photoresist is removed except for where the gate electrode is required.

Finally, the exposed areas of the metal are removed by chemical etching that selectively etches the metal and underlying oxide layers below. The etch step does not significantly affect the much thicker $0.3\mu m$ SiO_2 regions.

f) *Source and Drain Doping*, Figure 6.30g).
In order to dope the source and drain regions of the MOSFET n-type, a *self-aligned* doping process that is particularly ingenious and effective is used. An n-type dopant such as phosphorus or arsenic is introduced using ion beam implantation. The spatial accuracy positioning the source and drain regions precisely at the edges of the gate electrode exceeds that of lithographic processes. The gate material acts as a mask thereby shielding ions from passing through it into either the SiO_2/HfO_2 or the silicon beneath.

The source and drain doping level is high, on the order of $10^{18}\,\text{cm}^{-3}$. The original p-type wafer doping concentration is approximately two orders of magnitude lower and makes only a small contribution to the net doping. The thick field oxide and the poly-silicon gate are barriers to the dopant.

During the ion implantation step, the energetic ions entering the silicon collide with silicon atoms in the silicon lattice and generate point defects such as interstitial silicon atoms and silicon site vacancies. In order to restore the crystal quality in the implanted regions a *rapid thermal anneal* is performed. This anneal allows the silicon to return toward its crystalline equilibrium structure which represents the lowest energy configuration. By optimizing the temperature and duration of this anneal, a desired degree of crystalline perfection can be reached. The conditions for the anneal depend on the dopant species. For example, in the case of boron in silicon, an anneal at 1000 °C for 10 seconds is more than sufficient. Spatial redistribution of the dopant by thermal diffusion must be carefully limited and this sets an upper bound to the anneal time and temperature.

This high temperature anneal can affect the threshold voltage of the MOSFET. This is because the interface between the gate oxide, in particular hafnium oxide, and the gate metal may be permanently altered by the anneal, thereby changing the value of $q\Phi_m$ in Figure 6.4. This may occur since trapped interface charge also exists at this interface which is dependent on details of the bonding condition. In addition, the amorphous hafnium oxide may start to crystallize which further influences threshold voltage and can cause a variation in V_T from device to device on a given wafer. A solution to this difficulty is called the *gate last* approach in which a sacrificial gate material, used during source and drain ion implantation to achieve self-alignment, is removed after the high temperature anneal and then a new gate material is grown. This approach can be extended to include final growth of the hafnium oxide until after the high temperature process as well. The penalty of the gate last approach is process complexity and slightly larger feature sizes, but it allows even low melting point gate metals such as aluminum to be used. Aluminum gate metal was assumed in Example 6.3.

g) Insulator growth (Figure 6.30h)

A layer of SiO_2 is grown using LPCVD. The decomposition of Tetraethylorthosilicate (TEOS) occurs easily at low temperatures of 300–500 °C. A low growth temperature prevents unwanted diffusion and damage to the existing structure beneath. The reaction is $Si(OC_2H_5)_4 \rightarrow SiO_2 +$ gas phase reaction products.

h) Contact openings (Figure 6.30i)

A third photolithographic step is used to selectively remove the silica layer. This photolithographic step opens the areas in which contacts to the transistor are to be made. Chemical or plasma etching selectively exposes bare silicon at the source and drain regions and tungsten in the gate contact areas.

i) Contact metallization (Figure 6.30j)

Metallization is formed at the gate, source, and drain regions. This requires a fourth photolithographic step. The metallization may be patterned to provide connections to adjacent devices in an integrated circuit which are not shown. Metals such as tungsten, copper, and aluminum may be used.

6.13   CMOS STRUCTURES

High density, low power integrated circuits including processors, random access memories, and a wide range of other logic chips depend on *complementary* MOS (CMOS). The building block for CMOS is an n-channel MOSFET and a p-channel MOSFET connected together as shown in Figure 6.31. This circuit is a logic inverter which can be configured and modified to form flip flops, logic gates, memory cells, line drivers, and registers.

When the voltage V_{in} is low (i.e. close to V_{ss}) the n-channel MOS transistor is in accumulation, disconnecting V_{ss} from V_{out}. The p-channel MOS transistor is in inversion, a low resistance state, connecting V_{dd} to V_{out} through the channel. V_{out} therefore registers V_{dd}.

When, however, V_{in} is high (i.e. close to V_{dd}), the p-channel MOS transistor is in a high resistance state, disconnecting V_{dd} from V_{out} and the n-channel MOS transistor is in a low resistance state, connecting V_{ss} to V_{out}. Now, V_{out} registers V_{ss}.

In short, the outputs of the p-channel MOS and n-channel MOS transistors are complementary such that when the input is low, the output is high, and when the input is high, the output is low. No matter what the input is, the output is never left floating. The CMOS circuit output is the logic inverse of the input.

Each MOSFET within the CMOS circuit requires virtually no input current to sustain a logic state due to the insulating property of its gate oxide. This ensures extremely low power consumption if there is no switching. Logic gates can therefore be connected sequentially or in parallel, and CMOS logic with billions of transistors is feasible.

Switching speed of a given CMOS circuit is determined by the current available from the output of the previous CMOS circuit as well as the combined capacitance of the gates of the given CMOS circuit. Power consumption of CMOS logic depends on the number of switching cycles per second because current flow occurs briefly during switching.

The process steps required to achieve CMOS chips are a straightforward extension to the steps just described for the n-channel MOSFET. Figure 6.32 shows the cross section of a silicon substrate onto which an n-channel and p-channel MOSFET are formed. In order to form the p-channel MOSFET an n-channel well is required.

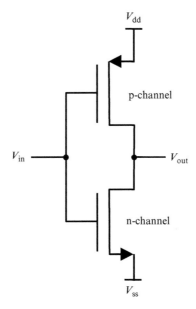

FIGURE 6.31 CMOS circuit, the logic input is connected to the gate of an n-channel MOSFET as well as the gate of a p-channel MOSFET. Except during switching, only one of the MOSFET pair is in its "on" state for which an inversion layer is formed while the other MOSFET is turned off. V_{dd} is the positive logic voltage connected to a power supply and V_{ss} is ground.

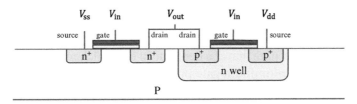

FIGURE 6.32 Cross section of simplified CMOS structure. An n-type well is formed in the p-type silicon to realize the p-channel MOSFET.

6.14 THRESHOLD VOLTAGE ADJUSTMENT

Values of V_{dd} have been decreasing over the years as MOSFET channel lengths decreased. See Table 6.1.

TABLE 6.1 Showing channel lengths and V_{dd} values versus year.

Approximate year	1990	1995	2000	2005	2010	2015
Channel length L	1 μm	0.5 μm	0.25 μm	100 nm	50 nm	25 nm
Power supply voltage V_{dd}	5 V	3.3 V	2.5 V	1.5 V	1.25 V	1 V

From Figure 6.31 it is clear that the threshold voltages of both the n-channel and the p-channel MOSFET must be set to allow for optimal switching. The n-channel MOSFET of Example 6.3 was shown to have threshold voltage of $-0.64V$. However if this n-channel MOSFET was incorporated into a CMOS circuit its threshold voltage should be positive and set at approximately $V_{dd}/2$ or 0.5 volts for a 25 nm gate length MOSFET. As V_{in} rises from low to high, the n-channel MOSFET turns on as the p-channel MOSFET turns off at approximately $V_{dd}/2$.

In order to increase the threshold voltage of a n-channel MOSFET to a desired value, ion implantation is used. If a negatively charged Group III dopant is implanted in the silicon just below the semiconductor–oxide interface of an n-channel MOSFET, the threshold voltage will rise because the dopant charge contributes to the interface charge term Q_i/C_i in Equation 6.13. This negative charge repels electrons and therefore a more positive gate voltage is required for electrons to form an inversion layer. Ion implantation is particularly useful because it can be used even after partial fabrication of the MOSFET.

EXAMPLE 6.4

Estimate the dose of boron ions required to increase the threshold voltage of the MOSFET of Example 6.3 to 1.25 volts. Boron is implanted at 15 kV through a metal gate of thickness 60 nm as well as the HfO_2 insulator layer. Assume for simplicity that the stopping power of the aluminum gate, the HfO_2, and the silicon are the same.

Solution

Referring to Figure 6.29, note that at a depth of 70 nm which matches the combined thickness of the 60 nm aluminum gate plus the 10 nm HfO_2, the implant at 15 kV has a range of 70 nm. This means that only the portion of the implant deeper than 70 nm dopes the silicon, whereas the implant shallower than 70 nm remains trapped in the oxide layer or the metal layer. We will assume that only boron ions within the silicon contribute to trapped interface charge.

From Figure 6.29, the boron concentration drops from just above $1 \times 10^{18} \, cm^{-3}$ right at the silicon surface to about $2 \times 10^{17} \, cm^{-3}$ at a depth of 70 nm into the silicon or 140 nm below the gate surface. Assume Figure 6.29 is applicable for the gate metal and for the oxide in addition to silicon. Approximating the implant within 70 nm of the silicon surface to a constant concentration of, say, $5 \times 10^{17} \, cm^{-3}$, we obtain an area doping density of $5 \times 10^{17} \, cm^{-3} \times 70 \, nm = 3.5 \times 10^{12} \, cm^{-2}$

From Example 6.3, $C_i = 2.21 \times 10^{-6} \, Fcm^{-2}$. Now we obtain

$$\Delta V_T = \left| \frac{\Delta Q_i}{C_i} \right| = \frac{1.6 \times 10^{-19} \, coul \times 3.5 \times 10^{12} \, cm^{-2}}{2.21 \times 10^{-6} \, Fcm^{-2}} = 0.25 \, V$$

This is for an implant dose of 10^{13} atoms cm^{-2}. We desire an adjustment in threshold voltage from

−0.64 V to 1.25 V which corresponds to $\Delta V_T = 1.89$ V. The estimated required dose is therefore $\dfrac{1.89 \text{ V}}{0.25 \text{ V}} \times 10^{13}$ atoms cm^{-2} $= 7.6 \times 10^{13}$ atoms cm^{-2}

The MOSFET is converted from a depletion mode device to an enhancement mode device.

Several approximations and assumptions have been made. One of them is to treat the implanted ions as being at the silicon surface. To be more accurate, the ions implanted deeper would affect depletion charge Q_d and could also influence ϕ_F. In practice, manufacturers use a combination of both more advanced numerical modeling and experimental results to refine their ion beam implant parameters.

Similarly, ion implantation can be used to adjust the threshold voltage of p-channel MOSFETs which would normally be set to approximately $V_{dd}/2$ in a CMOS circuit. If positive implanted charge is needed then phosphorous ions can be used in place of boron ions.

6.15 TWO-DIMENSIONAL ELECTRON GAS

In Chapter 2 the allowed wavefunctions for electrons in a three-dimensional infinite-walled potential box of dimensions a,b, and c were derived using Schrödinger's equation. The resulting electron energy levels from Equation 2.30 are

$$E = \frac{\hbar^2 \pi^2}{2m}\left[\left(\frac{n_x}{a}\right)^2 + \left(\frac{n_y}{b}\right)^2 + \left(\frac{n_z}{c}\right)^2\right] \tag{6.17}$$

Each possible wavefunction and each corresponding allowed electron energy is obtained by specifying positive integers (n_x, n_y, n_z). The three-dimensional reciprocal lattice space of Figure 2.12 is a convenient way of viewing all these possible integer sets.

Using Equation 6.17 we see that electrons having energy values below energy level E must occupy states inside a spherical octant of radius reciprocal space.

$$r = \sqrt{\frac{2mE}{\pi^2 \hbar^2}}$$

This model can be applied to better understand the electrons in the inversion layer of a p-channel MOS capacitor at the nanoscale. See Figure 6.33. For the purposes of this discussion a coordinate system will be used with the inversion layer of length L and width W in the x-y plane, and thickness δ specified along the z-axis. Therefore $a = L$, $b = W$, and $c = \delta$.

Note that the shape of the potential along the z-axis is triangular; however, this makes solving for the electron states much more mathematically complex.[2] We will assume an infinite-walled rectangular potential well of thickness equal to the inversion layer thickness.

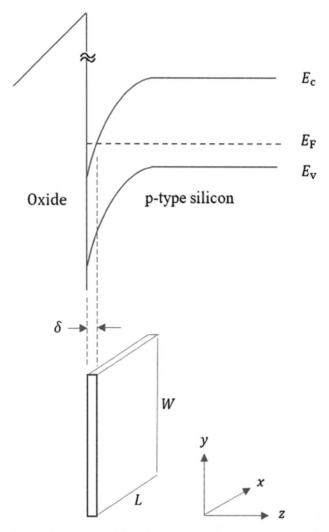

FIGURE 6.33   Shows the inversion layer of a MOS capacitor. A coordinate system defines the length L and width W of the inversion layer in the x-y plane. The triangular shape of the potential is approximated by a rectangular well of thickness δ along the z-axis.

Inversion layer thickness δ was calculated in Example 6.2 near threshold voltage and a value of 3.9 nm was obtained. The reciprocal space lattice of Figure 2.12 is redrawn in Figure 6.34 for an inversion layer of a MOS capacitor. Thickness $\delta = 3.9$ nm is assumed and we will follow Example 6.3 in which channel dimensions of $L = W = 100$ nm were given. What makes this situation interesting is the small thickness of the box in which inversion layer electrons exist relative to the length and width of the box. The unit cells of the reciprocal space lattice now have a much larger dimension along the vertical axis than along the other axes.

The dependence of available electron states on energy can now be visualized from Figure 6.34. The red spherical shell octant of radius r represents a constant energy

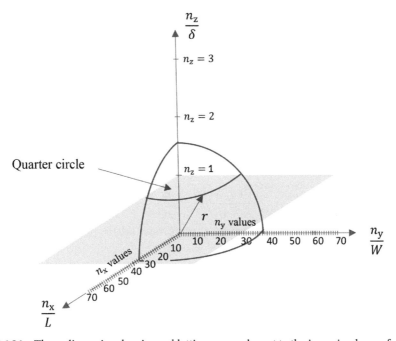

FIGURE 6.34 Three-dimensional reciprocal lattice space relevant to the inversion layer of a MOS capacitor with $L = 100\,\text{nm}$, $W = 100\,\text{nm}$ *and* $\delta = 3.9\,\text{nm}$. A plane normal to the vertical axis containing reciprocal lattice points with $n_z = 1$ is shown. The red spherical shell octant of radius r represents an equal energy surface. It encloses reciprocal lattice points with $n_z = 1$ that lie inside a quarter circle.

surface with energy E and inverse lattice points inside the shell correspond to smaller energies. There can be no allowed states when $r < \dfrac{1}{\delta}$.

If $r = \dfrac{1}{\delta}$ then

$$E = \frac{\hbar^2 \pi^2 r^2}{2m} = \frac{\hbar^2 \pi^2}{2m\delta^2}$$

As we consider higher electron energy states the spherical octant grows in radius over range $\dfrac{1}{\delta} < r < \dfrac{2}{\delta}$. It now begins to enclose lattice points in the red-shaded $n_z = 1$ horizontal plane as shown in Figure 6.34. The portion of this plane enclosed by the spherical shell octant is a quarter circle. From Equation 6.17

$$\frac{2m}{\hbar^2 \pi^2} E - \frac{1}{\delta^2} = \left(\frac{n_x}{a}\right)^2 + \left(\frac{n_y}{b}\right)^2$$

which is the equation of a circle of area $\pi \left(\dfrac{2m}{\hbar^2 \pi^2} E - \dfrac{1}{\delta^2} \right)$ provided $E \geq E_1$ as expected.

The number of available electron states $n_{n_z = 1}(E)$ that exist in this circle up to energy E can be approximated by dividing the circle area by the area $\dfrac{1}{L} \times \dfrac{1}{W}$ associated with each reciprocal lattice point. Therefore

$$n_{n_z=1}\left(E\right)=\frac{2}{4LW}\left|\frac{\pi\left(\frac{2m}{\hbar^2\pi^2}E-\frac{1}{\delta^2}\right)}{\frac{1}{L}\times\frac{1}{W}}\right|=\frac{m}{\pi\hbar^2}E-\frac{\pi}{2}\frac{1}{\delta^2}$$

A factor of 2 for spin and a factor of $\frac{1}{4}$ to restrict the area of the circle to positive values of n_x and n_y are included. Sample area LW appears in the denominator since the calculation is on a per-unit-area basis.

The density of states turns out to be the constant

$$\frac{dn_{n_z=1}\left(E\right)}{dE}=\frac{m}{\pi\hbar^2}$$

Now, consider a still larger energy range of reciprocal lattice points. If $r=\frac{2}{\delta}$ then

$$E=\frac{\hbar^2\pi^2r^2}{2m}=\frac{2\hbar^2\pi^2}{m\delta^2}$$

If radius r of the sphere octant exceeds $\frac{2}{\delta}$ then it will enclose some reciprocal lattice points for which $n_z=2$. The density of states arising from the lattice points on this plane must be the same as that obtained from the plane at $n_z=1$ because the two-dimensional density of states for a specific value of n_z is independent of n_z or δ.

A similar analysis can be made every time the sphere octant radius extends to subsequent planes with higher values of n_z. A plot of the density of states versus E is shown in Figure 6.35. Note that the effective electron mass or hole mass is used.

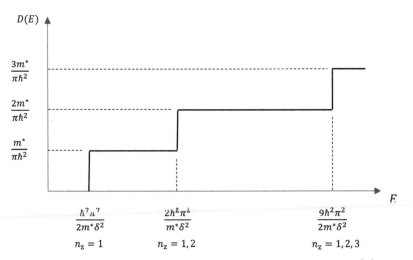

FIGURE 6.35   Density of states of a MOSFET channel in inversion. Note that $D\left(E\right)$ is a function that grows in steps as additional planes of inverse lattice points along the $\frac{n_z}{\delta}$ axis contribute to the density of states. The appropriate effective mass m^* for the specific band in the semiconductor is used as shown. This is known as a two-dimensional electron gas, although it is important to realize that it is based on a three-dimensional model.

EXAMPLE 6.5

The inversion layer of a MOS capacitor with a HfO_2 gate oxide having relative dielectric constant 25 and thickness 10 nm is treated as a three-dimensional infinite-walled potential box of thickness $\delta = 3.9$ nm. Show the position of the Fermi energy on a plot of $D(E)$ versus E and hence determine the highest value of n_z. Assume that the gate voltage exceeds threshold voltage by

a) 0.2 volts
b) 6 volts

Solution: $C_i = 2.21 \times 10^{-6}$ Fcm^{-2} which is the same as in Example 6.3.

a) $|Q_n(x)| = C_i(V_G - V_T) = 2.21 \times 10^{-6}$ F cm^{-2} $\times 0.2$ V $= 4.42 \times 10^{-7}$ C cm^{-2}

Therefore

$$n_{0.2 \text{ volts}} = \frac{4.42 \times 10^{-7} \text{ C cm}^{-2}}{1.6 \times 10^{-19} \text{ C}} = 2.76 \times 10^{12} \text{ cm}^{-2}$$

If we only consider $n_z = 1$, the number of available states is the area under the first density of states step in Figure 6.35 or

$$n_{n_z=1} = \frac{m^*}{\pi \hbar^2} \times \frac{3\hbar^2 \pi^2}{2m^* \delta^2} = \frac{3\pi}{2\delta^2} = 3.1 \times 10^{13} \text{ cm}^{-2}$$

At $V_G - V_T = 0.2$ V the electrons present therefore have energies not much beyond the lowest allowable energy of $\dfrac{\hbar^2 \pi^2}{2m^* \delta^2}$.

b) $|Q_n(x)| = C_i(V_G - V_T) = 2.21 \times 10^{-6}$ F cm^{-2} $\times 6$ V $= 1.33 \times 10^{-5}$ C cm^{-2}

Therefore

$$n_{6 \text{ volts}} = \frac{1.33 \times 10^{-5} \text{ C cm}^{-2}}{1.6 \times 10^{-19} \text{ C}} = 8.31 \times 10^{13} \text{ cm}^{-2}$$

Counting all available states in which $n_z = 1$ or $n_z = 2$, the area under the second density of states step is added to that under the first step and

$$n_{n_z=1,2} = 3.1 \times 10^{13} \text{ cm}^{-2} + \frac{2m^*}{\pi \hbar^2} \times \frac{5\pi^2}{2m^* \delta^2} = 1.34 \times 10^{14} \text{ cm}^{-2}$$

There are therefore enough available states with $n_z = 1$ and $n_z = 2$ to accommodate this surface charge density for $V_G - V_T = 6$ V.
Red lines are shown in the figure below at the approximate energies of the highest filled states for parts a) and b).

(continued)

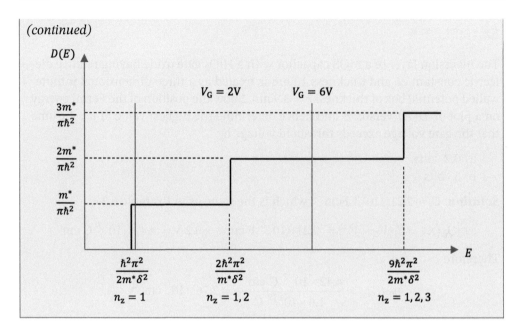

(continued)

In Chapter 1, Example 1.9, the first five wavefunction for a one-dimensional finite-walled box were calculated using Schrödinger's equation. Based on these results, the lowest energy wavefunction along the z-axis of the potential well of Figure 6.33 for $n_z = 1$ consists of an even wavefunction with a single maximum at the center of the well. Electron probability density is the square of wavefunction magnitude and it therefore follows that electron density is also a maximum at the center of the potential well. For the second energy state, an odd wavefunction is obtained which produces two maxima. Figure 6.36 shows simplified results of computer modeling of the inversion layer of a MOSFET channel which are consistent with the predictions of Example 1.9.

Poisson's equation predicts that electron density will be a maximum at the silicon surface. This changes to maxima slightly below the silicon surface once we include quantum well analysis. Combined Poisson-Schrödinger modeling is therefore required to obtain more advanced understanding of MOSFET inversion layers.

6.16   MODELING NANOSCALE MOSFETS

The equations we have derived in this chapter become less accurate when channel lengths of MOSFETs decrease into the low nanoscale range. More sophisticated three-dimensional modeling becomes necessary. Key issues include the following.

1. The "infinite planes of charge" assumption in Section 6.3 needs modification because the length and width of the channel are no longer large compared to channel thickness.

   As a result, three-dimensional modeling of the depletion region and inversion layer are required for more accurate characterization of MOSFET devices. Then the quantum well considerations of Section 6.15 must also be considered.

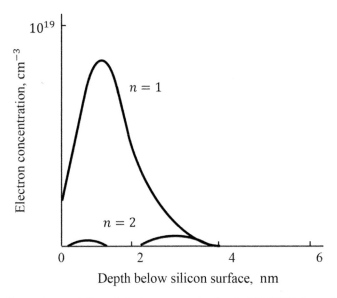

FIGURE 6.36   Computer modeling of electron concentration in MOSFET channel versus depth. A single maximum exists for the ground state and two maxima exist for the higher energy state. Adapted from L´opez-Villanueva et al., 1997.

2. In an n-channel MOSFET with a positive drain voltage, the depletion region near the drain will be wider than the depletion region near the source. This is because the junction between the n^+ drain and the underlying p-type semiconductor is more strongly reverse-biased than for the source. Since this depletion region extends both vertically and laterally into the p-type region forming the channel, it will reduce the effective width of the channel. For short channel devices where channel lengths are in the nanometer range, the result of this is to narrow the effective width of the channel dramatically. Now, even if the gate potential is set to a value below the expected threshold voltage, electrons from the source can pass through the very short effective channel, accelerate through the depletion region and reach the drain. This current is known as *subthreshold channel current*. The phenomenon is known as *drain-induced barrier lowering*. For gate voltages below the threshold voltage, the potential barrier to electron flow that normally exists at the source-channel n-p junction is lowered. See Problem 6.10.

3. Mobility is assumed constant through the channel of the MOSFET in Section 6.8. However for short channel lengths, the transverse electric field can be high enough to cause velocity saturation to occur which affects conductance and saturation transconductance. See Section 2.15. This is the case for nanoscale MOSFETs.

4. Due to high carrier velocity, degradation of the gate oxide can occur. In n-channel devices, high velocity or "hot" electrons flowing through the channel can enter the oxide layer and damage or disrupt oxide bonds. This influences Q_i

and can lead to unacceptable variations in threshold voltage. Hafnium oxide dielectrics reduce electric fields in the insulator layer and mitigate this effect.

5. Hot electrons in the channel can also enter the depletion region and create electron-hole pairs by a collision process. These unwanted carriers can lead to leakage currents that must be managed by careful device design.

6. If mechanical strain is introduced into the silicon forming a MOSFET channel, channel mobility can change. The mobility now depends on the direction of current flow relative to the strain direction. Advantageous strain that increases channel mobility and hence transconductance can be engineered into nanoscale MOSFETs by using silicon-germanium alloys to modify lattice constants.

In Chapter 9, advanced MOSFET designs that are heavily dependent on three dimensional modelling are introduced.

6.17   FLASH MEMORY

The concept of flash memory, which is a nonvolatile type of memory based on the MOSFET, was invented in 1967 at Bell Labs, but was further developed and successfully commercialized in 1987 by Toshiba. It has come to dominate portable digital storage. Finger-size flash memory sticks can now store a terabyte of data or more.

Each memory cell comprises a MOSFET except that the transistor has two gates instead of one. A floating gate (FG) is inserted between the control gate (CG) and the channel as shown in Figure 6.37. The FG is surrounded by a dielectric. This vertical structure makes efficient use of the silicon wafer and allows for very high memory cell density.

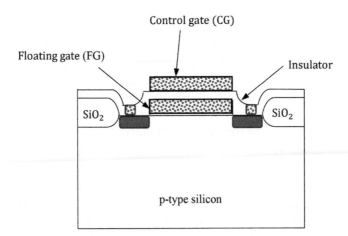

FIGURE 6.37   A flash memory cell is formed by a MOSFET with two stacked gate electrodes. The lower gate electrode is a floating gate (FG) that is surrounded by an oxide layer, typically SiO$_2$. It is not connected to any external electrode. The upper gate is the control gate (CG) which is externally connected.

Electrons placed on the FG are trapped or stored, because the FG is electrically isolated from both the silicon and the CG.

If the FG is charged with electrons, this stored charge reduces the electric field at the silicon surface that is normally produced by applying a positive voltage to the CG. The apparent threshold voltage of the MOSFET measured at the CG is thereby increased compared to the case of an uncharged FG.

A digital "0" or "1" is stored by the presence or absence, respectively, of stored charge on the FG.

In order to read the state of the stored charge, a very specific voltage is applied to the CG such that, in the absence of stored charge on the FG, the MOSFET channel is formed and current can flow between source and drain. In the presence of negative stored FG charge, however, this channel is not formed. The stored information can therefore be read out from the memory cell by applying a small voltage between source and drain and detecting a lack of drain current for a digital "0" or the presence of drain current for a digital "1." This method of reading the digital state of the cell does not affect the information since the charge stored on the FG is not affected by the read process. As a result, flash memory can be read a virtually unlimited number of times without negatively affecting the integrity of the stored information.

In order to change the stored information in the cell, charge must be supplied to, or removed from, the FG. Charge supply is achieved by causing hot channel electrons to tunnel through the oxide layer from the channel to the FG. See Figure 6.38. Channel

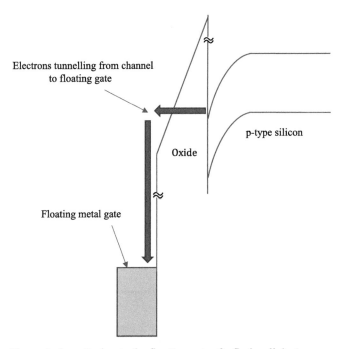

FIGURE 6.38 Charge is deposited onto the floating gate of a flash cell during a programming step. Electrons tunnel from the channel to the floating gate due to high electron velocity in the channel in addition to a large positive potential difference between the floating gate and the channel.

electrons are accelerated initially by a potential difference applied between source and drain. There is a probability that these electrons scatter into the oxide layer and electrons may tunnel directly to the FG where they are trapped.

Electrons may also be removed from the FG by a tunneling process. By applying a negative voltage to the CG relative to the MOSFET source terminal, a sufficiently high electric field is produced between the n^+ source region and the FG to enable electrons to tunnel from the FG to the source.

Gradual degradation during the write operation occurs due to the very high electric field ($\simeq 10^6$ Vcm^{-1}) experienced by the oxide during tunneling. The electrons can damage the oxide and interface layers by breaking atomic bonds over time leading to dangling bonds. The electrically insulating properties of the oxide are compromised, eventually allowing electrons to leak from the floating gate into the oxide, increasing the likelihood of data loss after ten thousand to one hundred thousand write cycles. To mitigate the risk of data loss, intelligent control hardware and software can detect the degradation of blocks of flash cells and can reallocate data to working areas of the chip.

These tunneling processes, although essential for the function of flash memory, are a serious issue for a normal MOSFET. As gate lengths decrease in successive generations of MOSFETs, decreases in gate oxide thickness also occur. Even with HfO$_2$ gate oxide, electron tunneling between the silicon and the gate electrode during MOSFET operation must be analyzed and minimized to prevent unwanted gate current and unwanted heat generation in high density integrated circuits. The last section of this chapter examines tunneling in detail.

6.18 TUNNELING

In Section 1.9 an electron incident on a potential step was discussed. We will now extend this to address electron tunneling.

Assume that a beam of electrons with kinetic energy $E < U_0$, moving from left to right, is incident on a narrow potential barrier of height U_0 and width L as shown in Figure 6.39. In this situation tunneling may occur.

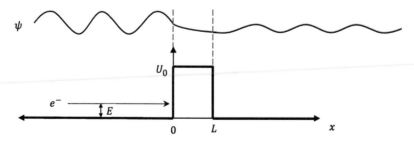

FIGURE 6.39 An electron traveling wave is incident at kinetic energy $E < U_0$ on a potential energy barrier of width L and height U_0. A portion of the wave energy is transmitted through the barrier in the form of a transmitted wave and the rest of the wave energy is reflected. Inside the barrier the electron wavefunction exhibits exponential decay.

For $x < 0$,

$$\psi_1(x) = Ae^{ikx} + Be^{-ikx}$$

where

$$k = \sqrt{\frac{2mE}{\hbar^2}} \tag{6.18}$$

For $0 \le x \le L$

$$\psi_2(x) = Ce^{-\beta x} + De^{\beta x}$$

where

$$\beta = \sqrt{\frac{2m(U_0 - E)}{\hbar^2}} \tag{6.19}$$

For $x > L$

$$\psi_3(x) = Fe^{ikx} + Ge^{-ikx}$$

We can conclude that $G = 0$ because no left-traveling electron wave arrives from the right side of the potential barrier.

For the electron wavefunction and its derivative to be continuous at $x = 0$ we obtain

$$A + B = C + D \tag{6.20}$$

and

$$ik(A - B) = \beta(D - C) \tag{6.21}$$

For the electron wavefunction and its derivative to be continuous at $x = L$ we obtain

$$Ce^{-\beta L} + De^{\beta L} = Fe^{ikL} \tag{6.22}$$

and

$$\beta\left(De^{\beta L} - Ce^{-\beta L}\right) = ikFe^{ikL} \tag{6.23}$$

The *tunneling probability* T will now be defined as the ratio of probability densities of arriving and departing traveling waves which is expressed as

$$T = \frac{F^* e^{-ikx} Fe^{ikx}}{A^* e^{-ikx} Ae^{ikx}} = \frac{F^* F}{A^* A} = \left(\frac{F}{A}\right)^* \frac{F}{A} \tag{6.24}$$

Dividing the four Equations 6.20, 6.21, 6.22, and 6.23 by A reduces the number of unknowns to four, namely $\dfrac{B}{A}, \dfrac{C}{A}, \dfrac{D}{A}, \dfrac{F}{A}$. Solving for $\dfrac{F}{A}$ and substituting the solution into Equation 6.24 the tunneling coefficient obtained is

$$T = \frac{1}{\cosh^2(\beta L) + \frac{1}{4}\left(\frac{\beta}{k} - \frac{k}{\beta}\right)^2 \sinh^2(\beta L)}$$

$$= \frac{1}{1 + \frac{U_0^2}{4E(U_0 - E)}\sinh^2(\beta L)}$$

See Problem 6.12 . In region 2 the characteristic decay length of the exponentially decaying wavefunction is $\frac{1}{\beta}$. If this decay length is much shorter than the barrier width L then $\frac{1/\beta}{L} \ll 1$ and therefore $\beta L \gg 1$. In this case $\sinh\beta L \simeq \frac{1}{2}e^{\beta L}$ and, since the second term in the denominator dominates,

$$T \cong 16\frac{E}{U_0}\left(1 - \frac{E}{U_0}\right)e^{-2\beta L} \tag{6.25}$$

This tunneling probability will now be applied to calculate the tunneling current density of electrons from the inversion layer of a MOSFET through the oxide to the gate. See Figure 6.40.

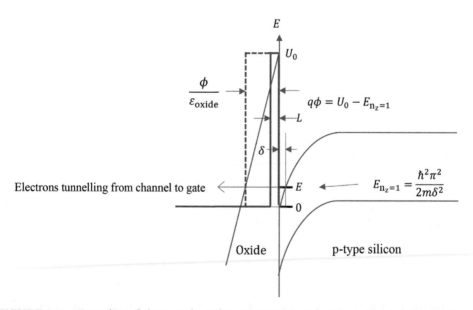

FIGURE 6.40 Tunneling of electrons from the inversion layer of a p-type silicon MOS capacitor through the oxide layer. The potential energy barrier in the oxide is approximated as a rectangular barrier of energy $q\phi$ for inversion layer electrons injected from the lowest energy level $E_{n_z=1}$ in a rectangular well of thickness δ. Effective barrier thickness is determined by either L or $\frac{\phi}{\varepsilon_{oxide}}$, whichever is smaller.

To simplify the calculation of tunneling current, a series of initial assumptions will be made:

1. The triangular potential well containing the inversion layer is approximated by a square well of thickness δ as in Section 6.15.
2. The electron concentration in the inversion layer is modeled as a two-dimensional electron gas. The electron concentration will be sufficiently small such that only quantum states within the first step of the density of states plot $E_{n_z=1}$ as shown in Figure 6.35 may be occupied. Tunneling is a one-dimensional phenomenon and the energy of the electrons incident on the barrier is properly determined by their momentum *perpendicular* to the barrier. It is therefore correct to assign all electrons the energy $E_{n_z=1}$ as shown in Figure 6.35 even though their total energy may lie between $E_{n_z=1}$ and $E_{n_z=2}$. Momentum components parallel to the plane of the energy barrier may be neglected.
3. The potential barrier formed by the oxide is assumed to be rectangular. In reality the oxide potential energy barrier shape is triangular for thicker oxides in which the electrons tunnel into the conduction band of the oxide and barrier thickness is $\dfrac{\phi}{\varepsilon_{\text{oxide}}}$. For very thin oxides, however, tunneling directly through the oxide may take place, in which case the barrier shape is trapezoidal and of thickness L. See Figure 6.40. In either case, a rectangular barrier of the relevant thickness is assumed.

The electric field in the oxide layer will be defined as $\varepsilon_{\text{oxide}}$. From Figure 6.40

$$\phi q = U_0 - E_{n_z=1}$$

Using Equation 6.19,

$$\beta = \sqrt{\frac{2m\left(U_0 - E\right)}{\hbar^2}} = \sqrt{\frac{2mq\phi}{\hbar^2}}$$

Using Equation 6.25 and assuming $E \ll U_0$

$$T \cong 16\frac{E}{U_0}\left(1 - \frac{E}{U_0}\right)e^{-2\beta L} \cong 16\frac{E}{U_0}e^{-2\beta L}$$

Now, for thin oxides provided $L < \dfrac{\phi}{\varepsilon_{\text{oxide}}}$,

$$T_{\text{thin}} \cong 16\frac{E}{U_0}e^{-2\sqrt{\frac{2mq\phi}{\hbar^2}}L} \tag{6.26a}$$

For thick oxides where $L > \dfrac{\phi}{\varepsilon_{\text{oxide}}}$ then, from Figure 6.40, a triangular barrier of width $\dfrac{\phi}{\varepsilon_{\text{oxide}}}$ is relevant and

$$T_{\text{thick}} \cong 16 \frac{E}{U_0} e^{-2\sqrt{\frac{2mq}{\hbar^2}} \frac{\phi^{\frac{3}{2}}}{\varepsilon_{\text{oxide}}}} \tag{6.26b}$$

The charge per unit area in the inversion layer Q_n can be expressed as the number of electrons per unit volume n in the inversion layer of thickness δ yielding

$$n = \frac{Q_n}{q\delta} \tag{6.27}$$

Current density from the channel incident on the oxide is

$$J_I = \frac{1}{2} nqv = \frac{1}{2} nq \frac{p}{m^*} \tag{6.28}$$

where p is the electron momentum in the direction normal to the channel. The electron exists in a lowest-energy standing electron wave along the x-axis in Figure 6.39 (equivalent to the z-axis in Figure 6.33). This standing wave is the sum of a forward and a backward traveling wave. Only half the electrons should therefore be counted which justifies the factor of $1/2$.

The required momentum is given by

$$p = \sqrt{2m^* E}$$

The relevant energy level for electrons incident on the barrier as shown in Figure 6.35 is

$$E_{n_z=1} = \frac{\hbar^2 \pi^2}{2m^* \delta^2}$$

Now,

$$p = \sqrt{2m^* E_{n_z=1}} = \frac{\hbar \pi}{\delta} \tag{6.29}$$

and combining Equations 6.27, 6.28, and 6.29,

$$J_I = \frac{1}{2} nq \frac{p}{m^*} = \frac{1}{2} \frac{Q_n}{\delta^2} \frac{\hbar \pi}{m^*}$$

The complete expression for the tunneling current density using Equation 6.26a for thin oxides is

$$J_{\text{thin}} = J_I T_{\text{thin}} = 4 \frac{Q_n}{\delta^4} \frac{\hbar^3 \pi^3}{\left(m^*\right)^2 U_0} e^{-2\sqrt{\frac{2mq\phi}{\hbar^2}} L} \tag{6.30a}$$

and using Equation 6.26b for thick oxides it is

$$J_{\text{thick}} = J_I T_{\text{thick}} = 4 \frac{Q_n}{\delta^4} \frac{\hbar^3 \pi^3}{\left(m^*\right)^2 U_0} e^{-2\sqrt{\frac{2mq}{\hbar^2}} \frac{\phi^{\frac{3}{2}}}{\varepsilon_{\text{oxide}}}} \tag{6.30b}$$

EXAMPLE 6.6

A silicon MOSFET has p-type doping $N_a = 2 \times 10^{16} \text{cm}^{-3}$. The silicon-to-insulator barrier height for electrons is $\phi = 3.1$ eV. The inversion layer is treated as a three-dimensional infinite-walled potential box of thickness $\delta = 3.9$ nm. Channel length is 100 nm and channel width is 100 nm. You may assume $\phi_{MS} = 0$ and $Q_i = 0$.

a) Find the tunneling current density through the oxide layer when the gate voltage is 0.2 volts above threshold voltage for a SiO_2 dielectric of thickness 1.2 nm.
b) Repeat a) for a dielectric based on HfO_2 of thickness 7 nm. Note that per unit area capacitance C_i is close to equivalent to that for a) due to the dielectric constant ratio between SiO_2 and HfO_2.
c) Find the time needed to transfer 100 electrons across the oxide for a MOSFET with a gate area of $100 \text{nm} \times 100 \text{nm}$ for the dielectric of part a). This is also relevant to programming and storage in a nanoscale Flash memory cell that typically stores under 1000 electrons.
d) Repeat c) for the dielectric of b).

Solution:

$$C_i = \frac{\epsilon_0 \epsilon_r}{d} = \frac{8.85 \times 10^{-14} \text{ Fcm}^{-1} \times 3.9}{1.2 \times 10^{-7} \text{cm}} = 2.88 \times 10^{-6} \text{ Fcm}^{-2} = 2.88 \times 10^{-2} \text{ F m}^{-2}$$

$$Q_n = -C_i (V_G - V_T) = -2.88 \times 10^{-6} \text{ Fcm}^{-2} \times 0.2V$$
$$= -5.75 \times 10^{-7} \text{ C cm}^{-2} = -5.75 \times 10^{-3} \text{ C m}^{-2}$$

$$N_{0.2 \text{ volts}} = \frac{5.75 \times 10^{-7} \text{ C cm}^{-2}}{1.6 \times 10^{-19} \text{ C}} = 3.6 \times 10^{12} \text{cm}^{-2} = 3.6 \times 10^{16} \text{m}^{-2}$$

From Example 6.5 it is clear that these electrons will exist in quantum states under the first step in Figure 6.35 because $N_{0.2 \text{ volts}}$ is much less than 3.1×10^{13} cm^{-2} which is the total number of available states available between $n_z = 1$ and $n_z = 2$. From Example 6.2,

$$W_m = 2.19 \times 10^{-5} \text{ cm}$$

$$Q_d = -qN_a W_m = -1.6 \times 10^{-19} \text{ C} \times 2 \times 10^{16} \text{ cm}^{-3} \times 2.19 \times 10^{-5} \text{ cm}$$
$$= -7.01 \times 10^{-8} \text{ C cm}^{-2}$$

Considering all the charge in the semiconductor,

$$\varepsilon_s = \frac{|Q_d + Q_n|}{\epsilon_s} = \frac{7.01 \times 10^{-8} \text{ C cm}^{-2} + 5.75 \times 10^{-7} \text{ C cm}^{-2}}{8.85 \times 10^{-14} \text{ Fcm}^{-1} \times 11.8} = 6.18 \times 10^5 \text{ Vcm}^{-1}$$

$$E_{n_z=1} = \frac{\hbar^2 \pi^2}{2m^* \delta^2} = \frac{\left(1.05 \times 10^{-34} \text{ Js}\right)^2 \times \pi^2}{2 \times 1.08 \times 9.11 \times 10^{-31} \text{Kg} \times \left(3.9 \times 10^{-9} \text{m}\right)^2} = 3.63 \times 10^{-21} \text{ J}$$

(continued)

(*continued*)

$$q\phi = 3.1 \text{ eV} \times 1.6 \times 10^{-19} \text{ J} = 4.96 \times 10^{-19} \text{ J}$$

$$U_0 = E_{n_z=1} + \phi = \frac{\hbar^2 \pi^2}{2m^* \delta^2} + \phi = 3.63 \times 10^{-21} \text{ J} + 4.96 \times 10^{-19} \text{ J} = 5.00 \times 10^{-19} \text{ J}$$

Now,

$$J_I = \frac{Q_n}{\delta^2} \frac{\hbar \pi}{m^*} = \frac{5.75 \times 10^{-3} \text{ C m}^{-2}}{\left(3.9 \times 10^{-9} \text{ m}\right)^2} \frac{1.05 \times 10^{-34} \text{ Js} \times \pi}{1.08 \times 9.11 \times 10^{-31} \text{ Kg}} = 1.27 \times 10^{11} \text{A m}^{-2}$$

In the oxide,

$$\varepsilon_{\text{oxide}} = \frac{\epsilon_s}{\epsilon_{\text{oxide}}} \varepsilon_s = \frac{11.8}{3.9} \times 6.18 \times 10^5 \text{ Vcm}^{-1} = 1.87 \times 10^6 \text{ Vcm}^{-1}$$

From Figure 6.40,

$$L = \frac{3.1 \text{V}}{\varepsilon_{\text{oxide}}} = \frac{3.1 \text{V}}{1.87 \times 10^6 \text{ Vcm}^{-1}} = 1.66 \times 10^{-6} \text{cm} = 16.6 \text{nm}$$

a) This is much wider than the 1.2 nm oxide thickness and therefore the oxide barrier will be trapezoidal and Equation 6.26a is relevant for determining tunneling probability. Now,

$$T_{\text{thin}} \cong 16 \frac{E}{U_0} e^{-2\sqrt{\frac{2mq\phi}{\hbar^2}}L} = 16 \frac{3.63 \times 10^{-21} \text{ J}}{5.00 \times 10^{-19} \text{ J}}$$

$$\exp\left[-2\sqrt{\frac{2 \times 9.11 \times 10^{-31} \text{ Kg} \times 4.96 \times 10^{-19} \text{ J}}{\left(1.05 \times 10^{-34} \text{ Js}\right)^2}} \times 1.2 \times 10^{-9} \text{m}\right] = 4.24 \times 10^{-11}$$

$$J_{\text{thin}} = J_I T_{\text{thin}} = 1.27 \times 10^{11} \text{A m}^{-2} \times 4.24 \times 10^{-11} = 5.37 \text{ Am}^{-2}$$

b) The same value of C_i as in part a) will be assumed since the increase in relative dielectric constant from 3.9 to 25 is approximately offset by the change in thickness from 1.2nm to 7 nm. Again using Equation 6.26a,

$$T_{\text{thin}} \cong 16 \frac{E}{U_0} e^{-2\sqrt{\frac{2mq\phi}{\hbar^2}}L} = 16 \frac{3.63 \times 10^{-21} \text{ J}}{5.00 \times 10^{-19} \text{ J}}$$

$$\exp\left[-2\sqrt{\frac{2 \times 9.11 \times 10^{-31} \text{ Kg} \times 4.96 \times 10^{-19} \text{ J}}{\left(1.05 \times 10^{-34} \text{ Js}\right)^2}} \times 7 \times 10^{-9} \text{m}\right] = 1.04 \times 10^{-56}$$

$$J_{\text{thin}} = J_I T_{\text{thin}} = 1.27 \times 10^{11} \text{A m}^{-2} \times 1.04 \times 10^{-56} = 1.32 \times 10^{-45} \text{ Am}^{-2}$$

c) For gate area of $100\,\text{nm}\times100\,\text{nm}$

$$I = JA = 5.37 \text{ Am}^{-2} \times \left(100\times10^{-9}\right)^2 \text{ m}^2 = 5.37\times10^{-14}\,\text{A}$$

$$= \frac{5.37\times10^{-14} \text{ C}}{\text{s}} = \frac{5.37\times10^{-14} \text{ C}}{\text{s}} \times \frac{1\,\text{q}}{1.6\times10^{-19} \text{ C}} = 3.4\times10^5 \text{ qs}^{-1}$$

d)
$$T_{100} = \frac{100 \text{ q}}{3.4\times10^5 \text{ qs}^{-1}} = 3.0\times10^{-4}\text{s}$$

$$T_{100} = 3.0\times10^{-4} \text{ s} \times \frac{4.24\times10^{-11}}{1.04\times10^{-56}} = 1.2\times10^{42}\text{s} \cong 4\times10^{34} \text{ years}$$

The benefit of the high K dielectric is obvious. If the electron concentration in the inversion layer is high enough such that quantum states under the second step in Figure 6.35 are also occupied, then tunneling needs to be considered from states with two distinct energy levels having quantum number $n_z = 1$ or $n_z = 2$. See Problem 6.13.

An *equivalent oxide thickness* (EOT) parameter is used to characterize the thickness of SiO_2 required to reproduce the capacitance per unit area of an alternative high K dielectric. Figure 6.41 shows the historical thickness of SiO_2 which decreased over the years and then remained at approximately 2 nm due to tunneling effects in about 2000. This tunneling problem is clearly illustrated in Example 6.6. The introduction of an HfO_2 dielectric has allowed the EOT to be significantly reduced while tunneling remains manageable. For example, a 5 nm thick HfO_2 dielectric that does not allow significant tunneling has an EOT of below 1 nm.

The *Fowler-Nordheim Equations* for field emission are well known equations that treat electron field emission over a triangular potential barrier. The standard Fowler-Nordheim equation assumes a metal-vacuum system with a large vacuum electric field that causes electron tunneling over a triangular energy barrier of height $q\phi$ equal to the work function of the metal. It is

$$J_{\text{Fowler-Nordheim}} = \frac{q^3\varepsilon^2}{8\pi hq\phi} \exp\left[-\frac{4}{3}\sqrt{\frac{2mq}{\hbar^2}}\frac{\phi^{\frac{3}{2}}}{\varepsilon}\right]$$

This equation has a very similar exponent to the exponent in Equation 6.30b) that we derived. The Fowler-Nordheim equation results from a more accurate treatment of the triangular barrier tunneling coefficient. In addition, the incident electron current from electrons in the conduction band of a metal are assumed. These electrons are treated over a distribution of incident energies rather than by assuming just one value of E as we have done.

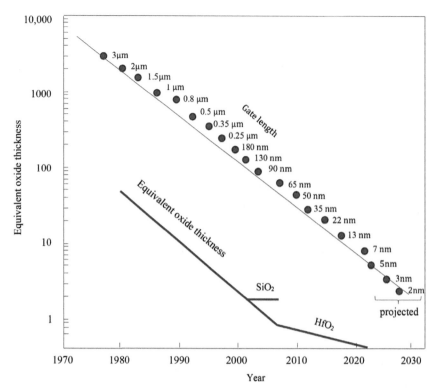

FIGURE 6.41   Equivalent oxide thickness (EOT) versus year. EOT reduction followed gate length reductions until approximately 2000 and then stopped and remained above approximately 2 nm due to tunneling current. The introduction of high K material HfO$_2$ has allowed further reductions in EOT to well below 1 nm.

6.19 SUMMARY

6.1 The planar MOSFET controls the flow of current through a channel that is near the surface of the semiconductor wafer by means of a gate electrode insulated from the channel. This structure facilitates high density device integration better than other transistor types.

6.2 MOSFET operation relies on the creation of an inversion layer forming a conducting channel between source and drain by means of the gate electrode. Channel doping is opposite in type compared to the source and drain regions which prevents current flow in the absence of channel inversion.

6.3 Using Gauss's law or Poisson's equation, the total net charge as a function of gate voltage in a MOS capacitor can be derived and plotted. This enables a quantitative understanding of accumulation, depletion, and inversion conditions.

6.4 The concept of the Debye length is derived. It allows for the thickness of an accumulation layer to be predicted. The inversion layer thickness can be estimated at its point of onset by considering an equilibrium between diffusion and drift currents in a direction perpendicular to the channel.

6.5 Gate capacitance in a MOS capacitor is determined by the thickness and dielectric constant of the dielectric material between the gate and the channel. In addition, an inversion layer may exist below this dielectric layer. In this case, capacitance is calculated from two series-connected capacitances, but this changes depending on the frequency at which the capacitance is measured.

6.6 MOSFET threshold voltage is dependent on pre-existing charge that exists at the interface between the channel and the dielectric layer adjacent to the channel. The addition of fixed charges to this interface by means of ion implantation allows MOSFET threshold voltage to be adjusted and controlled.

6.7 Since carriers flow very close to the semiconductor surface in the channel, surface scattering effects must be considered. Interface roughness effects at the semiconductor/insulator interface play a role in mobility reduction. In addition, phonon scattering of electrons and holes is modified near a semiconductor surface because bulk phonon modes and surface phonon modes differ.

6.8 A simplified model allows MOSFET transistor characteristics to be predicted. The most important parameter is the transconductance specified in terms of $dI\,/\,dV$ where dI is a change in drain current brought about by a change in gate voltage dV.

6.9 Moore's law has predicted MOSFET size reduction as a function of year since the 1970s. It cannot last forever but it has been a reliable roadmap for the semiconductor industry for half a century.

6.10 A range of innovative techniques in photolithography have been the key enabler of MOSFET size reduction. Extreme UV lithography using a wavelength of 13.5 nm is required for the smallest MOSFET structures having gate lengths in single digit nanometer length scales.

6.11 Ion beam implantation relies on knowledge of the average or expected depth range of high energy ions that enter MOSFET structures during fabrication. The range and the straggle are parameters that characterize this. These parameters depend on the species being implanted and their incident kinetic energy as produced by an ion implanter.

6.12 The steps associated with efficient MOSFET fabrication include self-aligned implants. This technique uses the gate electrode as a mask to precisely control the regions of the semiconductor being implanted to form the source and drain regions.

6.13 To realize low power consumption logic gates in digital integrated circuits, an n-channel and a p-channel MOSFET are fabricated and connected in pairs resulting in CMOS integrated circuits.

6.14 The calibrated adjustment of threshold voltage using ion beam implantation can be modeled using knowledge of the ion beam dose and depth profile.

6.15 The reduction of carrier depth in a MOSFET channel results in the need to apply a two-dimensional electron gas model. Only a specific number of carriers can exist in a channel due to the limited availability of quantum states within the applicable energy range.

6.16 Modeling nanoscale MOSFETs requires considerations that go beyond basic theory. These may include mobility effects due to surface carrier scattering, surface phonons, and lattice strain. In addition, drain-induced barrier lowering, and hot carrier effects must be considered.

6.17 Flash memory relies on a quantity of charge trapped in a floating gate. This charge determines the threshold voltage of a MOSFET channel and can be detected by measuring channel current without affecting the floating gate. Charge on the floating gate can be altered by tunneling current allowing electron flow across the oxide potential barrier surrounding the floating gate or by hot carriers that overcome an oxide potential barrier.

6.18 As MOSFETs decrease in size, the gate insulator is also reduced in thickness leading to the issue of tunneling current between gate and channel. Tunneling can be modeled using Schrödinger's equation applied to a potential barrier. To simplify the problem, a square potential barrier can be used in place of a triangular barrier leading to an approximation of the accepted expression for Fowler-Nordheim tunneling. This describes tunneling current through an oxide layer from a metal electrode as a function of applied electric field.

PROBLEMS

6.1 Show that the electric field in the metal gate of a MOSFET is zero by considering all the sources of charge in the MOS capacitor. See Section 6.3. Hint: In order to correctly treat the metal, the metal surface charge must lie just *outside* the metal surface.

6.2 Derive Poisson's equation starting from Gauss's law.

6.3 (a) Use Gauss's law to find the electric field in all 5 regions of the following hypothetical charge distribution. Each region is infinite in extent in the x-y plane.
(b) Plot the electric field as a function of z.
(c) Find and plot the electric potential as a function of z. Assume that the silicon surface adjacent to Region 5 is at potential $V = 0$.

Region 1, vacuum
Region 2, thickness 1.0 µm, silicon, positive charge density of magnitude ρ coul cm^{-3}. Assume the silicon is depleted of mobile charge carriers.
Region 3, thickness 3.0 µm, neutral silicon. Assume the silicon is depleted of mobile charge carriers.
Region 4, thickness 1.0 µm, silicon, negative charge density of magnitude ρ coul cm^{-3}. Assume the silicon is depleted of mobile charge carriers.
Region 5, vacuum

6.4 $g_{m(sat)}$ and g appear to be equal. Using Figure 6.12, explain why this is not necessarily true.

6.5 Derive Equation 6.4 and explain the connection to previous equations.

6.6 Derive Equation 6.7 by integrating the previous equation.

6.7 (a) Find the maximum depletion width, minimum capacitance, and threshold voltage for an ideal MOS capacitor with a 10 nm thick SiO_2 oxide layer on p-type silicon with $N_a = 1 \times 10^{16}$ cm^{-3}. Ignore trapped charges.

(b) Repeat a) if the dielectric is HfO_2 having the same thickness.

(c) How much negative, ion-implanted, trapped charge at the silicon–HfO_2 interface would be required to raise the threshold voltage of the MOSFET of b) by one volt?

(d) What ion species is suitable for the implant?

(e) At approximately what kinetic energy should the implant be performed for the ion of part d) ?

6.8 A MOS capacitor is formed on p-type silicon with doping of 7×10^{15} cm^{-3} and a positive trapped interface charge of 3×10^{10} charges cm^{-2} exists at the dielectric–silicon interface. The MOS capacitor has a gate dielectric of hafnium oxide 15 nm in thickness.

(a) Find the maximum depletion width.

(b) Find the minimum gate capacitance.

(c) Find the threshold voltage.

(d) Make a sketch of the capacitance versus voltage. Show numerical values for important values on both axes.

6.9 A MOSFET is made using the MOS capacitor of Problem 6.8. It has a channel length of 100 nm and a channel width of 100 nm. Assume room temperature conditions.

(a) Is the MOSFET a depletion mode or an enhancement mode device?

(b) What is the inversion layer thickness at the onset of inversion?

(c) Find conductance g if $V_G - V_T = 0.5$V. Assume V_D is very small. Use Figure 6.12.

(d) What is the minimum gate voltage at which electrons start to occupy quantum states with quantum number $n_z = 2$ if we approximate the channel to be a two-dimensional electron gas?

(e) Repeat d) if $n_z = 3$.

6.10 Look up literature explaining drain-induced barrier lowering in detail. Summarize it using your own words in 1-2 pages. Include energy diagrams showing the potential barrier for electrons and how it depends on drain voltage and channel width for an n-channel MOSFET.

6.11 Look up literature on strained silicon used to enhance the mobility in MOSFET channels. Describe the methods used to fabricate strained silicon channels and describe the performance improvements that are achieved. Present a summary in your own words in 1-2 pages. Include relevant figures.

6.12 Using Equations 6.20, 6.21, 6.22, 6.23, and 6.24, find the tunneling coefficient T.

6.13 A silicon MOSFET has p-type doping $N_a = 2 \times 10^{16}$ cm^{-3}. The silicon-to-insulator barrier height for electrons occupying the lowest energy quantum states of the inversion layer electron gas is $\phi = 3.1$ eV. The inversion layer is treated as a three-dimensional infinite-walled potential box of thickness $\delta = 3.9$ nm. Channel length is 100 nm and channel width is 100 nm. You may assume $\phi_{MS} = 0$ and $Q_i = 0$.

(a) Find the tunneling current density through the oxide layer when the gate voltage is 6 volts above threshold voltage for a SiO_2 dielectric of thickness 1.2 nm.

Hint: Add the contributions to tunneling current from the two discrete electron energy values relevant to the tunneling direction from quantum states under the first two steps of Figure 6.35. See Examples 6.5 and 6.6.

(b) Repeat a) for a dielectric based on HfO_2 of thickness 7 nm. Note that per-unit-area capacitance C_i is close to equivalent to that for a) due to the dielectric constant ratio between SiO_2 and HfO_2.

(c) Find the time needed to transfer 100 electrons across the oxide for a MOSFET with a gate area of $100 \, nm \times 100 \, nm$ for the dielectric of part a). This is also relevant to programming and storage in a nanoscale Flash memory cell that typically stores under 1000 electrons.

(d) Repeat c) for the dielectric of b).

6.14 (a) For the calculation in Example 6.6 a), show that $\sinh \beta L \cong \frac{1}{2} e^{\beta L}$ which justifies the use of Equation 6.25.

NOTES

1. We are effectively subdividing the total thickness of the volume enclosed by the Gaussian box into infinitesimally thin layers of thickness dx along the x-axis and are integrating or adding a potential difference across each thin layer to find the resulting electric field through the final Gaussian box face. This is a one-dimensional case of the divergence theorem which, when applied to Gauss's law, leads to Poisson's equation.

2. The solutions to Schrödinger's equation for a triangular well are in the form of Airy functions. Airy functions are solutions to differential equations of the form $\frac{d^2 y}{dx^2} - xy = 0$.

RECOMMENDED READING

1. Streetman, B.G. et al. (2014). *Solid State Electronic Devices*, 7e. Pearson. ISBN 978-0133356038.

2. Smith, B.W. et al. (eds.) (2020). *Microlithography: Science and Technology*, 3e. CRC Press. ISBN 978-1439876756.

The Quantum Dot

CONTENTS

Objectives

1. Introduce the materials, the structures, and the function of a quantum dot.
2. Describe various top-down and bottom-up synthesis methods for quantum dot formation.

Fundamentals of Semiconductor Materials and Devices, First Edition. Adrian Kitai.
© 2023 John Wiley & Sons Ltd. Published 2023 by John Wiley & Sons Ltd.
Companion Website: www.wiley.com/go/kitai_fundamentals

3. Calculate bandgap increase due to quantum confinement.
4. Consider the exciton Bohr radius versus the quantum dot radius.
5. Introduce vibronic transition concepts and the Stokes shift.
6. Introduce both core-shell epitaxial passivation and organic surface passivation approaches.
7. Introduce quantum dot charging and Auger processes.
8. Explain the blinking effect.
9. Introduce concepts of quantum dot functionalization and bio-compatibility.

7.1 INTRODUCTION AND OVERVIEW

The quantum dot has emerged in the current century as an important industrial application of semiconductors. Quantum dots have unique optoelectronic properties that form the basis for backlighting in high performance mass-produced liquid crystal displays because they efficiently produce backlighting with primary colors of high color purity. Quantum dots are finding growing application in biological imaging and labeling. Research and development is underway for quantum dot applications in diverse fields from solar cells to quantum computing.

The use of quantum dots predates the understanding of their nanoscale structure and the associated physics. In the early twentieth century, CdS and CdSe were incorporated into silicate glasses to achieve red-yellow glass colors. It was not until the 1980s, however, that quantum dot physics was understood. The industrial synthesis of high performance semiconductor quantum dots was developed even more recently.

In a size range that straddles bulk semiconductors and molecular semiconductors, the quantum dot exhibits properties that can be understood from a combination of quantum physics, semiconductor physics, and radiation theory.

The quantum dot is often approximately spherical in shape, although it can also have faceted surfaces. It is composed of 100–10,000 atoms of an inorganic semiconductor material. This constitutes a number range larger than a typical small molecule but smaller than what we would regard as a bulk material. The diameter of a spherical quantum dot is in the range from 1 nm to 20 nm.

The ability to synthesize quantum dots with tight control of size and purity has been a key enabling factor. Quantum dots may be formed as entities within an inorganic crystalline or amorphous surrounding. This surrounding is known as the *matrix* material. They may also be formed as inorganic semiconductor precipitates with an organic matrix, in which case organic *ligands* surround and extend away from the quantum dot surface allowing for dispersion in liquids and polymers. Ligands also serve to stabilize surface dangling bonds and they maintain a degree of separation between neighboring quantum dots, allowing them to function as independent entities.

A *core-shell* structure is commonly employed to further optimize quantum efficiency. In this construction a spherical core semiconductor material of energy gap E_g is frequently surrounded by a semiconductor shell layer having energy gap E'_g that is usually larger than E_g. This helps to confine electrons and holes to the core region and away from the quantum dot surface.

Quantum dots are almost exclusively used as *fluorescent* materials in which light is absorbed and re-emitted.

Quantum dot functionality may include charge transfer between quantum dots and adjacent charge transport materials. This is an active area of research and development because it enables electrical excitation of quantum dot light emission which results in quantum dot electroluminescence.

Properties that quantum dots exhibit that set them apart from many other fluorescent materials include

1. high extinction coefficients. The *extinction coefficient*, also known as *attenuation coefficient* quantifies the loss in intensity of a light beam passing through, in our context, a composite material containing many quantum dots. A high extinction coefficient arises from efficient light absorption into the quantum dots. The absorption coefficent α that was introduced in Chapter 2 for bulk semiconductors is a similar constant to the extinction coefficient, where the latter includes additional light losses due to optical scattering.

2. high quantum yield. In the context of fluorescence, quantum yield is the number of emitted photons divided by the number of incident photons.

3. minimal photobleaching. *Photobleaching* refers to a loss of quantum yield over time. It usually can be correlated with the number of photon absorption events that have occurred and the resulting loss in quantum yield. It is caused by a degradation mechanism such as bond damage or structural re-configuration of a material.

4. size-tunability of absorbance and emission wavelengths. This is a fundamental attribute of quantum confinement which will be further discussed in Section 7.4.

5. a broad excitation window. The *excitation window* refers to the wavelength range of illumination over which light absorption and hence the fluorescence can occur.

6. narrow emission peaks. Due to the band structure of the semiconductor used in quantum dots, the spectrum of the emission peak of a quantum dot is narrow which leads to high color purity.

7. low interference. Multiple quantum dots can be used together with minimal undesirable interference due to neighboring quantum dots.

8. low toxicity. In biological use, depending on composition, toxicity may be less than that from conventional organic dyes.

9. amenability to functionalization. Quantum dots may be *functionalized* with a wide variety of bio-active agents or other agents. This involves organic ligands that attach to, and extend away from, the quantum dot.

7.2　QUANTUM DOT SEMICONDUCTOR MATERIALS

The most important materials used in quantum dots are direct gap semiconductors due to the efficient radiative recombination that they provide. Table 7.1 summarizes a set of Group III-V, Group II-VI, Group IV-VI, and Group IV semiconductor materials and their key properties. A few materials having indirect gaps are included.

Quantum dots have a very high surface-to-volume ratio compared to bulk materials and a significant fraction of atoms are surface atoms. Chemical stability as well as the availability of suitable synthesis methods allow for narrow quantum dot size distributions and high fluorescence quantum yields.

TABLE 7.1　Relevant semiconductor materials and their associated parameters.

Group	Name	Bandgap (eV)	Relative dielectric constant	Crystal structure	Bandgap type
III-V	GaN	3.2	8.9	Zincblende	Direct
	GaN	3.4	8.9	Wurzite	Direct
	GaP	2.26	11.1	Zincblende	Indirect
	GaAs	1.43	13.2	Zincblende	Direct
	InN	1.97	9.3	Zincblende	Direct
	InP	1.35	12.4	Zincblende	Direct
	InAs	0.36	14.6	Zincblende	Direct
	InSb	0.18	17.7	Zincblende	Direct
II-VI	ZnO	3.8	8.1	Wurzite	Direct
	ZnS	3.6	8.9	Zincblende	Direct
	ZnS	3.8	8.3	Wurzite	Direct
	ZnSe	2.7	9.2	Zincblende	Direct
	ZnTe	2.25	10.4	Zincbende	Direct
	CdS	2.42	8.9	Zincblende	Direct
	CdS	2.53	8.28, 8.64 (c-axis)	Wurzite	Direct
	CdSe	1.73	8.75, 9.25 (c-axis)	Wurzite	Direct
	CdTe	1.51	10.2	Zincblende	Direct
IV-VI	PbS	0.29	17.3	FCC	Direct
IV	Si	1.11	11.8	Diamond	Indirect
	Ge	0.67	16	Diamond	Indirect
	SiC	3.2	10	Hexagonal (4H)	Indirect

7.3   SYNTHESIS OF QUANTUM DOTS

The technique for synthesis of quantum dots may involve a *top-down* or a *bottom-up* approach.

In top-down synthesis, a semiconductor material may be deposited to form a layer on a substrate in the thickness range of 1 nm to 20 nm. Lithography and chemical etching methods are used to selectively remove the semiconductor material, leaving behind well-defined quantum dot semiconductor islands of size 1 nm to 20 nm. This approach offers a high degree of flexibility in the design of nanostructured systems. Many shapes of quantum dots with precise separation and periodicity may be realized with this technique.

Nanoscale patterned etching is the key enabling process in this first approach. Dry plasma etching (see Section 6.12) with patterning via optical lithography may be used to form close-packed arrays of quantum dots. Semiconductor material is selectively removed to leave behind the desired quantum dots.

Focused ion beam (FIB) techniques may also be used for etching quantum dots with extremely high spatial precision. Highly focused beams from a molten metal source such as gallium may be used directly to selectively *sputter* the surface of the semiconductor layer. For sputtering, ion beam energy is in the range of 10 KeV which is lower than the ion beam energies discussed in Section 6.11 required for ion beam implanatation. In this lower energy range, surface substrate atoms are removed (sputtered) by ion collisions. The ion beam must be moved to cover the regions of the semiconductor film to be removed.

Another alternative to the optical lithographic techniques discussed in Chapter 6 is *electron beam lithography*. This is a *direct-write* approach to patterning in which an electron beam is positioned under computer control to create a spatially selective exposure pattern. Resist materials that are optimized for exposure by a highly focused beam of electrons, rather than photons, may be used. Electron beams offer high spatial resolution due to the small de Broglie wavelengths of energetic electrons, See Chapter 1, Example 1.2.

Top-down synthesis methods are slow compared to bottom-up methods and are therefore more suitable for research than for production. Whereas lithography is cost effective and necessary for high performance nanometer scale semiconductor devices discussed in detail in Chapter 6, the production of large numbers of nominally identical quantum dots can be achieved far more economically using bottom-up approaches.

Self-assembly of nanostructures in material may be enabled by *molecular beam epitaxy* (MBE). This leads to an example of a physical bottom-up growth method.

In MBE growth, atoms arrive at a substrate within a vacuum chamber and the growth of a semiconductor film occurs as the atoms condense slowly on the substrate. MBE has been used to deposit semiconductor layers and to grow elemental, compound or alloy semiconductor materials on a heated substrate, often under ultrahigh vacuum conditions. The basic principle of the MBE process is evaporation from a *Knudsen* source (a small container of a solid that is heated, releasing atoms through an aperture) to form a beam of atoms or molecules that condense on a substrate.

Growth is usually *epitaxial*. In epitaxial growth, provided the substrate is a single crystal with an appropriate lattice constant, the lattice structure of the substrate is replicated in the film because arriving atoms come to rest at thermodynamically stable locations that are consistent with the underlying crystal structure.

If the lattice constant of the substrate differs from the lattice constant of the growing film, growth modes can occur which result in the spontaneous formation of quantum dots due to the onset of island growth. Three growth modes shown in Figure 7.1 include the *van der Merwe* mode growth of a smooth layer. This is the growth mode expected for lattice-matched materials. For materials that are not perfectly lattice-matched, the layer is able to accommodate the strain associated with the lattice mismatch and smooth layers are formed over a limited thickness range.

The *Stranski-Krastanov* mode of growth initially forms a smooth layer; however, as the layer thickness increases, strain energy builds up. This energy can be released by the spontaneous formation of islands and hence quantum dots.

In *Volmer-Weber* mode growth, small islands form on the substrate that may constitute quantum dots. The bonding energy of the adatoms with the substrate surface atoms (interfacial energy) is weaker than adatom-adatom bonding due to a lattice mismatch.

MBE has been used to self-assemble quantum dots from III-V semiconductors using the lattice mismatch approach. For example, an InAs epitaxial layer on a GaAs substrate has a 7% lattice mismatch as shown in Figure 2.34 and undergoes the Stranski-Krastanov mode growth with the spontaneous formation of quantum dots. See Figure 7.2.

Chemical vapor deposition is another method to form thin films. See Section 6.12. Quantum dots can also self-assemble using this growth type. Precursors are introduced in a chamber at a particular pressure and temperature and they diffuse to the heated substrate and react to form a film. Gas-phase byproducts desorb from the substrate and are removed from the chamber. InGaAs and AlInAs quantum dots have been synthesized using chemical vapor deposition which formed due to lattice mismatch on GaAs substrates.

Of the many bottom-up growth methods, the most industrially important approach involves a *precipitation reaction* in which a particle is induced to grow into a quantum dot usually from a liquid precursor. The precipitation process involves nucleation followed by the limited growth of nanoparticles. Homogeneous nucleation occurs when solute atoms or molecules combine and reach a critical size without the assistance of a pre-existing solid interface. By varying factors including temperature and the chemistries of the solution and precipitate, quantum dots of the desired size, shape, and composition can be achieved.

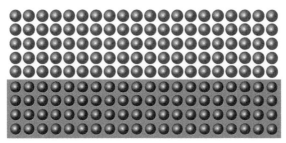

Frank–van–der–Merwe mode

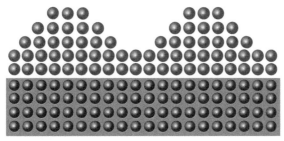

Stranski–Krastanov mode

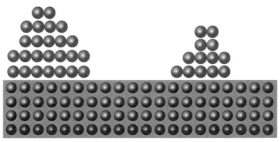

Volmer–Weber mode

FIGURE 7.1 a) Van der Merwe growth mode. Epitaxial growth causes atoms to replicate substrate lattice structure. Atoms being deposited attach preferentially to surface sites resulting in atomically smooth, continuous layers. This layer-by-layer growth is two-dimensional, indicating that complete films form prior to the growth of subsequent layers. b) Stranski Krastanov growth mode. Initially, Van der Merwe growth proceeds; however, the spontanteous formation of islands takes place later. A lattice-mismatch between substrate and the growing film can be responsible for this change in growth type: as the epitaxial later thickness increases, strain energy also rises due to the lattice mismatch. This strain energy can be released by island formation. c) Volmer-Weber growth mode. Island formation occurs immediately. In this growth mode, bonding between arriving atoms that lie on the growth surface (adatoms) is stronger than bonding between adatoms and substrate atoms. Source: David W / Wikimedia commons / Public Domain.

Sol-gel techniques represent a first type of precipitation reaction and have been used to synthesize a wide range of nanoparticles including quantum dots. The sol (nanoparticles dispersed in a solvent by Brownian motion) may be prepared using a Group II metal precursor (generally the metal in the form of an alkoxide, an acetate,

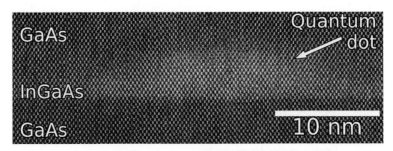

FIGURE 7.2 An atomic resolution image of a single indium gallium arsenide (InGaAs) quantum dot embedded within gallium arsenide (GaAs). Image is acquired using scanning transmission electron microscopy. The quantum dot is formed by $Ga_xIn_{1-x}As$ growth by MBE on a GaAs substrate. Depending on the indium concentration, strain energy increases during growth and quantum dots will spontaneously form due to the Stranski-Krastanov mode growth. The covering GaAs material is grown by MBE after the formation of the quantum dots. Magnunor / Wikimedia commons / Public Domain.

or a nitrate) in an acidic or basic medium. The three main steps in this process are hydrolysis, condensation (sol nucleation), and growth (gel formation). In brief, the metal precursor *hydrolyzes*, or reacts with water in the medium and condenses to form a sol, followed by polymerization to form a gel or network of precipitated particles. This method has been used to synthesize II-VI quantum dots such as CdS and ZnO. For example, ZnO quantum dots have been prepared by mixing solutions of Zn-acetate in alcohol and sodium hydroxide, followed by controlled reaction in air. The process is simple, cost-effective, and suitable for scale-up. The main disadvantages of the sol-gel process include a broad size distribution and a high concentration of defects.

A second type of precipitation reaction, the *microemulsion* process, is a convenient method for synthesizing quantum dots at room temperature. The processes can be categorized as either a *normal* microemulsion process (oil-in-water) or as a *reverse* microemulsion process (water-in-oil). In some cases, other polar solvents such as alcohol may be used instead of water. The reverse micelle process for synthesizing quantum dots involves the two immiscible liquids (for example polar water and nonpolar long-chain alkane) being stirred together to form an emulsion. Nanometer-scale water droplets dispersed in *n*-alkane solutions can be achieved using surfactants which reduce the surface tension of the alkane in which the water is dispersed. Surfactants are ligands that comprise a hydrophilic group on one end and and a hydrophobic group on the opposite end. See Figure 7.3. Numerous tiny droplets called micelles are formed in the continuous oil medium. These micelles are thermodynamically stable and can act as "nanoreactors." A vigorously stirred micellar suspension leads to a continuous exchange of reactants introduced for quantum dot formation due to dynamic collisions. Growth of the resultant quantum dots is limited by the size of the micelle which is controlled by the molar ratio of water and surfactant.

A third type of precipitation reaction has proven to be the most effective method for synthesizing quantum dots having a uniform particle size distribution and

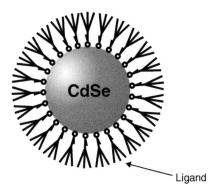

FIGURE 7.3 Ligands have a polar (hydrophilic) termination allowing binding to the ionic CdSe quantum dot semiconductor at one end and a nonpolar (hydrophobic) termination at the other end allowing suspension in a nonpolar medium.

excellent luminous efficiency. It comprises a high temperature injection method in which a low-temperature precursor solution is rapidly injected into a high temperature solution. *Hot-injection* II-VI quantum dot synthesis is widely used in production for synthesizing quantum dots having a narrow size distribution. This narrow size distribution is known as a *monodisperse* size range. CdS, CdSe, and CdTe quantum dots are examples of materials that may be grown by this method.

During hot injection synthesis, a condition above a saturation concentration (super-saturation) is achieved in which quantum dot particles are nucleated. At the beginning of this procedure, only one of the precursors (either anion or cation) is mixed with the solvents and ligands in a reaction vessel, followed by heating to high temperature in the range of $200°C - 300°C$. The surfactant molecules or ligands are available to coat precipitates that will be formed in subsequent steps. They prevent the resulting quantum dots from agglomerating.

Next, a solution containing the other precursor is swiftly injected into this hot solution, resulting in nucleation by supersaturation. The growth stage now proceeds through diffusion-controlled growth in which the larger particles in the reaction solution grow more slowly than the smaller ones. Hence, the average particle size increases over time and quantum dots with a narrow size distribution are achieved. The precursors that do not immediately participate in nucleation react with the surface of the already-nucleated quantum dots, causing quantum dots to grow. When the supersaturation is high, small particles grow rapidly and relatively large particles grow slowly due to the difference in surface area. A homogeneous diffusion-controlled growth is observed across the solution. Since larger quantum dots grow more slowly than the smaller ones, a size-focusing effect occurs and, as a result, particles become uniform in size. Thereafter, when the concentration of the precursor in the container falls below a critical concentration, the *Ostwald ripening* process takes place, in which slightly smaller particles contract as their surface atoms are released and the larger-sized particles grow, thereby increasing the size distribution of the precipitates. Before

Ostwald ripening occurs, the temperature of the reactor may be lowered to terminate the chemical reaction, thereby synthesizing quantum dots having a substantially uniform size. The hot-injection method is particularly effective because it offers a high level of control over the size of the particles as well as their size distribution by allowing a rapid nucleation separated from the growth stage. By varying the temperature, the concentration of the surfactants, and the reaction time, it is possible to obtain monodisperse quantum dots having a desired size. The high temperature process results in excellent crystal quality of the growing II-VI semiconductor material. See Figure 7.4.

Due to the high surface-to-volume ratio of quantum dots, electronic states associated with the surface (surface states) are important to consider. The energy of these states usually lies near the middle of the quantum dot bandgap. See Section 2.20. Therefore the surface states can trap charge carriers leading to non-radiative recombination of electron-hole pairs. As a result, surface states can dramatically lower quantum yield. Approximately 15% of the atoms in a 5 nm CdS quantum dot are at the surface. Dangling bonds arise predominantly from unsatisfied bonds at the surface, but may also be caused by non-stoichiometry and voids. The crystal quality of the semiconductor comprising the quantum dot is therefore important. Strategies to minimize or eliminate surface dangling bonds will be discussed in Section 7.8.

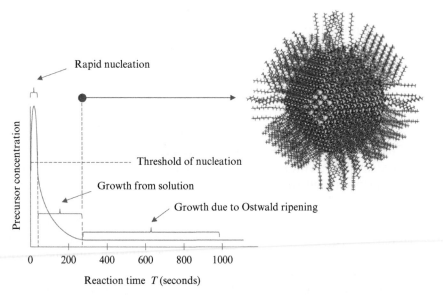

FIGURE 7.4 Quantum dots formed by the hot injection method. Either the cation or the anion of the desired quantum dot material is heated along with the ligand-based surfactant in a container. At reaction time $T = 0$ the complimentary ion in the form of a precursor complex is rapidly injected into the container leading to supersaturation of the nucleating species. The ligands surround the growing nuclei and prevent agglomeration. Inset (wikimedia) depicts the quantum dot with surfactant ligands that would appear after a reaction time of $\cong 260$ seconds.

7.4   QUANTUM DOT CONFINEMENT PHYSICS

The size-tunability of absorption and emission wavelengths is a key property of quantum dots. This can be modeled using quantum and semiconductor physics.

The *effective mass approximation model* is frequently used to predict the size dependence of the quantum dot bandgap. The starting point for this model is the band model for a bulk semiconductor forming the quantum dot. On a plot of energy E versus wavenumber k, parabolic valence and conduction bands are assumed with characteristic electron effective mass m_e^* and hole effective mass m_h^* as described in Chapter 2. An important point is that the parabolic bands are *not* continuous functions, but are instead a set of points, each having a distinct value of k, as a result of the finite dimension of the semiconductor crystal. This was covered in detail for a one-dimensional case in Section 2.7 and Example 2.1. It was covered again in three dimensions in Section 2.10.

The minimum energy level in a parabolic direct gap conduction band represents the lowest allowed energy state in the conduction band for a semiconductor of infinite dimensions in which wavenumber k approaches zero. There is no specific lower bound for wavenumbers k_x, k_y, and k_z in a three-dimensional semiconductor crystal of infinite size because the spacing between k values becomes infinitely small as size approaches infinity. As the size of the semiconductor is reduced, the spacing between k-values increases.

It is simplest to approximate the quantum dot as a box-shaped semiconductor sample shown in Figure 7.5 with dimensions $a = b = c$ and to assume infinitely high potential energy barriers at the box surfaces.

From Equation 2.30, the allowed values of k in each dimension and the possible electron energies for a free electron in the box are

$$E = \frac{\hbar^2}{2m}(k_x^2 + k_y^2 + k_z^2) = \frac{\hbar^2\pi^2}{2m}\left[\left(\frac{n_x}{a}\right)^2 + \left(\frac{n_y}{b}\right)^2 + \left(\frac{n_z}{c}\right)^2\right] \tag{7.1}$$

where n_x, n_y, and n_z are integers.

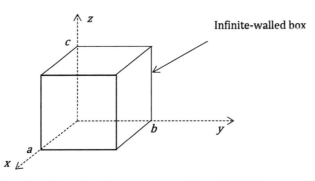

FIGURE 7.5   A box-shaped semiconductor quantum dot with infinite potential walls is used to obtain the allowed quantum states for an electron.

Equation 7.1 yields energy values approaching zero if dimensions a, b, and c approach infinity. It is useful to think of a very large semiconductor sample being contained in a three-dimensional box of almost infinite dimensions. This was the effective assumption for the various band models presented in Sections 2.2 to 2.5.

If the semiconductor box is a small cube having edge length a, the lowest allowed values of k_x, k_y, and k_z in Equation 7.1 are $k = k_x = k_y = k_z = \dfrac{\pi}{a}$ when $n_x = n_y = n_z = 1$. As a result, the lowest available energy state is

$$E = \frac{3\hbar^2 k^2}{2m} = \frac{3\hbar^2 \pi^2}{2ma^2}$$

This parabolic relationship between E and k applies to states in an infinite-walled potential well with $U = 0$. It needs to be modified to be consistent with the electron states near the edges of energy bands. The parabolic, isotropic dispersion relation from Section 2.6 is directly applicable. The appropriate electron effective mass needs to be used in place of the electron rest mass. For the conduction band,

$$E_{e,cube}^1 = \frac{3\hbar^2 \pi^2}{2m_e^* a^2}$$

where $E_{e,cube}^1$ is defined as the energy difference between the bottom of the conduction band and the lowest allowed state.

The same considerations can be used to determine the lowest energy for a hole in the valence band relative to the top of the valence band and

$$E_{h,cube}^1 = \frac{3\hbar^2 \pi^2}{2m_h^* a^2}$$

This is illustrated in Figure 7.6.

The effective quantum dot bandgap as a result of quantum confinement is therefore now

$$E_{g_{QD}} = E_g + \frac{3\hbar^2 \pi^2}{2m_e^* a^2} + \frac{3\hbar^2 \pi^2}{2m_h^* a^2} = E_g + \frac{3\hbar^2 \pi^2}{2a^2 \mu}$$

where μ, the reduced mass expressed as

$$\frac{1}{\mu} = \frac{1}{m_e^*} + \frac{1}{m_h^*}$$

is used for convenience.

An additional energy consideration arises from the formation of an exciton in quantum dots. In bulk inorganic semiconductor materials, excitons are unstable at room temperature as discussed in Chapter 4. However the size of a quantum dot is similar to an exciton and therefore the exciton is prevented from dissociation due to the potential barrier at the surface of the quantum dot.

We can include the ground state energy of the exciton (the exciton binding energy) which will lower the effective quantum dot bandgap. The quantum dot effective bandgap now becomes

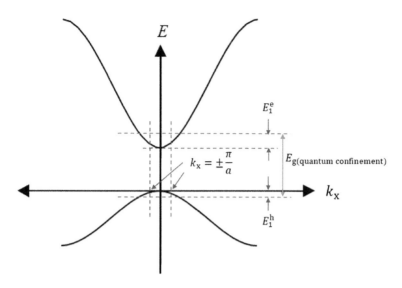

FIGURE 7.6 The conduction and valence bands for a semiconductor can be approximated as parabolic near the band edges. Although continuous black lines are shown, the bands are actually a set of discrete allowed points. For the small cube of edge length a, the smallest allowed value of k_x is π/a. If this standing wave quantum state is written as a sum of two travelling waves $e^{\pm ikx}$ then $k = \pm \frac{\pi}{a}$. The y and z dimensions are not shown but must be included to obtain the quantum confinement energy gap.

$$E_{g_{QD}} = E_g + \frac{3\hbar^2\pi^2}{2a^2\mu} - \frac{\mu q^4}{8\epsilon_0^2\epsilon_r^2 h^2}$$

where the negative exciton binding energy term was discussed in Section 4.5. This assumes that the exciton is smaller than the quantum dot, but that the quantum dot is small enough to restrict the exciton to its lowest energy or its ground state.

EXAMPLE 7.1

Estimate the effective bandgap of a quantum dot made from CdSe. Assume the shape of the quantum dot is a cube of edge length 8 nm with $m_e^* = 0.13m$ and $m_h^* = 0.3m$ Include both the quantum size effect and the exciton model.

Solution:

Reduced mass μ is determined as

$$\mu = \left(\frac{1}{m_e^*} + \frac{1}{m_h^*}\right)^{-1} = m\left(\frac{1}{0.13} + \frac{1}{0.3}\right)^{-1} = 9.11\times10^{-31}\,\text{Kg}\times9.07\times10^{-2} = 8.26\times10^{-32}\,\text{Kg}$$

From Table 7.1 for CdSe, $\epsilon_r \cong 9$ and $E_g = 1.73\text{eV}$

(continued)

(continued)

$$E_{gQD} = E_g + \frac{3\hbar^2 \pi^2 n^2}{2a^2 \mu} - \frac{\mu q^4}{8\epsilon_0^2 \epsilon_r^2 h^2} = 1.73\text{eV} + \frac{3 \times \left(1.05 \times 10^{-34}\,\text{Js}\right)^2 \pi^2}{2 \times \left(8 \times 10^{-9}\,\text{m}\right)^2 \times 8.26 \times 10^{-32}\,\text{Kg}}$$

$$- \frac{8.26 \times 10^{-32}\,\text{Kg} \times \left(1.6 \times 10^{-19}\,\text{C}\right)^4}{8 \times \left(8.85 \times 10^{-12}\,\text{Fm}^{-1}\right)^2 \times 9^2 \times \left(6.62 \times 10^{-34}\,\text{Js}\right)^2}$$

$$= 1.73\text{eV} + 3.08 \times 10^{-20}\,\text{J} - 2.43 \times 10^{-21}\,\text{J} = 1.73\text{eV} + 0.19\text{eV} - 0.015\text{eV} = 1.91\text{eV}$$

The effective bandgap of the quantum dot increases from 1.73eV to 1.91 eV. The change is dominated by the confinement effect.

Whereas Equation 7.1 assumes a cube-shaped quantum dot, quantum dots made by precipitation reactions as depicted in Figures 7.3 and 7.4 are approximately spherical in shape because surface area and hence the energy of formation during chemical synthesis is minimized in this case. It is therefore necessary to model quantum confinement based on a spherical geometry.

We will first find the lowest available energy state for an electron in a spherical region with potential energy $U = 0$ for radii between 0 and R. The energy barrier for an electron will be assumed to be infinite at radius R.

Schrödinger's equation must be written using spherical polar coordinates r, θ, ϕ in order to obtain solutions to this spherically symmetric problem. Equation 1.27 is the Schrödinger equation written in terms of the radial coordinate r assuming that the wavefunctions being sought are independent of θ and ϕ. This equation

$$-\frac{\hbar^2}{2mr}\frac{\partial^2}{\partial r^2}\left[r\psi(r)\right] + U(r)\psi(r) = E\psi(r)$$

becomes

$$\frac{-\partial^2}{\partial r^2}\left[r\psi(r)\right] = \frac{2mE}{\hbar^2}r\psi(r)$$

for $r < R$ because $U = 0$.
Letting $\Omega(r) = r\psi(r)$ we have

$$-\frac{\partial^2}{\partial r^2}\Omega(r) = \frac{2mE}{\hbar^2}\Omega(r)$$

which has general solution

$$\Omega(r) = A\sin(kr) + B\cos(kr)$$

where $k = \sqrt{\dfrac{2mE}{\hbar^2}}$ and therefore

$$\psi(r) = \frac{A\sin(kr)}{r} + \frac{B\cos(kr)}{r}$$

In order to avoid an infinite solution at $r = 0$ we require that $B = 0$. In addition, the boundary condition $\psi(R) = 0$ ensures that the wavefunction is continuous at $r = R$. Hence,

$$\psi(r) = \frac{A\sin(kr)}{r}$$

where, for positive integer values n, $k = \dfrac{n\pi}{R}$.

Coefficient A is evaluated by normalizing $\psi(r)$:

$$\int_0^{2\pi} \int_0^{\pi} \sin\theta \int_0^{\infty} r^2 |\psi(r)|^2 \, dr d\theta d\phi = \int_0^{\infty} 4\pi r^2 \psi^2(r) dr = \int_0^{R} 4\pi A^2 \sin^2\left(\frac{n\pi}{R}r\right) dr = 1$$

Solving this we obtain $A = \sqrt{\dfrac{1}{2\pi R}}$ and

$$\psi(r) = \sqrt{\frac{1}{2\pi R}} \frac{\sin\left(\dfrac{n\pi}{R}r\right)}{r}$$

The lowest available state occurs when $n = 1$ in which case

$$k = \frac{\pi}{R} = \sqrt{\frac{2mE}{\hbar^2}}$$

Solving for the eigenenergy, we obtain

$$E = \frac{\hbar^2 k^2}{2m} = \frac{\hbar^2 \pi^2}{2mR^2}$$

In order to apply this minimum allowed energy to the band edges for the quantum dot semiconductor we again invoke the parabolic, isotropic dispersion relation to obtain

$$E_{e,\text{sphere}}^1 = \frac{\hbar^2 k^2}{2m_e^*} = \frac{\hbar^2 \pi^2}{2m_e^* R^2}$$

and

$$E_{h,\text{sphere}}^1 = \frac{\hbar^2 k^2}{2m_h^*} = \frac{\hbar^2 \pi^2}{2m_h^* R^2}$$

The resulting effective energy gap for the spherical quantum dot including the exciton binding energy is

$$E_{g_{QD}} = E_g + \frac{\hbar^2 \pi^2}{2\mu R^2} - \frac{\mu q^4}{8\epsilon_0^2 \epsilon_r^2 h^2}$$

The energy gap of a spherical quantum dot will now be compared to the result for the cubic quantum dot in Example 7.1.

EXAMPLE 7.2

Repeat 7.1 using the spherical quantum dot model in place of the cubic model and compare results.

Solution: The sphere radius will be $R = 4$ nm which is half the dimension of the cube in Example 7.1.

$$E_{gQD} = E_g + \frac{\hbar^2 \pi^2}{2\mu R^2} - \frac{\mu q^4}{8\epsilon_0^2 \epsilon_r^2 h^2} = 1.73\,\text{eV} + \frac{\left(1.05 \times 10^{-34}\,\text{Js}\right)^2 \pi^2}{2 \times \left(4 \times 10^{-9}\,\text{m}\right)^2 \times 8.26 \times 10^{-32}\,\text{Kg}}$$

$$- \frac{8.26 \times 10^{-32}\,\text{Kg} \times \left(1.6 \times 10^{-19}\,\text{C}\right)^4}{8 \times \left(8.85 \times 10^{-12}\,\text{Fm}^{-1}\right)^2 \times 9^2 \times \left(6.62 \times 10^{-34}\,\text{Js}\right)^2}$$

$$= 1.73\,\text{eV} + 4.11 \times 10^{-20}\,\text{J} - 2.43 \times 10^{-21}\,\text{J} = 1.73\,\text{eV} + 0.26\,\text{eV} - 0.015\,\text{eV} = 1.98\,\text{eV}$$

This result is larger than the solution to Example 7.1. This is not surprising since the volume of the sphere is smaller than that of the cube. Increased confinement is expected to increase the energy gap.

Two additional considerations relate to small quantum dots. Firstly, if the quantum dot radius is smaller than the exciton radius, the electron and hole are forced to be closer to each other, on average, compared to the case of an exciton in a bulk semiconductor due to the energy barrier imposed by the quantum dot. Some quantum dot semiconductor materials such as PbS and InSb have large relative dielectric constants ϵ_r and/or small values of reduced mass μ. See Table 7.1 and Appendix 6. We can therefore expect exciton radius a_0 to be large due to the expression for exciton radius from Section 2.14

$$a_0 = \frac{4\pi\epsilon_0 \epsilon_r \hbar^2}{\mu q^2} \tag{7.2}$$

Detailed modeling of effective quantum dot energy gap in the case when quantum dot radius is smaller than exciton radius is beyond the scope of this book.

Secondly, for smaller quantum dots, the parabolic energy band assumption inherent in the effective mass approximation model may become invalid. This is because in our modeling of quantum dot bandgap, we assumed an effective mass which is consistent with conduction and valence bands that are well approximated by parabolas in E versus k space, and this parabolic assumption is not accurate except near band edges as can be seen from the band diagrams in Figure 2.17. From Figure 7.2, a smaller quantum dot which leads to larger values of $k = \dfrac{\pi}{a}$ leads also to larger values of E_1^e and E_1^h and energy band character further away from band edges may no longer be parabolic.

In summary, for small quantum dots, modeling can lead to errors in both the quantum confinement term *and* the exciton binding energy term.

One aspect of quantum dot excitons we have not considered in spin. Quantum dot excitons may be singlet excitons or triplet excitons. The importance of singlets versus triplets arises from the large energy spacings between excited quantum dot states. Normally only one excited electron exists in a quantum dot and its spin direction is therefore important, whereas in bulk semiconductors, large numbers of very closely spaced excited energy levels exist having both spin directions. During photoluminescence, the excited quantum dot will normally be in a singlet state since the incoming photon that generated the exciton did not carry a magnetic moment. See Section 4.6.

It is also possible, however, for the excited quantum dot state to be a triplet state. Triplet states are slightly lower in energy than the singlet states. One means by which triplet states can be formed is through electroluminescence in which excitation does not involve a photon. This is further discussed in the context of molecules for organic light emitting devices in Section 8.11.

7.5   FRANCK-CONDON PRINCIPLE AND THE STOKES SHIFT

The absorption and emission spectra of a quantum dot differ from each other. Consider a quantum dot that is optically excited with monochromatic light of wavelength $\lambda_{excitation}$ leading to fluorescence. The fluorescence is found to have an emission spectrum rather than just a single emission wavelength. The maximum intensity of the emission spectrum occurs at a wavelength that is longer compared to $\lambda_{excitation}$.

An absorption spectrum also exists. It can be determined by varying the excitation wavelength $\lambda_{excitation}$ while plotting emission intensity at a fixed emission wavelength which is normally at or near the peak of the emission spectrum. Observed emission and absorption spectra for a CdSe quantum dot are shown in Figure 7.7.

The *Stokes shift* is the difference in the position of the maximum intensity between the emission spectrum and the absorption spectrum. It can be expressed in terms of energy, wavenumber or wavelength.

Section 4.8 describes the absorption spectrum of a bulk, direct gap semiconductor by considering a range of energy values available for electrons and holes in the relevant energy bands. The joint density of states leads to an increasing absorption

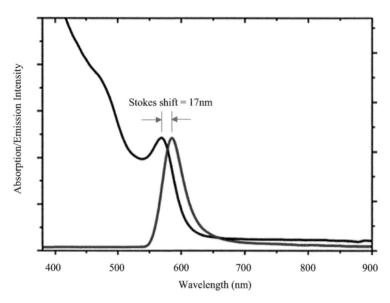

FIGURE 7.7 The absorption spectrum for a CdSe quantum dot is shown in black. The emission spectrum is shown in red. The peak of the emission spectrum for the quantum dot occurs at a longer wavelength than the peak of the absorption spectrum. The change in the peak is the Stokes shift expressed in terms of wavelength. Mohamed et al., 2019. / Springer Nature.

coefficient as wavelength decreases. Figure 4.12, the examples in Figure 5.5 and the absorption spectrum in Figure 7.7 all show increasing absorption toward shorter wavelengths as the joint density of states increases.

The joint density of states and the probability of occupancy of these states also explain the emission spectrum of a bulk direct gap semiconductor in Section 4.8. Quantum dot emission, however, involves only *one* electron transition energy $E_{g(\text{quantum confinement})}$ from the lowest state in the conduction band to the highest available state in the valence band shown in Figure 7.6. Another explanation is therefore needed for the broadening of the emission spectrum of Figure 7.7. In addition, the absorption spectrum in Figure 7.7 shows an unexplained local maximum, again with some broadening, where the Stokes shift in measured.

In order to explain the origin of the Stokes shift, the broadening of the emission spectrum and the local maximum in the absorption spectrum, it is necessary to consider vibrations of the atoms in a quantum dot. The atoms are connected to each other with bonds that can be modelled as springs. This mass-spring system is subject to thermal vibration. In bulk semiconductors, lattice vibrations or phonons have been discussed in Chapter 2 in the context of their effects on carrier scattering and carrier recombination.

Lattice vibrations in a quantum dot need to be treated differently from those in bulk materials due to the small number of atoms and their associated bonds. Consider the simplest case of just two atoms and one bond between them in a hypothetical diatomic molecule with a covalent bond. The energy diagram for this situation is shown in Figure 7.8. The equilibrium bond length r_0 is defined at the minimum bond energy. In compression, with lengths smaller than r_0, repulsive forces exceed attractive forces, and vice versa for dilation when lengths are larger than r_0.

Now consider what happens to the diatomic molecule after it absorbs a photon. A valence electron is excited into a higher energy state leaving a missing electron in the covalent bond. This weakens the bond between the two atoms due to the lack of the full complement of valence electrons.

Two scenarios are possible. In the first case the molecule breaks apart due the absorption of the photon. This is an example of *photodissociation* which constitutes a chemical change in the system. Many important natural and industrial chemical reactions involve photodissociation.

Of interest here is the second case in which the bond weakens, but the molecule does not dissociate. The result is a modified plot of potential energy versus internuclear distance as shown in Figure 7.9.

Quantum dots have more than two atoms. Nevertheless, the concepts illustrated in Figure 7.9 are still relevant. This can be understood by the sharing of a missing electron among a collection of bonds. Every bond has a probability at any instant in time of missing an electron and the net effect is that every bond behaves as an incomplete bond.

In a bulk semiconductor, the number of atoms is very large and the effect of a hole or holes on the atomic lattice is insignificant because the hole concentration is normally orders of magnitude smaller than the number of atomic bonds per unit volume. In an excited quantum dot, however, the number of bonds is very limited and the sharing of even one hole among these bonds leaves them sufficiently incomplete, on average, to influence the resulting atomic positions.

It the case of a quantum dot, it is not appropriate to use the label on the horizontal axis of Figure 7.9. Instead, the shared hole in the quantum dot causes a small change

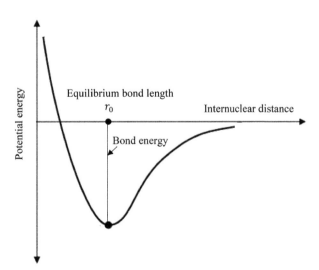

FIGURE 7.8 Bond energy versus bond length for an atomic bond. The equilibrium bond length occurs at minimum bond energy.

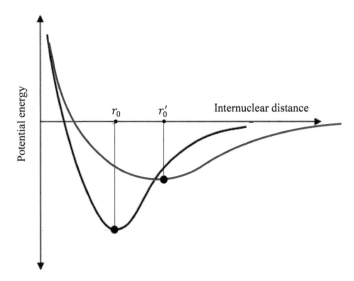

FIGURE 7.9   Bond energy versus bond length for an atomic bond of an photoexcited diatomic molecule (red curve) compared to the same molecule before excitation (black curve). The bond energy decreases after photoexcitation and the internuclear distance (bond length) increases.

in the relative positions of the quantum dot atoms. The horizontal axis is now labeled as a *configurational coordinate*. A simple example of a reconfiguration would be a slight dilation of the quantum dot due to a slight increase in the average bond length. Other more complex reconfigurations are also possible. Smaller quantum dots with fewer atoms and fewer bonds would be expected to exhibit more of this effect than larger quantum dots. A small quantum dot may have a diameter of $1 - 2$ nm and contain only approximately 100 atoms.

The equilibrium bond length shown in Figures 7.8 and 7.9 is defined at the energy minimum on the vertical axis. In reality, vibrations of the atoms are always present, and a series of possible vibrational energies or vibrational modes exists. The equilibrium bond length is now the average bond length of the vibrating bond. This concept explains thermal expansion of almost all solids in which an increase in temperature causes an increase in average bond length. A higher degree of asymmetry in the shape of the bond in Figure 7.8 leads to a higher coefficient of thermal expansion.

The processes involved in the *Franck-Condon* principle also rely on these atomic vibrational energies. They will initially be described with simplifications, and then the quantum mechanical treatment that provides more insight into the physics involved will follow.

Consider a small quantum dot. In Figure 7.10 the electron levels in the quantum dot that include vibrational energies are shown. The vibrational energies of the nuclei are added onto the electronic energy gap from Figure 7.6 which explains why the red and black curves are vertically separated in Figure 7.10 but not in Figure 7.9. The vertical axis now represents the sum of the energy of the electron and a set of possible

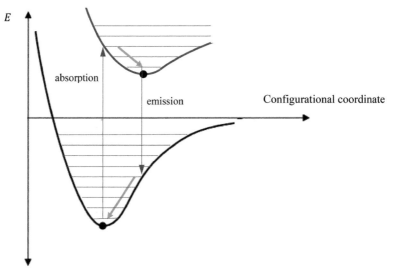

FIGURE 7.10   Electron energy levels in a quantum dot including vibrational energy levels are shown. Bond vibrations having various vibrational amplitudes are shown as horizontal lines. The curves are offset along the vertical energy axis to show the change in electron energy between the unexcited quantum dot (electron in the valence band) and the exciton or excited quantum dot (electron in the conduction band). The absorption of a photon that raises an electron from the lowest vibrational level of the quantum dot before excitation is shown by a red arrow. The electron making a vertical transition will end up in a vibrational level of the excited quantum dot. Before recombination, the excited quantum dot will rapidly thermalize (lose vibrational energy to its surroundings) and the excited electron will reach the lowest vibrational level of the excited quantum dot as indicated by the upper blue arrow. Then, when the radiation event takes place, the electron makes a second electronic transition. If this transition is also vertical then the ground state quantum dot ends up in a vibrational state of the unexcited quantum dot before thermalizing again as shown by the lower blue arrow. The energy difference between the absorption process and the emission process explains the Stokes shift.

vibrational energies of the atomic nuclei. In this picture, the electron state, in either the valence or the conduction band, possesses an additional vibrational energy since the nuclei of the atoms are moving. The temperature effect on average bond length that leads to thermal expansion in solids is evident since the midpoint of each vibration energy line yields an increasing configurational coordinate as vibrational energies increase. This is particularly evident in the case of the excited quantum dot.

Justification for adding the vibrational energies to the electron energies is based on the *Born-Oppenheimer approximation* which states that the electron motion and the nuclear motion can be considered separately. This means that the electron energies and wavefunctions can be calculated by assuming the nuclei are static, and then the additional energy of the moving nuclei can be added later. The rationale for this is the large ratio of the mass of an average-sized atomic nucleus to the electron mass. This ratio is on the order of 10^4 which means that electrons move much faster than nuclei. The relatively slow-moving nuclei have a negligible effect on the instantaneous interactions of a given nucleus with surrounding electrons which determine the electron wavefunctions and energy states of electrons. As an analogy, we usually neglect the orbital or rotational motions of the earth when we study the effect of gravity on small projectiles because the resulting errors are very small.

Now consider what happens to the quantum dot after it absorbs a photon as shown in Figure 7.10. A valence band electron is excited into the conduction band leaving a hole in the valence band. At the start of a photoluminescence process, the unexcited quantum dot exists in a low energy vibrational state near the bottom of the black curve. From a classical viewpoint, some minimal vibrational energy is always present due to thermal vibrations consistent with ambient temperature. We will assume that the quantum dot is in thermal equilibrium with its surroundings. At room temperature the average vibrational energy is $kT = 0.026\,\text{eV}$. The reason that the vibrational states shown in Figure 7.10 are discrete will be covered in Section 7.6.

The absorption of the incoming photon is extremely fast. The process occurs in a time frame of about 10^{-15} seconds. The photon is annihilated when its energy is transferred to the electron making the transition from valence band to conduction band. Unlike the opposite process of radiation described in Section 4.4 in which a superposition state forms an oscillating dipole and produces energy that builds up over time to eventually create a photon, the process of photon absorption cannot gradually remove energy from a photon. The photon represents the smallest quantum of electromagnetic energy and therefore the photon either does exist, or does not exist. At the instant in time that the photon ceases to exist, its entire energy must have transferred to the electron in order to satisfy energy conservation. There is no classical picture that can allow us to visualize exactly how this energy transfer takes place.

The natural frequency range of phonons depends primarily on the masses of the atomic nuclei and the effective spring constants of the bonds between them. The well-known mass-spring model describes a mass m attached to one end of a spring of spring constant K resulting in a resonance frequency $\omega = \sqrt{\dfrac{K}{m}}$. If numerical values of K and m relevant to atoms in a crystalline solid are used, values of resonant frequency are found to typically lie in the range of 1–20 THz ($1\,\text{THz} = 10^{12}\,\text{Hz}$). The duration of a cycle of one atomic vibration is therefore on the order of $10^{-12}\,\text{s}$ or $10^{-13}\,\text{s}$ which is two or three orders of magnitude larger than the duration of the photon absorption process. See Problem 7.5.

There is therefore insufficient time for the configuration of the atoms to significantly change during the very fast photon absorption process. This means that the absorption event in Figure 7.10 must occur along an almost vertical line as shown. The third vibrational level of the excited quantum dot results in the example shown. From a classical viewpoint, we can consider that the atoms in the excited state of the quantum dot at the maximum amplitude of the third vibrational level exist in a configuration that happens to match the configuration of the atoms in the unexcited quantum dot. The result is that there is excess atomic vibrational energy that inherently results from the absorption of the photon. The vibration of the excited quantum dot is initiated by the photon absorption, but the vibrations occur after the absorption event is completed. The amplitude of this vibration depends on the extent to which changes in equilibrium bond lengths and atomic positions are caused by the creation of the exciton.

The vibrational energy in the excited quantum dot will very rapidly be dissipated. The vibrational level will relax to, or near to, its lowest vibrational level as shown by the upper blue arrow. Using thermodynamic terminology, heat flows out of the quantum dot and is dissipated by the quantum dot surroundings. The time scale for this to occur is typically on the order of the period of vibration which is 10^{-12} s or 10^{-13} s.

The next step in the fluorescence process involves radiative recombination of the excited electron-hole pair constituting the exciton in the quantum dot. As a result of this radiation process, the electron in Figure 7.10 drops down to a vibrational level of the unexcited quantum dot and a photon is released. The downward transition during emission is shown as a vertical arrow in Figure 7.10. Justification for this transition being a vertical transition will be discussed in detail in Section 7.6.

The emitted photon clearly has a lower energy than the photon that was originally absorbed. The difference between the absorption energy and the emission energy explains the Stokes shift. Figure 7.10 shows that the decrease in electron transition energy ends up in the form of vibrational energy or heat. Vibrational energies in both the unexcited and the excited quantum dot are involved.

The spectral widths of the peaks in the absorption and emission spectra shown in Figure 7.7 can also be explained now. Figure 7.11 shows that additional transitions could occur due to the close spacing of vibrational modes. Details of the shape of the absorption and emission spectra depend on the probabilities of the various transitions. The two transitions shown in Figure 7.10 should be viewed as the most probable transitions that coincide with the spectral peaks in Figure 7.7, and the added transitions in Figure 7.11 also occur, but with lower probabilities, that determine the shapes of the spectra.

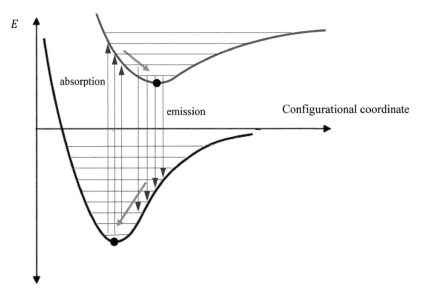

FIGURE 7.11   A set of possible transitions exists during both absorption and emission due to the close energy spacing of the vibrational modes in a quantum dot.

7.6 THE QUANTUM MECHANICAL OSCILLATOR

Justification for the downward transition being a vertical arrow in Figure 7.10 is more complicated than in the case of the upward arrow because, unlike the photon absorption process, the radiative recombination process is not virtually instantaneous relative to the period of atomic vibrations. Radiative recombination occurs over a time frame that is, on average, much *longer* than the vibrational cycle of the atomic vibrations. In Example 4.1 this time frame is on the order of 10^{-8} seconds compared to an atomic vibrational cycle time of 10^{-12} s or 10^{-13} s. The radiative recombination process involves dipole radiation in which the expected value of the position of the electron oscillates, electric and magnetic fields form, and a photon results. The electron is in a superposition state composed of both the excited state and the ground state of the electron.

It is first necessary to understand the vibrations of atoms from a quantum mechanical perspective. Then the radiative transition will be viewed as a *vibronic transition*. The concept of the vibronic transition also provides a much better understanding of the probabilities of the various transitions in Figure 7.11, and these probabilities can then be calculated.

Quantum mechanics concepts introduced in Chapter 1 focus on electrons; however, these concepts are also applicable to other particles including protons, neutrons, and atomic nuclei. The uncertainty principle $\Delta x \Delta p \geq \frac{\hbar}{2}$ is still valid even though the mass of an atomic nucleus is 3–5 orders of magnitude larger than electron mass.

Consider a single atom of mass m inside a solid. In a simple one-dimensional model, the bonds between this atom and its n nearest neighbors may be modeled as n springs, each obeying Hooke's law, and each having spring constant K_n. We can express the total net force on the atom as a function of its displacement about its equilibrium position as $F = -Kx$ where K is the combined spring constant that results by suitably adding the forces due to each of the n springs.

The potential energy of an atom nucleus moved a distance x from its equilibrium position is

$$U_n(x) = \int_0^x |F|(x)dx = \int_0^x Kx_0 dx_0 = \frac{K}{2}x^2$$

This parabolic potential energy versus position dependence is the basis for the quantum mechanical oscillator. Potential energy in Figure 7.8 is well approximated by a parabola except for the case of large atomic vibrations.

In a classical oscillator, the resonant frequency is

$$\nu = \frac{\omega}{2\pi} = \frac{1}{2\pi}\sqrt{\frac{K}{m}}$$

and therefore

$$U_n(x) = \frac{K}{2}x^2 = \frac{(2\pi\nu)^2 m}{2}x^2$$

For convenience, we will continue to express the potential energy of the atom in terms of the classical resonant frequency. Substituting the potential energy into Schrödinger's equation yields

$$\frac{-\hbar^2}{2m}\frac{d^2\psi(x)}{dx^2} + \frac{(2\pi\nu)^2 m}{2}x^2\psi(x) = E\psi(x)$$

or

$$\frac{d^2\psi(x)}{dx^2} + \left[\frac{2mE}{\hbar^2} - \left(\frac{2\pi m\nu}{\hbar}\right)^2 x^2\right]\psi(x) = 0$$

Let $\alpha = 2\pi m\nu / \hbar$ and $\beta = 2mE / \hbar^2$. Now,

$$\frac{d^2\psi(x)}{dx^2} + (\beta - \alpha^2 x^2)\psi(x) = 0 \tag{7.3}$$

An exponential solution is a good trial solution for this type of differential equation. Motivated by the existence of x^2 in the equation, we will try a solution of the form

$$\psi(x) = A_0 e^{-\gamma x^2}$$

The negative sign in the exponent ensures that the function decays to zero for large values of $|x|$.

Substituting this trial solution into Equation 7.3, the first term is

$$\frac{d^2\psi(x)}{dx^2} = \frac{d^2}{dx^2}\left[A_0 e^{-\gamma x^2}\right] = A_0\left[-2\gamma + 4\gamma^2 x^2\right]e^{-\gamma x^2} \tag{7.4}$$

By examining this and the second term of Equation 7.3, a solution is available provided $\alpha = \beta = 2\gamma$. From the definitions of α and β, the energy eigenvalue is now seen to be

$$E = \frac{\beta\hbar^2}{2m} = \frac{\alpha\hbar^2}{2m} = \frac{2\pi m\nu}{\hbar}\frac{\hbar^2}{2m} = \frac{h\nu}{2}$$

and the associated eigenfunction is

$$\psi(x) = A_0 e^{-\frac{\pi m\nu}{\hbar}x^2} \tag{7.5}$$

This is a Gaussian function of x which is plotted as the lowest curve in Figure 7.12. It represents the ground state of the quantum mechanical oscillator.

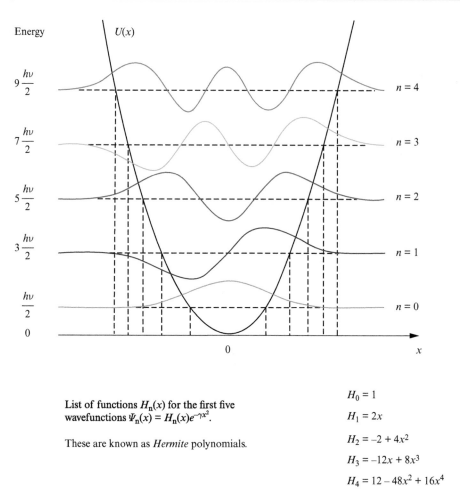

List of functions $H_n(x)$ for the first five wavefunctions $\Psi_n(x) = H_n(x)e^{-\gamma x^2}$.

These are known as *Hermite* polynomials.

$H_0 = 1$

$H_1 = 2x$

$H_2 = -2 + 4x^2$

$H_3 = -12x + 8x^3$

$H_4 = 12 - 48x^2 + 16x^4$

FIGURE 7.12 The first five wavefunctions and eigenenergies of the quantum mechanical oscillator that are solutions to Equation 7.3. The potential energy curve $U(x)$ is parabolic. It is scaled such that its width coincides with the amplitudes on the x-axis of a classical oscillator having the energies shown on the vertical axis. In contrast to the case of a square potential well, the eigenenergies are equally spaced.

The energy value $E = \dfrac{h\nu}{2}$ is called the *zero-point energy* which exists for the atoms in the system even at a temperature of absolute zero. It is the lowest allowed energy and it is the energy of the ground state of the atomic vibration. The motion of the atom is now no longer regarded as a classical vibration. Instead, only the probabilities of finding the atom at various positions along the x-axis may be known. These probabilities are determined from probability density $|\psi|^2$. The zero-point energy turns out to be on the same order of magnitude as kT at room temperature. See Problem 7.6. Note that this Gaussian wavefunction was used to obtain the Uncertainty Relation in Appendix 2.

Other valid solutions to the Schrödinger equation exist having higher energies. A set of polynomial functions known as *Hermite polynomials* that multiply $e^{-\gamma x^2}$ are found for these solutions. The first five of these are shown in Figure 7.12, and the discrete vibrational energy levels in Figures 7.10 and 7.11 are the associated energy eigenvalues. The eigenfunctions and eigenvalues are numbered with integer quantum numbers n. The general expression for any eigenvalue is

$$E_n = (2n+1)\frac{h\nu}{2}$$

The spacing between energy eigenvalues is uniform and equal to $h\nu$. The zero-point energy corresponds to the case with $n = 0$. See Problem 7.7.

As vibrational quantum number n increases, and also as mass m increases, the loops and nodes of the wavefunctions along the x-axis become more numerous and more spatially compressed. At the same time, the amplitudes of the wavefunctions near the limits of $+x$ and $-x$ increasingly grow relative to the amplitudes further away from the limits. The result is that the probability density function from the quantum mechanical result gradually merges with that for the classical sinusoidal result for vibrations.

All the atomic nuclei in a quantum dot can be modeled as quantum mechanical oscillators. There is vibrational coupling between bonded atoms which means that various traveling and standing waves that we describe as phonons exist a in an infinite lattice of atoms. In a quantum dot, the traveling waves would soon suffer reflections at the quantum dot surfaces resulting in standing waves that are not propagating. Therefore Figures 7.10–7.12 can be applied to the lattice of atoms in a quantum dot. The same concepts are also valid for atoms in molecules.

7.7 VIBRONIC TRANSITIONS

Consider an electronic dipole transition that emits a photon from an excited quantum dot at room temperature. An electron in the electronic energy level of the vibrating exciton state of the quantum dot transitions to an available electronic energy level in the un-excited and vibrating quantum dot. This transition corresponds to the downward arrow in Figure 7.10.

The electronic dipole transition was treated in Section 4.4 in the absence of atomic vibrations. During the transition, the electron is in a superposition of initial state wavefunction ψ_m and final state wavefunction $\psi_{m'}$. During the superposition state this results in an expected value of the position of the electron $\langle r \rangle_s (t)$ where

$$\langle r \rangle_s (t) = 2ab\langle \phi_m \mid r \mid \phi_{m'} \rangle \cos(\omega_{mm'}t)$$

oscillates as a function of time. The amplitude of this oscillation will be denoted A_{dipole} where

$$A_{dipole} = 2ab\langle \phi_m \mid r \mid \phi_{m'} \rangle = 2ab\int_V \phi_m^* r \phi_{m'} dr$$

Here, ϕ_m and $\phi_{m'}$ are the spatial parts of the initial and final electronic wavefunctions as described in Section 4.4. The matrix element is the integral $\int_V \phi_m^* r \phi_{m'} dr$ and the oscillator strength of the dipole transition is proportional to the square of the matrix element as shown in Equation 4.6.

In order to clearly identify the above wavefunctions as electronic wavefunctions they are now rewritten using the subscript "e" as

$$2ab\langle \phi_{m_e} |r_e| \phi_{m_e'}\rangle = 2ab \int_V \phi_{m_e}^* r_e \phi_{m_e'} dr_e \tag{7.6}$$

and the position coordinate of the electron is written as r_e. The integral is over all space which requires three spatial coordinates; however, for simplicity the integrand shows just one spatial variable r_e. The oscillator strength which determines the photon emission rate is proportional to

$$\left| \int_V \phi_{m_e}^* r_e \phi_{m_e'} dr_e \right|^2$$

We have seen before in Section 4.6 how the Schrödinger equation for two electrons, with individually labeled position coordinates, can be constructed. The combined wavefunction of the two particles was taken to be the product of the wavefunction of each particle. Then the Schrödinger equation was solved by the method of separation of variables and the expected single particle Schrödinger equation for each particle was duly obtained. Motivated by this approach, a spatial wavefunction ϕ_c that combines both the electronic and the vibrational wavefunctions may be written

$$\phi_c = \phi_{m_e} \phi_{n_v}$$

which is the product of the electronic spatial wavefunctions ϕ_{m_e} (ignoring vibrations) and the vibrational spatial wavefunctions ϕ_{n_v} now written with the subscript "v". Each wavefunction has its own set of spatial coordinates as in Section 4.6, although the interaction between the two entities (one atomic and one electronic) is very different from the case of the two electrons in Section 4.6.

The combined wavefunction ϕ_c in the form of the product of electronic and vibrational wavefunctions may be substituted into the time-independent Schrödinger equation. The combined equation may then be separated into two equations that become the vibrational Schrödinger equation and the electronic Schrödinger equation. The technique of separation of variables used for this is the same as that detailed in Section 4.6.

The combined Schrödinger equation takes the form

$$-\frac{\hbar^2}{2m}\frac{d^2}{dr_e^2}\left[\phi_{m_e}\phi_{n_v}\right] - \frac{\hbar^2}{2m}\frac{d^2}{dr_n^2}\left[\phi_{m_e}\phi_{n_v}\right] + (U_e + U_n)\phi_{m_e}\phi_{n_v} = E_T\phi_{m_e}\phi_{n_v}$$

where U_n is the potential energy of the atomic nuclei and U_e is the potential energy for the electron, respectively and the vibrational coordinate is labeled r_n.

After separation of variables, the resulting equations are the electronic Schrödinger equation with eigenenergy E_e

$$-\frac{\hbar^2}{2m}\frac{d^2}{dr_e^2}\phi_{m_e} + U_e\phi_{m_e} = E_e\phi_{m_e}$$

and the atomic oscillator Schrödinger equation with vibrational eigenenergy E_n

$$-\frac{\hbar^2}{2m}\frac{d^2}{dr_n^2}\phi_{n_v} + U_n\phi_{n_v} = E_n\phi_{n_v}$$

See Problem 7.9.

The total system energy E_T now becomes the sum

$$E_T = E_e + E_n$$

This confirms that if the combined wavefunction is taken to be the product of the electronic and vibrational wavefunctions, then the total energy is the sum of the eigenenergies of the separated Schrödinger equations. This is consistent with the Born-Oppenheimer approximation.

The combined wavefunction $\phi_c = \phi_{m_e}\phi_{n_v}$ will now be used to calculate a *vibronic* matrix element that extends Equation 7.6 to include both the electronic and the vibrational wavefunctions.

To see the effect of the nuclear wavefunction on the amplitude of the electronic oscillation we take the expected position of the electron relative to a fixed reference point as $r_e + r_n$. This appropriately adds the atomic position relative to the fixed reference point to the electronic position relative to the atomic position. The amplitude of the electron oscillation during the superposition state is now

$$A_{vibronic\,dipole} = 2ab\langle\phi_c|r_e + r_n|\phi_c'\rangle$$
$$= 2ab\langle\phi_{m_e}\phi_{n_v}|r_e + r_n|\phi_{m_e'}\phi_{n_v'}\rangle$$
$$= 2ab\int_V \phi_{m_e}^*\phi_{n_v}^* (r_e + r_n)\phi_{m_e'}\phi_{n_v'}\,dr_e dr_n$$
$$= 2ab\int_V \phi_{m_e}^*\phi_{n_v}^* (r_e)\phi_{m_e'}\phi_{n_v'}\,dr_e dr_n + 2ab\int_V \phi_{m_e}^*\phi_{n_v}^* (r_n)\phi_{m_e'}\phi_{n_v'}\,dr_e dr_n$$

At this point, we will apply the *Condon approximation* which states that the electronic integrals can be approximated as independent of the atomic positions, again justified by the large mass ratio involved. Therefore

$$A_{vibronic\,dipole} \cong 2ab\int_V \phi_{n_v}^*\phi_{n_v'}\,dr_n \int_V \phi_{m_e}^* (r_e)\phi_{m_e'}\,dr_e$$
$$+ 2ab\int_V \phi_{m_e}^*\phi_{m_e'}\,dr_e \int_V \phi_{n_v}^* (r_n)\phi_{n_v'}\,dr_n$$

(7.7)

The integral

$$2ab \int_V \phi_{m_e}^* \phi_{m_e'} dr_e$$

in the second term is zero. This is due to the orthogonality of wavefunctions $\phi_{m_e}^*$ and $\phi_{m_e'}$ that are solutions to Schrödinger's equation and that are therefore eigenstates of the Hermitian operator $\hat{H}$. See Section 1.14.

The surviving first term of Equation 7.7 contains the familiar electronic dipole oscillation amplitude $2ab \int_V \phi_{m_e}^* (r_e) \phi_{m_e'} dr_e$ which is identical to Equation 7.6. However, this is now multiplied by the factor

$$\int_V \phi_{n_v}^* \phi_{n_v'} dr_n$$

This factor is known as the *vibrational overlap integral*, also called the *Franck–Condon integral*. Unlike the electronic integral $\int_V \phi_{m_e}^* \phi_{m_e'} dr_e$, we cannot make the assumption that it is zero because $\phi_{m_e}^*$ and $\phi_{m_e'}$ are vibrational wavefunctions associated with two distinct potential energy functions shown in Figure 7.9. This means that they are not subject to an orthogonality relation.

The Franck-Condon integral is a measure of the degree to which the vibrational wavefunctions of the ground and excited state overlap spatially. The oscillator strength of the electronic transition therefore depends on the square of the Franck-Condon integral which is known as the *Franck-Condon factor*

$$\left| \int_V \phi_{n_v}^* \phi_{n_v'} dr_n \right|^2$$

Figure 7.13 shows an example of vibrational wavefunctions superimposed on the energy diagram of Figure 7.11 that now illustrates the degree of overlap of vibrational wavefunctions during electronic transitions. The transitions for which the Franck-Condon factor is maximum will dominate. In a vertical transition to a specific vibrational level where the overlap integral is a maximum the transition is most likely, and therefore of maximum oscillator strength. This explains why the emission processes in Figures 7.10 and 7.11 are shown as vertical transitions. Partial overlaps for which the Franck-Condon factors are smaller also have a probability of occurring, leading to broadening of the emission spectrum.

In Figure 7.13, the potential energy of the atomic positions are more asymmetric to better approximate bond potential energy. This asymmetry implies non-parabolic potentials that are known as *Morse potentials*. We will not treat these in detail in this book.

The Franck-Condon factor is also relevant to the absorption process in Figures 7.10 and 7.11. This is because a vertical transition into several vibrational states of the excited quantum dot can occur. The Franck-Condon factor predicts the probabilities of each vibrational state being accessed and the shape of the absorption spectrum can

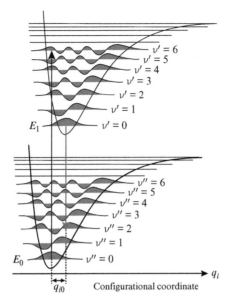

FIGURE 7.13 The Franck-Condon principle favors vertical transitions in both absorption and emission. The vibrational wavefunctions are shown in the diagram for a range of vibrational energies. Maximum overlaps in vibrational wavefunctions yield the highest oscillator strength and occur for vertical transitions. Weaker overlaps involving a range of vibrational levels determine the width of the absorption and emission spectra. Spectral broadening occurs since there are probabilities that absorption and emission events can populate a range of vibrational energy levels due to the range of vibrational wavefunctions.that overlap to varying degrees. The probability of populating each vibrational level is predicted by the Franck-Condon factor in either emission or absorption. Samoza / Wikimedia commons / CC BY-SA 3.0.

thereby be determined. Numerical methods are commonly used to perform the overlap integrals.

7.8 SURFACE PASSIVATION

Semiconductor surfaces tend to pin the Fermi level to surface energy levels (surface states) formed by dangling bonds that lie within the bulk energy gap. These surface states behave as traps that enable non-radiative EHP recombination. See Section 2.20. In quantum dots, surface state traps are of great consequence since excited electrons and holes are in close proximity to a surface. Without taking appropriate measures, all recombinations could occur non-radiatively at surfaces. Surface passivation is therefore critical.

A perfectly passivated surface of a quantum dot has all dangling bonds saturated and therefore surface states are eliminated. In this case, excitons have a high probability of producing the desired radiative recombination events.

Consider CdS as an example of a II-VI semiconductor quantum dot. Negative surface S ions (anions) form dangling sulphur bonds and positive surface Cd ions (cations) form dangling cadmium bonds. Because electronegativity generally increases when moving from the left side to the right side of the periodic table, a sulphur⁻ ion can be expected to attract surface electrons more strongly than a Cd ion. The dangling bond electrons of sulphur are therefore more bound than the dangling bond electrons of Cd. Whereas approximately mid-gap surface states form from a Group IV semiconductor such as silicon, CdS produces a wider range of surface trap energies including lower-lying states typically closer to the valence band (traceable to danging sulphur bonds) to higher energy states typically closer to the conduction band (traceable to dangling cadmium bonds). The elimination of dangling bonds requires the elimination of both cation and anion dangling bonds.

In Figures 7.3 and 7.4, organic ligands involved in the synthesis process of solution-processed spherical quantum dots effectively terminate a percentage of surface dangling bonds. These ligands are often referred to as *capping agents*. To understand this in more detail, consider the oleic acid ligands shown in Figure 7.14.

Oleic acid is a molecule described as a fatty acid that occurs naturally in numerous animal and vegetable fats and oils. It is an example of a quantum dot ligand. Most of the oleic acid molecule comprises a carbon chain that resembles a hydrophobic oil molecule. One end of the molecule, however, has two oxygen atoms, one of which is in the form of a hydroxide, that allows it to chemically bond to the quantum dot. During the bonding reaction, the hydrogen ion from the hydroxide is released and the oxygen can then bond to the surface of the quantum dot. See Figure 7.15. At this point the ligand becomes an anion with a net negative charge, and is referred to as an *oleate ion* due to the loss of the H^+ ion.

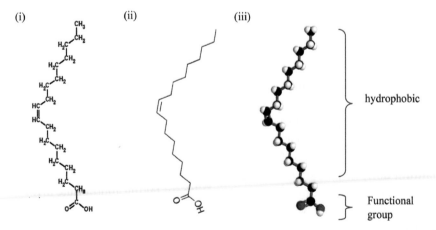

FIGURE 7.14   The oleic acid molecule with chemical formula $C_{18}H_{34}O_2$ presented in three ways: i) labeled structure, ii) chemical symbol, iii) 3D rendering. Most of the molecule is formed from a hydrocarbon backbone that is hydrophobic. At one end, there is a functional OH group that facilitates bonding. The hydrogen bonded to one oxygen can be released and the available oxygen dangling bond is then able to bond to a quantum dot surface atom.

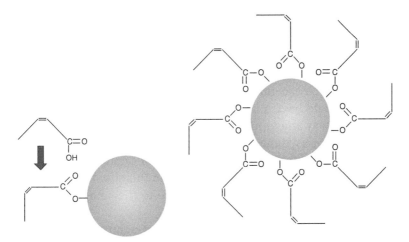

FIGURE 7.15 After bonding, the oleic acid becomes an oleate ion due to the release of H^+. The number of oleate ions that can terminate one quantum dot is limited by the number of available dangling bonds and the packing density of the ligands. A properly functioning ligand-terminated quantum dot has no net charge. It is typically surrounded by an organic matrix such as a polymer that forms weak but stable van der Waals bonds with the hydrophobic tails of the ligands.

Organic ligands are generally unable to pack together closely enough to match the separation of neighboring surface atoms of the quantum dot. Many of the organic capping molecules are bent, such as the oleic ion ligand, or are branched as depicted in Figure 7.3. As a result, coverage of surface atoms with organic capping molecules tends to be *sterically hindered* meaning that only partial saturation of dangling bonds is achieved.

Another issue is the need for the simultaneous passivation of both anionic and cationic surface sites using such capping agents. This is very difficult since chemically stable attachment points for ligands tend to be site-specific which precludes the simultaneous passivation of all surface atoms.

The bonding character of organic ligands with cations and anions is often weak due to the availability of ligand electrons dedicated to participate in bonding. Usually only a single bond exists between a ligand and the quantum dot. The energetically demanding quantum dot environment in which incident radiation excites fluorescence means that such ligand bonding is prone to failure, exhibiting *photoinstability*.

A second and much more satisfactory approach to passivation of the quantum dot surface is the use of a core-shell structure. A crystalline inorganic semiconductor shell is epitaxially grown onto the core. If the shell semiconductor has a larger bandgap than that of the core, then confinement of electrons and/or holes is possible. The quantum efficiency of quantum dots is increased by a defect-free shell. When the shell lattice constant does not match the lattice constant of the core, lattice strain results. This lattice strain can be elastically accommodated if the shell is sufficiently thin. Dislocations must be avoided since they cause dangling bonds which would undermine the passivation function of the shell. Ligands then surround the shell to further passivate the core-shell structure.

A core-shell quantum dot with an idealized *Type* I *heterojunction* is shown in Figure 7.16a). The Type I heterojunction produces energy band offsets that simultaneously raise the available energies in the shell for an electron in the conduction band and for a hole in the valence band. Limiting the shell thickness to two monolayers allows lattice strain to be accommodated without forming dislocations. In practice, the shell material should have a lattice parameter within approximately 10% of the core to encourage epitaxy and minimize strain.

In Figure 7.16b) an asymmetric case is shown in which the conduction band offset is small compared to the valence band offset. For example, a CdSe quantum dot core may be passivated with a CdS shell resulting in this situation. Electron confinement is weak compared to hole confinement due to the smaller conduction band offsets.

An additional cause of weak electron confinement compared to hole confinement is the relative mass ratio between the electron and the hole. From Section 4.5 and

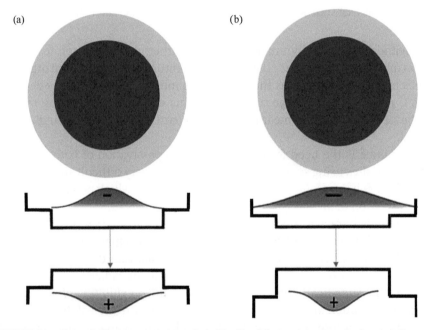

FIGURE 7.16 Core-shell quantum dot. a) A idealized heterojunction is formed from a core semiconductor and a shell semiconductor having a larger bandgap than that of the core. Provided the shell conduction band is higher in energy for electrons and the shell valence band is higher in energy for holes compared to core conduction and valence band levels, respectively, this is defined as a Type I heterojunction. It forms an energy well with finite energy barriers confining both carriers to the core region. The confinement of carriers is indicated by the blue and red shaded regions that correspond to radial wavefunction probability density. Recombination occurs due to a transition from the core conduction band to the core valence band as indicated. b) In reality, the band offsets and degrees of confinement in the conduction and valence bands are often highly asymmetric. This is often called a quasi-Type II quantum dot.

Appendix 4 the expected radius values of electron and hole in an exciton are shown to be inversely proportional to their effective masses. In the case of CdSe, from Appendix 6, $\frac{m_h^*}{m_e^*} = 3.5$ which predicts that the expected electron radius is 3.5 times larger than the expected hole radius.

As a result of these two considerations the electron in a CdS/CdSe core/shell quantum dot extends through both the core and the shell whereas the hole is strongly confined to the core.

Surrounding the shell layer with organic ligands allows these quantum dots to be suspended in a polymer matrix. Another important function of the ligands surrounding each quantum dot is to avoid agglomeration by keeping a minimum separation between adjacent quantum dots.

From Appendix 6 the core and shell lattice constants are within 4% for a CdSe/CdS core/shell quantum dot. Another example of a Type I heterojunction core shell quantum dot that is free from the toxic heavy metal cadmium is the InP/ZnS core/shell quantum dot. From Appendix 6 the the core and shell lattice constants are within 8%. See Figure 7.17.

The *Type* II *heterojunction* core-shell quantum dot is also of interest. In a Type II heterojunction the electron and the hole are confined to opposing portions of the

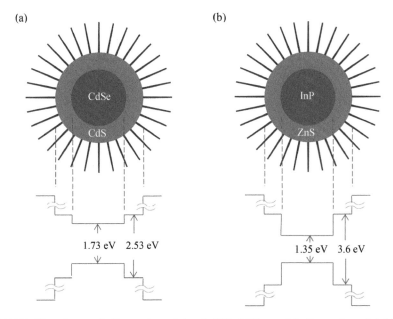

FIGURE 7.17 Type I core-shell quantum dots. a) CdSe/CdS core/shell quantum dot. b) InP/ZnS core/shell quantum dot. Bandgaps shown are bulk values and increase due to quantum confinement allowing a range of visible light emission wavelengths. Ligands have wide effective bandgaps and are electric insulators due to their carbon chains that have tightly bound electrons.

core-shell structure. This will happen if the heterojunction conduction and valence band offsets do not favor confinement of both the electron and the hole in the same region.

Examples of Type II quantum dots include CdTe/CdSe (core/shell) or CdSe/ZnTe (core/shell). These are shown in Figure 7.18. The magnitudes and directions of conduction and valence band offsets between two compound semiconductors that determine whether a Type I or a Type II heterojunction forms are controlled by details of charge distribution at the heterojunction and are difficult to model and predict without experimental data.

Oscillator strengths and hence quantum yields of Type II quantum dots tend to be lower than in Type I quantum dots due to the more limited overlap between electron and hole spatial distributions. Nevertheless, in Type II heterojunction quantum dots, a wider range of emission wavelengths is possible because the emission spectrum is

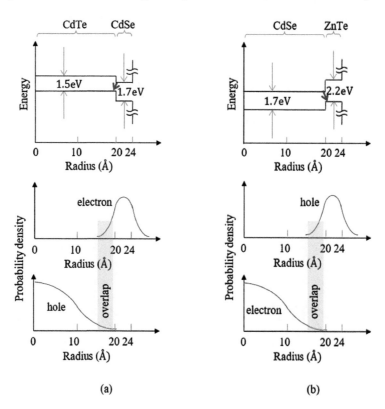

(a) (b)

FIGURE 7.18 Type II heterojunction quantum dots using CdTe/CdSe core/shell and CdSe/ZnTe core/shell semiconductors. a) Electron confinement in the shell and hole confinement in the core occur in the CdTe/CdSe core/shell example. b) Electron confinement in the core and hole confinement in the shell occur in the CdSe/ZnTe core/shell example. In either case there is some spatial overlap of the electron and the hole probability densities because the heterojunction potential barriers are finite. This spatial overlap (yellow regions) allows dipole radiation owing to a non-zero transition matrix element. Note that the energy separation between electron and hole are smaller than the effective bandgaps of either the core or the shell resulting in infrared emission wavelengths. Red arrows indicate the radiative transitions. Bulk bandgaps are shown; however, effective bandgaps are larger due to quantum confinement. Adapted from Kim et al., 2003.

controlled by band offsets across the heterojunction as indicated by the red arrows in Figure 7.18. In these examples, emission spectra are observed in the 800–900 nm range whereas both core and shell effective bandgaps are in the visible wavelength range. This facilitates more flexibility in choices of semiconductor materials that can optimize lattice matching between core and shell. In addition, semiconductor materials that minimize toxicity may be available.

For the Type I quantum dot in Figure 7.16b) the substantial asymmetry in the confinement between electron and hole causes a decreased spatial overlap compared to the idealized case of Figure 7.16a). Because of this, these quantum dots are often referred to as *quasi-Type II* quantum dots.

7.9   AUGER PROCESSES

Exciton recombination may occur via an Auger process. The Auger process requires very close-range coulomb interaction between carriers. As discussed in Section 5.9 this process becomes significant in semiconductor devices such as LEDs at high carrier injection levels when close-range carrier-to-carrier interactions can occur.

The substantial carrier confinement inherent in quantum dots makes the Auger process more probable compared to the case in bulk semiconductors. Provided the additional carrier required for an Auger process to occur exists, and a suitable energy state for this third carrier to occupy also exists, energy may be conserved during a recombination event and no radiation will be observed.

Consider the case of a quantum dot with four carriers comprising two electrons and two holes known as a *biexciton* as shown in Figure 7.19). A biexciton can form if one additional electron-hole pair is created in an already-excited quantum dot. For example, if optical excitation is sufficiently intense, the first exciton may not have had a chance to decay before an additional exciton is created. Two electrons and two holes may occupy the same energy level because there are two possible spin states. The biexciton has interesting potential applications in lasers as well as for the production of entangled photons used for quantum information and quantum computing. See Problem 7.10.

The existence of biexcitons may be detected experimentally by measuring the fluorescence decay rate of an ensemble of quantum dots. As shown in Figure 7.19, an intense flash of light is incident upon an ensemble of CdSe/CdS core/shell quantum dots at time $t = 0$. The existence of biexcitons may be detected by a higher fluorescence radiative decay rate for the quantum dots because the probability of a recombination event is now four times higher than for a single exciton. Both the electron and the hole concentrations within the quantum dot have doubled. Since the decay rate is four times higher, the characteristic decay time is four times shorter. See Figure 7.19.

Figure 7.20a) shows an example of Auger events that might take place in a biexciton. The second electron is excited during the recombination of one electron-hole pair. No radiation is produced, and instead, energy is conserved since this electron is ejected from the quantum dot. In some cases the second electron could also become

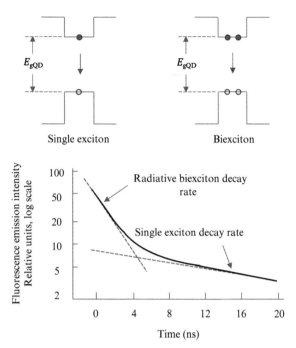

FIGURE 7.19   Expected single exciton and biexciton fluorescence intensity versus time for an ensemble of CdS/CdSe core/shell quantum dots following a flash of light at time $t = 0$. A decay time four times shorter than for the single exciton case is expected for the biexciton quantum dot since both the hole and the electron concentrations are doubled. The biexciton decay is only observed with high intensity optical excitation during the short time interval where biexcitons dominate. Auger recombination processes which can occur for a biexciton are neglected and only radiative recombination is assumed. Additional coulombic effects that influence the emission spectrum associated with the biexciton are also neglected in this simplified picture. Adapted from Bae et al., 2013.

trapped in a higher energy level that exists outside the semiconductor region, but within the ligand region, of the quantum dot.

After this Auger process, the quantum dot is positively charged due to the remaining second hole. If this positively charged quantum dot is now excited radiatively, another electron-hole pair is generated without altering the remaining hole. This is known as a *trion*. There is a chance that the trion could recombine radiatively resulting once again in the positively charged quantum dot; however, an Auger process is likely due to the availability of the extra hole.

The Auger process would excite the extra hole into a higher energy level. In the example of Figure 7.20b) this hole becomes trapped at a surface bond of the quantum dot core and results in ligand dissociation. In order to return to normal operation, this quantum dot would need to acquire a single electron from its surroundings to become

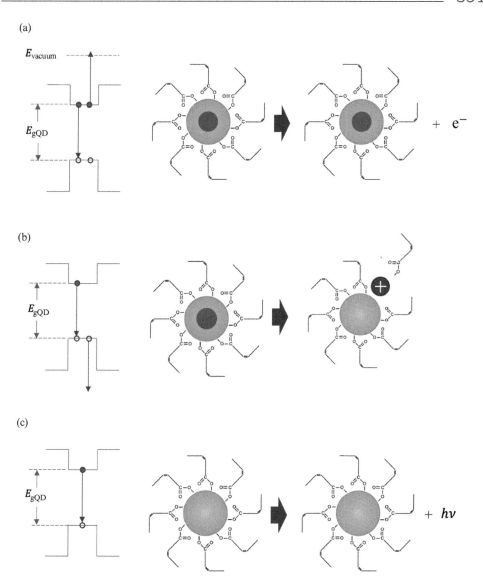

FIGURE 7.20 Examples of Auger events for a quantum dot. a) A quantum dot with a biexciton undergoes an Auger event involving two electrons and one hole. One electron-hole pair recombines, but in place of energy conservation by dipole radiation, the second electron is excited to a higher energy level such that system energy is conserved. If the second electron is sufficiently energetic, the work function can be exceeded resulting in a free electron and a positively charged quantum dot. b) If this positively charged quantum dot absorbs a photon it becomes a trion. In this example, another Auger process takes place. One electron-hole pair recombines, but in place of energy conservation by dipole radiation, the associated hole is excited to a high energy level. The hole is excited to the surface of the quantum dot resulting in an incomplete surface bond that detaches a ligand. c) If the quantum dot from b) were to gain one electron from its surroundings it would become neutral, thereby restoring the ligand bond and restoring radiative operation of the quantum dot.

neutral. The normal operation of the neutral quantum dot following the absorption of a new photon is shown in Figure 7.20c).

Charged and hence inactive quantum dots are frequently observed in what is known as the "blinking" effect. Individual quantum dots are observed to go dark and to then function again periodically due to Auger carrier ejection followed by a subsequent charge transfer that neutralizes the quantum dot allowing it to function normally again.

A more detailed understanding of the mechanisms involved in Auger processes may be obtained by measuring radiative decay times for quantum dot fluorescence. If we now include Auger processes, the decay rate of a CdS/CdSe core/shell quantum dot will change dramatically. Figure 7.21 shows the *observed* fluorescence decay of an ensemble of quantum dots. Since the biexciton enables Auger processes, the observed fluorescence decay rate is much higher than expected from the radiative decay assumption of Figure 7.19. Auger processes can also result from intermediate trion states that are produced during the decay process as illustrated in Figure 7.20. The

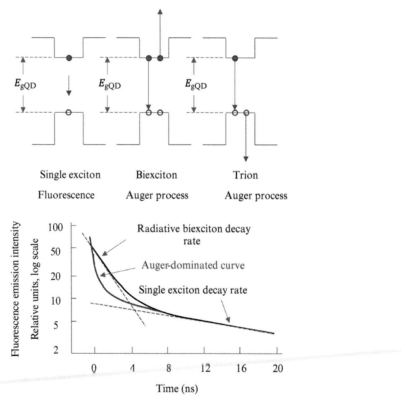

FIGURE 7.21 The observed radiative decay (red curve) for an ensemble of CdS/CdSe core/shell quantum dots following a high intensity flash of light at time $t = 0$. A decay rate much higher than expected for a biexciton is observed due to Auger processes. These Auger processes could occur in the biexciton or in a trion as shown. Adapted from Bae et al., 2013.

degree to which the observed decay rate for short times after the light flash exceeds the expected biexciton radiative decay rate shown in Figure 7.19 is a measure of the relative rate of Auger recombination compared to radiative recombination.

In Figure 7.21 the core/shell CdSe/CdS quantum dots are fabricated with an abrupt heterojunction. It is also possible to fabricate a *graded heterojunction* quantum dot as shown in Figure 7.22 in which an intermediate layer of the ternary alloy semiconductor $CdSe_xS_{1-x}$ is inserted between the CdSe core and the CdS shell. The value of x decreases from 1 to 0 as the radius increases through the graded junction region. The observed fluorescence intensity decay curve of the graded heterojunction quantum dot (green curve) is shown in comparison with the results from Figures 7.19 and 7.21. The results indicate that Auger processes have been substantially suppressed by the introduction of the graded heterojunction.

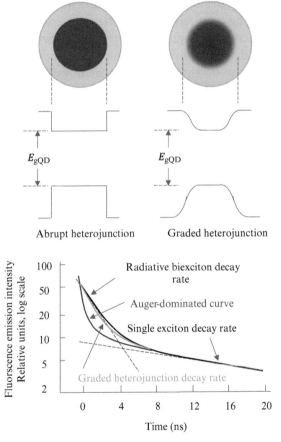

FIGURE 7.22 The observed radiative decay (green curve) for an ensemble of graded alloy heterojunction $CdSe/CdSe_xS_{1-x}/CdS$ core/shell quantum dots following a high intensity flash of light at time $t = 0$. A decay rate much closer to the expected radiative biexciton decay rate of Figure 7.19 is observed which indicates that the graded heterojunction effectively suppresses exciton processes. Adapted from Bae et al., 2013.

An explanation for the mitigation of Auger events due to graded heterojunctions requires rather complex quantum mechanical modeling. For a dipole transition, the transition matrix element determines the transition rate. For an Auger process, the concept of a transition matrix element also exists and is referred to as *the electronic transition matrix element of the interparticle Coulomb interaction*. It depends on the wavefunction overlap integral between the initial and the final states of the system for the Auger process. Details of such modeling are beyond the scope of this book; however, some insight can be obtained by analogy with classical mechanics of a multi-body system.

Consider a multi-body system in which one or more elastic collisions occur. Classical mechanics states that the system must satisfy momentum and energy conservation. In the case of a quantum dot, the combined initial momentum and initial energy of the electron(s) and hole(s) that participate in the Auger process is likewise conserved following the interparticle coulomb interaction.

Consider specifically the requirement of momentum conservation in the case of an abrupt heterojunction core-shell quantum dot. Within a few femtoseconds following the initiation of the Auger coulomb interaction, the relevant charge carrier being ejected will exist with high probability in the close vicinity of the vertical energy band offset of the abrupt heterojunction. Figure 7.23 shows the evolution of the wavefunction of one charge carrier (an electron in this example) with time as predicted by advanced modeling.

This brief spatial confinement (small Δx) causes a large uncertainty in the particle's momentum (large Δp) due to the uncertainty principle. This large uncertainty in momentum relaxes the requirement for the conservation of momentum for the Auger process, and this instant in time, which is only some femtoseconds after the start of

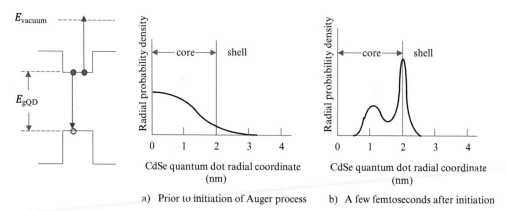

a)   Prior to initiation of Auger process   b)   A few femtoseconds after initiation

FIGURE 7.23   Radial probability density for a trion undergoing an Auger process in an abrupt heterojunction core-shell quantum dot with a core radius of 2 nm. a) Before initiation of the Auger event the electron being ejected is in the expected excitonic state. b) At time t_0 (a few femtoseconds after initiation) the probability density of the electron is momentarily strongly localized at the core-shell boundary. Subsequently, the ejected electron escapes from the the quantum dot. Adapted from Vaxenburg et al., 2015.

the coulomb interaction, represents the time after which the confined particle wave-function disintegrates. The Auger process can now continue to completion.

If the energy band offset of the core-shell quantum dot is graded then the band offset is spatially delocalized. As a result, carrier localization is minimized and carrier momentum remains better defined. It is now less likely that momentum conservation can be satisfied. The Auger process is substantially forbidden.

Commercially available quantum dots that are optimized for the highest available quantum efficiencies consist of core-shell structures synthesized with graded hetero-junctions. Quantum yields of over 95% can be achieved in volume production and the blinking effect is minimized due to Auger process suppression. One key application is liquid crystal display backlighting. Blue light from GaInN light emitting diodes excites fluorescence in a mixture of green-emitting and red-emitting quantum dots. The resulting superposition of blue, green, and red spectral peaks produces *triluminous* white light as shown in Figure 7.24. If this light source is used as the backlight of a color liquid crystal display, a wide color gamut is achieved by the display.

Quantum dots have also enabled very high performance displays which are based on a blue organic light emitting pixel array. Color conversion using quantum dots results in a full color display. Each full color pixel is composed of three blue subpixels. A red emitting subpixel and a green emitting subpixel are constructed from blue sub-pixels coated with red-emitting and green-emitting quantum dots, respectively. These quantum dots are excited by the blue light. The third blue subpixel is not coated.

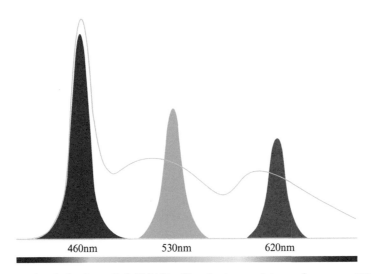

FIGURE 7.24   Blue light from GaInN LEDs illuminates a mixture of green-emitting and red-emitting quantum dots. The spectrum of the remaining blue light in combination with the green and red quantum dot fluorescence produces triluminous white light that is ideally suited for full color liquid crystal display backlighting. Wide color gamut liquid crystal displays are enabled. The less ideal spectrum of alternative phosphor-based fluorescent materials is also shown for comparison. If blue light from an organic blue-emitting light emitting diode display is used in place of blue light from inorganic LEDs, a high performance full color quantum-dot-augmented organic light emitting diode display is enabled. Organic light emitting diodes are introduced in Chapter 8.

While red, green, and blue organic light emitting diodes are available, the individual colors age at different rates thereby causing changes in display color characteristics over time. It is also challenging to lithographically pattern the three distinct organic light emitting device materials required to produce the three colors due to their chemical and moisture sensitivity. The use of red and green quantum dots which are relatively easy to lithographically pattern overcomes both of these issues.

The resulting organic light emitting diode/quantum dot display eliminates light leakage inherent in liquid crystal displays. Extremely high contrast ratios between lit and dark pixels are achieved. In addition, the use of color filters to separate white light into the primary colors is also eliminated which dramatically improves power efficiency. Televisions using this technology are now commercially available.

7.10   BIOLOGICAL APPLICATIONS OF QUANTUM DOTS

Quantum dots are becoming an important diagnostic tool for discerning cellular function at the molecular level. Their high brightness and their long-lasting, size-tunable, narrow band luminescence set them apart from conventional fluorescent dyes. Quantum dots are being developed for a variety of biologically-oriented applications such as for fluorescent markers used to track drug delivery into living tissue and for disease detection.

To provide fluorescence detection, the quantum dot must be reliably attached to the species of interest which may constitute a drug or some other biological entity such as a cell. Ligands used in biocompatible quantum dots can be engineered to have functional groups not only for binding to the quantum dot, but also for binding to adjacent biomolecules. See Figure 7.25. This allows the incorporation of quantum dots into biological systems to be highly tailored and hence very versatile. If quantum dots are tagged to a specific drug that is being tested, the fluorescence of these quantum dots allows the precise location of the drug to be followed in living tissue. If quantum dots are tagged to specific cells, then detailed mechanisms of cancerous tumor formation can be studied.

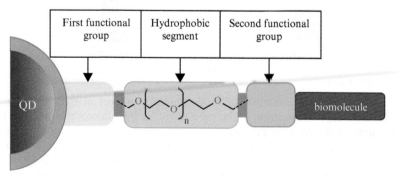

FIGURE 7.25   Biocompatible quantum dots may be designed with ligands that have a second functional group that attaches to relevant molecules such as water, proteins, and other biomolecules.

Heavy metals such as Cd are toxic for biological systems and therefore alternative quantum dot materials are used. Silicon, carbon, and sulphur-based quantum dots have been developed. In some cases the materials used do not function as direct-gap semiconductors. The Type II quantum dot in Figure 7.18 is one example of this approach because it does not involve recombination via the bandgap of either the core or the shell material. Weaker oscillator strengths in these biological applications may be acceptable, whereas in lighting applications, the higher performance available from direct gap semiconductors in Type I core-shell quantum dots is required.

Type II quantum dots may also be of interest because of the reduced electron-hole wavefunction overlap. The wavefunction overlap between the hole and electron is small compared to Type I heterojunction quantum dots leading to weaker oscillator strengths. As a result, Type II quantum dots exhibit extended exciton lifetimes which can be beneficial for tracking the quantum dots in biological systems in real time.

Infrared emission available from Type II quantum dots can cause infrared light absorption and heating in cells. This provides opportunities for therapeutic cancer tumor destruction.

The toxicity of heavy metals such as cadmium or indium used in quantum dots can be problematic for biological applications. Strategies to avoid this problem are being investigated. The use of Type II heterojunction core-shell quantum dots allows a wider range of semiconductor types including those with indirect gaps to be considered.

7.11   SUMMARY

7.1 The semiconductor quantum dot spatially confines an electron and a hole to a size on the scale of 2 to 20nm in all three dimensions. This confinement causes the effective bandgap of the semiconductor to increase due to quantum confinement. This phenomenon can be understood and modeled by applying Schödinger's equation to a three-dimensional quantum well.

7.2 The electron and hole form an exciton. The exciton binding energy also impacts the effective bandgap. The exciton radius needs to be compared to the radius of the quantum dot for more detailed modeling.

7.3 Quantum dots exhibit efficient fluorescence if they are formed from direct gap semiconductors. Group II-VI direct gap semiconductors such as CdSe and group III-V direct gap semiconductors such as InP are examples of industrially important quantum dot materials.

7.4 The synthesis of quantum dots can be top-down or bottom-up. Top down synthesis is costly and it typically requires patterning of materials using lithographic or direct-write methods. Bottom-up synthesis allows for the spontaneous formation of monodisperse quantum dots. Precipitation reactions are important industrially exploited approaches to bottom-up synthesis.

7.5 In addition to electronic confinement effects, vibrations of the atoms comprising the quantum dot play a role in the observed absorption and emission spectra. This leads to an explanation of the Stokes shift and the Franck-Condon effect. The vibrations of atoms can be introduced using classical physics; however, a more accurate understanding requires a quantum-mechanical treatment. This modeling results in eignenenergies and eigenstates predicted by the quantum harmonic oscillator model. Quantum dot fluorescence is now described in terms of vibronic absorption and emission.

7.6 Due to the extremely high surface-to-volume ratio in quantum dot materials, surface states are of critical importance. Surface passivation is required for efficient fluorescence. Approaches to surface passivation include core-shell structures and organic ligands that surround the quantum dot. Type I and Type II quantum dot heterostructors can be understood using band diagrams.

7.7 Due to the inherent confinement of carriers in quantum dots, Auger processes have a high probability of occurring. This leads to a blinking effect that decreases the quantum efficiency of quantum dot fluorescence. The effect is strongly influenced by quantum dot charging. Mitigation strategies include graded core-shell quantum dot structures.

7.8 By means of ligand chemistry, quantum dots can be preferentially bonded to specific sites thereby tagging areas of interest in biological systems. Toxicity of quantum dots can be mitigated by an expanded range of semiconductors available in Type II heterojunction core-shell quantum dots.

PROBLEMS

7.1 Estimate the effective bandgap of a quantum dot made from a CdS core that is 5 nm in diameter. Include both the quantum size effect and the exciton model. Look up appropriate parameters as needed.

7.2 Look up literature on a core-shell quantum dot with a CdSe core. Confirm the shell material that is used. Find evidence of the Stokes shift on measured and reported excitation and emission spectra. Show the spectra graphically.

7.3 Based on the observed Stokes shift for Problem 7.2, sketch a configurational coordinate diagram that is consistent with this Stokes shift showing the energies of the absorption and emission transitions on the y-axis and the configurational coordinate on the x-axis.

7.4 Show, by calculation, that the Bohr radius of the exciton in InSb is approximately 60nm. Look up literature results for the size range of successfully grown monodisperse InSb quantum dots and compare to the Bohr radius.

7.5 Estimate the frequency of vibration of a carbon-carbon bond using $\omega = \sqrt{\dfrac{K}{m}}$. The value of spring constant K is 150 N/m. In what part of the electromagnetic spectrum does this vibrational frequency lie?

7.6 Using the value of ω from Problem 7.5, compare $\dfrac{h\nu}{2}$, the approximate zero point energy for vibrations in diamond, to kT at room temperature.

7.7 Look up and summarize the full derivation of harmonic oscillator wavefunctions and eigenvalues. See, for example, Eisberg et al. in Recommended Reading.

7.8 Find journal articles on the blinking effect in quantum dots. Summarize what you read in your own words. Discuss this in the context of the content of Section 7.9.

7.9 Starting from

$$-\frac{\hbar^2}{2m}\frac{d^2}{dr_e^2}\left[\phi_{m_e}\phi_{n_v}\right]-\frac{\hbar^2}{2m}\frac{d^2}{dr_n^2}\left[\phi_{m_e}\phi_{n_v}\right]+(U_e+U_n)\phi_{m_e}\phi_{n_v}=E_T\phi_{m_e}\phi_{n_v}$$

perform separation of variables. Hint: Separation of variables was used in Section 4.6 in the case of two electrons. The steps are quite analogous.

7.10 Find literature on entangled photons. What are they? How may they be produced? How may they be exploited for quantum computing?

7.11 Find a plot of effective bandgap energy versus radius for well-known direct gap quantum dot semiconductors that include PbS, PbTe, CdSe, and InP. PbS and PbTe can be sized to achieve a specific visible photon emission wavelength; however, this is more difficult compared to the use of CdSe or InP. Not considering the issue of toxicity, what makes this challenging for PbS or PbTe based on the plot? Visible light has bandgaps between 1.8 eV and 3.1 eV.

RECOMMENDED READING

1. Manders, J.R. et al. (2016). Quantum dots for displays and solid-state lighting. Book chapter. In: *Materials for Solid State Lighting and Displays* (ed. A. Kitai). Wiley, ISBN:9781119140580.

2. Amita, R. et al. (ed.) (2022). *Quantum Dots: Fundamentals, Synthesis and Applications*, 1e. Elsevier, ISBN 0128241535.

3. Eisberg, R. et al. (1991). *Quantum Physics of Atoms, Molecules, Solids, Nuclei, and Particles*, 2e. Wiley, ISBN 978-0471873730.

CHAPTER 8

Organic Semiconductor Materials and Devices

<table>
<tr><td colspan="3">CONTENTS</td></tr>
</table>

Fundamentals of Semiconductor Materials and Devices, First Edition. Adrian Kitai.
© 2023 John Wiley & Sons Ltd. Published 2023 by John Wiley & Sons Ltd.
Companion Website: www.wiley.com/go/kitai_fundamentals

Objectives

1. Understand conjugated polymers as distinct from polymers with saturated bonding.
2. Understand the mechanism of electric conductivity in conjugated polymers.
3. Explain absorption and emission processes as based on the molecular exciton.
4. Introduce polymer OLED devices and their structures.
5. Explain polymer OLED operation with reference to HOMO and LUMO levels.
6. Discuss materials used for polymer OLED devices and the band model of the polymer OLED.
7. Introduce small molecule OLEDs and their device structures.
8. Describe anode and cathode materials for small molecule OLEDs.
9. Discuss hole injection and transport layers as well as electron injection and transport layers in small molecule OLEDs.
10. Describe light emission in small molecule OLEDs including host materials and fluorescent and phosphorescent dopants.
11. Introduce the basic organic solar cell structure and discuss carrier dissociation at interfaces as well as the band model for the heterojunction solar cell.
12. Describe the advantages of more advanced organic solar cell structures including the bulk heterojunction and examples of self-organization.
13. Introduce hybrid perovskite materials as applied to solar cells.
14. Introduce the organic field effect transistor.

8.1   INTRODUCTION TO ORGANIC ELECTRONICS

In the twentieth century the plastics revolution enabled low cost manufacturing of new lightweight products for a wide range of applications from molded vehicle components to food wrapping film to thermal foam insulation building products. The basic attributes of plastic that drove this revolution include its low cost, ease of manufacture, mechanical flexibility, and toughness.

Organic semiconductor devices now form the basis of an ongoing revolution in electronics in the twenty-first century. The field of *organic electronics* is developing rapidly.

Organic materials having well-defined electrical properties are being deployed for optoelectronic devices such as organic light emitting diodes (OLEDs). OLEDs are now displacing the use of liquid crystal displays in consumer electronics such as cellphones, tablets, and televisions.

The replacement of inorganic solar cells with organic materials is challenging due to the requirement of long term exposure to the sun. A new hybrid inorganic/organic solar cell perovskite material system is the subject of intensive development. Although more akin to an inorganic semiconductor, this hybrid material rivals silicon in terms of solar cell efficiency.

Transistors can also be made using organic materials. Organic field effect transistors that operate much like the MOSFET, can be deposited on flexible plastic sheets to enable flexible circuitry.

8.2   CONJUGATED SYSTEMS

Organic materials generally can be classified into *small molecule* organic materials and *large molecule* organic materials. Small molecule materials, often referred to as *oligomers*, contain only a few repeat units, or *mers*, per molecule, whereas large molecule materials, referred to as polymers, contain many mers per molecule. Typical polymer molecules contain hundreds or thousands of mers per molecule.

Electrically conductive organic molecules as electronic materials were understood in the 1970s, and they now constitute a new family of organic semiconductors that have the essential properties needed to make electronic devices. Conductive oligomers/polymers, however, have electrical properties that are orders of magnitude different from typical inorganic semiconductors and they may be processed at much lower temperatures.

In order to understand the properties of these materials we need to review their molecular structure and understand the origin of the energy levels for electrons and holes as well as the mechanisms for electron and hole transport. We will compare these to inorganic materials to highlight the distinctive properties that must be considered in order to make use of conjugated polymers in semiconductor devices.

Conventional polymers are *saturated*, which means that the valence electrons of carbon atoms in a carbon chain are fully utilized in bonding and four atoms are bonded to each carbon atom. The simplest example of a saturated polymer is polyethylene, as shown in Figure 8.1. In polyethylene a carbon atom has six electrons, which normally occupy energy states $1s^2 2s^2 2p^2$.

For the carbon incorporated in polyethylene the four electrons in the $2s^2 2p^2$ shell achieve a lowest energy state in a *hybridized* sp^3 configuration that combines the spatial characters of the s-orbital and the p-orbital resulting in what are known as σ *bonds*. The bonding symmetry of a carbon atom in polyethylene is very close to tetrahedral, and the carbon bonds are *stereospecific*, meaning that the carbon bond angles are quite rigid and approach the tetrahedral bond angle of 109°. The well-known

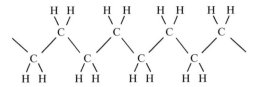

FIGURE 8.1 The molecular structure of polyethylene. Each carbon has four nearest neighbors and forms four bonds. Polyethylene is an insulator and has a wide energy gap in the ultraviolet energy range. Each carbon atom has an almost perfect tetrahedral bond symmetry even though it bonds to both carbon and hydrogen nearest neighbors.

tetrahedrally symmetric bonding character of carbon in inorganic diamond crystals is an example of the stereospecific nature of carbon bond angles.

The simplest conductive polymer is polyacetylene. Its structure is shown in Figure 8.2. In this polymer there are three nearest neighbors leading to a different type of hybridization, which is denoted sp^2p_z. In this hybridization there is a lone electron called the π electron and three other electrons. These three other electrons take part in σ bonds with the three nearest neighbors. The π electron, however, does not play a role in bonding and is left "dangling" at each carbon atom. Polymers having this configuration are referred to as *conjugated*.

Since a chain of closely spaced carbon atoms is created, there are as many π electrons as carbon atoms in the chain, and there is a spatial overlap of these individual π electrons. The π electron at each atom interacts with all the other π electrons in the chain to form a number of energy bands. These bands are collectively referred to as *π-sub-bands*. Electrical conductivity also arises as a result of these π electron sub-bands since the electrons they contain become delocalized. As with inorganic semiconductors, the Pauli exclusion principle forces the π electrons to occupy a range of energy levels leading to the formation of bands.

In addition we can regard such a conjugated molecule as a periodic potential in which the π electrons exist, which means that the band concepts of Chapter 2 would predict energy bands and energy gaps. The length scale of one period in the periodic potential depends on the length of one mer on the polymer chain. Figure 8.3 shows a band

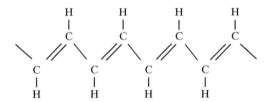

FIGURE 8.2 Polyacetylene is the simplest conjugated molecule. It is often thought of as a chain of single bonds alternating with double bonds although actually all the bonds are equal and are neither purely single nor purely double. It consists of a carbon chain with one hydrogen atom per carbon atom. Since only three of the four valence electrons of carbon are used for bonding, one π electron per carbon atom is available for electrical conduction and becomes delocalized along the carbon chain.

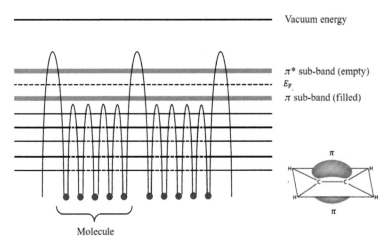

FIGURE 8.3   Energy levels and bands in a few closely spaced organic molecules. Note the small energy barriers caused by the intramolecular bonding and the larger energy barriers caused by the intermolecular bonding. The conducting π electrons are delocalized and lie above and below the σ bonds as depicted for a simple ethylene molecule in the inset.

diagram of a molecular solid containing a few molecules. The small potential energy barriers exist between atoms or mers within a molecule, and the large barriers represent the separation between molecules. Intramolecular bonds are normally covalent or conjugated bonds, and intermolecular bonds are much weaker van der Waals bonds.

In addition to the periodicity of the molecule repeat units modeled by the periodic potential energy $U(x)$ depicted in Figure 2.1, the overall dimension of the conjugated system, or *conjugation length,* can be viewed as the effective length of the molecule over which electron delocalization occurs. This conjugation length acts as the length L of a one-dimensional box in which the electron is confined, and the simplest model of the molecule is therefore somewhat analogous to Example 2.1. The longer the conjugation length the smaller the energy level spacing. In Chapter 9, more sophisticated atomic scale modeling methods are introduced. These methods rely on using Schrödinger's equation to obtain the properties of electrons in structures containing small numbers of atoms such as molecules. This modeling can be used for molecules that have structures other than simple linear repeating units.

The number of electrons relative to the number of available states in the π sub-bands determines the electrical nature of the polymer. If the π sub-bands are either filled or empty then the material will be semiconducting and if the highest filled band is partly filled then the material will be metallic. The energy difference between the highest occupied π sub-band and the lowest unoccupied π sub-band determines the energy gap in the case of a semiconducting polymer. The unoccupied π bands are commonly referred to as π^* bands.

A wide range of large molecules exist that have semiconducting properties. A number of these are shown in Figure 8.4. Note that with the exception of polyacetylene, the remaining molecules have ring elements containing five or six atoms per ring.

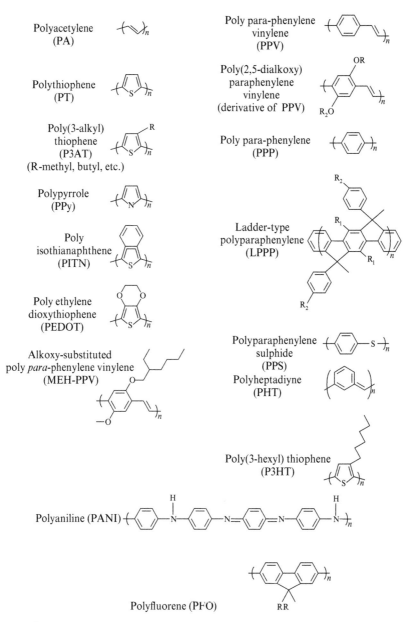

FIGURE 8.4 [1] Molecular structures of well-known conjugated polymers. Many molecules contain a combination of linear and ring-type structures, the simplest example being poly paraphenylene vinylene (PPV). The mer of each of these conjugated polymers is shown. Reprinted from Li, Z., Meng, H., Organic Light-Emitting Materials and Devices, 157444-574X. Copyright (2006).

The simplest ring element is the benzene ring having six carbon atoms and six delocalized π electrons, one from each atom. Six-sided rings are also the building blocks of graphene, a conductive material. It is therefore not surprising that ring-containing molecules are prominent among conducting polymers.

The macroscopic electrical conductivity of an organic material depends both upon the conduction within a molecule (*intramolecular* conduction) as well as the ability for charges to transfer from molecule to molecule (*intermolecular* conduction). Intermolecular conduction is strongly dependent on the proximity of molecules to each other and their orientation to each other. Crystalline materials are generally more densely packed and generally achieve higher electrical conductivities than amorphous materials; however, both amorphous and crystalline phases are candidate materials. Crystalline organic semiconductors often contain crystalline domains in an amorphous matrix.

Carrier mobility values for organic semiconductors typically span the range from 10^{-7} cm^{2}V^{-1}s^{-1} to about 1 cm^{2}V^{-1}s^{-1}. Note that these values are much lower than a typical inorganic semiconductor like silicon with mobility on the order of 10^3 cm^{2}V^{-1}s^{-1}. There are a number of interesting opportunities offered by molecular semiconductors:

(a) *Chemical properties* can be modified by changing the side-groups on the molecules. For example, in Figure 8.4, PPV is insoluble in organic solvents. Since PPV contains a benzene ring it can be modified by adding groups to the ring, and soluble molecules are obtained, which are referred to as *PPV-derivatives*. One example of a soluble PPV derivative is shown in Figure 8.5. A soluble PPV derivative allows a polymer with the electrical functionality of PPV to be formulated in a solution, deposited as a liquid and then baked to release the solvent. This is a cost-effective device preparation technique that wastes very little material.

(b) *Electronic and optical properties* can be modified by changing side-groups on the molecule. For example, PPV can be modified to become MEH-PPV shown in Figure 8.4 by adding the indicated groups to the benzene ring of PPV. Figure 8.6 shows PPV and derivatives that are frequently used in polymer OLEDs. Side groups modify the energy gap of PPV as shown allowing absorption or emission wavelengths to be tuned to desired values, which is of particular relevance to optoelectronic applications.

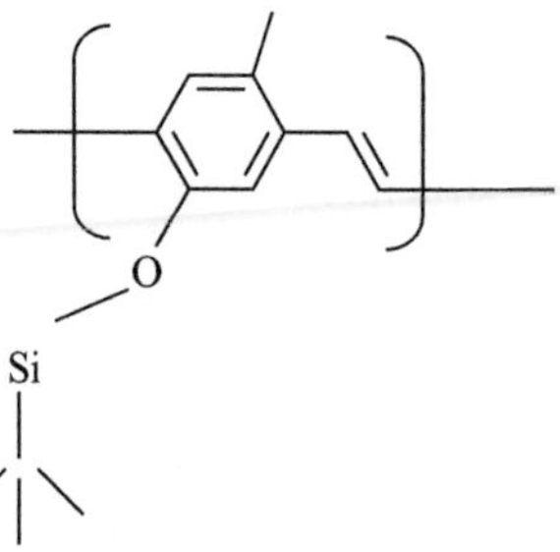

FIGURE 8.5   Poly para-phenylene vinylene (PPV) derivative forming a silicon-substituted soluble polymer.

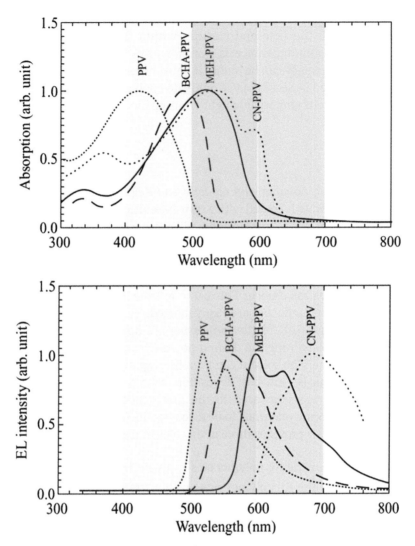

FIGURE 8.6 Absorption and emission of poly para-phenylene vinylene (PPV) and PPV derivatives. The energy gap determines the upper wavelength range of absorption as well as the lower wavelength range of emission. Here energy gaps from 1.9 eV ($\cong$ 640 nm) to 2.5 eV ($\cong$ 500 nm) result in these spectra. After Li, Z., Meng, H., Organic Light-Emitting Materials and Devices, 157444-574X, Taylor & Francis, 2006.

The absorption process excites an electron using a photon from the highest occupied π sub-band to the lowest unoccupied π^* sub-band; the emission process generates a photon when the electron falls back to its original band. An exciton is formed and is then annihilated during emission as discussed in detail in Chapter 4. The *photoluminescence efficiency* of this absorption and emission process can be in the range of 10% to over 90%, which makes these materials suitable for high-efficiency optoelectronic device applications.

Electroluminescence efficiency is lower than photoluminescence efficiency due to spin associated with the hole and electron within the exciton. The maximum efficiency under electroluminescence is normally 25% of the photoluminescence efficiency, although efficiency can be higher if phosphorescence occurs. These optical processes were discussed in Section 4.7. Conjugated polymers are suitable candidates for both organic solar cells and organic light emitting diodes.

8.3　POLYMER OLEDS

The structure of a basic polymer *organic light emitting diode* (OLED) is shown in Figure 8.7. A glass substrate is coated with a transparent conductor such as indium tin oxide (ITO). The ITO must be as smooth as possible because the subsequent conjugated polymer layer is generally under 100 nm thick. An EL polymer layer is deposited on the anode layer. The thickness of the conjugated polymer layer must be minimized due to the low electrical conductivity of the material. Finally a low-workfunction cathode layer completes the device. This cathode layer is composed of Group I or Group II metals or compounds that are easily ionized.

Deposition techniques of *solution processing* are applicable to most electronic polymers. A solution containing the desired polymer can be formed as a liquid and spread over the substrate. The solvent can then be evaporated away leaving a layer of the desired polymer. This is a low-cost deposition technique and is a key advantage of polymer organic materials compared to the more expensive deposition methods required for small molecule materials to be discussed in Section 8.4.

The EL polymer layer provides three main roles in the polymer OLED:

(a) Near the cathode the EL polymer acts as an electron transport layer.
(b) Near the anode the EL polymer acts as a hole transport layer.
(c) The EL polymer provides for the recombination of holes and electrons via molecular exciton formation and annihilation with the consequent emission of light.

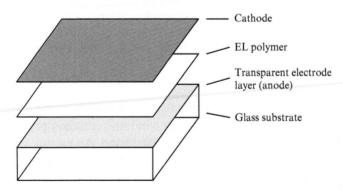

Cathode

EL polymer

Transparent electrode layer (anode)

Glass substrate

FIGURE 8.7　Structure of basic polymer OLED consisting of a glass substrate, a transparent ITO anode layer, an EL polymer layer, and a low workfunction cathode layer.

We can understand the ability of one layer to serve all these functions by understanding the workfunction difference between the two electrodes. Assume a cathode
and an anode material are placed in vacuum, are electrically isolated from each other
and are not connected to any voltage source. The anode has a high workfunction, typically in the range of 5 eV, and the cathode has a low workfunction in the range of
1–2 eV. This means that for an electron to leave the surface of the anode and enter
vacuum requires about 5 eV but an electron leaving the cathode requires only about
1–2 eV to enter the same vacuum. This may be illustrated by defining a vacuum energy
E_{vac} for both anode and cathode as shown in Figure 8.8.

Since the electrons leaving the anode and cathode leave from their Fermi levels
(highest occupied energy levels) it is clear that we cannot draw a diagram simultaneously showing aligned Fermi levels and a constant value of E_{vac}. This problem is
resolved since an electric field is generated in the vacuum giving rise to a potential
gradient between the anode and cathode that accounts for the difference in work
function. A built-in charge is present on each electrode to create this electric field: the

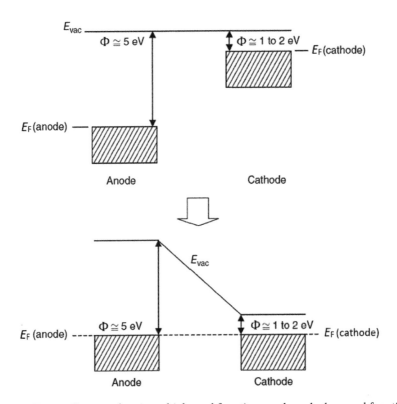

FIGURE 8.8   Energy diagram showing a high workfunction anode and a low workfunction
cathode. With a constant vacuum energy the Fermi levels cannot be aligned (upper diagram) and
this diagram is therefore not an equilibrium diagram, which invalidates the concept of Fermi
energy. To resolve this problem an electric field forms between anode and cathode and an
equilibrium diagram (lower diagram) having aligned Fermi energies is the result. This is the result
of charge transfer.

anode carries a *negative* charge and the cathode carries an equal and opposite *positive* charge. For the charges to build up, there must be an opportunity for charge to transfer between anode and cathode; however, once the appropriate charges are present there will be no further charge transfer and equilibrium is maintained. Without the opportunity to transfer charge equilibrium can never be reached.

We can now return to the case of the EL polymer layer sandwiched between anode and cathode. The low carrier concentration in the EL polymer layer ($\leq 10^{14}$ cm^{-2}) combined with its low carrier mobility means that we can regard this layer as a virtual insulator sandwiched between the anode and cathode electrodes. There will be a built-in electric field within the polymer caused by the workfunction difference between the two metals, as shown in an ideal form in Figure 8.9a, which is drawn for equilibrium conditions. A slope in the upper π^* band and the lower π band is present, which is a consequence of the electric field present in the EL polymer. The finite but small conductivity of the EL polymer is sufficient to transfer the necessary charge for equilibrium to be reached.

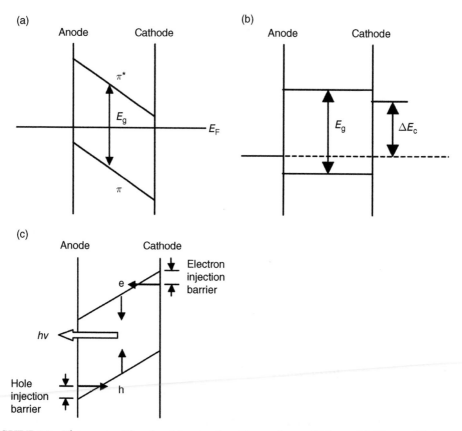

FIGURE 8.9   The upper π^* band and lower π band in a polymer EL layer. (a) The equilibrium condition. (b) The flat-band condition in which a positive voltage is applied to the anode. (c) Device in forward bias in which holes and electrons are injected and form molecular excitons, which annihilate to generate photons. The resulting hole injection barrier and electron injection barrier are shown.

If a voltage equal to the difference between the two workfunctions is applied across the anode and cathode the flat-band condition is reached. The anode is biased positive with respect to the cathode. At this voltage, which is the flat-band voltage, the EL polymer energy bands are horizontal and no electric field exists. The applied voltage causes additional charges to be added to the anode and cathode that cancel out the built-in charges. The flat-band condition is shown in Figure 8.9b. The difference between the conduction band edges of the two electrodes, ΔE_c, is shown. The flat band concept was introduced in Chapter 6.

The vacuum workfunctions of the anode and cathode materials are actually modified by the details of the electrode-EL polymer interfaces, which also affect the flat-band voltages. This modification of workfunctions is caused by trapped charges at the interfaces and other impurity and diffusion effects that occur there. A similar set of issues was described in Section 3.9 for inorganic metal-semiconductor contacts, although the polymer EL material offers particularly high diffusivity even at room temperature and is therefore more prone to the incorporation of undesired species.

If the anode voltage is made still more positive then the situation in Figure 8.9c results. Here the electrons from the cathode tunnel across the potential "spike" at the cathode edge of the π^* band and are injected into the EL polymer. The electric field is in the opposite direction to that shown in Figure 8.9a, and electrons that enter the π^* band start drifting toward the anode. It is important to minimize the energy barrier seen by cathode electrons as they are injected into the band. The electrons will enter the lowest energy state in this π^* band and we refer to this as the *lowest unoccupied molecular orbital* (LUMO). Note that this is analogous to E_c, the bottom of the conduction band, in inorganic semiconductors. Generally a difference between cathode Fermi energy and the LUMO energy of 0.2 eV or less is suitable.

Proper modeling of electron current flowing across this potential barrier also requires the application of thermionic emission physics, which was discussed in detail in Section 3.9. In practice both thermionic emission and tunneling are often involved.

At the same time, holes can be injected into the π band from the anode. It is more correct to describe this as the formation of holes at the anode-EL polymer interface due to the transfer of electrons from the π band to the anode. These holes are swept by the electric field toward the cathode in the highest available π band energy level, which is referred to as the *highest occupied molecular orbital* (HOMO), and the holes have an opportunity to combine with π^* band electrons to form excitons. Upon annihilation of these excitons, light is emitted.

Cathode metals require the use of low-ionization-energy elements. Examples of cathode materials include a thin barium-based, calcium-based or lithium-based layer in contact with the active polymer. This layer may be coated with a final aluminum capping layer to offer some protection and to improve sheet conductivity. Since Group I and Group II metals are very easily oxidized and react vigorously with water, care must be taken to protect them from the atmosphere. OLED devices in the laboratory must be handled and tested in a very dry inert gas environment requiring the use of a glove box.

Anode materials are generally transparent to allow light to leave the active layers, and indium tin oxide (ITO) is popular. There are three challenges with ITO, however:

(a) ITO is a polycrystalline film and has naturally occurring surface roughness, which can be comparable in scale to the thickness of the EL polymer layer (20–100 nm). This can cause OLED failure due to short circuits.

(b) Indium is highly mobile in EL polymers and can migrate from the anode into the active region.

(c) The ITO surface workfunction is highly unstable and inhomogeneous. There are many dangling bonds on the ITO surface, which need to be hydroxide-terminated. The surface of the ITO contains Sn atoms that substitute for In atoms and add to the chemical complexity.

To prevent these difficulties an additional conductive polymer anode layer is commonly inserted between the ITO and the EL layer. This layer has a HOMO level close to the HOMO level of the EL layer to promote hole injection. It also acts as a planarization layer to create a smooth surface for the EL polymer layer. As a result of adding a polyaniline (PANI) anode layer (Figure 8.4) an increase in the life of a PPV-based polymer OLED from hundreds of hours to over 10,000 hours has been obtained.

Typical operating characteristics of a polymer OLED are illustrated in Figure 8.10. Both current and luminance are shown as a function of applied voltage. Below a threshold voltage of about 1.8 V, current does not flow and there is no light output. Both current and luminance rise rapidly as voltage is increased and carriers are able to be injected into the LUMO and HOMO levels. The luminance is essentially proportional to the current over several orders of magnitude.

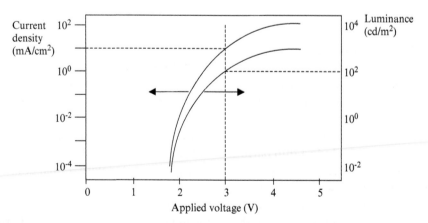

FIGURE 8.10 Typical luminance–voltage (L–V) and I–V characteristics of a polymer OLED. A well-defined threshold voltage is observed due to the sharp onset of carrier injection from the electrodes across the potential barriers at the electrode-EL polymer interfaces. Note the similarity between the shapes of the current and luminance curves.

The performance of OLEDs is frequently expressed in terms of cd A^{-1}. In Figure 8.10 at an applied voltage of 3 V the current density is 10 mA cm^{-2} = 100 A m^{-2} and the luminance is 100 cd m^{-2}. This means that the operating efficiency at 3 V is (100 cd m^{-2})/(100 A m^{-2}) = 1.0 cd A^{-1}. Polymer OLEDs with operating efficiency values of well over 10 cd A^{-1} have been achieved. In addition, very high peak luminance values of up to $\cong 10^6$ cd m^{-2} may be obtained if large transient currents in the range of 100 A cm^{-2} are applied.

The most challenging aspect of OLED performance has been operating lifetime, which normally is defined as the operating time under constant current conditions for the OLED to decrease in light emission intensity to half its initial intensity. Purity and interface stability are two requirements of practical OLED materials. It is found that lifetime can be well correlated to the total charge density $\rho_{accumulated}$ passing through the OLED during its lifetime. This total charge densitycan be determined for an OLED subjected to a time-varying current density by integrating the current density over time yielding

$$\rho_{accumulated} = \int_0^T J(t) \, dt$$

Typical lifetime values of $\rho_{accumulated}$ greater than 10^5 C cm^{-2} are achievable in a sealed, dry environment at room temperature.

8.4 SMALL-MOLECULE OLEDS

In addition to polymer EL materials, *small molecule* organic materials can also be used to fabricate efficient LED and solar cell devices. Examples of small molecules with useful optoelectronic properties are shown in Figure 8.11. Small-molecule materials are generally not soluble and are deposited by vapor deposition methods. Using a vacuum chamber, thin films of these materials can be grown by heating a source pellet or powder of the material causing it to evaporate onto a substrate. Using an appropriate chamber design, deposited films with good thickness uniformity and high purity can be obtained; however, there is significant material waste due to the evaporation process. The small-molecule materials in Figure 8.11 all contain ring structures and have delocalized electrons for high intramolecular conduction. Intermolecular conduction is also critical for bulk conductivity to exist.

There are several requirements of the small-molecule materials:

(a) They must not decompose during thermal evaporation in the vapor deposition process. A source pellet of the material is heated to temperatures in the range from 150°C to over 400°C in a vacuum chamber to create a high enough vapor pressure for the molecules to evaporate and condense on a cool substrate positioned a set distance away from the source material.

(b) The deposited films must be of high quality and purity. Typical thicknesses are in the range from 5 to 200 nm.

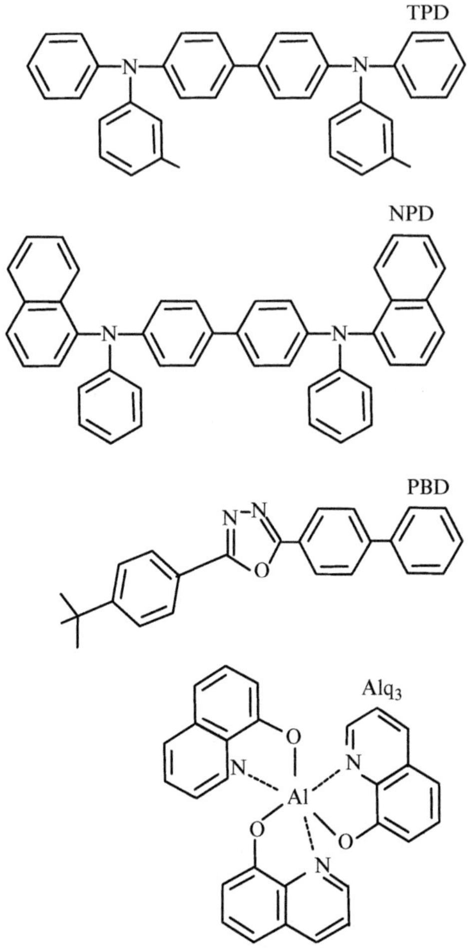

FIGURE 8.11 Small organic molecules used for small-molecule OLED devices. Hole-transporting materials are TPD and NPD. Electron-transporting materials are PBD and Alq$_3$. Reprinted from Li, Z., Meng, H., Organic Light-Emitting Materials and Devices, 157444-574X.

(c) The glass transition temperature T_g should be high enough to prevent crystallization under the conditions of normal operation. Typically T_g should be over 85°C. Crystallization can lead to intermolecular interactions that decrease quantum efficiency.

Unlike polymer OLEDs, small-molecule devices typically use separate electron- and hole-transporting materials and are therefore more analogous to inorganic p-n junction devices. A basic structure of a small-molecule OLED is shown in Figure 8.12. A *hole transport layer* (HTL) and a separate *electron transport layer* (ETL) form a junction sandwiched between anode and cathode electrodes.

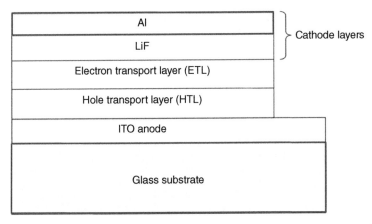

FIGURE 8.12 *Small-molecule OLED structure.* The OLED includes a transparent substrate, transparent ITO anode, hole transport layer (HTL), electron transport layer (HTL), and cathode. HTL materials such as TPD or NPD and electron transport materials such as Alq$_3$ or PBD are suitable. A popular cathode is a two-layer Al/LiF structure as shown.

It is also possible to introduce a *light emitting material* (LEM) positioned between the HTL and ETL, which is optimized for radiative recombination efficiency and in which the holes and electrons form excitons. In these materials a number of additional energy transfer processes are often involved, which are described in more detail in Sections 8.11–8.14.

The finally process involves photon outcoupling. There are internal reflections that limit the escape cone of the photons. This is similar to the situation in inorganic LEDs; however, the refractive index n of the active organic layers is lower than for inorganic semiconductors, which allows better outcoupling. A typical value of $n \cong 1.5$ allows 20–30% outcoupling through the glass substrate.

The small-molecule OLED with separate HTL and ETL materials is more efficient than single-layer designs. This is true since the relevant layers can be individually optimized for carrying one type of charge carrier. Recombination near the interface of the HTL and ETL layers is achieved. This prevents recombination from occurring too close to the anode and cathode layers, which could result in quenching luminescence and causing non-radiative decay of the excitons. Even better performance can be obtained by adding the light emitting material as well as an anode *hole injection layer* (HIL). The cathode is shown to include an *electron injection layer* (EIL), which is composed of LiF in Figure 8.12. An OLED structure containing these layers is shown in Figure 8.13.

The operation of the small-molecule OLED can be better understood with reference to the band diagram shown in Figure 8.14. Upon application of an electric field electrons are injected by means of the EIL and ETL while holes are injected by means of the HIL and HTL. These holes and electrons meet in the LEM and form excitons there. Optimum performance is obtained when the same number of active holes and electrons enter the LEM per unit time, otherwise excess carriers of one type or the other are unable to participate in the recombination process, which causes efficiency

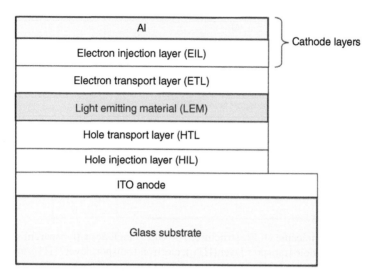

FIGURE 8.13   A more optimized small-molecule OLED structure includes an electron injection layer, a hole injection layer, and a light emitting material. The cathode includes the electron injection layer.

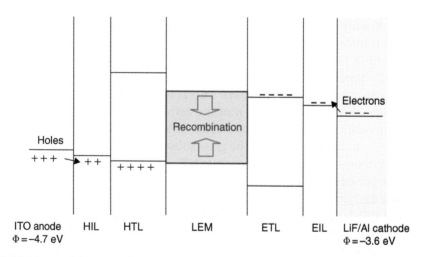

FIGURE 8.14   Band diagram of a small-molecule OLED showing LUMO and HOMO levels for the various layers of the device. The band diagram is drawn without a bias applied. The accepted workfunctions of anode (ITO) and cathode (LiF/Al,), which are 4.7 eV and 3.6 eV respectively, are shown.

loss. This is a current balance condition between holes and electrons similar to that discussed in the context of inorganic LEDs in Section 5.9.

In operation, holes are injected by means of the anode into the HTL and electrons are injected from the cathode into the ETL. With a suitable applied voltage a sufficient electric field exists for the holes and electrons to drift across their respective layers. The mobilities of these layers are important; compared to inorganic p-type and n-type materials mobility values are much lower. A typical hole mobility for small-molecule HTL materials is 10^{-3} cm^{2}V^{-1}s^{-1}, and a typical electron mobility for small-molecule ETL materials is 10^{-5} to 10^{-4} cm^{2}V^{-1}s^{-1}. These values are sufficient provided the layers are very thin; layer thicknesses in the range of only 10–200 nm are typical.

Next, exciton formation occurs by coulombic attraction of the holes and electrons as they encounter each other. In the absence of a LEM, this normally occurs within the transport layers. Since the HTL typically has a higher mobility, the ETL is likely the layer in which excitons are formed. Photon emission from these excitons is normally limited to singlet excitons that decay very rapidly. The excited state has a very short spontaneous lifetime typically in the range of 10 ns due to the strong dipole strength of the radiating exciton. This means that the radiation originates very close in location to where the exciton was created.

There are a large number of small molecules that have been investigated for small-molecule OLEDs, and new molecules with specific attributes continue to be researched. Sections 8.5–8.14 present selected examples of molecular materials designed for various layers in small-molecule OLEDs as well as their most important properties.

8.5 ANODE MATERIALS

The ITO anode offers ease of patterning and good stability. As with polymer OLEDs, the surface smoothness of both the substrate and the ITO layer is important. A surface roughness below 2 nm is generally required. ITO has a high workfunction ($\varphi > 4.1$ eV) allowing it to inject holes efficiently. The transparency of the ITO is a result of its wide bandgap of over 4 eV.

There are challenges associated with the use of ITO for small-molecule OLEDs, some of which were also noted for polymer OLEDs. ITO has a resistivity of $\cong$ 2×10^{-4} Ω cm, which limits the current flow through an ITO layer and results in unwanted voltage drops along the anode conductors, which is particularly problematic if narrow anode rows or columns are required for an OLED display. ITO is inevitably rough due to its polycrystallinity. It has a chemically active surface that can cause migration of indium into subsequent polymer layers. The ITO workfunction is sensitive to the cleaning process used to prepare the ITO for subsequent processing. ITO is a brittle inorganic film that is not ideal on polymer substrates because it normally requires high temperatures to deposit (200–400°C) and can crack due to mechanical and thermal deformation on polymer substrates. The development of alternative transparent conductive materials to ITO is an active field.

8.6 CATHODE MATERIALS

Unlike anode materials, cathode materials are generally not transparent, which provides a wider range of materials choices. They must provide high conductivity and low workfunction, and good adhesion to the underlying organic layers. Stability is also important and is highly dependent on packaging.

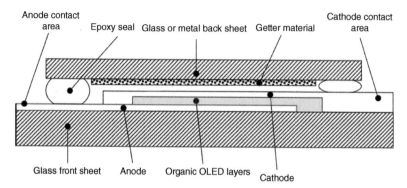

FIGURE 8.15 OLED package includes front and back sheets, epoxy seal material on all edges, sacrificial desiccant or getter material, cathode and transparent anode having cathode contact areas for external connections. The rate of moisture penetration must be calculated to ensure a specified product life.

Challenges associated with cathodes include ease of oxidation, which is a consequence of the low-workfunction materials that require easily ionized Group I and Group II metals. There is also a tendency for these cathode layers to cause chemical reduction of adjacent organic layers. As with polymer OLEDs a two-layer cathode is popular, and the LiF/Al structure is widely used in which the aluminum protects the reactive LiF layer and also provides improved sheet conductivity. The formation of LiF preserves the low workfunction of the cathode but reduces its tendency to oxidize as a result of reaction with the organic EIL material. LiF/Al cathodes have a workfunction of 3.6 to 3.8 eV.

The cathode layer is usually the most reactive layer in the OLED in the presence of oxygen or water. OLED devices must be protected with encapsulation. In practice glass /metal or glass/glass sheets cover and seal OLED devices on the front and back respectively. An epoxy edge seal between front and back sheets is used; however, moisture can slowly diffuse through this polymer layer. Inside the OLED package a sacrificial *getter* material may be used to scavenge moisture or oxygen. An OLED package is shown in Figure 8.15.

Encapsulation suitable for flexible substrates has been developed and has enabled folding cellphone displays and roll-up displays. See Problem 8.7.

8.7   HOLE INJECTION LAYER

The HIL acts to improve the smoothness of the anode surface due to the native ITO roughness. It also facilitates efficient hole injection and long-term hole injection stability. The polymer-metal interface is very complex due to charges that are trapped there as a result of dangling bonds and metal atoms that react with the organic layer. Materials used for the HIL include copper phthalocyanine or CuPc, which is a widely used *porphyrinic metal complex*. Its molecular structure is shown in Figure 8.16.

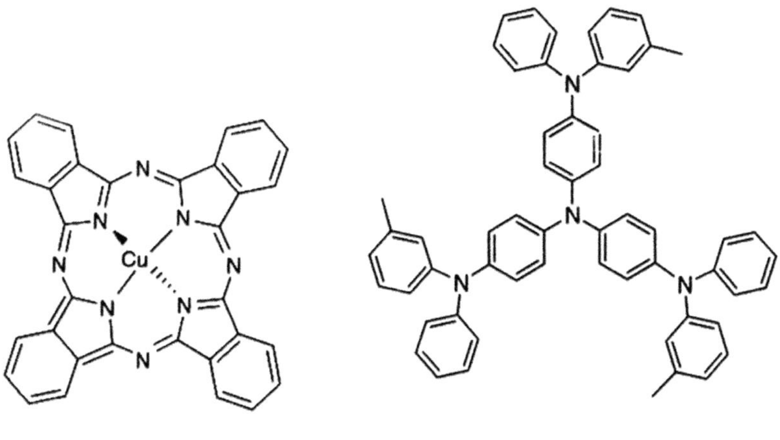

CuPc

m-MTDATA

FIGURE 8.16 Copper phthalocyanine, or CuPc, a widely used metal complex used for the HIL in small molecule OLEDs. Another HIL molecule is m-MTDATA, also shown. Chemical Structure reproduced from Organic light-emitting materials and devices, ed. by Z. Li and H. Meng 9781574445749 (2007) Taylor and Francis.

This material improves efficiency and life in OLEDs; however, the precise mechanisms for the improvements are controversial. CuPc may reduce the hole injection barrier, which is equivalent to the hole injection barrier illustrated in Figure 8.9c for polymer OLEDs. In addition CuPc has a particularly good wetting characteristic on ITO, which may contribute to better interface strength. CuPc has good thermal stability and can reduce the dependence of the hole injection barrier on the cleaning procedure used on the ITO layer. CuPc is a semiconductor, which gives it sufficient electrical conductivity to carry current between the anode and the HTL.

Another important aspect of OLED performance relates to the current balance between electron and hole currents. CuPc may in some cases also improve this current balance and therefore the efficiency of OLED devices. This is highly dependent on the subsequent layers and particularly the electron injection and transport materials used in the OLED. Hole current generally tends to be in excess compared to electron current in OLEDs due to the higher hole mobility in organic materials. This suggests that CuPc may controllably decrease hole current in some cases. CuPc and phthalocyanines in general are *ambipolar*, meaning that they can act as both hole and electron conductors although the electron mobility is one to two orders of magnitude less than the hole mobility. Sometimes these layers are called *electron-blocking* layers.

A challenge in the use of CuPc is the material's tendency to crystallize over time. The crystallization process is thermally driven. This may effectively contribute to limiting the ultimate device lifetime. Other materials that are useful in forming desirable ITO interfaces are under investigation and include fluorocarbon and organosiloxane materials.

8.8 ELECTRON INJECTION LAYER

Pure and easily ionized cathode metals such as Ca or Ba can be used as the low-work-function electron injection layer; however, these metals are highly unstable with gaseous oxygen and water molecules as well as with the organic material used in the electron transport layer. Compounds of such metals that reduce their reactivity and instability while retaining their desired low workfunction include LiF, CsF, Li_2O, and Na_2O. Although these are insulators they can be deposited to a thickness of only one or two monolayers and therefore really function to establish the desired interfacial chemistry and a desired trapped interface charge density. This allows the electron injection barrier illustrated in Figure 8.9 to be established.

Other complexes that include alkali metals or alkaline earth metals incorporated in organic molecules are also candidate materials for the electron injection layer. These include *lithium-quinolate* complexes, as shown in Figure 8.17. The development of materials based on organic molecules is of interest due to their inherent compatibility with the other organic layers.

8.9 HOLE TRANSPORT LAYER

The basic requirement of the HTL is good hole conductivity. In conjugated polymers, hole conductivity arises through conjugated bonding, and in small-molecule hole transport materials the same mechanism applies, combined with the transfer of charge between HTL molecules. As shown in Figure 8.11, TPD and NPD are popular hole conductors consisting of small molecules containing six-carbon rings with conjugated bonds allowing intramolecular hole transport. TPD and NPD are members of a family of compounds known as *triarylamines.*

Triarylamines were developed for xerography in the 1970s and are well-developed photo-conductive materials. Here the electrical conductivity is controlled by the

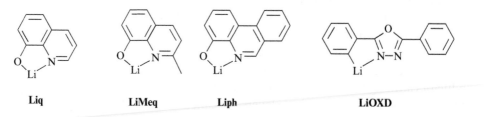

| Liq | LiMeq | Liph | LiOXD |

FIGURE 8.17 Organo-metallic complexes may also be used for the electron injection layer. Examples are shown consisting of some lithium-quinolate complexes. Liq, LiMeq, Liph, and LiOXD. Chemical Structure reproduced from Organic light-emitting materials and devices, ed. by Z. Li and H. Meng 9781574445749 (2007) Taylor and Francis.

density of mobile charge carriers that are generated by illumination of the triaryl-amine in the Xerox process.

Both TPD and NPD are commonly applied to OLEDs due to their modestly high hole mobilities in the range of 10^{-3} to 10^{-4} cm^2/Vs. A significant challenge is their low-temperature crystallization, which progresses slowly at typical device operation temperatures of 30–40°C. This causes the materials to become mechanically unstable and device stability is compromised.

Another group of triarylamines includes the hole conductors triphenylamine (TPA) and TPTE, shown in Figure 8.18. OLEDs employing these materials may be operated continuously at temperatures of 140°C without breakdown since they do not crystallize readily. A number of other triarylamines are being studied also. One additional key requirement for efficient OLED devices using them is the size of the energy barrier at the interface of the HTL and the HIL, which must be small enough to result in an efficient OLED. Hence the HOMO level of the HTL should be within a fraction of an electron-volt from the anode energy band. In Figure 8.14 the conduction band of ITO is shown at 4.7 eV below a vacuum reference level and the HOMO levels of TPD and NPD are suitable, being close to 5 eV below the vacuum level.

In addition to the triarylamines, another family of hole transport materials consists of the *phenylazomethines*. Four examples of phenylazomethines are shown in Figure 8.19. Upon mixing these phenylazomethines with metal ions, such as Sn ions, the resulting metal complexes can form good HTL materials with thermal stability and highly efficient injection of holes. These complexed materials have HOMO levels in the range of −5.2 eV to −5.4 eV (5.2–5.4 eV below the vacuum level), which contributes to their good efficiency.

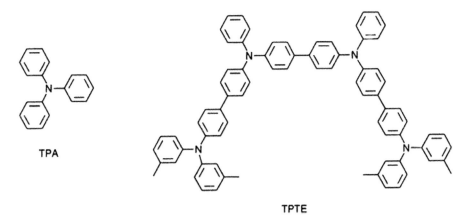

FIGURE 8.18   Two further examples of hole-conducting triarylamines include TPA (triphenylamine) and TPTE (a tetramer of TPA). TPTE enables high-temperature OLED operation without crystallization. Chemical Structure reproduced from Organic light-emitting materials and devices, ed. by Z. Li and H. Meng 9781574445749 (2007) Taylor and Francis.

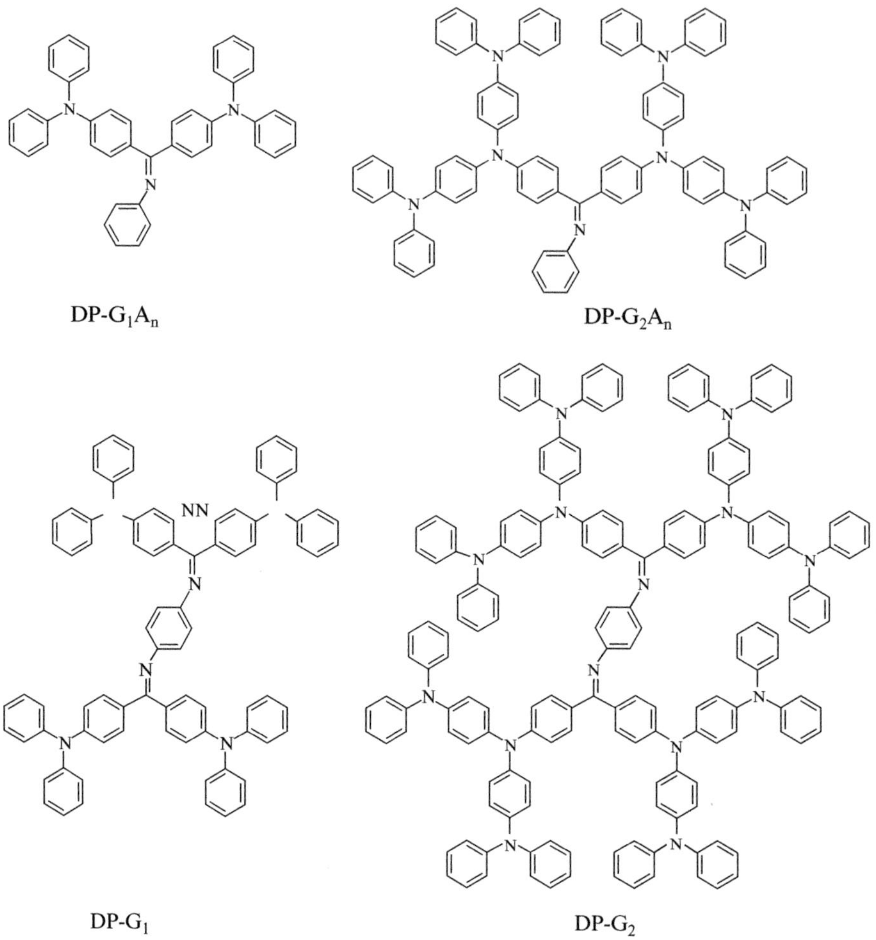

DP-G$_1$A$_n$

DP-G$_2$A$_n$

DP-G$_1$

DP-G$_2$

FIGURE 8.19 Phenylazomethines are formed by various arrangements of nitrogen-terminated six-carbon rings. These phenylazomethine molecules are thermally stable and are complexed with metal ions such as Sn ions introduced in the form of SnCl$_2$ molecules to form the HTL material. Chemical Structure reproduced from Organic light-emitting materials and devices, ed. by Z. Li and H. Meng 9781574445749 (2007) Taylor and Francis.

8.10   ELECTRON TRANSPORT LAYER

Materials for the electron transport layer (ETL) have been investigated intensively and several families of candidate materials are known. Intermolecular transport occurs by electron hopping, and a LUMO level that is similar in energy to the work-function of the cathode and the electron-conducting level in the EIL is required, as shown in Figure 8.14.

The ETL should have a mobility of at least 10^{-6} cm^{2}V^{-1}s^{-1}, which is one to two orders of magnitude smaller than the mobility range of HTL materials. Improving this low mobility has been one key target of the intensive investigation of these materials. Insufficient electron mobility in the ETL means that in many cases holes that enter the light emitting layer (LEM) will not encounter electrons and will therefore continue until they reach the ETL before they recombine. Since the ETL is not optimized for high recombination efficiency, a lower device efficiency can result. ETL is also often oxidized by hole conduction, in which electron loss and the subsequent degradation of the ETL material occur due to holes that enter the material. This is a major degradation mechanism for small-molecule OLEDs.

The crystallization temperature of the ETL should be high enough to retain the amorphous structure during device operation at the operating temperature. Generally a glass transition temperature should be above 120°C.

The ability of the ETL to withstand long-term exposure to the applied electric field is essential. Since the ETL has a lower mobility and therefore lower conductivity than the HTL a larger voltage drop and hence a larger electric field drops across the ETL. The molecules in the ETL should not lose multiple electrons by field ionization. However, they need to permit the flow of one electron at a time and reversibly change charge state by a single electron charge as the electron enters and leaves a given molecule.

Finally, the ETL material must be able to be processed and coated with high interface stability and with sufficient layer uniformity and quality.

The most common and most successful ETL is Alq_3 as shown in Figure 8.11. This is an example of a *metal chelate* material. An Al^{3+} ion at the center of the molecule is surrounded by three side-groups called *quinolines*. Alq_3 has a glass transition temperature of over 172°C and an electron mobility of 1.4×10^{-6} cm^{2}V^{-1}s^{-1}. The LUMO level is -3.0 eV below the vacuum level, which is a good match to the cathode workfunction of 3.6–3.8 eV in the case of LiF/Al.

The concern regarding the low mobility of Alq_3 is lessened since Alq_3 also functions well as a LEM. This means that an oversupply of holes can penetrate a relatively thick ETL and recombine radiatively within this combined LEM/ETL. The emission wavelength in this material has been extensively studied and various substitutions may be made to modify the emission characteristics while retaining the electron transport properties. Examples of substitutions will be discussed in the context of light emitting materials discussed in the next section.

Another important class of ETL materials is the group of *oxadiazoles*. In Figure 8.11 the molecule PBD is an example of an oxadiazole having a LUMO level of -2.16 eV, which permits a high device efficiency. Unfortunately the materials have low glass transition temperatures of about 60°C, although this can be increased by making larger molecules that resemble groups of two or four PBD molecules connected to each other forming a new molecule with a linear or a star shape, respectively. The most serious difficulty associated with the use of oxadiazoles, however, is the tendency of the excited states of the molecules to be unstable resulting in short device lifetimes.

Other potential ETL materials include various molecules containing double-bonded C=N groups, which are known as *imines*. These include TPBI, ATZL, and TPQ, as shown in Figure 8.20. TPBI has an electron mobility in the range of 10^{-6} to

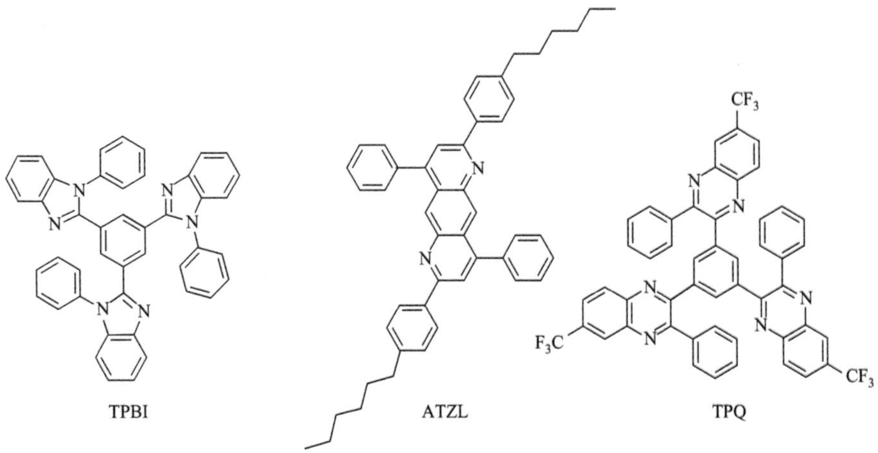

FIGURE 8.20   TPBI, ATZL, and TPQ are members of imine-based molecules which are candidate electron transport layer (ETL) materials as well as light emission materials. Other candidate ETL materials include C_{60}. See Section 8.16. Chemical Structure reproduced from Organic light-emitting materials and devices, ed. by Z. Li and H. Meng 9781574445749 (2007) Taylor and Francis.

10^{-5} cm^{2}V^{-1}s^{-1}, which is slightly higher than Alq$_3$. It has a LUMO level of -2.7 eV. TPQ has even higher mobility of 10^{-4} cm^{2}V^{-1}s^{-1} and good thermal stability.

8.11   LIGHT EMITTING MATERIAL PROCESSES

A key material for successful OLED operation, the LEM must be amenable to a highly controllable deposition technique such as vacuum deposition. It also requires the capability to transport both holes and electrons to enable the recombination of these carriers. Moreover, it must effectively allow for the creation of excitons and their decay to generate photons, it must remain stable at the electric fields needed to transport the holes and electrons, and the migration of molecules must be minimized for device stability.

In OLED operation, electrons injected from the cathode and holes injected from the anode combine to form molecular excitons, which were discussed in Chapter 4. The spin of the electron and hole are generally random because there is no spin preference when electrically generated electrons and holes form excitons. This means that the exciton population will occur with a 25% chance of being a singlet exciton and a 75% chance of being a triplet exciton. The singlet exciton can recombine by a dipole emission process in which case fluorescence will occur; however, the triplet exciton is spin-forbidden to emit radiation. The maximum quantum yield is therefore expected to be 25%.

The emission color of the OLED is ultimately determined by the LEM, and in many cases molecules are modified by changing side-groups, for example, to achieve a specific desired emission color. The molecules used of the LEM must then both transport carriers and fluoresce efficiently.

It is possible for mixtures of two or more molecules to be used within the LEM for the layer to provide the various required functions. In such mixtures a solid solution of the component molecular materials is usually preferred and segregation of the components is avoided. This is commonly referred to as *molecular doping* of one molecular material by another molecular material.

A common process used in the LEM layer is the *host-guest* energy transfer process which relies on molecular doping. Here, a wide bandgap host molecule is excited due to the electrons and holes arriving at the LEM and it can efficiently transfer its energy to a lower bandgap fluorescent guest molecule. Typically guest molecules are strongly diluted and comprise in the range of 0.1% to 10% of the LEM.

Since host molecules dominate the LEM host-guest system, most of the excitons form on host molecules. Host singlet excitons generally transfer to guest molecules to create guest singlet excitons through Förster energy transfer. Host triplet excitons generally transfer to become guest triplet excitons through Dexter energy transfer which does not rely on a dipole interaction. See Section 4.7. There are advantages to this energy transfer process which are as follows:

(a) The transfer process can create emission at wavelengths needed for a variety of emission colors without changing the host material. By this means, the host can define the energy levels that optimize the device band structure and charge transport while the guest molecules allow independent selection of emission wavelengths with high quantum efficiency.

(b) By limiting the concentration of guest molecules, energy transfer between guest molecules can be minimized. This can enhance quantum yield since coupling between adjacent fluorescent molecules can give rise to undesired inter-molecular interactions forming a radiationless energy-loss mechanism or *channel*, a process known as *aggregation-quenching*.[2]

(c) A range of specialized guest molecules have been developed that are able to overcome the 25% quantum yield limit to approach a quantum yield of closer to 100%. These guest molecules enable *triplet harvesting* and are further discussed in Section 8.14.

Host-guest transfer is illustrated in Figure 8.21. S_0^H and S_0^G refer to the ground singlet states of the host and guest respectively. S_1^H and S_1^G refer to excited singlet states of the host and guest respectively. T_1^H and T_1^G refer to the excited triplet states of the host and guest respectively. When a host molecule is excited to its S_1^H level the exciton may radiate and emit a photon by a dipole-allowed process and fluoresce. Alternatively the molecule could lose energy by collisions and the excited electron may return to the ground state and transfer its energy to heat, which is a non-radiative process. In addition the molecule may transfer energy to another molecule by Förster, Dexter, or radiative energy transfer, as shown. Unwanted energy transfer within either host or guest molecule through *intersystem crossing* from singlet to triplet excited states can occur but is normally forbidden in fluorescent molecules due to the requirement for angular momentum conservation. Once transferred, the energy can be radiated from

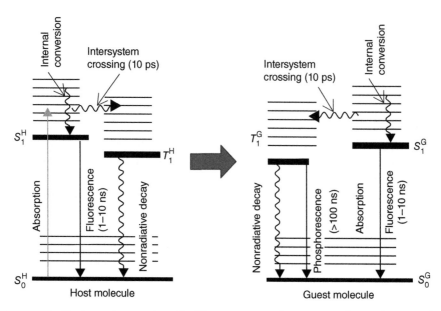

FIGURE 8.21　Host-guest energy transfer. The energy transfer can occur due to three possible processes, which may be Förster, Dexter, or radiative energy transfer. After Li, Z., Meng, H., Organic Light-Emitting Materials and Devices, 157444-574X, Taylor & Francis, 2006.

the guest molecule by a dipole-allowed fluorescence process, or it may radiate by phosphorescence. Phosphorescent guest molecules which rely on intersystem crossing are described in Section 8.14.

The transfer of energy from host to guest molecules may be a virtually complete transfer or it may involve only a fraction of the excited host molecules. As a result the measured emission spectrum may contain features characteristic of both host and guest emission. This is not desirable when saturated color coordinates are required for red, green and blue emission in full color OLED displays. In these cases full host-guest transfer is required. In contrast to this, white emitting OLED devices currently used for for lighting and television applications can benefit from a combined guest-host emission process which leads to a broader emission spectrum to approximate a desired white spectrum. An important requirement for efficient energy transfer is that the host exciton energy be higher than the guest exciton energy due to the Stokes shift. This difference in wavelength between absorption and emission peaks is detailed in Chapter 7 and is illustrated in Figure 7.7.

8.12　HOST MATERIALS

Suitable host materials must exhibit good electron and/or hole conduction to ensure the recombination of charge carriers and the effective formation of excitons. Their LUMO and HOMO levels must suit the guest molecules: Compared to the guest

LUMO level the host LUMO level should lie less deep in energy (closer to the vacuum level). Compared to the guest HOMO level the host HOMO level should lie deeper (further from the vacuum level) to ensure effective energy transfer. The host and guest molecules must exhibit good miscibility to maintain a stable solution without the tendency for precipitation, which will decrease energy transfer efficiency. Finally energy transfer processes should occur rapidly.

Among the simplest LEM hosts is the single molecule Alq$_3$, which also functions as an ETL material. Alq$_3$ emits in a band centered near 560 nm, which results a yellow-green emission color. Since the emission spectrum is not an appropriate red, green, or blue color for a full-color display it can be combined with other emission spectra to yield white emission for white-emitting OLED lamps. It can also be optically filtered to achieve red or green emission. Filtering does reduce efficiency, however, as unwanted parts of the emission spectrum are absorbed by the filter. The LUMO and HOMO levels of Alq$_3$ are −3 eV and −5.7 eV respectively, and its triplet energy is 2 eV. One modification of Alq$_3$ is BAlq, which has a LUMO level of −3 eV and a HOMO level of −5.9 eV. BAlq has a slightly higher triplet energy of approximately 2.2 eV compared to Alq$_3$, which makes it an ideal host for some red-emitting phosphorescent guest molecules, which will be described in the next section. The relevant LUMO and HOMO energy levels of Alq$_3$, BAlq as well as two other hosts that favor electron transport are shown in Figure 8.22.

In order to provide hole transport within the LEM, hole transport hosts are often combined with these electron transport hosts. A number of effective hole transport hosts are known. They include CPB, with HOMO and LUMO levels of −3 eV and −6 eV respectively, and a triplet level of 2.67 eV. CPB works well for red, yellow, and green triplet emitters, but the triplet level is not high enough for blue emission. For blue emission, CDBP has been shown to work well, with HOMO and LUMO levels of −3 eV and −7.3 eV, and triplet level of 3.0 eV. Both CPB and CDBP are shown in Figure 8.23.

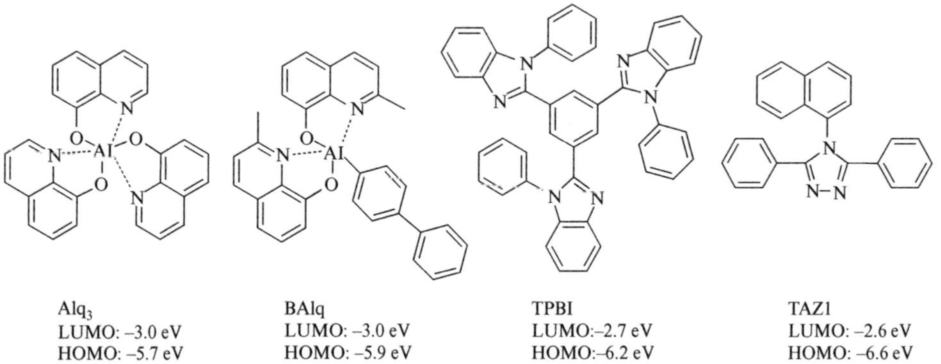

Alq$_3$
LUMO: −3.0 eV
HOMO: −5.7 eV

BAlq
LUMO: −3.0 eV
HOMO: −5.9 eV

TPBI
LUMO: −2.7 eV
HOMO: −6.2 eV

TAZ1
LUMO: −2.6 eV
HOMO: −6.6 eV

FIGURE 8.22 Electron transport hosts Alq$_3$, BAlq, TPBI, and TAZ1. Chemical Structure reproduced from Organic light-emitting materials and devices, ed. by Z. Li and H. Meng 9781574445749 (2007) Taylor and Francis.

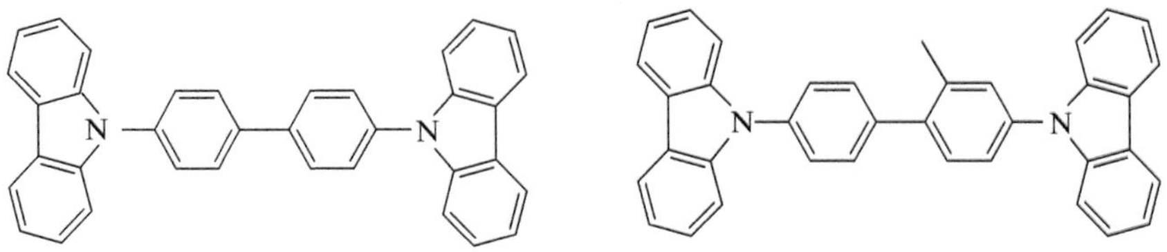

CBP 110
LUMO: –3.0 eV
HOMO: –6.3 eV
*T1: 2.67 eV

CDBP 111
LUMO: –3.0 eV
HOMO: –6.3 eV
*T1: 3.0 eV

FIGURE 8.23   Hole transport hosts CBP and CDBP. Chemical Structure reproduced from Organic light-emitting materials and devices, ed. by Z. Li and H. Meng 9781574445749 (2007) Taylor and Francis.

8.13 FLUORESCENT DOPANTS

The requirements for full-color display applications of OLEDs include red, green, and blue emitters with color coordinates close to the following values: for green emitters, $x = 0.3$ and $y = 0.6$; for red $x = 0.62$ and $y = 0.37$; for blue $x = 0.14$ and $y = 0.10$. Fluorescent dopants emitting with approximately these color coordinates are required.

An example of a green dopant is based on a *coumarin* dye molecule such as C-545TB, shown in Figure 8.24. This dopant yields saturated green emission with color coordinates $x = 0.3$, $y = 0.64$, a luminescent efficiency of 12.9 cd A^{-1}, a power efficiency of 3.5 lm W^{-1} at 20 mA cm^{-2}, and a brightness of 2585 cd m^{-2}. Another type of green dopant is DMQA, which is an example of a *quinacridone* molecule. DMQA achieves a luminescent efficiency of 21.1 cd A^{-1} and a luminance of over 88,000 cd m^{-2}. Coumarin-based C-545TB and quinacridone-based DMQA are shown in Figure 8.24.

Red fluorescent dopants have been developed that exhibit satisfactory color coordinates with good stability and efficiency based on the *arylidene* family of molecules. An example of a red fluorescent molecule is DCJPP, shown in Figure 8.25. There are large numbers of other candidate red fluorescent materials in this family; however, the less suitable ones suffer from a tendency to undergo unwanted chemical reactions. Some fluoresce with *y*-values of their color coordinates that are too large and orange-red emission results. Still others exhibit deep-red emission but have low quantum efficiencies.

Red fluorescent molecules based on other molecular families exist. For example, an *isophorone*-based red emitter, DCDDC (see Figure 8.25), has been used as a red emitter in OLEDs when dissolved in the host Alq_3. Emission from both Alq_3 and DCDDC is observed for small concentrations of dopant; however, if the DCDDC doping level is increased to above 2% concentration, only the red emission is observed due to a strong host-to-guest energy transfer. The emission peak is 630 nm from the DCDDC. At a 1% DCDDC concentration, which does somewhat compromise the red color, a peak luminance of 5600 cd m^{-2} at a voltage of 15 V with maximum efficiency of 1.6 lm W^{-1} is achieved.

There has been a great deal of effort invested in blue fluorescent molecules and suitable hosts. The challenge is that the short wavelength of emission in the range of

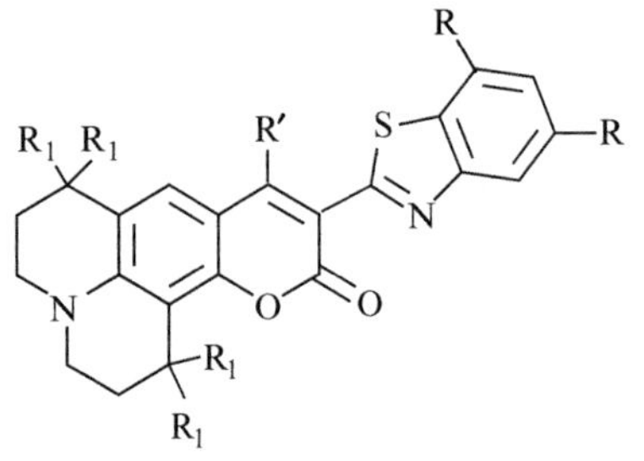

C-545TB ($R_1 = CH_3$, R = t-butyl; R′ = H)

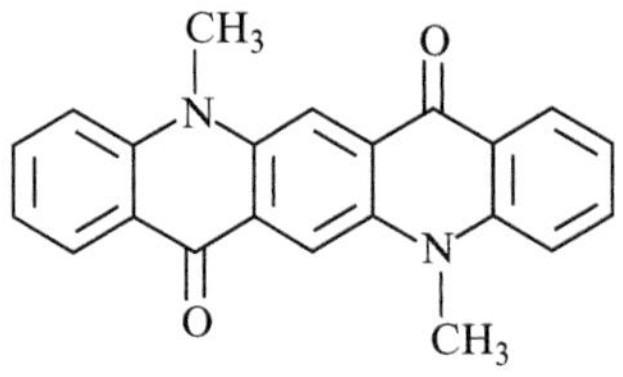

DMQA

FIGURE 8.24 Coumarin-based green fluorescent dopant C-545TB and quinacridone-based dopant DMQA. Chemical Structure reproduced from Organic light-emitting materials and devices, ed. by Z. Li and H. Meng 9781574445749 (2007) Taylor and Francis.

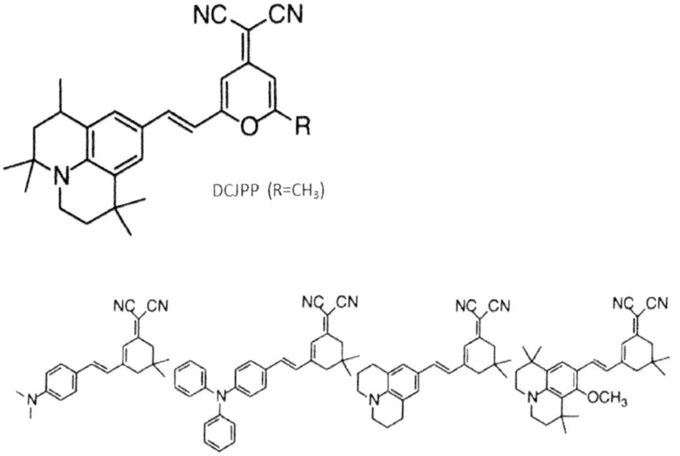

DCJPP ($R=CH_3$)

DCDDC1 DCDDC2 DCDDC3 DCDDC4

FIGURE 8.25 The red fluorescent molecule DCJPP derived from the arylidene family of molecules and four variations of red fluorescent molecule DCDDC derived from the isophorone family of molecules. Chemical Structure reproduced from Organic light-emitting materials and devices, ed. by Z. Li and H. Meng 9781574445749 (2007) Taylor and Francis.

450 nm required for blue color coordinates with y-values near 0.1 calls for high gaps between guest LUMO and HOMO levels and even higher gaps near 3 eV between suitable host LUMO and HOMO levels. Molecules with these properties exist but the resulting OLEDs have proven less stable than red and green emitters.

An example of a blue emitter host that is an arylene derivative is DPVBI, having HOMO and LUMO levels of −5.9 eV and −2.8 eV respectively. A suitable fluorescent guest that is also an arylene derivative is BCzVBI, with HOMO and LUMO levels of −5.4 eV and −2.42 eV. Luminance centered at 468 nm at 10,000 cd m^{-2} at an efficiency of 0.7–0.8 lm W^{-1} has been achieved (see Figure 8.26).

Also shown in Figure 8.26 is a candidate blue system from the *anthracene* family. The host is JBEM with HOMO and LUMO levels −5.8 eV and −2.8 eV respectively,

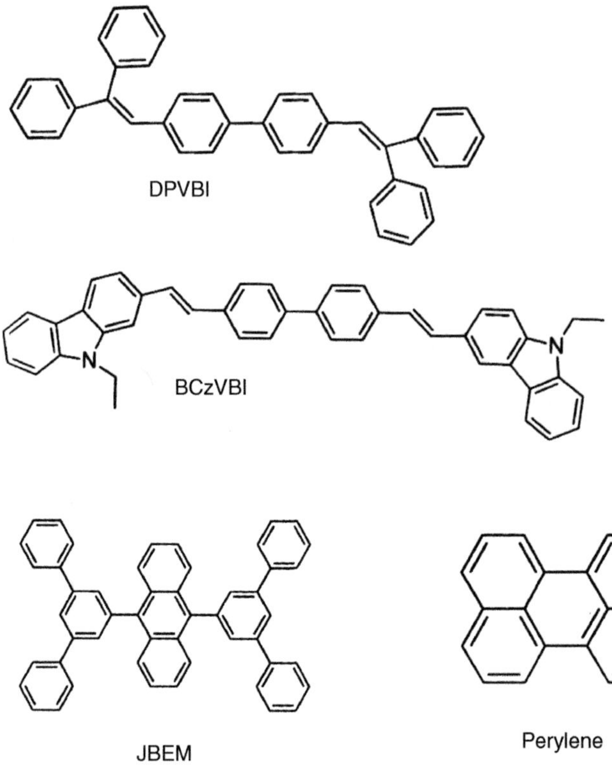

FIGURE 8.26 Arylene family host DPVBI and dopant BCzVBI. Also shown are anthracene family host JBEM and dopant perylene. Chemical Structure reproduced from Organic light-emitting materials and devices, ed. by Z. Li and H. Meng 9781574445749 (2007) Taylor and Francis.

and the guest is the well-known *perylene* molecule with HOMO and LUMO levels −5.3 eV and −2.5 eV respectively. Resulting OLED performance achieves 400 cd m^{-2} at a current density of 20 mA cm^{-2}, a maximum efficiency of 1.45 lm W^{-1} and color coordinates of $x = 0.24$ and $y = 0.21$. A half-life (life to half initial luminance) starting at 100 cd m^{-2} brightness of over 1000 hours can be obtained. Improvements to lower the y-component of the color coordinate and achieve pure blue emission can be realized by using other anthracene derivatives.

8.14 PHOSPHORESCENT AND THERMALLY ACTIVATED DELAYED FLUORESCENCE DOPANTS

In order to be competitive in commercial OLED devices, guest molecules in the LEM achieving quantum yields well beyond 25% are essential. The most successful approach to exceed a 25% quantum yield is to harvest the 75% triplet state exciton population using guest molecules known as *phosphorescent emitters*.

A heavy transition metal having an unfilled inner shell and hence non-zero angular momentum may be included as a core of an organic molecule. Such *organometallic* molecules allow for the transfer of angular momentum between the transition metal and the triplet states such that dipole radiation between triplet excitons and the singlet ground state can occur. High-efficiency phosphorescence can result, in which the 25% quantum yield limit associated with the 1:3 singlet-triplet population ratio can be overcome.

Among the highest efficiency phosphorescent emitters are the iridium organometallic complexes. They have a short triplet lifetime of 1–100 μs, which means that radiative recombination is assisted by triplet spin-orbit interaction with iridium in the molecule. Iridium, a transition metal with an unfilled inner shell and a net angular momentum, provides the needed spin-orbit interaction. Platinum organometallic molecules have also been developed that provide a pathway for angular momentum transfer, again through spin-orbit interactions. Both iridium and platinum are high atomic mass transition metals. Spin-orbit transfer increases as the transition metal atomic number increases.

The angular momentum transfer within organometallic phosphorescent molecules enables spin conservation from the triplet state to the ground singlet state, but not surprisingly, it simultaneously permits rapid energy transfer between the excited singlet state and the triplet state. In Figure 8.21 this transfer is shown as an intersystem crossing. The consequence is that virtually all singlet state fluorescence is replaced with intersystem energy transfer to the excited triplet state followed by phosphorescence. The observed emission spectra from guest phosphorescent molecules are therefore consistent with the energy gap between the excited triplet state T_1^G and the ground singlet state S_0^G. See Figure 8.27.

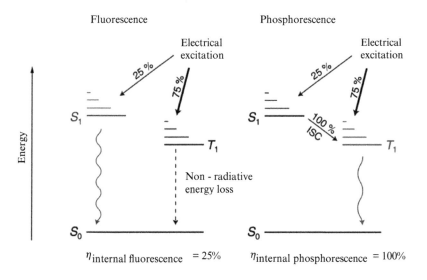

FIGURE 8.27 Scheme illustrating internal quantum efficiencies that can be achieved under electroluminescence based on conventional fluorescent emitters (left) and organometallic phosphorescent emitters (right). With 75% triplet exciton generation, the internal efficiency of conventional fluorescence in OLEDs is limited to about 25%. On the right side ISC (intersystem crossing) enables up to 100% quantum yield. Reineke et al., 2013 / With permission from American Physical Society.

Due to strong demand for more efficient battery-powered electronic devices, the range of well-studied phosphorescent dopant molecules continues to expand and they are heavily used in production OLED devices for cell phones and tablets. Of key interest are phosphorescent molecules providing emission of red, green, and blue for full color displays. Figure 8.28 shows a range of such molecules. Phosphorescent molecules having red and green emission are well developed and provide sufficiently

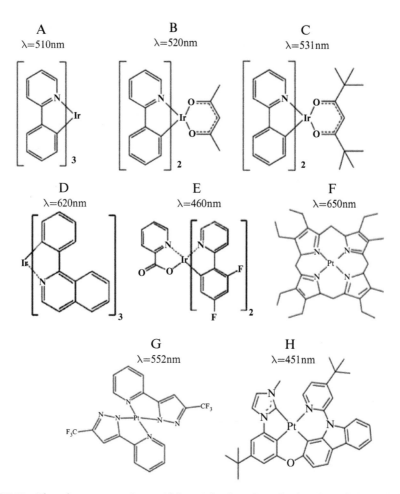

FIGURE 8.28   Phosphorescent emitters: A) fac tris(2-phenylpyridine) iridium (Ir(ppy)₃) B) bis(2-phenylpyridine) iridium(III) acetylacetonate (Ir(ppy)₂(acac)), C) bis(2-3,5-dimethylphenyl)-1 methylpyridine) iridium(III) (2,2,6,6-tetramethylheptane-3,5-diketonate) (Ir(3′,5′,4-mppy)₂(tmd)), D) tris(1-phenylisoquinolinato-C2,N)iridium(III) (Ir(piq)₃), E) bis[(4,6-difluorophenyl)-pyridinato-N,C2′] iridium(III) picolinate (FIrpic), F) (2,3,7,8,12,13,17,18-octaethyl-21H,23H-porphine) platinum(II) (PtOEP), G) bis[3-trifluoromethyl-5-(2-pyridyl)-1,2-pyrazolato] platinum(II) (Pt-A), H) [6-(1,3-dihydro-3-methyl-2H-imidazol-2-ylidene-κC²)-4-tert-butyl-1,2-phenylene-κC¹]oxy[9-(4-tert-butyltpyridin-2-yl-κN)-9H-carbazole-1,2-diyl-κC¹] platinum(II) (PtON7-dtb). With permission from Aziz H. (Kitai A.H. editor) Materials for Solid State Lighting and Displays, Wiley 2017.

stable performance in OLED displays for portable electronics; however, stability in blue emitting phosphorescent molecules is generally not sufficient and therefore blue emission generally relies on fluorescent guest molecules having a quantum yield of no more than 25%. In blue phosphorescent molecules the blue emission band arises from a singlet exciton that must extend into the UV or near-UV wavelength range before transferring by an intersystem crossing to the triplet state. Molecules providing these short wavelength transitions tend to be less stable.

An interesting alternative approach to phosphorescence that can also overcome the 25% singlet quantum yield limit is called the thermally activated delayed fluorescence (TADF) process. In TADF, the energy splitting between the singlet and lower energy triplet state is reduced to below 0.1eV. This reduction is achieved by using molecules in which there is reduced spatial overlap of ground state and excited state electrons. See Section 4.6 in which electron-electron coulomb interaction is discussed as the origin of this energy splitting. At room temperature $kT = 26 \times 10^{-3}$eV and therefore the opportunity to thermally stimulate triplet excitons to the singlet level is enabled. This thermal stimulation process is shown in Figure 8.29.

Note that the TADF process relies on reverse intersystem crossing (RISC). Figure 8.29 (right) shows how RISC is favored when singlet-triplet energy splitting is small. However as the splitting becomes smaller the fluorescence channel in the molecule weakens. A balance between these competing effects maximizes TADF. TADF has an advantage over phosphorescence in that there is no UV or near-UV transition involved in blue emitters. TADF materials are not as well developed as phosphorescent materials. See Problem 8.9.

Efficient white OLEDs are becoming increasingly important for both display and lighting applications. The host-guest LEM must therefore be optimized for maximum system efficiency. In a clever approach to engineer an efficient white LEM, host

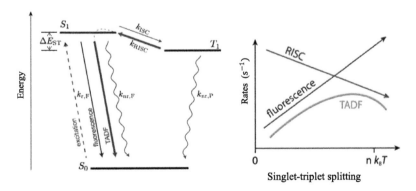

FIGURE 8.29   Left: Thermally-activated delayed fluorescence (TADF) emitter molecule. S_1 and T_1 denote the first excited singlet and triplet state, respectively. k_i represents various rates. ISC = intersystem crossing, RISC = reverse ISC, r = radiative, nr = non-radiative, F = fluorescence, P = phosphorescence. ΔE_{ST} indicates the singlet-triplet splitting. Right: Qualitative representation of how RISC and the fluorescence rate depend on the singlet-triplet splitting ΔE_{ST}. After Reineke S. (Kitai A.H. editor) Materials for Solid State Lighting and Displays, Wiley 2017.

molecule triplet excitons transfer their energy by Dexter energy transfer to guest molecules providing green and red emission. These guest molecules use phosphorescence and possibly TADF processes to maximize red and green quantum yield. Simultaneously, host singlet excitons either provide blue fluorescence directly or else they are harvested by energy transfer to excite singlet excitons in blue-emitting guest molecules yielding efficient blue fluorescence. In this way a blue phosphorescent dopant is not necessary. Luminous efficiencies of over 100 lm/W for white light are achievable. Due to the competitive nature of this industry, details of the LEM are often proprietary.

Red, green, and blue emission from white-emitting OLED material can be achieved using color filters. Color filtering is particularly effective when the white emitting LEM can produce well-defined red, green, and blue spectral emission ranges. This approach has enabled successful OLED television displays. The use of a single white LEM simplifies display manufacturing since patterning of red, green, and blue emitters is avoided. In addition the simultaneous achievement of long OLED lifetime and high quantum yield is facilitated since phosphorescence is only needed for green and red emission. In contrast to this, a dedicated blue LEM is normally limited to a quantum yield of 25% since triplet harvesting for blue emission is difficult to achieve in materials with sufficient stability for commercial applications.

The direct excitation of quantum dots by electrical means (*quantum dot electroluminescence*) is being intensively investigated. Device structure is similar to small molecule OLED devices except that the LEM consists of quantum dots. Achieving efficient and balanced electron and hole transfer to the quantum dots is challenging. See Problem 8.15.

8.15   ORGANIC SOLAR CELLS

The absorption of sunlight in a molecular organic semiconductor results in the formation of molecular excitons, and in accordance with the dipole oscillator and the discussion in Sections 4.6 and 4.7 the ground-state singlet is excited into an excited singlet molecular exciton. This exciton is localized to a single molecule, which, for small molecules, is on the nanometer length scale. Unless the exciton can be dissociated and its hole and electron extracted no current can result. In contrast to this, photon absorption in inorganic semiconductors used in solar cells results in separated holes and electrons that are free to flow independently of each other and hence directly contribute to current flow.

A key challenge in the development of organic solar cells is to overcome the exciton binding energy of optically generated holes and electrons. Once dissociated, charges can flow from molecule to molecule by a hopping process. Materials and device architectures designed to facilitate exciton dissociation are the key to successful organic solar cells.

The simplest organic solar cell structure is the single-layer device shown in Figure 8.30. Photons create molecular excitons in the organic semiconductor layer. Only the excitons at or near the anode-semiconductor or the cathode-semiconductor junctions have an opportunity to be dissociated. The dissociation is enabled by the electric field resulting from steps in the LUMO and HOMO levels at these interfaces. Only a small fraction of the light absorbing molecules exist at the interfaces. The majority of the generated excitons are never dissociated. A limited number of excitons up to about 10 nm away from these interfaces may diffuse by Förster or Dexter processes to reach the molecules at the interfaces. The energy level diagram for this structure is shown in Figure 8.31. A portion of the successfully dissociated carriers may immediately recombine because electrodes are not specifically optimized as an ETL or HTL.

An increase in the collection of charge may be accomplished by the *planar heterojunction* solar cell, which is shown in Figure 8.32. Its energy band diagram is shown in Figure 8.33. The introduction of an electron transport *acceptor* layer (ETL) and a

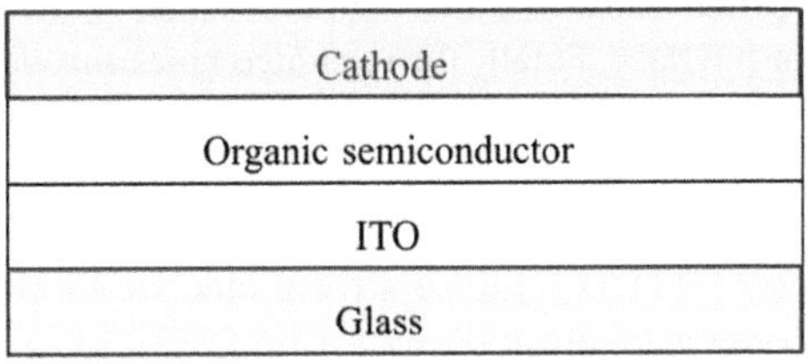

Cathode
Organic semiconductor
ITO
Glass

FIGURE 8.30 Single-layer organic solar cell consisting of a single organic semiconductor layer, a low workfunction cathode and a transparent anode. Device efficiency is well below 1%.

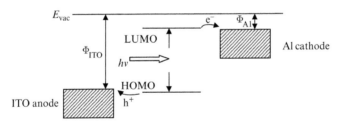

FIGURE 8.31 Energy level diagram for single-layer organic solar cell. The absorption of light creates excitons through the promotion of molecular electrons from the HOMO level to the LUMO level. In reality, exciton dissociation is not possible except for those excitons that are located at or near an electrode interface.

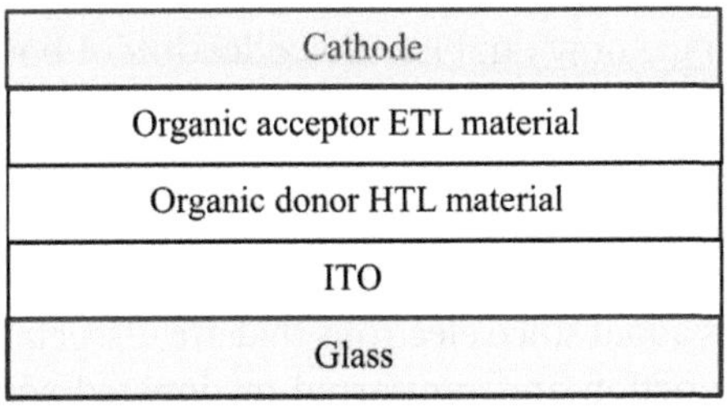

Cathode
Organic acceptor ETL material
Organic donor HTL material
ITO
Glass

FIGURE 8.32 Organic planar heterojunction solar cell structure showing donor and acceptor organic layers.

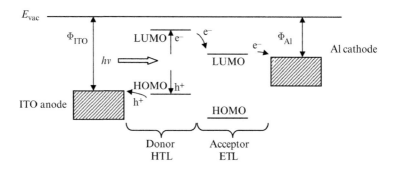

FIGURE 8.33 Heterojunction solar cell showing donor and acceptor LUMO and HOMO levels. Excitons are generated throughout the donor layer and these excitons are dissociated when they diffuse to the donor-acceptor interface. Finally the separated holes and electrons can drift to their respective electrodes.

hole transport *donor* layer (HTL) creates a strong electric field at the heterojunction interface that greatly enhances exciton dissociation there. In principle both the donor and acceptor layers can absorb photons and become populated with excitons. These excitons can then diffuse toward the heterojunction interface and dissociate there.

The sharp and narrow high field region at the heterojunction interface in organic solar cells may be contrasted with the inorganic p-n junction. The width of the depletion region in inorganic junctions is determined by the spatial extent over which carriers recombine to establish an equilibrium condition. At the organic interface in organic junctions, charge carriers in the HOMO and LUMO levels transfer from molecule to molecule by hopping and only minimal charge transfer occurs, leaving the equivalent of a depletion region of very small thickness. The potential difference between the HOMO and LUMO levels falls across a very small spatial range of dimension in the nanometer scale giving rise to a high electric field at the junction.

In practice, the donor layer is specifically designed to absorb photons and it therefore becomes populated with excitons. Since the mobility of holes is relatively higher than electrons, as we saw in Section 8.10, the donor HTL allows the holes to diffuse toward the heterojunction interface. The electrons will remain bound to the holes since exciton dissociation will not readily take place until the excitons reach the interface. This interface now enables the collection of both electrons and holes, which drift across their respective layers: Holes reach the ITO electrode through the hole-conducting layer, and electrons reach the cathode through the electron-conducting layer.

The terminology "donor" and "acceptor" used to describe the two layers forming the heterojunction comes about since electrons that are dissociated from the excitons in the donor layer at the junction are transferred or donated across this junction from donor molecules and are accepted by acceptor molecules in the acceptor layer. The holes from the dissociated excitons remain in the donor layer and drift to the anode. The terminology is a molecular analogue of the terms "donor" and "acceptor" applied to dopants used in inorganic semiconductors; however, the organic molecules donate and accept electrons to/from neighboring molecules rather than to/from energy bands. Since the acceptor layer becomes populated with electrons this layer needs to be an electron conductor and charge is carried in its LUMO level. Conversely the donor layer, being populated with holes, needs to be a hole conductor and these holes are carried in its HOMO level.

The thickness of the p-type donor layer is controlled by the diffusion length of the excitons which must reach the interface to be dissociated. A donor layer that is too thick will lower efficiency since a significant fraction of the generated excitons will recombine before they can reach the interface. A layer that is too thin will result in less absorption of light. A solar cell efficiency of only a few percent is achievable with the planar heterojunction design.

Since exciton diffusion lengths in organic materials are approximately 10 nm, the useful absorption depth in the donor layer is only about 10 nm, which means that incomplete absorption of sunlight limits the performance of the heterojunction solar cell. The thickness for virtually complete absorption of sunlight is closer to 100 nm in organic materials; however, a donor layer of this thickness would result in poor efficiency and most generated excitons would recombine without reaching the interface.

A successful approach to improving performance further is to arrange several interfaces within the light path of the incoming sunlight, and to make each donor layer thin enough to allow effective exciton diffusion to the nearest heterojunction interface. A portion of sunlight is absorbed in each thin donor layer and the remaining light can then continue to a subsequent layer. This approach relies on a *bulk heterojunction* layer that incorporates multiple donor and acceptor regions. The device structure is as shown in Figure 8.34.

The bulk heterojunction layer can be formed using a variety of nanostructures, and the development of techniques and materials for the achievement of these nanostructures has been a focal point in further improving organic solar cells. Bulk heterojunction solar cells have attained the highest efficiency levels available in organic solar cells that use a single organic absorption band to absorb light.

The length scale of the desired nanostructures is in the nanometer range, and it is very desirable to use *self-organization* of the organic materials to achieve a low-cost

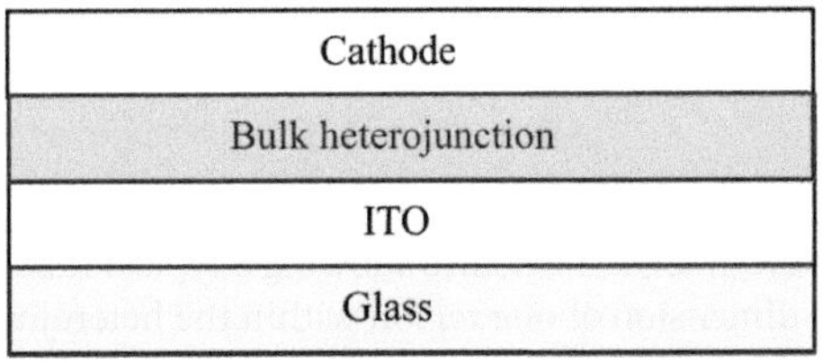

| Cathode |
| Bulk heterojunction |
| ITO |
| Glass |

FIGURE 8.34 Bulk heterojunction organic solar cell. A number of small (~10 nm) donor regions are organized within the bulk heterojunction layer and optimized to absorb sunlight and allow exciton diffusion to a nearby junction.

method to create these nanostructures. For example, the donor and acceptor organic materials can be mixed together and then deposited onto the solar cell substrate. If the mixed material segregates spontaneously under suitable conditions to form the desired bulk heterojunction then self-organization has been achieved. This dramatically lowers the cost of processing since submicron lithography and patterning techniques are avoided. Although these techniques are well known and highly developed for inorganic semiconductor device processing they are not cost effective for large-area solar cells.

An important requirement of bulk heterojunctions is to provide for the effective conduction of current away from the donor and acceptor regions and for the collection of this current by the electrodes. Two specific examples of heterostructures are shown in Figure 8.35. Figure 8.35a shows a morphology that limits the effectiveness of current collection because the donor and acceptor layers are not well connected to the electrodes. This is not an accurate representation of the morphology in real systems, but serves for illustration purposes. Figure 8.35b shows a more desirable structure since

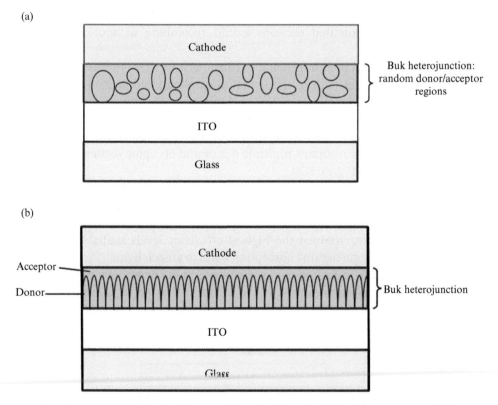

FIGURE 8.35 (a) Bulk heterojunction structure showing a typical random structure of donor and acceptor materials. The dimension of one region within the heterojunction is about 10 nm. The problem is the connectivity of these regions to their appropriate contact materials. (b) Bulk heterojunction of vertically oriented stripes of donor and acceptor materials that enables the donor material to be in contact with the ITO electrode and the acceptor layer to be in contact with the cathode. The acceptor layer could be made using vertically oriented carbon nanotubes.

the donor and acceptor materials are arranged to allow for effective connection to the electrodes. The achievement of organic layers that self-organize into optimal structures at low cost is an area of ongoing research.

Another important area of development is the use of multilayer organic solar cells in which various layers act to absorb different portions of the solar spectrum in a manner analogous to the inorganic multiple junction solar cells described in Section 5.6. This is particularly important for organic solar cells since the absorption bandwidth of a given organic material is small. The π and π^* bands are much narrower than the conduction and valence bands in inorganic semiconductors, which limits the absorption bandwidth.

The heterojunction structures we have discussed are capable of providing two absorption bands if excitons can be generated and harvested in both the donor and acceptor materials. The energy gaps of these two layers can be different and two absorption bands can be realized. A challenge associated with this is achieving high enough diffusion lengths in both the donor and acceptor layers and collecting carriers effectively. A single organic material having all the attributes needed specifically for an ideal acceptor, including good electron mobility, high optical absorption, and effective electron capture from the donor material, has not yet been found. Nevertheless numerous organic material blends and mixtures are being investigated to obtain multiple absorption bands in organic solar cell structures.

Efficiencies as high as 16% have been reached in the laboratory for bulk heterojunction organic solar cells. The attributes of low cost, low weight, and flexibility are key drivers behind this development.

8.16   ORGANIC SOLAR CELL MATERIALS

A common *thiophene*-based donor material is poly(3-hexylthiophene), or P3HT. It is soluble in several organic solvents, which makes it compatible with low-cost solution processing. Another donor material is poly(3,3′″ -didodecyl quaterthiophene), or PQT-12. Their molecular structures are shown in Figure 8.36. The optical absorption spectra of these compounds (Figure 8.37) show that their absorption bands are limited to the green and red parts of the solar spectrum.

Acceptor materials must provide good electron conductivity but optical absorption is not desired. A popular material is a C_{60} derivative. C_{60} is a fullerene, or a molecule composed entirely of carbon. The graphene-like surface of C_{60} allows effective electron transport through the delocalized electrons in the fullerene molecular orbitals, which result in the fullerene LUMO level. C_{60} as well as its derivative [6,6]-phenyl-C_{61} -butyric acid methyl ester (PCBM) are shown in Figure 8.38. The modification allows the fullerene derivative PCBM to be solution-processed, which reduces manufacturing costs, whereas C_{60} must be vacuum deposited.

Fullerenes are excellent acceptors since they have a LUMO level with electron energy well below the vacuum level and more specifically somewhat below the LUMO levels of a variety of donor molecules as required for solar cells. This is a requirement

(a)   (b)

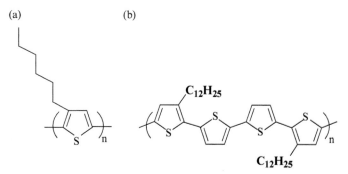

FIGURE 8.36   Molecular structures of a) poly(3-hexylthiophene) (or P3HT) and b) PQT-12. Reprinted with permission from Organic Electronics, Efficient bulk heterojunction solar cells from regio-regular-poly(3,3‴-didodecyl quaterthiophene)/PC70BM blends by P. Vemulamada, G. Hao, T. Kietzke and A. Sellinger, 9, 5, 661–666 Copyright (2008) Elsevier.

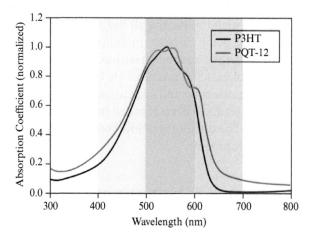

FIGURE 8.37   Absorption spectra of P3HT and PQT-12. After Organic Electronics, Efficient bulk heterojunction solar cells from regio-regular-poly(3,3‴-didodecyl quaterthiophene)/PC70BM blends. Adapted from Vemulamada et al, 2008.

(a)   (b)

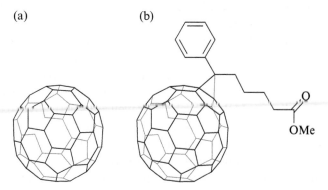

FIGURE 8.38   Molecular structures of the fullerene C_{60} and its derivative [6,6]-phenyl-C_{61}-butyric acid methyl ester (PCBM). Courtesy of Dr. J.H.G Owen and the Oxford University QIP-IRC.

of an acceptor molecule, as illustrated in Figure 8.33. Each C_{60} actually has the ability to accept multiple electrons due to the number of available vacancies in the LUMO level. Electron transport within a single C_{60} molecule is very fast and efficient. The electrons are metastable within the molecules and are therefore readily transferred to the solar cell electrode. Finally these fullerenes have an absorption spectrum that peaks in the ultraviolet part of the spectrum, which means that they exhibit only minimal absorption in the important parts of the solar spectrum, leaving the donor layer free to absorb the visible or infrared solar radiation.

The *carbon nanotube* is another acceptor material of considerable interest. A promising development is the use of carbon nanotubes as rod-like acceptors to form structures similar to that shown in Figure 8.35b. Nanotubes, being composed of rolled-up graphene sheets, have electronic properties that are similar to the fullerenes, allowing them to function effectively as acceptors. The achievement of bulk heterojunctions with oriented nanotubes is required. A carbon nanotube is shown in Figure 8.39.

An ongoing challenge is stability. Whereas silicon solar cells last for over 25 years, organic materials degrade rapidly in full sun conditions.

A more recent type of organic solar cell using perovskite materials has reached over 25% efficiency. Perovskite solar cells rely on organic molecules, but they function much more like inorganic solar cells. A single semiconductor material shown in Figure 8.40 is used to absorb sunlight, to generate electrons and holes, and to collect these carriers. The material is crystalline with Pb and I atoms but it also contains the organic molecule methylammonium (MA) at specific locations in the crystal lattice. The halide shown is iodine; however, other halides such as chlorine and bromine may be substituted for iodine.

The bandgap of the perovskite may be varied between 1.5 eV and 2.4 eV depending on the choice of the halide. There are two important defect types in this material comprising a lead atom vacancy that acts like an acceptor state near the valence band edge and a MA molecule interstitial which acts like a donor state near the conduction band. Depending on details of the preparation process used, the prevalence of either defect type can be controlled.

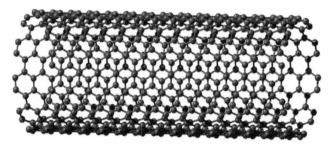

FIGURE 8.39   Carbon nanotube. Courtesy of Dr. J.H.G Owen and the Oxford University QIP-IRC.

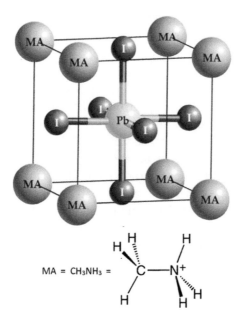

MA = CH₃NH₃ =

FIGURE 8.40 Perovskite solar cell material containing Pb, I, and MA (methylammonium molecule shown below) in a perovskite crystal lattice. Adapted from Yin et al., 2014.

Mobile charges in the perovskite material are predominantly present as free electrons and holes, rather than as bound excitons. Excitons in $CH_3NH_3PbI_3$ have a weak binding energy of less than 50 meV which means that most of them can be dissociated very rapidly into free carriers at room temperature. Unlike other organic semiconductor solar cells the MA molecule does not directly absorb light. Therefore no molecular exciton is formed and there is no need to have a junction to dissociate an exciton.

The electrons and holes produced in this material exhibit small effective masses resulting in high carrier mobilities ranging from approximately 25 cm^2 V^{-1} s^{-1} for electrons to 100 cm^2 V^{-1} s^{-1} for holes. These mobility values are much higher than the values of under 1 cm^2 V^{-1} s^{-1} achieved in typical organic semiconductors. A major feature of these perovskites is their direct gap band structure. This leads to a high efficiency thin film solar cell having long minority carrier diffusion lengths of 1μm or more. See Figure 8.41. Electron transport and hole transport layers are required to ensure suitable collection of electrons and holes from the perovskite. Due to the poor electron mobilities achievable in organic electron transport layers the inorganic semiconductor TiO_2 is used as the electron injection layer. A promising hole transport layer material is Spiro-MeOTAD, a small molecule organic, shown in Figure 8.42.

Another key aspect of these perovskites is their ease of synthesis. Low temperature solution processing on glass or polymer substrates is possible. Current challenges include long term perovskite stability and the reliance on lead which is toxic.

metal back contact
hole transport layer (Spiro-MeOTAD)
perovskite
electron transport layer (TiO₂)
transparent conductor (ITO)
glass or polymer substrate

FIGURE 8.41 Perovskite solar cell structure. Perovskite layer thickness is approximately 0.5 μm. TiO$_2$ is a suitable electron transport layer and Spiro-MeOTAD is an organic hole transport material.

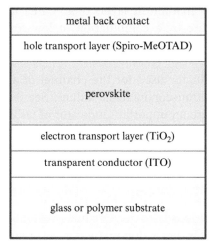

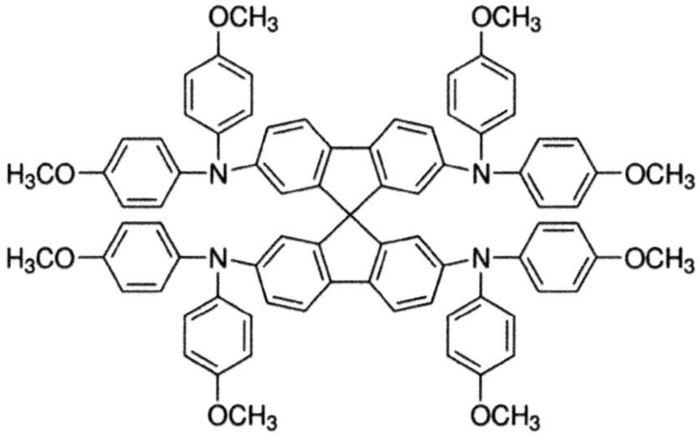

FIGURE 8.42 Spiro-MeOTAD hole transport layer used for perovskite solar cells.

Tandem solar cells may be formed using a perovskite cell as the upper cell and a crystalline silicon p-n junction solar cell as the lower cell. See Section 5.6. Such tandem cells have achieved a conversion efficiency of approximately 30%. Impediments to the commercialization of these potentially very cost effective tandem cells include, once again, the toxicity of lead and the degradation over time of the perovskite material.

8.17 THE ORGANIC FIELD EFFECT TRANSISTOR

A organic field effect transistor (OFET) may be achieved using organic materials that form structures that resemble the planar MOSFET of Chapter 6. The electrodes, the channel and the insulator may be realized using organic materials. One of the main technological attractions is that all the layers of an OFET can be deposited and patterned at room temperature by a combination of low-cost solution-processing or printing methods. This makes OFETs ideally suited for low-cost, large-area electronics on flexible substrates.

The organic semiconductor used for the channel of an OFET requires a high mobility to achieve good transconductance values. See Section 6.8. Therefore the achieved mobility values are an important indicator of OFET performance. Mobility values parallel to the plane of a deposited thin film are relevant to OFETs, whereas OLEDs and organic solar cells generally rely on electrical properties perpendicular to the thin film. OFET mobility trends for both polymer and small molecule organics are shown in Figure 8.33.

Organic small molecules can achieve high mobility values if they are formed with high crystallinity and a low defect density. Organic small molecule films can be grown by thermal evaporation. This deposition method gives organic small molecules great advantages in optimizing the morphology and uniformity of the thin film, thus suppressing unwanted device-to-device variations.

For p-type conductivity, pentacene is one of the most widely studied small molecules. See Problem 8.2. Other candidate materials include copper phthalocyanine CuPc shown in Figure 8.16. More recently rubrene and C8-BTBT have been developed as crystalline organic materials. See Figures 8.43 and 8.44.

Compared to p-type OFET channel materials, n-type organic conductors are less well developed due to the difficulty in stabilizing electrons in LUMO levels that are considerably higher in energy than carriers in the HOMO level. This higher electron energy causes n-type materials to be less chemically stable. These electrons are susceptible to being chemically bound by oxygen or hydroxides and encapsulation is necessary.

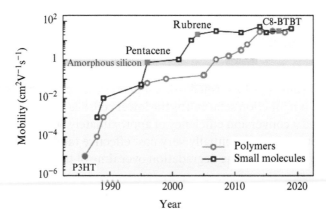

FIGURE 8.43 Since the 1980s mobility values achieved in polymer or small molecule thin films have increased by over six orders of magnitude. Mobility values now exceed the mobility of amorphous silicon, the latter being widely used for the thin film transistors in liquid crystal displays. Yan et al., 2021 / John Wiley & Sons.

(a) C8 BTBT

(b) Rubrene

FIGURE 8.44 a) C8-BTBT is a conducting polymer with the name [1] benzothieno[3,2-b] [1]-benzothiophene. It is a p-type air-stable semiconductor that can be deposited to form a thin film with high carrier mobility. Its exceptionally high carrier mobility of over $40\ cm^2V^{-1}s^{-1}$ is a result of spontaneous crystallization as well as the elongated structure of the molecule. b) Rubrene is also a p-type small molecule. Its crystal structures are rather complex and mobility values of up to $40\ cm^2V^{-1}s^{-1}$ are again achievable. Both molecules have simple shapes that promote crystallization.

Figure 8.45 shows a top-gate OFET device. It is closely related to the MOSFET structures of Chapter 6. A variety of alternative structures are possible. For example, the gate could be deposited before the channel and be situated underneath the channel. The source and drain contacts could be deposited above the channel. The insulator layer can be a polymer. Deposition methods that can be used for the required layers include thermal evaporation which works well for small molecule materials; however, solution coating techniques are also applicable to either small molecule or polymer materials provided the organic can be prepared in a solvent. Additional methods including dip-coating, spin-coating, inkjet printing, and screen printing are possible. See Problem 8.13.

Applications of OFETs include flexible displays and bio-electronics in the health care industry.

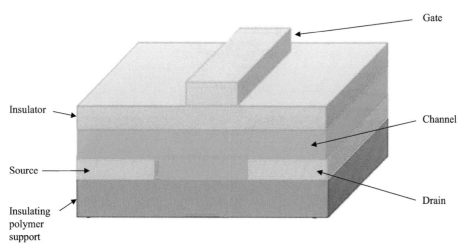

FIGURE 8.45 Top gate OFET device structure. The channel material should exhibit high mobility parallel to the plane of the channel. The best performance is available for p-type channels.

8.18 SUMMARY

8.1 The field of organic electronics is very active primarily due to the success of OLED displays. These displays are being increasingly used in cellphones, tablets, and televisions.

8.2 Electrical conductivity in molecular materials is made possible by conjugated bonding in which electrons are delocalized in π sub-bands. Electron transport within molecules occurs by intramolecular transport, and transport between molecules occurs by intermolecular hopping. Conjugated polymers allow for numerous modifications that can modify bandgaps as well as side-groups that facilitate solution processing.

8.3 In its simplest form the polymer OLED uses one electroluminescent polymer layer that acts as an electron transport layer near the cathode and as a hole transport layer near the anode. The EL polymer layer also provides for the recombination of electrons and holes that form molecular excitons within this layer. In an energy diagram an electric field forms between anode and cathode, which is the result of a charge transfer. Electrons from the cathode drift by means of the lowest unoccupied molecular orbital, or LUMO, while holes drift from the anode by means of the highest occupied molecular orbital, or HOMO. Molecular excitons and electroluminescence result.

8.4 The small-molecule OLED uses separate electron- and hole-transporting materials. A hole transport layer (HTL) and a separate electron transport layer (ETL) form a junction sandwiched between anode and cathode electrodes. The cathode electrode has a low workfunction to allow electrons to readily flow into the ETL. A light emitting material (LEM) positioned between the HTL and ETL is optimized for the formation of molecular excitons, and their radiative recombination efficiency is also optimized. Other layers may be added to further optimize device operation.

8.5 Anode materials are normally transparent to allow light emission, and indium tin oxide (ITO) is the most common material. Surface smoothness of both the substrate and the ITO layer is important. A surface roughness below 2 nm is generally required although ITO is inevitably rough due to its polycrystalline structure. ITO has a high workfunction (φ >4.1 eV) allowing it to inject holes efficiently. It has a chemically active surface that can cause migration of indium into adjacent polymer layers.

8.6 Cathode materials must provide high conductivity, a low workfunction, and good adhesion to the underlying polymer layers with long-term stability. Ease of oxidation and a tendency for these cathode layers to cause chemical reduction of adjacent polymer layers are challenges. A two-layer cathode is popular and the LiF/Al structure is widely used in which the aluminum protects the reactive LiF layer and also provides improved sheet conductivity. OLED devices must be protected with encapsulation. The rate of moisture penetration must be calculated to ensure a specified product life.

8.7 The hole injection layer (HIL) acts to improve the smoothness of the anode surface due to the native ITO roughness and facilitates efficient hole injection and long-term hole injection stability. The polymer-metal interface is very complex due to charges that are trapped there as a result of dangling bonds and metal atoms that react with the organic layer. Another important aspect of OLED performance relates to the current balance between electron and hole currents.

8.8 Pure and easily ionized metals such as Ca or Ba can be used as the low workfunction cathode electron injection layer; however, these metals are highly unstable with gaseous oxygen and water molecules as well as with the organic material used in the electron transport layer. Compounds of such metals that reduce their reactivity and instability while retaining their desired low workfunction include LiF. Although these are insulators they can be deposited to a thickness of only one or two monolayers. The development of materials based on organic molecules is of interest due to their inherent compatibility with the other organic layers.

8.9 The hole transport layer (HTL) must provide effective hole transport. TPD and NPD are popular hole conductors consisting of small molecules containing six-carbon rings with conjugated bonds allowing intramolecular hole transport. TPD and NPD are members of a family of compounds known as triarylamines and have hole mobilities in the range of 10^{-3} to 10^{-4} $cm^2\,V^{-1}\,s^{-1}$. A challenge is their low-temperature crystallization. OLEDs employing TPA and TPTE may be operated continuously at temperatures of 140°C since they do not crystallize readily.

8.10 The electron transport layer (ETL) provides electron transport. A LUMO level that is similar in energy to the workfunction of the cathode is required. The ETL should have a mobility of at least 10^{-6} cm^2 V^{-1} s^{-1}, which is one to two orders of magnitude smaller than the mobility range of HTL materials. Improving this low mobility has been a key target. Since the ETL has a lower mobility it must support a larger electric field, which can further compromise stability. The most common and most successful ETL is Alq$_3$.

8.11 Several processes occur to achieve light emission in the light emitting material (LEM). It must be able to transport both holes and electrons to enable the recombination of these carriers. It must effectively allow for the creation of excitons and their decay to generate photons. It must also remain stable at the electric fields needed to transport the holes and electrons, and the migration of molecules must be minimized for device stability. It is common for mixtures of two or more molecules to be used as light emitting materials. In the host-guest energy transfer process an excited host molecule can either directly produce radiation or transfer its energy to a guest molecule. This can overcome the 25% efficiency limit of singlet recombination.

8.12 Suitable host materials must exhibit good electron and/or hole conduction and their LUMO and HOMO levels must match the guest molecules. They must exhibit good miscibility to maintain a stable solution without the tendency for crystallization, which will decrease energy transfer efficiency. Finally, energy transfer processes to guest molecules should occur rapidly. Alq$_3$ is a simple host material that also functions as an ETL material. It emits at 560 nm, which is a yellow-green color, and it can be combined with other guest emission spectra to yield a white emission or it can be optically filtered to achieve red or green emission.

8.13 Guest materials may undergo fluorescence or photoluminescence. Fluorescent dopants emitting with suitable red, green, and blue color coordinates are required for full-color displays. In addition the HOMO and LUMO levels of the dopants must match the host materials. Guest phosphorescent dopants should have triplet energy levels smaller than the host to enable effective energy transfer. The guest dopant should remain soluble in the host to prevent unwanted segregation. An example of a green dopant is based on a coumarin dye molecule. Red fluorescent dopants have been developed that exhibit satisfactory color coordinates with good stability and efficiency based on the arylidene family of molecules. For blue emission, hosts from the anthracenes combined with the guest molecule perylene may be used.

8.14 Guest molecules based on iridium organometallic complexes have proven the most effective phosphorescent molecules. They have a short triplet lifetime of 1–100 μs. Radiative recombination is assisted since the normally forbidden radiation from the triplet exciton is somewhat allowed due to spin-orbital interaction in the molecule. This alters the spin states and renders high-efficiency phosphorescence. Numerous phosphorescent emitters are well studied and some have been commercialized in OLEDs for battery-powered devices that require high efficiency. Thermally activated delayed fluorescence guest molecules provide another channel for triplet harvesting. In these molecules singlet and triplet exciton levels are close in energy to allow for thermal excitation.

8.15 In the organic solar cell, absorption of sunlight creates molecular excitons that are localized to a single molecule, which is generally on the nanometer length scale. Unless the exciton can be dissociated and its hole and electron extracted, no current can result. A key challenge is to overcome the localization and pairing in the form of excitons of optically generated holes and electrons. Materials and device architectures designed to facilitate exciton dissociation enable organic solar cells. The single-layer solar cell relies on the differing workfunctions between cathode and anode to generate an electric field high enough to collect these charges. Better performance may be achieved in heterojunction and bulk heterojunction devices in which excitons are dissociated at heterojunctions between donor and acceptor materials.

8.16 The thiophene P3HT and fullerene C_{60} are simple examples of a donor material and acceptor material respectively. Acceptor materials must provide good electron conductivity, but optical absorption is not desired. Electron transport within a single C_{60} molecule is very fast and efficient. The electrons are metastable within the molecules and are therefore readily transferred to the solar cell electrode. Finally, these fullerenes have an absorption spectrum that peaks in the ultraviolet part of the spectrum, which means that they exhibit only minimal absorption in the important parts of the solar spectrum, leaving the donor layer free to absorb the visible or infrared solar radiation. The carbon nanotube is another acceptor material of considerable interest.

8.17 Perovskite solar cells have rapidly progressed and now achieve over 20% efficiency. They are based on a perovskite crystal containing an organic methylammonium molecule. The working principle is similar to that of an inorganic solar cell and stable excitons are not formed at room temperature. An organic hole transport layer and an inorganic electron transport layer are required. The direct gap in the perovskite enables this film construction and a solution-based deposition process can minimize manufacturing costs. Stability and toxicity are ongoing challenges.

8.18 Organic field effect transistors allow for flexible and low-cost electronics using roll-to-roll manufacturing methods. Such devices are of interest for flexible displays and bio-compatible electronics. Both small molecule and polymer organic materials have been developed to form high mobility channels using thin film deposition techniques.

PROBLEMS

8.1 The well-known firefly produces light by a process called bioluminescence. Find more information on the specific organic molecules in the firefly and the way they are excited.

8.2 The benzene series is shown below:

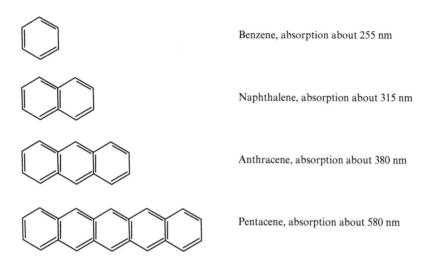

Benzene, absorption about 255 nm

Naphthalene, absorption about 315 nm

Anthracene, absorption about 380 nm

Pentacene, absorption about 580 nm

(a) Look up and reproduce absorption spectra for the four members of the benzene series shown.

(b) Pentacene is a well-studied organic semiconductor. Look up its melting point.

(c) What values of mobility have been measured for electrons in pentacene? The mobility is a measure of electron flow through the material, which requires both intramolecular flow enabled by the delocalization of the π-band electrons as well as intermolecular flow by hopping. A high degree of crystallization of pentacene is very important to obtain high values of mobility. Explain.

(d) Pentacene is not stable in air. What reaction occurs? How does that affect its feasibility for commercial flexible electronics?

8.3 Although OLEDs have been most successful in display application, they are also available as lighting products.

(a) Write a 1-2 page summary based on internet research on commercially available OLED lamps.

(b) What negative attributes of OLEDs versus inorganic LEDs have made it challenging for OLEDs to penetrate the lighting market?

(c) Based on your answer to a), what specific attributes of OLEDs are being leveraged to differentiate OLED lighting products in the lighting market?

8.4   There are commercial applications of both organic and inorganic-based pixel arrays. Inorganic LED arrays are used for the large display market that comprises outdoor displays and billboards viewed from distances of tens of metres, whereas OLED pixel arrays are used in markets for smaller display in portable electronics and television.

    (a) Make a list of strong points and weak points for each technology in the context of the basic requirements of display markets including the following attributes

       cost

       brightness

       operating life

       color gamut

       power consumption (efficiency)

    (b) Based on your answers to a), rationalize the choice of technology (organic versus inorganic) for the two markets.

8.5   A full-color OLED display contains red, green, and blue pixels. In order to generate white light the relative luminance values of these three pixel colors are as follows:

red: 30%

green: 60%

blue: 10%

The brightness specification of the OLED display is measured by the maximum *average* luminance that it can produce in a white screen. Assume that 25% of the display area is red pixels, 25% is green pixels, and 25% is blue pixels. A final 25% of the display area is the spaces between pixels, which do not emit light.

    (a) For an average white luminance of 500 cd m^{-2} find the required luminance values for the red pixels, the green pixels, and the blue pixels.

    (b) A required specification for this OLED display in terms of lifetime depends on the application. For television, 60,000 hours to half luminance is a common requirement. Estimate the lifetime requirement of a cell phone display, and comment on the applicability of current OLED technology for these two applications.

    (c) The peak emission wavelengths of the red, green, and blue pixels are approximately 615 nm, 540 nm, and 470 nm respectively. Assuming, for ease of calculation, that the emission in these three colors is monochromatic (light of a single wavelength), find the radiative power per unit area from each pixel for the result of (a). *Hint*: This requires a conversion from photometric to radiometric units, which can be obtained from the plot of luminous efficacy in Chapter 4. You may assume a lambertian source.

    (d) For the result of (c) find the relative photon emission rate from each of the three colors in photons per unit area per second. You may assume a lambertian source.

8.6 The structures of OLED devices shown in Figures 8.12 and 8.13 will not yield good contrast because of the high reflectivity of the aluminum rear electrode. In a display application, ambient light will reflect from the rear electrode and be re-emitted. This reflected light may exceed the light emitted by the OLED material and result in poor display contrast. In order to resolve this problem a *circular polarizer* may be placed in front of the OLED. The circular polarizer will prevent the reflection of ambient light off a *specular reflector* such as the aluminum rear electrode. The OLED device will therefore appear black unless it is electrically excited.

(a) Find reference material on polarizers and briefly explain both linear polarization and circular polarization of electromagnetic radiation. How does the circular polarizer prevent the re-emission of ambient light from a specular reflector placed behind the polarizer? *Hint*: Ambient light must first pass through the circular polarizer in one direction before reflecting off the specular reflector and leaving through the circular polarizer in the opposite direction.

(b) The circular polarizer will also attenuate the desired light emission from the OLED. If OLED electroluminescence is not polarized, what maximum fraction of the generated OLED light can pass through the circular polarizer?

(c) The OLED devices of Figures 8.12 and 8.13 may be simplified from an optical viewpoint as in the following figure, which shows the light emission from the OLED device with an aluminum rear electrode:

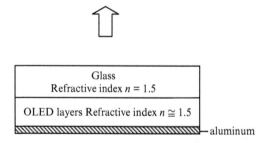

(d) A viewing cone will be formed by considering the critical angle of OLED light emission that is generated in the OLED layer, travels through the glass and is incident on the front surface of the glass. At incident angles higher than the critical angle this light will be reflected back into the device off the front surface of the glass. Find the critical angle relative to the normal of the glass. What fraction of generated light can leave the OLED device? *Hint*: See Section 5.10.

(e) How might the fraction of generated OLED light leaving the device be increased? *Hint*: Use concepts of optical outcoupling in Chapter 5 to discuss this.

8.7 The promise of flexible displays has been a major driver for OLED development. Since OLED active layers as well as OLED cathode materials are highly moisture and oxygen sensitive it is necessary to encapsulate the devices. Flexible polymer sheets are not impermeable since water or oxygen molecules can diffuse between the polymer molecules due to their weak van der Waals intermolecular bonding. Polymer sheets therefore do not offer protection from the atmosphere to OLED layers deposited on them, resulting in rapid degradation. To remedy this, a series of inorganic thin films may be deposited on a polymer sheet with thin organic layers between them to keep them separated. The concept is that the inorganic layers, being impermeable to moisture and oxygen, will render the sheet impermeable. Since there are inevitable cracks in any given inorganic layer, the use of multiple inorganic layers will force unwanted molecules to take a very tortuous pathway to pass through the sheet as illustrated below:

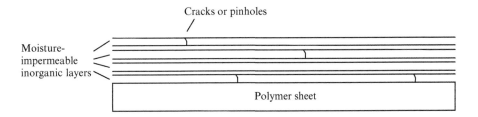

(a) Look up information on multilayer moisture barrier technology for flexible OLED devices. *Hint*: Search using the following keywords: multilayer OLED moisture protection.
(b) Find one or two quoted maximum moisture penetration rates allowable for OLED devices in units of grams per square metre per day.
(c) Flexible OLED devices are on the market. Search for information on foldable and rollable OLED displays that have been commercialized.

8.8 A typical OLED display emits light as a lambertian source. This means that the luminance of the OLED display is independent of the angle between the observer and the normal axis to the plane of the OLED display. See Section 5.10.

If the lambertian OLED display provides a luminance of 100 cd m^{-2}, has a light emitting surface area of 100 cm^2, and requires an electrical input power of 1 watt,

(a) Show that the total luminous flux emitted by the display is 3.14 lumens. Hint: Use spherical polar coordinates. Integrate the amount of light emitted over the solid angle range as follows: 1) Obtain the luminous intensity in candelas emitted as a function of angle by multiplying the luminance by the OLED surface area in square meters subtended by the viewer at each viewing angle. 2) To obtain the total luminous flux in lumens (lm), integrate the luminous intensity over the entire range of solid angles that is viewable.

Luminous intensity may be expressed as lm Sr^{-1}. Note the definition of luminous flux in Section 4.9. Use spherical polar coordinates.

(b) Find the luminous efficiency of the OLED display in lm W^{-1}. Compare this to luminous efficiency values for inorganic LEDs in Figure 5.11. Discuss the significance of this comparison.

8.9   Write a review of recent literature on TADF guest molecules and the degree to which this process has become commercially relevant. Show structures of specific TADF molecules. Have stable blue-emitting TADF molecules been developed? How to they compare with blue emitting phosphorescent molecules in terms of color coordinates, quantum yield, and operational stability? Are blue emitters for practical OLED displays still forced to rely on fluorescent molecules with a maximum of 25% quantum yield?

8.10   Although the organic solar cell has developed rapidly since 2001, it is well behind inorganic technology in terms of maturity. A major challenge is the stability of organic solar cells. Traditional solar farm installations and rooftop installations demand 20- to 30-year lifetimes with not more than a 20% reduction in performance. This specification will not be easy to meet using organic solar cells. There are applications of organic solar cells that require much shorter lifetimes.

(a) Given the inherent flexibility and light weight of organic solar cells, describe some existing applications that take advantage of the special attributes of organic solar cells and that are also less demanding in terms of efficiency and longevity.

(b) Now add some potential *new* applications of organic solar cells to your list. Be creative. New applications are often those that force new technologies to mature quickly and become mainstream since they fill such market niches better than anything else.

8.11   Research is underway on organic-inorganic solar cell structures in which the bulk heterojunction has an inorganic acceptor material combined with an organic donor material. For example, the HTL may be solution-processed MEH-PPV and the ETL may be ZnO. The motivation for this is the availability and stability of inorganic electron conductors. An example of such as structure is shown below:

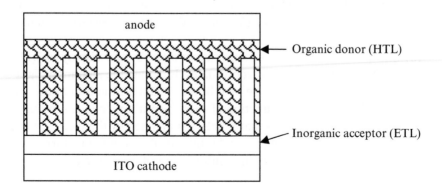

(a) What challenges are there with this approach? *Hint*: Think carefully about what needs to happen right at the inorganic-organic junction, and what could go wrong at the junction.

(b) Search for more information on recent developments in this field. Use the following keywords: organic solar cells; organic/inorganic heterojunctions.

(c) Organic electron-conducting acceptor materials tend to be more moisture and oxygen sensitive than hole-conducting donor materials. Why?

(d) An OLED can also be made using a combination of organic and inorganic materials for the electron-hole injection and recombination processes. Provide and explain an example based on an internet search What challenges are inherent in this approach?

8.12 Write a review of recent literature on perovskite solar cells. Are replacements for lead available for the perovskite layer? How do their costs or projected costs compare with silicon solar cells? How stable are perovskite solar cells? What steps can be taken to improve stability? Have they been commercialized?

8.13 Seek online examples of the solution deposition of OFET channel materials by the following methods: dip-coating, spin-coating, inkjet printing and screen printing. Briefly describe each method and its advantages.

8.14 Look up and show the molecular structures of the poly para-phenylene vinylene (PPV) and PPV derivatives shown in Figure 8.6.

8.15 Search for an up-to-date review paper on quantum dot electroluminescence. What are the benefits of this approach? In two pages, summarize the achieved performance levels. Has is been commercialized? What are the target applications?

NOTES

1. Full molecular names such as Poly(2,5-dialkoxy) paraphenylene vinylene will not generally be listed in this chapter but may be found in Suggestions for Further Reading, Z. Li et al.

2. It should be noted that work on intermolecular interaction has led to developments in LEM materials in which molecules fluoresce efficiently as aggregates. This is known as *aggregation-induced emission* but it is currently at the research stage and the host-guest approach dominates current high performance OLEDs.

SUGGESTIONS FOR FURTHER READING

1. So, F. (2010). *Organic Electronics – Materials, Processing, Devices, and Applications*. CRC Press.

2. Li, Z. and Meng, H. (2007). *Organic Light Emitting Materials and Devices*. CRC Press.

3. Sun, S.S. and Dalton, L.R. (2008). *Introduction to Organic Electronic and Optoelectronic Materials and Devices*. CRC Press.

4. Kitai, A.H. (ed.) (2017). *Materials for Solid State Lighting and Displays*. Wiley.

CHAPTER 9

One- and Two-Dimensional Semiconductor Materials and Devices

CONTENTS

Objectives

1. Motivate the need for atomic scale modeling.
2. Introduce the linear combination of atomic orbitals.
3. Illustrate the application of the linear combination of atomic orbitals to an ionized hydrogen molecule.
4. Introduce and illustrate modeling using density functional theory.
5. Present transition metal dichalcogenides as a prototypical two-dimensional semiconductor.
6. Present examples of multigate MOSFETs and their advantages over planar devices.

Fundamentals of Semiconductor Materials and Devices, First Edition. Adrian Kitai.
© 2023 John Wiley & Sons Ltd. Published 2023 by John Wiley & Sons Ltd.
Companion Website: www.wiley.com/go/kitai_fundamentals

9.1   INTRODUCTION

This chapter is intended to illustrate a few emerging one- and two-dimensional semiconductor materials and devices. It also provides a very basic introduction to quantum-mechanical modeling strategies that are suited to these materials systems.

An understanding of the electronic properties of semiconductors using the band model introduced in Chapter 2 has been applied to semiconductor devices with dimensions from the bulk down to those of the quantum dot. In the case of the smallest quantum dots, however, limitations have been noted for the application of band theory. Some alternative type of nanoscale modeling is required here as well as for even smaller systems.

Emerging nanoscale semiconductor materials and devices exist in which at least one semiconductor dimension is only an atom to a few atoms thick. In fact, molecules introduced in Chapter 8 are also examples of structures whose size scale is beyond the reach of a band model.

In this chapter, useful approaches for modeling materials having a limited number of atoms in one or more dimensions are presented. One approach is the *Linear Combination of Atomic Orbitals* (LCAO) and a second, more recently developed approach is *Density Functional Theory* (DFT).

These modeling techniques generally require the use of a computer. Our focus will be on the very basic physical and mathematical concepts upon which the modeling is based. These techniques may be applied to gain detailed insights into molecular materials and other nanoscale materials including, for example, nanowires and nanotubes.

After this, properties of some important materials will be presented and discussed. Reference materials listed in Recommended Reading will provide further information on these topics.

The validity of the modeling approaches introduced in this chapter has improved and continues to improve. These improvements are the result of increasingly sophisticated computations and a large body of work spanning almost a century in comparing modeling results to measured properties. The field of *computational materials science* has evolved and is now an established discipline.

The most well-studied atomically thin two-dimensional material is graphene; however, it is a semi-metal and not a semiconductor. Graphene consists of a robust, covalently bonded carbon atom layer only a single atom in thickness that may be weakly bonded to adjacent layers by van der Waals bonds.

Atomically thin two-dimensional *semiconductor* materials exist that resemble graphene in terms of having atomically thin layers weakly bonded to adjacent layers by van der Waals bonds. One of the most promising classes of these materials is based on members of the transition metal dichalcogenide family of compounds. This chapter provides an introduction to transition metal dichalcogenide materials, their electronic and optical properties, and their potential applications.

In spite of the development of these fascinating new classes of materials, silicon continues to be the industrially important material used for mass-produced state-of-the-art transistors in integrated circuits. As shown in Chapter 6, the length scale of silicon MOSFETs is approaching atomic dimensions. The diamond unit cell of silicon has a unit cell dimension of 5.43 Å which means that a silicon MOSFET with a channel length of 2 nm is only four unit cells in length. In addition, state-of-the-art MOSFETs are no longer planar devices as described in Chapter 6 and important three-dimensional MOSFET structures than can scale to atomically small length scales will be presented and discussed.

9.2   LINEAR COMBINATION OF ATOMIC ORBITALS

In order to understand the basic approach taken in LCAO theory, consider an example comprising the simplest possible system containing more than one atom. This is the positively ionized hydrogen gas molecule comprising one electron and two protons. It is a stable molecular ion and it is easily produced in a gas plasma.

In Chapter 1, the ground state wavefunction of the hydrogen atom

$$\psi(r) = \frac{1}{\sqrt{\pi}} a_0^{-\frac{3}{2}} e^{-\frac{r}{a_0}}$$

was found from Schrödinger's equation written using radius as the spatial variable in spherical polar coordinates.

Consider two protons separated by a distance R and positioned on a coordinate system as shown in Figure 9.1.

The wavefunction is unknown for the electron in this system of the two protons labeled Proton A and Proton B. However we do know the electron wavefunction for one proton. We will propose that the electron wavefunction for the pair of protons be approximated by the superposition of two single proton wavefunctions

$$\psi_s = a\psi_A + b\psi_B \tag{9.1}$$

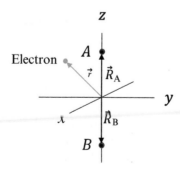

FIGURE 9.1   Two protons separated by distance R are located on the z-axis at distances $\frac{R}{2}$ above and below the xy plane. Their positions are defined using vectors $\vec{R}_A$ and $\vec{R}_B$ where $\left|\vec{R}_A\right| = \left|\vec{R}_B\right| = \frac{R}{2}$. The electron is at a position defined by vector $\vec{r}$.

where

$$\psi_A = \frac{1}{\sqrt{\pi}} a_0^{-\frac{3}{2}} e^{-\frac{|\vec{r}-\vec{R}_A|}{a_0}}$$

and

$$\psi_B = \frac{1}{\sqrt{\pi}} a_0^{-\frac{3}{2}} e^{-\frac{|\vec{r}-\vec{R}_B|}{a_0}}$$

are the normalized electron wavefunctions relevant to one electron and each individual proton. This assumption is the central idea in the appropriately named method of linear combination of atomic orbitals. Justification for this approach is based on the idea that electrons will have a considerable probability of being in the close vicinity of either one of the nuclei.

In order to find the values of coefficients a and b we will seek an expression for the total electron energy E_e of the system and will then minimize this energy to be consistent with thermodynamic equilibrium. The result will predict the molecular electron energy levels and will also enable the prediction of the value of the proton separation R.

No eigenenergy is available because the proposed wavefunction is a superposition state which is not an eigenstate of Schrödinger's equation. In order to calculate system energy, the expectation value of energy E_e will be calculated. See Section 1.14. We obtain

$$E_e = \frac{\int \psi_s^* \widehat{H} \psi_s dV}{\int \psi_s^* \psi_s dV} \tag{9.2}$$

The denominator is required because ψ_s has not been normalized.

The Hamiltonian consists of the kinetic energy and potential energy terms

$$\widehat{H} = \widehat{T} + \widehat{U}$$

See Section 1.13. It may be written using Equation 1.26 as

$$\widehat{H} = \widehat{T} - \left[\frac{qQ}{4\pi\varepsilon_0 |\vec{r} - \vec{R}_A|} + \frac{qQ}{4\pi\varepsilon_0 |\vec{r} - \vec{R}_B|} \right]$$

where the charge on each proton is $Q = q$. A term for nuclear repulsion is not included and will be introduced later.

The denominator of Equation 9.2 may be expanded as

$$\int \psi_s^* \psi_s dV = \int (a\psi_A + b\psi_B)^* (a\psi_A + b\psi_B) dV$$

$$= a^2 \int \psi_A^2 \left(|\vec{r} - \vec{R}_A| \right) dV + b^2 \int \psi_B^2 \left(|\vec{r} - \vec{R}_B| \right) dV + 2ab \int \psi_A \left(|\vec{r} - \vec{R}_A| \right) \psi_B \left(|\vec{r} - \vec{R}_B| \right) dV$$

Note that ψ_A and ψ_B are real wavefunctions which allows the complex congugation to be ignored. The first two terms reduce to $a^2 + b^2$ because ψ_A and ψ_B are normalized. The integral in the third term is unknown and will be denoted as I_{AB}. Therefore

$$\int \psi_s^* \psi_s dV = a^2 + b^2 + 2ab I_{AB} \qquad (9.3)$$

The numerator of Equation 9.2 may also be expanded as

$$\int \psi_s^* \widehat{H} \psi_s dV = \int (a\psi_A + b\psi_B)^* \widehat{H} (a\psi_A + b\psi_B) dV$$
$$= a^2 \int \psi_A^* \widehat{H} \psi_A dV + b^2 \int \psi_B^* \widehat{H} \psi_B dV + ab \int \psi_A^* \widehat{H} \psi_B dV + ab \int \psi_B^* \widehat{H} \psi_A dV$$

By symmetry, the last two integrals must be identical and therefore

$$\int \psi_s^* \widehat{H} \psi_s dV = a^2 H_{AA} + b^2 H_{BB} + 2ab H_{AB}$$

where

$$H_{AA} = \int \psi_A^* \widehat{H} \psi_A dV$$

$$H_{BB} = \int \psi_B^* \widehat{H} \psi_B dV$$

and

$$H_{AB} = \int \psi_A^* \widehat{H} \psi_B dV = \int \psi_B^* \widehat{H} \psi_A dV$$

Also by symmetry $H_{BB} = H_{AA}$

From Equation 9.2

$$E_e = \frac{\left(a^2 + b^2\right) H_{AA} + 2ab H_{AB}}{a^2 + b^2 + 2ab I_{AB}} \qquad (9.4)$$

We are now ready to minimize energy E_e. Equation 9.4 is a function of variables a and b. At a minimum value of this function $\dfrac{\partial E_e}{\partial a} = 0$ and $\dfrac{\partial E_e}{\partial b} = 0$

Performing the derivatives we obtain

$$\frac{\partial E_e}{\partial a} = \frac{2a^2 b H_{AA} I_{AB} + 2b^3 H_{AB} - 2b^3 H_{AA} I_{AB} - 2a^2 b H_{AB}}{\left(a^2 + b^2 + 2ab I_{AB}\right)^2} = 0$$

and

$$\frac{\partial E_e}{\partial b} = \frac{2ab^2 H_{AA} I_{AB} + 2a^3 H_{AB} - 2a^3 H_{AA} I_{AB} - 2ab^2 H_{AB}}{\left(a^2 + b^2 + 2ab I_{AB}\right)^2} = 0$$

The numerators of the above equations must be zero. After dividing by common terms and rearranging,

$$a^2 H_{AA} I_{AB} + b^2 H_{AB} = b^2 H_{AA} I_{AB} + a^2 H_{AB}$$

and

$$b^2 H_{AA} I_{AB} + a^2 H_{AB} = a^2 H_{AA} I_{AB} + b^2 H_{AB}$$

It is now clear that the two equations are identical and reduce to a single equation, whereas two equations are required to solve for the two variables a and b. Nevertheless, the ratio $\frac{a}{b}$ can be found by dividing either equation by b^2. After rearranging terms we obtain

$$\frac{a^2}{b^2}(H_{AA} I_{AB} - H_{AB}) = H_{AA} I_{AB} - H_{AB}$$

which can only be true if $a = b$ or if $a = -b$.

There are now two distinct superposition wavefunctions ψ_s which can be expressed in terms of a new constant $c = |a| = |b|$ as

$$\psi_{s+} = c(\psi_A + \psi_B)$$

or as

$$\psi_{s-} = c(\psi_A - \psi_B)$$

The constant c can be found by normalizing the superposition wavefunctions. For ψ_{s+} using Equation 9.3 we have

$$\int \psi_s^* \psi_s dV = a^2 + b^2 + 2ab I_{AB} = \int \psi_{s+}^* \psi_{s+} dV = 2c^2(1 + I_{AB}) = 1$$

and therefore

$$c = \frac{1}{\sqrt{2(1 + I_{AB})}}$$

and for ψ_{s-}

$$\int \psi_s^* \psi_s dV = a^2 + b^2 + 2ab I_{AB} = \int \psi_{s-}^* \psi_{s-} dV = 2c^2(1 - I_{AB}) = 1$$

$$c = \frac{1}{\sqrt{2(1 - I_{AB})}}$$

The normalized wavefunction ψ_{s+} may now be written using Equation 9.1 as

$$\psi_{s+} = \frac{1}{\sqrt{2(1 + I_{AB})}} \left[\frac{1}{\sqrt{\pi}} a_0^{-\frac{3}{2}} e^{-\frac{|\vec{r} - \vec{R}_A|}{a_0}} \right] + \frac{1}{\sqrt{2(1 + I_{AB})}} \left[\frac{1}{\sqrt{\pi}} a_0^{-\frac{3}{2}} e^{-\frac{|\vec{r} - \vec{R}_B|}{a_0}} \right] \quad (9.5a)$$

This is called a *bonding orbital* wavefunction.

The normalized wavefunction ψ_{s-} may be written as

$$\psi_{s-} = \frac{1}{\sqrt{2(1 - I_{AB})}} \left[\frac{1}{\sqrt{\pi}} a_0^{-\frac{3}{2}} e^{-\frac{|\vec{r} - \vec{R}_A|}{a_0}} \right] - \frac{1}{\sqrt{2(1 - I_{AB})}} \left[\frac{1}{\sqrt{\pi}} a_0^{-\frac{3}{2}} e^{-\frac{|\vec{r} - \vec{R}_B|}{a_0}} \right] \quad (9.5b)$$

This is called an *antibonding orbital* wavefunction.

Plots of Equations (9.5a) and (9.5b) are shown in Figure 9.2.

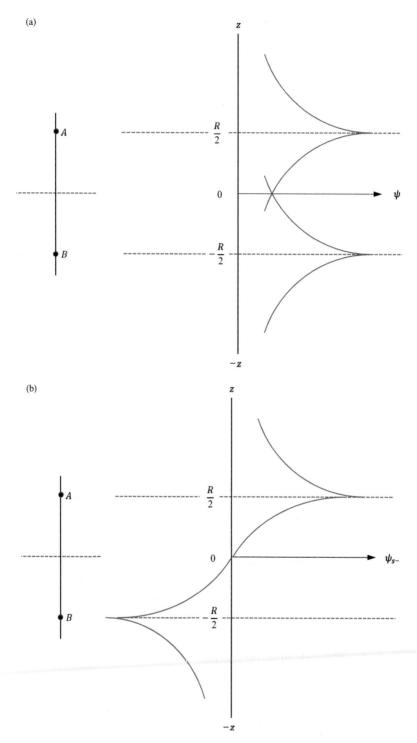

FIGURE 9.2 Orbital wavefunction amplitudes for the hydrogen molecular ion plotted as a function of position along the z axis. a) Bonding orbital wavefunction. b) Antibonding orbital wavefunction.

In order to determine the equilibrium internuclear separation predicted by the LCAO model, the minimum system energy needs to be determined. The system energy for the case of a bonding orbital can be expressed using Equation 9.4 as

$$E_+ = \frac{H_{AA}(R) + H_{AB}(R)}{1 + I_{AB}(R)} + \frac{q^2}{4\pi\varepsilon_0 R}$$

where the second term is the energy due to the coulomb repulsion between the two protons. The change in potential energy for an electron and a proton separated by a distance r relative to the electron and proton infinitely far apart was derived in Equation 1.26 as

$\Delta U = -\dfrac{q^2}{4\pi\varepsilon_0 r}$. By making the result positive, this change in potential energy becomes correct for two protons.

In the case of an antibonding orbital the analogous result is

$$E_- = \frac{H_{AA}(R) - H_{AB}(R)}{1 - I_{AB}(R)} + \frac{q^2}{4\pi\varepsilon_0 R}$$

where $H_{AA}(R)$, $H_{AB}(R)$, and $I_{AB}(R)$ are integrals that must be performed. These integrals are worked out in Appendix 7. The results are shown in Figure 9.3.

The experimentally determined dissociation energy of the molecular ion is 2.79 eV versus the LCAO result of 1.76 eV. The experimentally determined bond length is 1.06Å versus the LCAO result of 1.32Å. The LCAO predictions are not very accurate but they provide a profound quantum mechanical understanding of the bonding and antibonding electron states. In the stable molecular ion, only the bonding state is occupied. In a neutral hydrogen molecule, only the bonding state is occupied except that there are now two electrons having opposing spin states.

The relevant energy diagram known as a *correlation diagram* used by chemists is shown in Figure 9.4.

One can view this molecule as having an energy gap due to the separation between the occupied bonding orbital and the unoccupied antibonding orbital. According to the Figure 9.3 the energy gap is approximately 7 eV at the equilibrium bond distance which is very large from a semiconductor viewpoint.

The LCAO approach can be extended to more than two atoms. Atoms having different nuclear charges can be modeled, and orbitals that are not spherically symmetric such as p, d, and f orbitals can also be used for LCAO modeling. If LCAO modeling is applied to more complicated molecules then estimates of the relevant molecular energy gaps suitable for organic semiconductors can be obtained. The LCAO-derived correlation diagrams for molecules define the LUMO and HOMO energy levels that were discussed in Chapter 8.

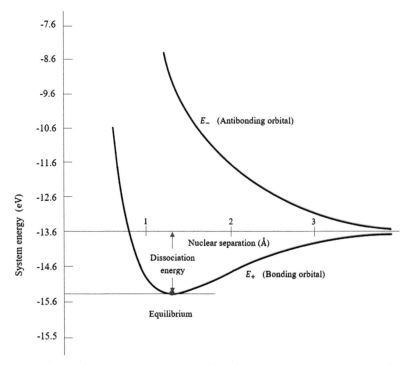

FIGURE 9.3   Bonding orbital system energy and antibonding orbital system energy plotted as a function of internuclear separation R. As $R \to \infty$ the system energy approaches the negative of the Rydberg constant (-13.6 eV) because the electron would have to choose one proton to form a hydrogen atom having dissociation energy 13.6 eV. The bonding orbital curve drops below this value by 1.76 eV indicating that there is an equilibrium bond length for which sharing the electron between both protons yields a system energy that is lower in energy than -13.6 eV by an additional 1.76 eV. The molecule dissociation energy of 1.76 eV at an equilibrium bond length of 1.32 Å as indicated by the double arrow.

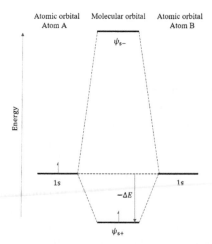

FIGURE 9.4 H_2^+ correlation diagram. The spin direction of the electron is indicated by the short arrow and the dissociation energy equal to $\left| -\Delta E \right|$ is indicated by the long arrow. The short arrow is shown populating the bonding state but not the antibonding state. This allows for the stable molecular ion. In a neutral hydrogen molecule, both spin up and spin down directions would be indicated populating the bonding state. In a He_2 molecule, the four electrons would fill both the bonding and the antibonding states. This molecule is not stable.

9.3   DENSITY FUNCTIONAL THEORY

One critical aspect of modeling that was not addressed in Section 9.2 is the effect of more than one electron. For example, if a helium atom were to be modeled it would require a second electron. It is not possible to define the location of the charge of one electron in order to write down the potential energy function $U(x,y,z)$ needed to derive the wavefunction of a second electron. Both electrons occupy regions of space with a spatial distribution of probabilities that can't be derived based on considering only one electron at a time. This difficulty underlies the quantum mechanical modeling of all materials with two or more electrons.

Density Functional Theory (DFT) builds on quantum mechanical models designed to treat the effects of multiple electrons on the quantum state of an additional electron, the electron of interest, in a material system.

Early modeling that originated in the 1920s focused on multielectron atoms in which inner shell electrons could be viewed as electrons that partly shield the nuclear charge from the outermost electron in the atom. By modifying the effective nuclear charge, multielectron atoms could be approximately modeled using simple modifications of hydrogen atom wavefunctions.

Subsequently, the extension of these early models allowed for the modeling of materials containing many atoms and multiple electrons. The availability of computers in conjunction with the theoretical framework of DFT has enabled unprecedented progress in the modeling of materials systems starting from the 1960s. Today, after extensive refinement and further development, DFT is a recognized and widely used tool.

Fundamental to DFT is the idea that the electrons other than the specific electron being modeled can be approximated as a spatially dependent charge density. This eliminates the complexity of knowing the precise location and wavefunction of each other electron at all times because only a spatial distribution function of the combined density of the other electrons is required.

The central challenge in DFT is the *functional* (function of a function) that makes use of this electron density in order to formulate the appropriate Schrödinger equation such that the properties of the electron under study are most accurately obtained as a solution.

To provide some insight we can write

$$\widehat{H}\psi = \left[\widehat{T}+\widehat{U}_n+\widehat{U}_e\right]\psi = \left[\sum_{i=1}^{N_e}\left(\frac{-\hbar^2}{2m}\nabla_i^2\right)+\sum_{i=1}^{N_n}U_n\left(\hat{r}_i\right)+\sum_{i=1}^{N_e}U_e\left(\hat{r}_i,\hat{r}_j\right)\right]\psi = E\psi$$

which is the many-electron time-independent Schrödinger equation. Here the atomic nuclei are assumed to be stationary resulting in a potential energy term $\sum_{i=1}^{N_n}U_n(\hat{r}_i)$ from all N_n atomic nuclei in the system. An example of $N_n = 2$ was used in Section 9.2 for LCAO modeling. The kinetic energy operator $\widehat{T}=-\dfrac{\hbar^2}{2m}\nabla^2$ introduced in

Section 1.13 now includes contributions from a total of N_e electrons and becomes $\sum_{i=1}^{N_e} \left(\dfrac{-\hbar^2}{2m} \nabla_i^2 \right)$. The final term $\widehat{U}_e = \sum_{i=1}^{N_e} U_e\left(r_i, r_j\right)$ requires knowledge of the location and the interaction associated with all the other electrons in the system. Due to this term, a solution to this equation is impossible to obtain directly. The wavefunction ψ is a many-electron wavefunction of the N_e electrons in the system and it therefore includes the spatial locations r_i of each electron and is written as $\psi(\hat{r}_1, \hat{r}_2 ... \hat{r}_{N_e})$. The analysis of a two-electron wavefunction for a two-electron time-independent Schrödinger equation was undertaken in Section 4.6 which illustrates details of the approach.

In DFT, the term $\widehat{U}_e$ is not present. In its place, the spatially dependent electron density $n(\hat{r})$ is used. In terms of the many electron wavefunction,

$$n(\hat{r}) = N_e \int d^3\hat{r}_2 ... \int d^3\hat{r}_{N_e} \psi^*(\hat{r}, \hat{r}_2, ... \hat{r}_{N_e}) \psi(\hat{r}, \hat{r}_2, ... \hat{r}_{N_e})$$

This equation states that if the many-electron wavefunction is known then the electron density is also known.

Under equilibrium conditions, the converse is valid. This means that *the many electron ground state wavefunction may be uniquely derived if the spatially dependent electron density is known*. This fact is quite surprising and it represents a theoretical breakthrough obtained in the 1960s that gives DFT a solid theoretical underpinning. It underlies the validity of using $n(\hat{r})$ in DFT.

The *Kohn–Sham equation* is the Schrödinger-like equation of a fictitious system of non-interacting particles (typically electrons) that generates the same electron density as a given system of interacting particles. The Kohn–Sham equation is defined by a local effective potential in which the non-interacting electrons move. It is this equation that links the electron density to the many-electron ground state wavefunction.

It may now be stated that the ground state wavefunction ψ is a unique functional of $n(\hat{r})$.

As a result, it becomes possible to calculate observables for the ground state of the system. For example, the expectation value of the ground state energy E_0 which is the lowest energy state of the system is

$$E_0 = E(n_0(\hat{r})) = \left\langle \psi(n_0) \left| \widehat{T} + \widehat{U}_n + \widehat{U}_e \right| \psi(n_0) \right\rangle$$

where $n_0(\hat{r})$ is the electron density of the ground state of the system and the ground state wavefunction ψ is a functional of n_0.

The functional that provides ψ from n should include electron–electron repulsion and should also include the exchange interaction resulting from indistinguishability.

In order to obtain the ground state energy of the system, the function

$$E(n) = T(n) + U_e(n) + \int U_n(\hat{r}) n(\hat{r}) d^3\hat{r}$$

must be minimized by taking the derivative of $E(n)$ with respect to $n(\hat{r})$ and then setting it to zero. A similar strategy was also used in Section 9.1 for LCAO modeling. The integral term in the above equation is defined by the charges and locations of the atomic nuclei. Functionals $T(n)$ and $U_e(n)$ are called *universal functionals*.

Observables other than energy may likewise be evaluated.

Through years of comparing DFT modeling results to measured properties, refinements to the functional $U_e(n)$ that more accurately take both electron coulomb interactions as well as exchange interactions into account continue to develop. Computer programs that allow the use of DFT modeling with sophisticated functionals and energy minimization algorithms have evolved to become widely available modeling tools.

DFT models the ground state of a system and it does not directly allow non-equilibrium states to be determined. It excels at predicting lattice structures, bonding lengths, lattice parameters, elastic properties, and other fundamentals that are based on the lowest energy configuration of the system under study. Unlike the LCAO approach, DFT provides accurate results for the ground state but is not normally able to determine, for example, the energy gap of a semiconductor which depends on excited states of the system. More recent commercially available DFT-based software is designed to overcome these limitations by making alterations to the functional or by the inclusion of additional terms. This allows non-equilibrium electron state modeling to be achieved.

9.4 TRANSITION METAL DICHALCOGENIDES

Since the discovery of graphene in 2004, a resurgence of scientific interest in other atomically thin materials has taken place. This can be explained by advances in sample preparation and optical and electrical characterization. As a result, an improved understanding of these materials has emerged allowing them to be considered for engineering applications.

The transition metal dichalcogenide (TMDC) family of naturally layered materials is an example of a graphene-like material. Unlike graphene, TMDCs are semiconductors. Like graphene, chemically bonded layers of atomic scale crystalline unit cells are weakly connected in a third dimension by van der Waals bonds allowing them to be used as lubricants due to the ability of the layers to slide relative to each other.

Figure 9.5 shows the structures of TMDC materials MoS_2 and WS_2. Layer thickness in Figure 9.5 is in the range of 3–3.5Å making these truly atomically thin materials even though two planes of the chalcogen are involved. Mechanical measurements performed on single-layer MoS_2 show that it is 30 times as strong as steel and can be deformed by up to 11% before breaking. This makes MoS_2 one of the strongest semiconducting materials with the potential for deployment onto flexible substrates.

In addition to the sulphide TMDCs shown in Figure 9.5, analogous TMDCs in which selenium or tellurium replaces sulphur also exist. Single-layer TMDCs exhibit

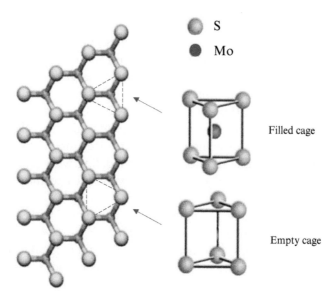

S
Mo

Filled cage

Empty cage

FIGURE 9.5   Structure of a single layer of MoS$_2$. Triangular cages formed by sulphur atoms each contain one molybdenum atom. Interspersed with these cages are empty triangular cages. The thickness of the layer is 3Å. Several views are shown of this structure. Adjacent layers are connected by weak van der Waals bonds.

direct semiconductor bandgaps and are of interest for applications in optoelectronics. The surfaces of a single TMDC layer have no dangling bonds and they therefore possess dramatic advantages compared to other crystalline semiconductors that require bond terminations to achieve surface passivation. TMDCs may be supported on a wide range of substrates that are not lattice matched and that may even be amorphous.

As the number of layers increases and bulk-like behavior develops, TMDCs exhibit indirect gap properties. Measurement of photoluminescence quantum yield is a simple way to observe this direct gap–indirect gap transition: A factor of 10^4 decrease in photoluminescence quantum yield is observed when monolayer MoS$_2$ is stacked to form bulk MoS$_2$ which is explained by the need for phonon participation in bulk indirect gap fluorescent transitions.

The band diagram of TMDC materials MoS$_2$ and WS$_2$ are shown in Figure 9.6 for the case of both bulk and single layer material. These band diagrams are calculated by DFT methods.

Portions of the band diagram can be connected to specific atomic orbitals of the atoms making up the TMDC crystal. The LCAO approach to band theory involves the superposition of atomic orbitals. In TMDC materials, the relevant valence orbitals are p orbitals and d orbitals rather than the s orbitals used in Section 9.2. It is therefore expected that these p and d orbitals form the LCAO basis for the TMDC valence and conduction bands. The most relevant orbitals in band formation are the metallic (Mo, W) outermost d orbitals and the chalcogen (S) outermost p orbitals.

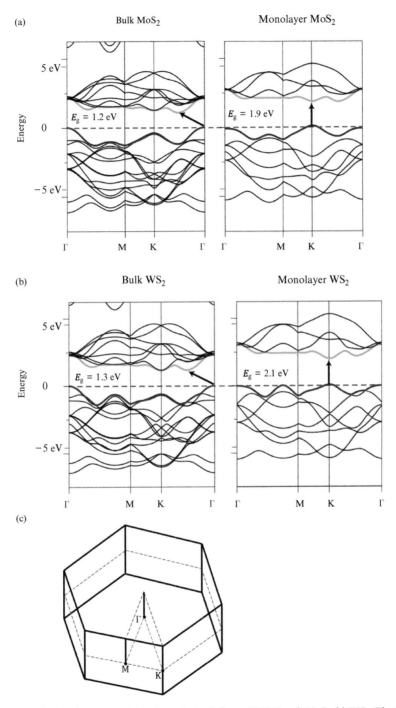

FIGURE 9.6 The band diagram for bulk and single layer TMDCs. a) MoS_2. b) WS_2. The band diagrams plot energy E versus wavevector along a variety of crystallographic directions that extend between defined labels Γ, M and K. These labels are shown in c) starting from the center of the unit cell at Γ and extending toward points M and K. The hexagons in c) are in the plane of a single layer TMDC. After T. Humberto et al., Novel hetero-layered materials with tunable direct band gaps by sandwiching different metal disulfides and diselenides, Adapted from Humberto et al, 2017.

For MoS_2, the band states at the K-point are mainly based on localized d orbitals on the Mo atoms situated in the middle of the S–Mo–S layer sandwiches. These orbitals are relatively unaffected by interlayer coupling.

In contrast to this, the states near the Γ-point are due to combinations of the anti-bonding p_z-orbitals on the S atoms as well as the d orbitals on Mo atoms. Since the S atoms are at the surface of a monolayer of TMDC, the band diagram at the Γ-point has a strong interlayer coupling effect.

As a result, with decreasing layer count, the states near the K-point are relatively unchanged but the energy gap shifts significantly from an indirect one to a larger, direct one. The changes in the MoS_2 and WS_2 bands shown in Figure 9.6 indicate that the band-gaps at the Γ-points increase in the monolayer cases compared to the bulk cases thereby enabling direct gap transitions at the K-points. All MoX_2 and WX_2 compounds undergo a similar indirect- to direct-bandgap transformation with decreasing numbers of layers, covering the bandgap energy range 1.1–1.9 eV, where X represents a chalcogen.

In addition to phonon scattering, coulomb scattering in monolayer TMDCs is caused by random charged impurities located within the TMDC layer or at its surfaces. This is the dominant scattering effect at low temperatures, as it is for graphene. Graphene placed on a dielectric material such as SiO_2 is known to experience carrier scattering due to surface phonons in the SiO_2. In freely suspended graphene, the primary scattering mechanism is due to out-of-plane flexural phonons. Freestanding MoS_2 has similar ripples to those observed in graphene. These may also contribute to scattering and mobility reduction.

Photoluminescence quantum yield is observed to be dependent on the type of physical support. Freely suspended single layer TMDC samples and samples placed on boron nitride substrates show higher quantum yield compared with samples placed on SiO_2 substrates.

The lack of a bandgap in graphene means that it cannot achieve a low off-state current, limiting its use as a MOS transistor channel. There is strong interest in nano-electronic materials with bandgaps to support high on/off current ratios having scalability to ever-smaller dimensions. As a result of their sizeable bandgaps and high mobilities, ultrathin TMDC materials can be used as the channel in MOS transistors with high on/off current ratios.

With bandgaps typically in the 1–2-eV range, the extreme thinness of TMDCs allows efficient control over switching and can help to reduce MOSFET power dissipation, the main limiting factor to transistor miniaturization.

Room-temperature mobility in TMDCs is 200–400 $cm^2V^{-1}s^{-1}$ and is primarily limited by phonon scattering. This range of values is not much different from carriers in silicon MOSFET channels.

An experimental TMDC MOSFET is shown in Figure 9.7. Drain current as a function of gate voltage for this device is shown in Figure 9.8 indicating excellent on/off current ratios of up to 10^7 with the channel providing n-type conduction. Room-temperature channel mobility is approximately 200 $cm^2V^{-1}s^{-1}$ in this case. Although III-V semiconductors with higher mobility values exist, the inherent lack of dangling surface bonds in TMDC devices is particularly attractive.

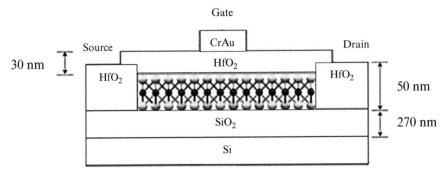

FIGURE 9.7 Transition metal dichalcogenide MOSFET having a channel composed of one monolayer of MoS_2. The gate dielectric is hafnium oxide and a silicon substrate supports the experimental device. The dimension of the MoS_2 channel is greatly exaggerated. It is actually only a few angstroms thick. Adapted from Jana et al, 2016.

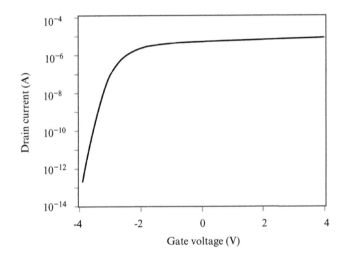

FIGURE 9.8 Drain current versus gate voltage for the transition metal dichalcogenide MOSFET of Figure 9.7. The on-off current ratio is 6–7 orders of magnitude which rivals silicon-based MOSFETs. The drain-source voltage is 0.5 volts. Adapted from Jana et al, 2016.

The top-gated geometry with a high-k dielectric improves performance as discussed for silicon MOSFETs in Chapter 6. Possible applications include flexible, transparent, 2D electronic circuits. The development of synthesis methods for obtaining large areas of MoS_2 is important for wafer-scale fabrication of devices. One way to prepare atomically thin layers of TMDCs that is not sufficiently well-controlled for reliable manufacturing is the use of simple mechanical exfoliation from bulk material using adhesive tape. This methodology was employed in the discovery of graphene. To achieve repeatable mass production of TMDC devices, a range of chemical vapor deposition growth methods by which transition metal atoms and the desired chalcogen react at the surface of a substrate are under development. The biggest on-going challenge with these growth methods is the achievement of defect-free layers.

9.5   MULTIGATE MOSFETS

In spite of the discovery and development of promising candidate materials such as TMDCs that could, in principle, replace silicon and allow for atomically thin MOSFET channels, silicon continues to be the semiconductor that is used for commercial nanoscale MOSFETs. This is driven by the well-understood processing methods and properties of silicon as well as the associated mature and installed production equipment base.

In Chapter 6 MOSFET structures with planar channels were discussed. Two new MOSFET structures known as *FinFETs* and *gate-all-around FETs* are now in volume production for high performance integrated circuits. These new nanostructures exhibit improved MOSFET performance and other advantages compared to what is achievable using planar channel MOSFETs.

Both FinFETs and gate-all-around FETs belong to a category of MOSFETs known as *multigate FETs*. This definition is somewhat misleading since, in most cases, there is only one electrically conductive gate element. The multigate FET category includes devices in which channel depletion is controlled by gates that act on multiple surfaces of the channel. The gate electrode may be continuous, but the effect of the gate electrode on the channel provides control that goes beyond what is achievable in a planar geometry.

As MOSFET size decreases, planar transistors increasingly suffer from undesirable short-channel effects, especially off-state leakage current, which increases the power required by the CMOS circuitry. See Section 6.16.

In a multi-gate device, because the channel can be depleted on more than one channel surface, better electrical control over the channel is achieved allowing more effective suppression of off-state leakage current. These advantages translate to lower power consumption. Nonplanar devices are also more compact than conventional planar transistors for a given channel cross-sectional area enabling higher transistor density and hence more transistors on a single die.

Figure 9.9 shows an example of a FinFET which is characterized by one or more vertical fins that form the channel. By extending the channel in the vertical direction, the following benefits are obtained:

(i)   The vertical fins allow for more channel cross-sectional area to be achieved than for a planar MOSFET with the same overall footprint. This enables a higher number of transistors per unit wafer area.

(ii) The subchannels are almost fully surrounded by the gate electrode. This enables a high transconductance, meaning that control of the channel state is achieved with smaller swings in gate voltage. See Section 6.8. A smaller required gate voltage range that can control the channel between depletion and strong inversion means lower operating voltages, smaller dimensions, less unwanted tunneling currents, and lower average power dissipation.

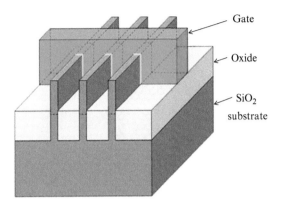

Gate

Oxide

SiO$_2$
substrate

FIGURE 9.9 A typical silicon FinFET transistor is distinguished from a planar MOSFET because the channel is broken up into multiple fin-like subchannels connected in parallel. This example has three subchannels, each of which is vertically aligned. Gate control is achieved on two sides as well as the upper surface of each subchannel. The device is categorized as a multigate transistor even though only one gate electrode exists.

A micrograph of a single FinFET fin cross section is shown in Figure 9.10. One weakness of the FinFET is the lack of gate control at the bottom of each subchannel fin. This bottom area is susceptible to leakage currents that limit the ratio between on-state and off-state channel currents.

Another architecture being introduced along with FinFETs for the smallest nodes (3 nm and lower) of MOSFET transistors is the gate-all-around FET (GAAFET) in which the channel is completely surrounded with gate electrode as shown in Figure 9.11. This is also categorized as a type of multigate transistor. The GAAFET has the potential to support the continued evolution of Moore's law because the dimensions of each subchannel can continue to decrease to the point that they are effectively nanowires having cross sections of atomic dimensions. In this case the silicon semiconductor channel material might even approach the atomic thickness range offered by TMDC materials.

The silicon-based GAAFET requires a rather innovative manufacturing strategy because the mobility of each subchannel must be as high as possible to achieve the highest possible transconductance as derived in Chapter 6. This implies high crystalline quality. To this end, the subchannels are fabricated by first growing a superlattice of alternating Si and Si$_{1-x}$Ge$_x$ epitaxial layers. It is possible to selectively etch away the Si$_{1-x}$Ge$_x$ material leaving behind a set of nanoscale silicon beams that are supported at the source and drain ends of the device only. Subsequently the gate insulator and the gate electrode are introduced as coatings that infill the voids resulting in the desired structure. See Figure 9.12.

New materials are being introduced for the gate metal. Cobalt, ruthenium, molybdenum, nickel, and some alloys are being developed for this purpose.

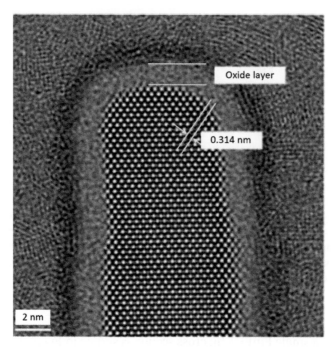

FIGURE 9.10   Cross section of a single fin of a FinFET fabricated from silicon. The picture was taken from a thin sample using scanning transmission electron microscopy (STEM). Individual silicon atoms are visible and the spacing of (111) planes in silicon is shown as 0.314 nm. The oxide layer is amorphous and the surrounding conductive layer is polycrystalline. From: Focused ion beam systems, Technical data, Data sheet, Hitachi, 2018.

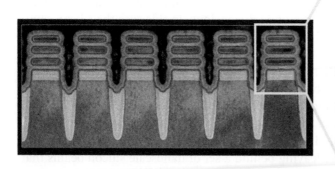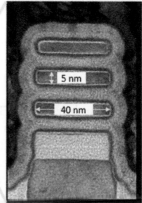

FIGURE 9.11   The Gate-all-around FET or GAAFET orients the channels horizontally rather than vertically as in the FinFET. In this micrograph cross section, gate electrode completely surrounds each of a series of three stacked subchannels. Channel control by the gate electrode is geometrically optimized, and as dimensions are reduced, each subchannel becomes effectively a nanowire or a nanosheet. Multiple subchannels enable a sufficiently large total drain current. Reprint Courtesy of IBM Corporation ©.

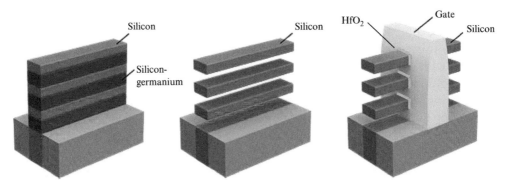

FIGURE 9.12 A method of formation of single crystal subchannels (nanosheets) relies on the initial growth of a stack of epitaxially grown alternating layers of silicon and silicon-germanium alloy semiconductor on a silicon substrate. The $Si_{1-x}Ge_x$ layers have a nominal lattice constant larger than that of silicon but are able to exist in a state of tensile compression due to their small thickness such that their lattice constant matches that of silicon. A selective etch allows the removal of the $Si_{1-x}Ge_x$ layers without damaging the silicon layers followed by the insertion of gate dielectric coatings and gate metallization. This results in a vertical stack of high-quality silicon subchannel nanosheets. As the technology is further developed these nanosheets are projected to become even smaller nanowires.

Modeling nanoscale multigate transistors requires two- and three-dimensional computer modeling that builds on the one-dimensional MOSFET modeling introduced in Chapter 6.

Models are based on a wide range of underlying theory including band theory, electrostatics, LCAO theory, DFT, Montecarlo simulations, and the materials science of crystalline defects. The following modeling resources are examples of commercially available modeling software that make meaningful optimization of nanoscale MOSFETs possible.

1. Contact resistance as a function of the atomic structure of metal–semiconductor interfaces
2. Doping-dependence of contact resistance
3. Band alignment and band offsets at interfaces
4. Schottky barrier calculations
5. Band structure calculations
6. Carrier transport simulations including phonon- and point defect-limited mobilities using quasi-random atomic configurations
7. Screening lengths and effects of metal and non-metal conductors
8. Mean free path and resistivities of metals and alloys
9. Surface and grain boundary scattering effects
10. Leakage and tunneling current effects
11. Electron-phonon scattering effects
12. DFT and LCAO simulations to simulate and predict carrier transport

The relevance and accuracy of modeling software is optimized by repeated updates based on the comparison between modeled and measured results. Cycle times of new nanoscale transistor device designs that previously took years to develop are now achieved within months.

In addition, software models also exist that predict structural aspects of nanoscale device fabrication. For example, the modeling of mechanically allowed force per unit area during FinFET semiconductor fin manufacturing impacts yield and is needed to avoid structural pattern collapse by using predictive atomic scale analysis to re-engineer material composition or process conditions.

The switch from excimer laser lithography to extreme UV lithography is being accelerated by the exacting requirements for finFET and GAAFET fabrication. Gate lengths in the range of just one nanometer are included in the future plans of major semiconductor device manufacturers.

9.6   SUMMARY

9.1 The combined effect of more than one atom on electron wavefunctions and energy levels may be predicted by the method of the Linear Combination of Atomic Orbitals (LCAO).
Such modeling becomes important for atomic-scale semiconductor materials and devices.

9.2 The LCAO method may be illustrated by applying it to an ionized hydrogen molecule.

9.3 More than one electron exists in practical nanoscale or atomic scale systems. In order to overcome the issue of not being able to precisely locate the other electrons when modeling the behavior of an additional electron, a spatially dependent charge density function can be used instead. The method by which this modeling approach works is called Density Functional Theory (DFT).

9.4 The Transition Metal Dichalcogenide (TMDC) family of crystalline materials is formed by a transition metal M such as Mo or W combined with a chalcogen X such as S, Se, or Te in the compound MX_2. Like graphene, these materials form a layered structure in which van der Waals bonds exist between atomically thin layers. TMDCs are semiconductors that are of interest as conductive channels for future FET devices or as direct gap optical materials.

9.5 Multigate FETs involve the control of channel conductivity by a structured gate electrode that acts on multiple surfaces of the channel. The result is an increase in transconductance, meaning that a large ratio of channel conductance can be achieved with smaller swings of gate voltage. This lowers overall power consumption. In addition, by utilizing a third dimension not exploited in planar MOSFETs, a higher density of devices can be achieved in integrated circuits. The continuing applicability of Moore's law is driving the adoption of these new transistor structures despite their fabrication challenges. The FinFET and the gate-all-around FET (GAAFET) are two important types of multigate FETs.

PROBLEMS

9.1 Molecular orbital theory is based on LCAO theory. Look up the biographies and scientific contributions of Friedrich Hund, Robert Mulliken, John C. Slater, and John Lennard-Jones. Molecular orbital theory was originally called the Hund-Mulliken theory. Write a page on each of these four pioneers of quantum chemistry and include a short description of each of their main contributions to quantum chemistry.

9.2 Look up the Hartree–Fock method in which the molecular orbitals are expanded in terms of an atomic orbital basis set. This led to the development of many computational quantum chemistry methods.

9.3 Look up ligand field theory and compare it to crystal field theory.

9.4 In Section 9.2, LCAO theory for two spherical atomic shells was worked out. LCAO theory can also be applied to p-orbitals which are atomic orbitals where the principal quantum number $n = 2$. Additional quantum numbers $l = 1$ and $m = 0$ are also defined resulting in atomic orbitals in the shape of lobes that depend on both θ and ϕ in spherical polar coordinates.

(a) Look up atomic p-orbitals and sketch the probability densities of these orbitals in spherical polar coordinates.

(b) Upon the application of LCAO theory on these p-orbitals, chemistry textbooks on molecular bonding show the following diagrams:

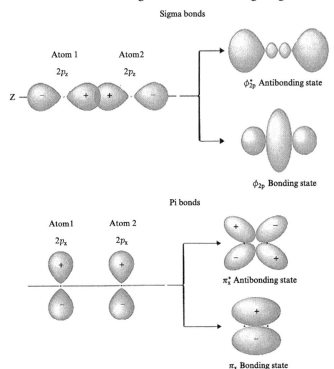

(a) Explain each diagram from the point of view of molecular orbitals that result from the application of the LCAO method to p-orbitals. Use chemistry reference materials as needed.

(b) In Chapter 8 organic semiconductor molecules were described in terms of their σ and π bonds. Describe these again in your own words in the context of your understanding of LCAO theory and your answers to part a).

9.5 (a) Look up and present a figure of the ground state electron density for C60 (buckyball) as determined by density functional theory.

(b) Look up computed versus experimental band gaps in elemental and binary semiconductors. Present a figure showing experimental bandgaps on the x-axis and computed bandgaps on the y-axis. Confirm that the computations are done based on a density functional theory approach. Comment on the validity of the computed results.

9.6 Look up DFT models of silicon vacancy-related defects in 4H-SiC. Show a figure of the crystal structure of this semiconductor. Localized spin states exist at these vacancies that have potential application to quantum computing. Explain in your own words how these localized spin states come about in the vicinity of the silicon vacancies. Show that the modeling is based on DFT.

9.7 For the micrograph of Figure 9.10, estimate the width of the fin. Also prove that the spacing between (111) planes in silicon is 3.14Å given that the unit cell dimension is 5.43Å.

9.8 Find high resolution micrographs showing the cross sections of thin slices state-of-the-art of fabricated GAAFETs. Determine the dimensions of the channels.

9.9 Seek and summarize up-to-date information on the fabrication methods and the size scales of state-of-the-art finFETs and GAAFETs.

9.10 Is Moore's law still on track? Seek and summarize up-to-date information on this topic.

9.11 Read Appendix 7. Perform the integral

$$\int \psi_B \left(\frac{1}{r_2}\right) \psi_A dV$$

from Appendix 7 and prove that

$$H_{AB} = -R_H I_{AB} - \frac{q^2}{4\pi\epsilon_0}\left(\frac{1}{a_0} + \frac{R}{a_0^2}\right)e^{-\frac{R}{a_0}}$$

9.12 Plot the final expressions in Appendix 7 for R and E_+ versus proton separation E_+ using a computer and compare the results to Figure 9.3. Find the minimum value of E_+ and compare it to the expected result.

RECOMMENDED READING

1. Harrison, W.A. (1989). *Electronic Structure and the Properties of Solids: The Physics of the Chemical Bond*. Dover Publications. ISBN 978–0486660219.

2. Giustino, F. (2014). *Materials Modelling Using Density Functional Theory: Properties and Predictions*. Oxford University Press. ISBN: 978–0199662449.

3. Sholl, D.S. and Steckel, J. A. (2009). *Density Functional Theory: A Practical Introduction*. John Wiley & Sons. ISBN: 978–0–470–37317-0.

4. Kaxiras, E. (2003). *Atomic and Electronic Structure of Solids*. Cambridge University Press. ISBN-13: 978–0521523394.

Appendix 1: Physical Constants

Boltzmann's constant	$k = 1.38 \times 10^{-23}\,\text{J K}^{-1} = 8.62 \times 10^{-5}\,\text{eV K}^{-1}$
Electron charge magnitude	$q = 1.6 \times 10^{-19}\,\text{C}$
Electron mass	$m = 9.11 \times 10^{-31}\,\text{kg}$
Permittivity of free space	$\epsilon_0 = 8.85 \times 10^{-12}\,\text{F m}^{-1} = 8.85 \times 10^{-14}\,\text{F cm}^{-1}$
Planck's constant	$h = 6.63 \times 10^{-34}\,\text{J s}$
kT at room temperature	$kT \cong 0.026\,\text{eV}$
Speed of light	$c = 3.00 \times 10^{8}\,\text{m s}^{-1}$

Fundamentals of Semiconductor Materials and Devices, First Edition. Adrian Kitai.
© 2023 John Wiley & Sons Ltd. Published 2023 by John Wiley & Sons Ltd.
Companion Website: www.wiley.com/go/kitai_fundamentals

Appendix 2: Derivation of the Uncertainty Principle

Consider a sinusoidal wave along the x-axis having amplitude a and wavelength λ. Its corresponding wavenumber k is defined as $k = \dfrac{2\pi}{\lambda}$. We can express the wave as

$$f_k(x) = ae^{ikx}$$

If an electron is described by this wave it will have infinite spatial extent. A spatially localized wave-packet may be obtained mathematically by adding a series of component sinusoidal waves together, each sinusoidal wave having a unique wavelength. This forms a Fourier series.

Consider the following sum of waves

$$f(x) = \sum_n a_n e^{ik_n x}$$

This may be written as an integral in the case of a continuum of wavelengths:

$$f(x) = \int_{-\infty}^{\infty} a(k) e^{ikx} dk \tag{A2.1}$$

Assume, for simplicity, that only a range of wavelengths of the component waves exists such that

$$a(k) = 1 \text{ for } k_0 - \frac{\Delta k}{2} \leq k \leq k_0 + \frac{\Delta k}{2}$$

and outside of this range $a(k) = 0$.

Now we obtain

$$f(x) = \int_{k_0 - \frac{\Delta k}{2}}^{k_0 + \frac{\Delta k}{2}} e^{ikx} dk$$

Fundamentals of Semiconductor Materials and Devices, First Edition. Adrian Kitai.
© 2023 John Wiley & Sons Ltd. Published 2023 by John Wiley & Sons Ltd.
Companion Website: www.wiley.com/go/kitai_fundamentals

$$= \Delta k \frac{\sin\left(\dfrac{\Delta k x}{2}\right)}{\dfrac{\Delta k x}{2}} e^{ik_0 x}$$

The envelope amplitude of this wave is now in the form of a sinc function. We will approximate the width of the sinc function by setting $\dfrac{\Delta k x}{2} = \pm\pi$ because $\sin(\pm\pi) = 0$ and therefore $x = \pm\dfrac{2\pi}{\Delta k}$. Note that the width of this sinc function will determine the uncertainty in position Δx of the wave-packet and hence

$$\Delta x \cong \frac{4\pi}{\Delta k} \tag{A2.2}$$

But from Equation 1.3,

$$p = \frac{h}{\lambda} = \frac{hk}{2\pi}$$

and therefore

$$\Delta p = \frac{h}{2\pi}\Delta k \tag{A2.3}$$

Using Equation A2.2 we obtain

$$\Delta p = \frac{h}{2\pi}\frac{4\pi}{\Delta x} = \frac{2h}{\Delta x}$$

Now,

$$\Delta p \Delta x \cong 2h$$

This represents an estimate of the minimum value of the product of uncertainty in position Δx and uncertainty in momentum Δp of the particle.

A lower and properly minimized value of $\Delta p \Delta x$ can be derived if a series of waves having a Gaussian distribution of amplitudes is summed as a Fourier series.

Using the previous approach, we instead define $a(k)$ to be a Gaussian function in wavenumber space of the form

$$a(k) = \exp\left(-\frac{\left(k - k_0\right)^2}{2\sigma_k^2}\right) \tag{A2.4}$$

Substituting this into Equation A2.1 we obtain[1]

$$f(x) = \int_{-\infty}^{\infty} \exp\left(-\frac{\left(k - k_0\right)^2}{2\sigma_k^2}\right) e^{ikx} dk$$

This integral is well known, and after several steps covered in calculus textbooks the result is another Gaussian function in real space

$$f(x) = \frac{1}{\sigma_x \sqrt{2\pi}} \exp\left(-\frac{x^2}{2\sigma_x^2}\right)$$

for which the relationship between the standard deviations σ_k and σ_x is

$$\sigma_k \sigma_x = 1.$$

If we define $\Delta k = \dfrac{\sigma_k}{\sqrt{2}}$ and $\Delta x = \dfrac{\sigma_x}{\sqrt{2}}$ then we have[2]

$$\Delta k \Delta x = \frac{\sigma_k \sigma_x}{2} = \frac{1}{2}$$

Finally, using Equation A2.3, we obtain

$$\Delta p \Delta x = \frac{1}{2}\left(\frac{h}{2\pi}\right) = \frac{\hbar}{2}$$

This defines the lower limit of $\Delta p \Delta x$ and hence the momentum and position form of the Uncertainty Principle states that

$$\Delta p \Delta x \geq \frac{\hbar}{2}$$

We can also find an analogous form of the uncertainty principle relating to energy and time by considering a Fourier series of waves having a range of frequencies. Let function $a(\omega)$ be

$$a(\omega) = \exp\left(-\frac{(\omega - \omega_0)^2}{2\sigma_\omega^2}\right)$$

Note that this is a Gaussian function as in Equation A2.4 except that it is now written in frequency space instead of wavenumber space. A Fourier series of waves having a Gaussian distribution of amplitudes may be written by changing Equation A2.1 to an integral over frequency ω and we obtain[3]

$$f(t) = \int_{-\infty}^{\infty} a(\omega) e^{i\omega t} d\omega$$

$$= \int_{-\infty}^{\infty} \exp\left(-\frac{(\omega - \omega_0)^2}{2\sigma_\omega^2}\right) e^{i\omega t} d\omega$$

$$= \frac{1}{\sigma_t \sqrt{2\pi}} \exp\left(-\frac{t^2}{2\sigma_t^2}\right)$$

where the product of standard deviations σ_t and σ_ω is $\sigma_t\sigma_\omega = 1$. Defining $\Delta t = \dfrac{\sigma_t}{\sqrt{2}}$ and $\Delta\omega = \dfrac{\sigma_\omega}{\sqrt{2}}$ we obtain $\Delta\omega\Delta t = \dfrac{1}{2}$ as a minimum. Since $E = \hbar\omega$, the energy and time form of the Uncertainty Principle is obtained as

$$\Delta E\Delta t \geq \frac{\hbar}{2}$$

NOTES

1. We are including negative wavenumbers in the integral due to the limits of integration from negative infinity to positive infinity. The negative wavenumbers are not physically meaningful. Euler's formula $e^{ikx} = \cos kx + i\sin kx$ was used to replace trigonometric functions $\cos kx$ and $\sin kx$. Only positive wavenumbers would appear if the integrals were written using trigonometric functions, and the limits of integration would be from zero to infinity. The use of e^{ikx} allows the integral to be written in a more mathematically elegant form.

2. It would seem reasonable to define uncertainties Δk and Δx by their respective standard deviations. In Section 1.6, however, we will see that the probability of the existence of an electron is determined by the *square* of the amplitude of the wavepacket. If a Gaussian function is squared, the result is another Gaussian function having a standard deviation that decreases by a factor of $\sqrt{2}$ compared to the initial Gaussian function because the exponent doubles. Hence the relevant uncertainties in position and wavenumber are reduced by a factor of $\sqrt{2}$.

3. Only physically meaningful positive frequencies would appear if the integral were written using trigonometric functions. (See note 1).

Appendix 3: Derivation of Group Velocity

Consider two traveling waves that may have slightly different phase velocities. We will write the waves as

$$f_1(x,t) = e^{i(\omega_1 t - k_1 x)}$$

and

$$f_2(x,t) = e^{i(\omega_2 t - k_2 x)}$$

We define average values of ω and k for the two waves as

$$\omega = \frac{(\omega_1 + \omega_2)}{2}$$

and

$$k = \frac{(k_1 + k_2)}{2}$$

and deviations from average values $\Delta\omega$ and Δk as

$$\Delta\omega = \frac{(\omega_1 - \omega_2)}{2}$$

and

$$\Delta k = \frac{(k_1 - k_2)}{2}$$

If we add the two waves to obtain sum $f(x,t)$ we obtain

$$f(x,t) = f_1(x,t) + f_2(x,t) = e^{i(\omega_1 t - k_1 x)} + e^{i(\omega_2 t - k_2 x)}$$

$$= e^{i(\omega t + \Delta\omega t - kx - \Delta kx)} + e^{i(\omega t - \Delta\omega t - kx + \Delta kx)}$$

Fundamentals of Semiconductor Materials and Devices, First Edition. Adrian Kitai.
© 2023 John Wiley & Sons Ltd. Published 2023 by John Wiley & Sons Ltd.
Companion Website: www.wiley.com/go/kitai_fundamentals

$$= e^{i(\omega t - kx)}\left(e^{i(\Delta\omega t - \Delta kx)} + e^{i(-\Delta\omega t + \Delta kx)}\right)$$

$$= 2\cos(\Delta\omega t - \Delta kx)\, e^{i(\omega t - kx)}$$

The wave envelope is seen to have the function $\cos(\Delta\omega t - \Delta kx)$ which travels at group velocity

$$v_g = \frac{\Delta\omega}{\Delta k}$$

If the wave-packet is composed from many waves, then provided that $\dfrac{\Delta\omega}{\Delta k}$ is the same for each consecutive pair of waves in the Fourier series, the resulting group velocity of the entire wave-packet becomes

$$v_g = \frac{d\omega}{dk}$$

and this therefore defines the group velocity of a wave-packet. Note that if the plot of ω versus k is a straight line then $\dfrac{d\omega}{dk} = \dfrac{\omega}{k}$ which corresponds to dispersion-free propagation.

Appendix 4: Reduced Mass

Consider two masses m_1 and m_2 that are at rest and that are attracted to each other by an electrostatic force. A straight line joining the two masses shown in the figure can be divided into two vector portions r_1 and r_2.

The mutual attraction between the masses will cause a time-dependent force of magnitude $F(t)$ pulling the masses toward each other. The center of mass is defined as the location where the masses would meet after they accelerate toward each other under the influence of force $F(t)$.

Consider any differential interval of time dt after time t while the masses travel toward each other starting at time $t = 0$. The distance dr_1 traveled by mass m_1 will be

$$dr_1 = s_1(t)dt = \left[\int_0^t a_1(t)dt \right] dt = \frac{1}{m_1} \left[\int_0^t F(t)dt \right] dt \qquad (A4.1)$$

where $s_1(t)$ is the time-dependent speed and $a_1(t)$ is the time-dependent acceleration of mass m_1. By examination, the instantaneous speed and the distance traveled by each mass will be in inverse proportion to the respective mass. This means that the following ratio must be satisfied

$$r_2 = -\frac{m_1}{m_2} r_1$$

Also, from the figure,

$$r = r_1 - r_2$$

Fundamentals of Semiconductor Materials and Devices, First Edition. Adrian Kitai.
© 2023 John Wiley & Sons Ltd. Published 2023 by John Wiley & Sons Ltd.
Companion Website: www.wiley.com/go/kitai_fundamentals

Combining these two equations,

$$r = \left(1 + \frac{m_1}{m_2}\right)r_1 \tag{A4.2}$$

Since r_1 and r_2 are inversely proportional to each mass, if the acceleration continued for long enough, both masses would arrive at the center of mass at exactly the same time.

Solving Equation A4.1 for the force $F(t)$ yields the equation of motion for mass m_1

$$m_1 \frac{d^2 r_1}{dt^2} = F(t)$$

This equation can also be written in terms of the magnitude of r using Equation A4.2. We obtain

$$m_1 \frac{d^2 r_1}{dt^2} = m_1 \frac{d^2}{dt^2}\left[\left(1 + \frac{m_1}{m_2}\right)^{-1} r\right] = \mu_0 \frac{d^2 r}{dt^2} = F(t)$$

where reduced mass μ_0 is defined as

$$\frac{1}{\mu_0} = \frac{1}{m_1} + \frac{1}{m_2}$$

The equation of motion of mass m_1 is now

$$\mu_0 \frac{d^2 r}{dt^2} = F(t)$$

Since there is nothing unique about either mass, this proves that the distance r between m_1 and m_2 can be used in the equation of motion for either mass provided the reduced mass μ_0 is used.

Appendix 5: The Boltzmann Distribution Function

The classical distribution of energy in an ensemble of interacting atoms or molecules forms the basis for a quantum energy distribution function. In this section we will review the assumptions and the methodology to calculate the probability distribution function of the energies of an ensemble of atoms or molecules that is used as the basis for the calculation of the Fermi–Dirac distribution in Section 2.10.

THE BOLTZMANN DISTRIBUTION FUNCTION

Consider a system in thermal equilibrium containing a large number of identical entities. Each entity, which could be an atom or a molecule, will have some energy E with a certain energy distribution function. Energy from each entity can be exchanged with other entities by means of collisions between entities. These collisions are assumed to be elastic, meaning that the total system energy is not changed as a result of energy exchange. The purpose of this section is to find this energy distribution function.

The following summarizes the assumptions that apply to the system of entities:

(a) The system is in thermal equilibrium at temperature T.
(b) Entities in the system can exchange energy by means of perfectly elastic collisions. This means that the total system energy is not altered when a collision takes place.
(c) The total system energy is constant.
(d) All possible divisions of total system energy among the entities in the system occur with equal probability. There is no preferred division of energy among the entities.
(e) The entities are distinguishable. This means that we can identify any given entity and its energy at any time.

Fundamentals of Semiconductor Materials and Devices, First Edition. Adrian Kitai.
© 2023 John Wiley & Sons Ltd. Published 2023 by John Wiley & Sons Ltd.
Companion Website: www.wiley.com/go/kitai_fundamentals

Consider an example of this situation. Assume there are four atoms, and each atom can have energy levels of $0, \Delta E, 2\Delta E, 3\Delta E$. Now assume the total system energy is $3\Delta E$. Possible divisions of energy are as listed in the following cases:

1. 3 atoms at $E = 0$ and 1 atom at $E = 3\Delta E$
2. 2 atoms at $E = 0$, 1 atom at $E = \Delta E$, and 1 atom at $E = 2\Delta E$
3. 1 atom at $E = 0$ and 3 atoms at $E = \Delta E$

These three cases do not cover all the possible divisions of total system energy, however. This is because in case 1 any one of the four atoms could be at energy $E = 3\Delta E$, and case 1 therefore really includes four distinct divisions of energy among the atoms. This is because the atoms are distinguishable.

In case 2, the atom at $E = 2\Delta E$ could be chosen from any one of the four atoms, leaving three atoms. Any one of the remaining three atoms could now be chosen for $E = \Delta E$. Finally the remaining two atoms would be at $E = 0$. Case 2 therefore results in $4 \times 3 = 12$ distinct divisions of energy among the atoms.

Finally in case 3, any one of the four atoms could be at energy $E = 0$, and case 3 therefore really includes four distinct divisions of energy among the atoms.

Now we can make a list of all 20 divisions and their probabilities of occurrence as shown in Table A5.1. For any one atom, we can now find the energy distribution by looking down the relevant column of the table and counting the number of entries having specific energies. For any one atom:

There is a chance of 1/20 that the atom is in energy state $E = 3\Delta E$.
There is a chance of 3/20 that the atom is in energy state $E = 2\Delta E$.
There is a chance of 6/20 that the atom is in energy state $E = \Delta E$.
There is a chance of 10/20 that the atom is in energy state $E = 0$.

We can plot this energy distribution function as shown in Figure A5.1. The probability of occurrence as a function of energy is the energy distribution function for the system.

TABLE A5.1   A list of all possible divisions of total system energy among the atoms in the system. Each division occurs with equal probability.

Arrangement	Atom 1	Atom 2	Atom 3	Atom 4	Probability of occurrence
1	0	0	0	$3\Delta E$	1/20
2	0	0	$3\Delta E$	0	1/20
3	0	$3\Delta E$	0	0	1/20
4	$3\Delta E$	0	0	0	1/20
5	0	0	ΔE	$2\Delta E$	1/20

Arrangement	Atom 1	Atom 2	Atom 3	Atom 4	Probability of occurrence
6	0	ΔE	0	$2\Delta E$	1/20
7	ΔE	0	0	$2\Delta E$	1/20
8	0	0	$2\Delta E$	ΔE	1/20
9	0	ΔE	$2\Delta E$	0	1/20
10	ΔE	0	$2\Delta E$	0	1/20
11	0	$2\Delta E$	0	ΔE	1/20
12	0	$2\Delta E$	ΔE	0	1/20
13	ΔE	$2\Delta E$	0	0	1/20
14	$2\Delta E$	0	0	ΔE	1/20
15	$2\Delta E$	0	ΔE	0	1/20
16	$2\Delta E$	ΔE	0	0	1/20
17	0	ΔE	ΔE	ΔE	1/20
18	ΔE	0	ΔE	ΔE	1/20
19	ΔE	ΔE	0	ΔE	1/20
20	ΔE	ΔE	ΔE	0	1/20

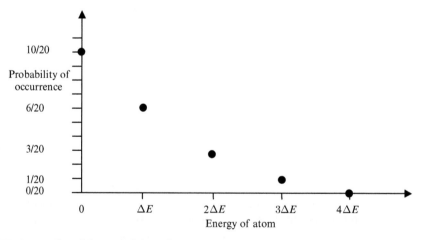

FIGURE A5.1   Plot of the probability of occurrence as a function of the energy for any atom in the system.

In Figure A5.2 an exponential function is now superimposed on the same data as shown in Figure A5.1. This exponential function is of the form:

$$p(E) = A\exp\left(-\frac{E}{E_0}\right)$$

(A5.1)

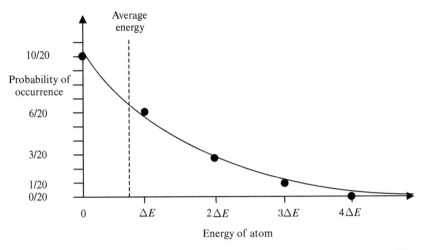

FIGURE A5.2   Plot of the probability of occurrence as a function of energy together with an exponential function as an approximation of the data. The average energy at $0.75\Delta E$ is also shown.

If we choose suitable parameters A and E_0 we can obtain a reasonable fit to the data.

If the same procedure using a much larger number of atoms than four is followed, then the number of data points increases rapidly and the ability of an exponential function of the form of Equation A5.1 to represent the data improves and becomes an excellent fit.

The significance of E_0 may now be understood by calculating the *average* energy of any atom in the system. This is done by adding all the possible energies and dividing by the number of divisions of energy. From Table A1.1, the average energy of any one atom is

$$\frac{(1\times3\Delta E)+(3\times2\Delta E)+(6\times\Delta E)+(10\times0)}{20}=\frac{15}{20}\Delta E=0.75\Delta E$$

This makes sense because the total system energy is $3\Delta E$ and there are four atoms each having average energy $0.75\Delta E$. The average energy is shown as a dashed line at $0.75\Delta E$ in Figure A3.2.

It is also possible to calculate the average energy based on the exponential function $p(E)$. This average is calculated again by adding all the possible energies and dividing by the number of divisions of energy; however, since $p(E)$ is a continuous function the addition becomes an integral and we obtain

$$E_{\text{average}}=\frac{\int_0^\infty Ep(E)\mathrm{d}E}{\int_0^\infty p(E)\mathrm{d}E}=\frac{\int_0^\infty E\exp\left(-\dfrac{E}{E_0}\right)\mathrm{d}E}{\int_0^\infty \exp\left(-\dfrac{E}{E_0}\right)\mathrm{d}E}$$

The numerator may be integrated by the method of integration by parts to obtain E_0^2 and the denominator is readily integrated as E_0. Hence we obtain

$$E_{average} = \frac{E_0^2}{E_0} = E_0$$

Therefore the average energy of one atom in a system with many atoms is E_0.

If the average energy of an atom belonging to a system in thermodynamic equilibrium at temperature T is defined as kT where k is Boltzmann's constant, then we have the famous Boltzmann distribution function

$$p(E) = A \exp\left(-\frac{E}{kT}\right)$$

which is used in Equation 2.22 in the derivation of the Fermi–Dirac distribution function.

In order to prove the validity of the exponential form of the Boltzmann distribution function, we will take a different approach. Consider atoms in the ensemble of particles. The atoms may transfer kinetic energy from one to the next. We will again assume the following:

(a) The atoms form a system in equilibrium.

(b) The total system energy is constant.

(c) No kinetic energy is lost when energy transfer occurs between atoms. Collisions are elastic.

(d) All possible divisions of the total system energy among the atoms occur with equal probability.

Consider two atoms that transfer energy between them. Atom 1 has energy E_1 and atom 2 has energy E_2. After a particular energy transfer occurs, atoms 1 and 2 end up with energies of E_1' and E_2' respectively. From assumption c) we can write

$$E_1 + E_2 = E_1' + E_2'$$

The likelihood of this energy transfer depends on the probabilities of the atoms having starting energies E_1' and E_2' respectively.

We now define $p(E)$ to be the probability of an atom having kinetic energy E. The likelihood of the energy transfer process is proportional to $p(E_1)p(E_2)$ because these are the necessary starting energies of the two atoms. From the requirement of equilibrium in assumption a) we can conclude that if the two atoms had energies E_1' and E_2' respectively, then the likelihood of a reverse energy transfer process between the two atoms resulting in energies E_1 and E_2 is also possible and must be equal in likelihood to that of the forward process.

Hence,

$$p(E_1)p(E_2) = p(E_1')p(E_2') \tag{A5.2}$$

Let $E_1' = E_1 + \Delta E$ and $E_2' = E_2 - \Delta E$. This ensures that energy is conserved.
Now Equation A5.2 becomes

$$p(E_1)p(E_2) = p(E_1 + \Delta E)p(E_2 - \Delta E)$$

This equation must work for all values of E_1 and E_2. If we examine the equation it becomes apparent **that the only functions $p(E)$ that can satisfy the equation are exponentials**. The most general exponential has the form

$$p(E) = Ae^{-\frac{E}{E_0}} \qquad\qquad (A5.3)$$

where A and E_0 are positive constants. The negative in the exponent ensures that the probability does not approach infinity at high energy values which is physically impossible.

Appendix 6: Properties of Semiconductor Materials

Fundamentals of Semiconductor Materials and Devices, First Edition. Adrian Kitai.
© 2023 John Wiley & Sons Ltd. Published 2023 by John Wiley & Sons Ltd.
Companion Website: www.wiley.com/go/kitai_fundamentals

Semi-conductor	Energy gap type	Energy gap E_g (eV)	Electron mobility μ_n (cm² V⁻¹ s⁻¹)	Hole mobility μ_p (cm² V⁻¹ s⁻¹)	Relative electron effective mass $\dfrac{m_n^*}{m}$	Relative hole effective mass $\dfrac{m_p^*}{m}$	Lattice constant (Å)	Relative dielectric constant ϵ_r	Density (g cm⁻³)
Si	Indirect	1.11	1350	480	1.08*	0.56*	5.43	11.8	2.33
Ge	Indirect	0.67	3900	1900	0.55*	0.37*	5.65	16	5.32
SiC	Indirect	2.86	500	–	0.6	1.0	3.08	10	3.21
AlP	Indirect	2.45	80	–	0.21	0.15*	5.46	9.8	2.40
AlAs	Indirect	2.16	1200	420	2.0	*	5.66	10.9	3.60
AlSb	Indirect	1.6	200	300	0.12	0.98	6.14	11	4.26
GaP	Indirect	2.26	300	150	*	*	5.45	11.1	4.13
GaAs	Direct	1.43	8500	400	0.067	*	5.65	13.2	5.31
GaN	Direct	3.4	380	–	0.19	0.6	4.5	8.9	6.1
GaSb	Direct	0.7	5000	1000	0.042*	*	6.09	15.7	5.61
InP	Direct	1.35	4000	100	0.077*	*	5.87	12.4	4.79
InAs	Direct	0.36	22 600	200	0.023*	*	6.06	14.6	5.67
InSb	Direct	0.18	100 000	1700	0.014*	0.4	6.48	17.7	5.78
ZnS	Direct	3.6	180	10	0.28	–	5.41	8.9	4.09
ZnSe	Direct	2.7	600	28	0.14	0.6	5.67	9.2	5.65
ZnTe	Direct	2.25	530	100	0.18	0.65	6.10	10.4	5.51
CdS	Direct	2.42	250	15	0.21	0.80	4.14	8.9	4.82
CdSe	Direct	1.73	800	–	0.13	0.3–0.45	4.30	10.2	5.81
CdTe	Direct	1.51	1050	100	0.10	0.37	6.48	10.2	6.20

*There is more than one value since both heavy holes and light holes may be involved. There may also be distinct direction-dependent values. If a value is given then it is averaged. See the scientific literature for details on each material and on how averages may be obtained.

Appendix 7: Calculation of the Bonding and Antibonding Orbital Energies Versus Interproton Separation for the Hydrogen Molecular Ion

From Chapter 9

$$I_{AB} = \int \psi_A \left(\left| \vec{r} - \vec{R}_A \right| \right) \psi_B \left(\left| \vec{r} - \vec{R}_B \right| \right) dV$$

For the purposes of this calculation, a spherical polar coordinate system will be initially centered on proton A as shown below.

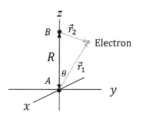

I_{AB} is known as an overlap integral because it calculates the degree of overlap of two normalized wavefunctions. I_{AB} provides a number in the range of 0 (no overlap) to 1 (full overlap). The limit of 1 can be understood because at this limit the two wavefunctions (real and normalized) are the same and $\int \psi_A \psi_A dV = 1$. In these new coordinates,

$$I_{AB} = \int \psi_A \left(\left| \vec{r}_1 \right| \right) \psi_B \left(\left| \vec{r}_2 \right| \right) dV$$

Fundamentals of Semiconductor Materials and Devices, First Edition. Adrian Kitai.
© 2023 John Wiley & Sons Ltd. Published 2023 by John Wiley & Sons Ltd.
Companion Website: www.wiley.com/go/kitai_fundamentals

Using the cosine law, $|\vec{r}_2| = \sqrt{|\vec{r}_1|^2 + R^2 - 2|\vec{r}_1|R\cos\theta}$

Defining for convenience $r_1 = |\vec{r}_1|$ and $r_2 = |\vec{r}_2|$ this becomes

$$r_2 = \sqrt{r_1^2 + R^2 - 2r_1 R\cos\theta}$$

Using spherical polar coordinates, the spatial integration is of the form

$$I_{AB} = \iiint \psi_A(r_1)\psi_B(r_2)r_1^2 \sin\theta\,d\phi\,d\theta\,dr_1$$

$$= \int_0^\infty \int_0^\pi \int_0^{2\pi} \frac{1}{\sqrt{\pi}} a_0^{-\frac{3}{2}} e^{-\frac{r_1}{a_0}} \frac{1}{\sqrt{\pi}} a_0^{-\frac{3}{2}} e^{-\frac{r_2}{a_0}} r_1^2 \sin\theta\,d\phi\,d\theta\,dr_1$$

$$= 2\pi \int_0^\infty \int_0^\pi \frac{1}{\sqrt{\pi}} a_0^{-\frac{3}{2}} e^{-\frac{r_1}{a_0}} \frac{1}{\sqrt{\pi}} a_0^{-\frac{3}{2}} e^{-\frac{\sqrt{r_1^2 + R^2 - 2r_1 R\cos\theta}}{a_0}} r_1^2 \sin\theta\,d\theta\,dr_1$$

To simplify notation, the subscript on r will be removed in the remaining steps.

$$I_{AB} = \frac{2}{a_0^3} \int_0^\infty \int_0^\pi e^{-\frac{r}{a_0}} e^{-\frac{\sqrt{r^2 + R - 2rR\cos\theta}}{a_0}} r^2 \sin\theta\,d\theta\,dr$$

A change of variable will now be used that replaces independent variable θ to simplify the expression.

Let $u = \sqrt{r^2 + R^2 - 2rR\cos\theta}$ in which case $u^2 = r^2 + R^2 - 2rR\cos\theta$.

Substituting the limits $\theta = 0$ to $\theta = \pi$ into this equation and simplifying, the limits of a new integral over variable u that replaces the integral over θ will extend from $u = |r - R|$ to $u = r + R$. The absolute magnitude is needed on the lower limit but not on the upper limit which is always positive.

Differentiating u with respect to θ, $2u\,du = 2rR\sin\theta\,d\theta$ and $\dfrac{u\,du}{R} = r\sin\theta\,d\theta$.

Now,

$$I_{AB} = \frac{2}{a_0^3 R} \int_0^\infty re^{-\frac{r}{a_0}} \int_{|r-R|}^{r+R} ue^{-\frac{u}{a_0}} du\,dr$$

Using integral tables or by integration by parts,

$$\int_{|r-R|}^{r+R} ue^{-\frac{u}{a_0}} du = (-ua_0 - a_0^2)e^{-\frac{u}{a_0}} \Big|_{|r-R|}^{r+R}$$

$$= a_0 \left[-(r + R + a_0)e^{-\frac{r+R}{a_0}} + (|r - R| + a_0)e^{-\frac{|r-R|}{a_0}} \right]$$

$$I_{AB} = \frac{2}{a_0^2 R} \int_0^\infty re^{-\frac{r}{a_0}} \left[-(r + R + a_0)e^{-\frac{r+R}{a_0}} + (|r - R| + a_0)e^{-\frac{|r-R|}{a_0}} \right] dr$$

If $r \leq R$ then $|r - R| = -r + R$ and if $r \geq R$ then $|r - R| = r - R$. Therefore

$$I_{AB} = \frac{2}{a_0^2 R} e^{-\frac{R}{a_0}} \left[\int_0^R \left(-r^2 - Rr - a_0 r \right) e^{-\frac{2r}{a_0}} dr + \int_0^R \left(-r^2 + Rr + a_0 r \right) e^{-\frac{2r}{a_0}} dr + \int_R^\infty \left(r^2 - Rr + a_0 r \right) e^{-\frac{2r}{a_0}} dr \right]$$

Using integral tables or by integration by parts, and after gathering terms,

$$I_{AB} = \left[\frac{R^2}{3a_0^2} + \frac{R}{a_0} + 1 \right] e^{-\frac{R}{a_0}}$$

Next, we will find

$$H_{AA} = \langle \psi_A | \widehat{H} | \psi_A \rangle = \int \psi_A^* \widehat{H} \psi_A dV = \int \psi_A \widehat{H} \psi_A dV$$

$$= \int \psi_A \left[\frac{-\hbar^2}{2m} \nabla^2 - \frac{q^2}{4\pi\epsilon_0} \left(\frac{1}{r_1} + \frac{1}{r_2} \right) \right] \psi_A dV$$

$$= \int \psi_A \left[\frac{-\hbar^2}{2m} \nabla^2 - \frac{q^2}{4\pi\epsilon_0 r_1} \right] \psi_A dV - \int \psi_A \frac{q^2}{4\pi\epsilon_0} \left(\frac{1}{r_2} \right) \psi_A dV$$

The first integral contains the Hamiltonian of the time-independent Schrödinger Equation for a single hydrogen atom operating on wavefunction ψ_A. Section 1.9 shows that this is equal to $-R_H \psi_A$ where Rydberg constant $R_H = 13.6$ eV. Therefore

$$H_{AA} = \int \psi_A (-R_H) \psi_A dV - \frac{q^2}{4\pi\epsilon_0} \int \psi_A \left(\frac{1}{r_2} \right) \psi_A dV$$

$$= -R_H \int \psi_A \psi_A dV - \frac{q^2}{4\pi\epsilon_0} \int \psi_A \left(\frac{1}{r_2} \right) \psi_A dV$$

$$= -R_H - \frac{q^2}{4\pi\epsilon_0} \frac{1}{\pi a_0^3} \int_0^\infty \int_0^\pi \int_0^{2\pi} \frac{1}{r_2} e^{-\frac{r_1}{a_0}} e^{-\frac{r_1}{a_0}} r_1^2 \sin\theta \, d\phi d\theta \, dr_1$$

$$= -R_H - \frac{q^2}{4\pi\epsilon_0} \frac{1}{\pi a_0^3} \int_0^\infty \int_0^\pi \int_0^{2\pi} \frac{1}{r_2} e^{-\frac{2r_1}{a_0}} r_1^2 \sin\theta \, d\phi d\theta dr_1$$

For the purposes of this integral, we will take advantage of the symmetry of the molecular ion and relocate the origin of the spherical polar coordinate system as shown below.

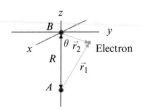

Now, $H_{AA} = -R_H - \dfrac{q^2}{4\pi\epsilon_0}\dfrac{1}{\pi a_0^3}\displaystyle\int_0^\infty\int_0^\pi\int_0^{2\pi}\dfrac{1}{r_2}e^{-\frac{2r_1}{a_0}}r_2^2\sin\theta\, d\phi d\theta dr_2$

Applying the cosine law $r_1 = \sqrt{r_2^2 + R^2 - 2r_2 R\cos\theta}$

$H_{AA} = -R_H - \dfrac{q^2}{4\pi\epsilon_0}\dfrac{1}{\pi a_0^3}\displaystyle\int_0^\infty\int_0^\pi\int_0^{2\pi}e^{-\frac{2\sqrt{r_2^2+R^2-2r_2R\cos\theta}}{a_0}}r_2\sin\theta\, d\phi d\theta dr_2$

To simplify notation, the subscript on r will be removed in the remaining steps

Let $u = \sqrt{r^2 + R^2 - 2rR\cos\theta}$ in which case $u^2 = r^2 + R^2 - 2rR\cos\theta$.

Differentiating with respect to θ, $2u\,du = 2rR\sin\theta\, d\theta$ and $\dfrac{u\,du}{R} = r\sin\theta\, d\theta$.

Integrating over $d\phi$ and appropriately changing the limits of the integral as in the calculation of I_{AB} we obtain

$$H_{AA} = -13.6\text{eV} - \dfrac{q^2}{4\pi\epsilon_0}\dfrac{2\pi}{\pi a_0^3 R}\int_0^\infty\int_{|r-R|}^{r+R}e^{-\frac{2u}{a_0}}u\,du\,dr$$

Using integral tables or by integration by parts

$$H_{AA} = -R_H - \dfrac{q^2}{4\pi\epsilon_0}\dfrac{2}{a_0^3 R}\dfrac{a_0}{2}\int_0^\infty\left[-e^{-\frac{2}{a_0}(r+R)}\left(r+R+\dfrac{a_0}{2}\right)+e^{-\frac{2}{a_0}|r-R|}\left(r-R+\dfrac{a_0}{2}\right)\right]dr$$

As before, if $r \le R$ then $|r - R| = -r + R$ and if $r \ge R$ then $|r - R| = r - R$. Therefore

$H_{AA} = -R_H - \dfrac{q^2}{4\pi\epsilon_0}\dfrac{1}{a_0^2 R}$

$$\left[e^{-\frac{2R}{a_0}}\int_0^\infty e^{-\frac{2r}{a_0}}\left(r+R+\dfrac{a_0}{2}\right)dr + e^{-\frac{2R}{a_0}}\int_0^R e^{\frac{2r}{a_0}}\left(R-r+\dfrac{a_0}{2}\right)dr + e^{\frac{2R}{a_0}}\int_R^\infty e^{-\frac{2r}{a_0}}\left(r-R+\dfrac{a_0}{2}\right)dr\right]$$

Using integral tables or by integration by parts, and after gathering terms,

$$H_{AA} = -R_H - \dfrac{q^2}{4\pi\epsilon_0}\left[\dfrac{1}{R}\left(\dfrac{1}{R}+\dfrac{1}{a_0}\right)e^{-\frac{2R}{a_0}}\right]$$

Next, due to symmetry,

$$H_{AB} = \langle\psi_A|\widehat{H}|\psi_B\rangle = \int\psi_A^*\widehat{H}\psi_B dV = \int\psi_A\widehat{H}\psi_B dV = \int\psi_B\widehat{H}\psi_A dV$$

$$= \int\psi_B\left[\dfrac{-\hbar^2}{2m}\nabla^2 - \dfrac{q^2}{4\pi\epsilon_0}\left(\dfrac{1}{r_1}+\dfrac{1}{r_2}\right)\right]\psi_A dV$$

$$= \int\psi_B\left(\dfrac{-\hbar^2}{2m}\nabla^2 - \dfrac{q^2}{4\pi\epsilon_0 r_1}\right)\psi_A dV - \int\psi_B\dfrac{q^2}{4\pi\epsilon_0}\left(\dfrac{1}{r_2}\right)\psi_A dV$$

The integrand of the first integral contains $\widehat{H}\psi_A$ for a hydrogen atom electron and from Schrödinger's Equation this yields $-R_H\psi_A$. Therefore

$$H_{AB} = -R_H \int \psi_B \psi_A dV - \frac{q^2}{4\pi\epsilon_0} \int \psi_B \left(\frac{1}{r_2}\right) \psi_A dV$$

The first integral is now the same as I_{AB} which is already calculated. The second integral can be calculated using virtually the same methods as were used in H_{AA}. See Problem 9.11. The result is

$$H_{AB} = -R_H I_{AB} - \frac{q^2}{4\pi\epsilon_0}\left(\frac{1}{a_0} + \frac{R}{a_0^2}\right)e^{-\frac{R}{a_0}}$$

Therefore

$$E_+ = \frac{H_{AA}(R) + H_{AB}(R)}{1 + I_{AB}(R)} + \frac{q^2}{4\pi\epsilon_0 R}$$

$$= \frac{-R_H - \frac{q^2}{4\pi\epsilon_0}\left[\frac{1}{R} - \left(\frac{1}{R} + \frac{1}{a_0}\right)e^{-\frac{2R}{a_0}}\right] - R_H I_{AB} - \frac{q^2}{4\pi\epsilon_0}\left(\frac{1}{a_0} + \frac{R}{a_0^2}\right)e^{-\frac{R}{a_0}}}{1 + I_{AB}(R)} + \frac{q^2}{4\pi\epsilon_0 R}$$

$$= -13.6eV - \frac{\frac{q^2}{4\pi\epsilon_0}\left[\frac{1}{R} - \left(\frac{1}{R} + \frac{1}{a_0}\right)e^{-\frac{2R}{a_0}}\right] + \frac{q^2}{4\pi\epsilon_0}\left(\frac{1}{a_0} + \frac{R}{a_0^2}\right)e^{-\frac{R}{a_0}}}{1 + \left[\frac{R^2}{3a_0^2} + \frac{R}{a_0} + 1\right]e^{-\frac{R}{a_0}}} + \frac{q^2}{4\pi\epsilon_0 R}$$

which is plotted as a function of R as the Bonding orbital curve in Figure 9.3 and

$$E_- = \frac{H_{AA}(R) - H_{AB}(R)}{1 - I_{AB}(R)} + \frac{q^2}{4\pi\epsilon_0 R}$$

$$= -13.6eV + \frac{-\frac{q^2}{4\pi\epsilon_0}\left[\frac{1}{R} - \left(\frac{1}{R} + \frac{1}{a_0}\right)e^{-\frac{2R}{a_0}}\right] + \frac{q^2}{4\pi\epsilon_0}\left(\frac{1}{a_0} + \frac{R}{a_0^2}\right)e^{-\frac{R}{a_0}}}{1 - \left[\frac{R^2}{3a_0^2} + \frac{R}{a_0} + 1\right]e^{-\frac{R}{a_0}}} + \frac{q^2}{4\pi\epsilon_0 R}$$

which is plotted as a function of R as the Antibonding orbital curve in Figure 9.3.

Index

Printed and bound by CPI Group (UK) Ltd, Croydon, CR0 4YY

16/04/2025

14658553-0004